ACS SYMPOSIUM SERIES 263

Resonances

In Electron–Molecule Scattering, van der Waals Complexes, and Reactive Chemical Dynamics

Donald G. Truhlar, EDITOR
University of Minnesota

Based on a symposium sponsored by
the Division of Physical Chemistry
at the 187th Meeting
of the American Chemical Society,
St. Louis, Missouri,
April 9–12, 1984

American Chemical Society, Washington, D.C. 1984

Library of Congress Cataloging in Publication Data

Resonances in electron-molecule scattering, van der Waals complexes, and reactive chemical dynamics.
(ACS symposium series, ISSN 0097-6156; 263)

"Based on a symposium sponsored by the Division of Physical Chemistry at the 187th Meeting of the American Chemical Society, St. Louis, Missouri, April 8-13, 1984."

Includes bibliographies and indexes.

1. Excited state chemistry—Congresses. 2. Collisional excitation—Congresses. 3. Van der Waals forces—Congresses.

I. Truhlar, Donald G., 1944- . II. American Chemical Society. Division of Physical Chemistry. III. Series.

QD461.5.R47 1984 541.2 84-16934
ISBN 0-8412-0865-4

PRINTED IN THE UNITED STATES OF AMERICA

ACS Symposium Series

M. Joan Comstock, *Series Editor*

FOREWORD

The ACS SYMPOSIUM SERIES was founded in 1974 to provide a medium for publishing symposia quickly in book form. The format of the Series parallels that of the continuing ADVANCES IN CHEMISTRY SERIES except that in order to save time the papers are not typeset but are reproduced as they are submitted by the authors in camera-ready form. Papers are reviewed under the supervision of the Editors with the assistance of the Series Advisory Board and are selected to maintain the integrity of the symposia; however, verbatim reproductions of previously published papers are not accepted. Both reviews and reports of research are acceptable since symposia may embrace both types of presentation.

CONTENTS

Preface .... **ix**

GENERAL DISCUSSIONS

1. **Roles Played by Metastable States in Chemistry** **3**
   Jack Simons

2. **Direct Variational Methods for Complex Resonance Energies** **17**
   C. William McCurdy

3. **Complex Energy Quantization for Molecular Systems** **35**
   R. Lefebvre

4. **Approximate Quantum Approaches to the Calculation of Resonances in Reactive and Nonreactive Scattering** **43**
   Joel M. Bowman, Ki Tung Lee, Hubert Romanowski, and Lawrence B. Harding

ELECTRON–MOLECULE SCATTERING AND PHOTOIONIZATION

5. **Resonances in Electron–Molecule Scattering and Photoionization** **65**
   B. I. Schneider and L. A. Collins

6. **Dynamics of Molecular Photoionization Processes** **89**
   D. L. Lynch, V. McKoy, and R. R. Lucchese

7. **Molecular Photoionization Resonances: A Theoretical Chemist's Perspective** **113**
   P. W. Langhoff

8. **Shape Resonances in Molecular Fields** **139**
   J. L. Dehmer

9. **Temporary Negative Ion States in Hydrocarbons and Their Derivatives** **165**
   K. D. Jordan and P. D. Burrow

10. **Negative Ion States of Three- and Four-Membered Ring Hydrocarbons Studied by Electron Transmission Spectroscopy** **183**
    Allison E. Howard and Stuart W. Staley

11. **Anion Resonance States of Organometallic Molecules** **193**
    Judith C. Giordan, John H. Moore, and John A. Tossell

12. **Vibrational–Librational Excitation and Shape Resonances in Electron Scattering from Condensed N_2, CO, O_2, and NO** **211**
    L. Sanche and M. Michaud

VAN DER WAALS COMPLEXES

13. **Vibrational Predissociation of Small van der Waals Molecules** **231**
    Robert J. Le Roy

14. **Complex-Coordinate Coupled-Channel Methods for Predissociating Resonances in van der Waals Molecules** 263
Shih-I Chu

15. **Vibrationally Excited States of Polyatomic van der Waals Molecules: Lifetimes and Decay Mechanisms** 289
W. Ronald Gentry

16. **Photodissociation of van der Waals Molecules: Do Angular Momentum Constraints Determine Decay Rates?** 305
M. P. Casassa, Colin M. Western, and Kenneth C. Janda

UNIMOLECULAR DYNAMICS

17. **Classical, Semiclassical, and Quantum Dynamics of Long-Lived Highly Excited Vibrational States of Triatoms** 323
R. M. Hedges, Jr., R. T. Skodje, F. Borondo, and W. P. Reinhardt

18. **The Intramolecular Dynamics of Highly Excited Carbonyl Sulfide (OCS)** 337
Michael J. Davis and Albert F. Wagner

BIMOLECULAR REACTIVE SYSTEMS

19. **Vibrationally Bonded Molecules: The Road from Resonances to a New Type of Chemical Bond** 353
Joachim Römelt and Eli Pollak

20. **Bimolecular Reactive Collisions: Adiabatic and Nonadiabatic Methods for Energies, Lifetimes, and Branching Probabilities** 375
Bruce C. Garrett, David W. Schwenke, Rex T. Skodje, Devarajan Thirumalai, Todd C. Thompson, and Donald G. Truhlar

21. **Atom–Diatom Resonances Within a Many-Body Approach to Reactive Scattering** 401
David A. Micha and Zeki C. Kuruoglu

22. **Resonances in the Collisional Excitation of Carbon Monoxide by Fast Hydrogen Atoms** 421
Lynn C. Geiger, George C. Schatz, and Bruce C. Garrett

23. **Resonant Quasi-periodic and Periodic Orbits for the Three-Dimensional Reaction of Fluorine Atoms with Hydrogen Molecules** 441
C. C. Marston and Robert E. Wyatt

24. **Resonance Phenomena in Quantal Reactive Infinite-Order Sudden Calculations** 457
Z. H. Zhang, N. Abusalbi, M. Baer, D. J. Kouri, and J. Jellinek

25. **Dynamic Resonances in the Reaction of Fluorine Atoms with Hydrogen Molecules** 479
D. M. Neumark, A. M. Wodtke, G. N. Robinson, C. C. Hayden, and Y. T. Lee

26. **Reactive Resonances and Angular Distributions in the Rotating Linear Model** 493
Edward F. Hayes and Robert B. Walker

Author Index 515

Subject Index 515

PREFACE

COLLISIONAL RESONANCES PROVIDE a unifying framework for the interpretation of phenomena in the areas of electron–molecule scattering, van der Waals complexes, and reactive chemical dynamics. Of these three areas, resonances have been studied for the longest time in electron–molecule scattering. Resonances in such processes can lead to enhancements of an order of magnitude or more in vibrational excitation and electronic excitation cross sections. As such, they have been widely studied for their effect on e-beam laser initiation, their role in energy degradation in planetary atmospheres and in magnetohydrodynamic energy generation in plasmas and other electron-rich systems, and their effect on transport in electronic devices and radiation chemistry. They also provide a mechanism for dissociative attachment, and they are of great fundamental interest from both a structural and a scattering-theory point of view.

Structurally, resonances provide information on metastable negative ions, negative electron affinities, orbital energies of unbound orbitals, and doubly excited electronic states. From the scattering theory point of view, a resonance is one of the clearest ways to test a quantum dynamical treatment, and it generally provides more definite and more sensitive tests of theory than nonresonant or background cross sections do. An important recent theoretical advance is the calculation of resonances by pseudo-bound-state techniques such as complex scaling and stabilization. This field has seen very rapid progress in the last few years because it is the cutting edge of a new way of thinking and calculating whereby dynamical processes are treated as much as possible by using well established structural methods and minimizing all nonessential scattering boundary conditions.

Another recent advance in electron–molecule resonances is their role in molecular autoionization and photoionization. Here they show up as an exit-channel effect. The increasing availability of synchrotron sources and the proliferation of high-resolution laser spectroscopic techniques are leading to expanded interest in these processes because of the necessity to interpret the resonance features for a greater variety of molecules of chemical interest. Electron–molecule shape resonances are also responsible for structure in inner-shell electron energy-loss spectra in the region around the core ionization threshold; acting as a final-state interaction, the same resonances explain characteristic features observed in X-ray absorption spectra in the region between the absorption edge and the EXAFS diffractive features.

Resonances in van der Waals systems has also been a very exciting area for several years now. The resonance model provides a way to interpret vibrational and rotational predissociation of molecular complexes such as the well studied van der Waals systems $He \cdot I_2$ and $(tetrazine)_2$ or the more recently studied hydrogen-bonded $(HF)_2$ and $(HF)_n$. Van der Waals complexes are being increasingly implicated in enhanced probabilities for low-energy or low-temperature vibrational energy transfer processes in bulk systems and in super-cold molecular beams. The now mature techniques of supersonic molecular beams and pulsed lasers are leading to increasingly detailed probes of the vibrational predissociation process. This process in turn is one of our most quantitative probes of state-selected vibrational–rotational–translational energy transfer mechanisms and the potential energy features responsible for such mechanisms. Photodissociation of van der Waals complexes is particularly relevant to many branches of chemistry because, much more so than conventional photodissociation of strongly bound species, it probes the weak attractive forces responsible for solvent bath effects. In addition, though, the process is sensitive to the strong repulsive forces that often dominate collisional energy transfer. Van der Waals resonances also provide test cases for current theories of mode-selective and statistical intramolecular energy redistribution.

Resonances in reactive dynamics is the newest field of the three, but potentially the most significant. The possible existence of a resonance in the $F + H_2$ reaction has generated much interest because this reaction has been studied by both molecular beam techniques and fully resolved IR chemiluminescence, in both cases as a function of translational energy. Furthermore, the most sophisticated and most generally applicable techniques of molecular collision theory are being applied to this reaction. Resonances are also of great current interest for the prototype chemical reaction $H + H_2$. Recent predictions of subthreshold resonances (which are related to the new subject of vibrationally adiabatic bound states on potential energy surfaces with no wells) are also of special interest. In addition to these bimolecular examples, resonances are providing an increasingly useful model for detailed interpretation of unimolecular dynamics above the dissociation threshold, as well as the reverse associations. A particularly significant result that is emerging in both bimolecular and unimolecular studies is the role of resonance states in determining branching ratios and product state distributions, that is, chemical and quantum-state specificity. Resonance effects in chemical reactions are also getting attention in fundamental studies because they are dramatic quantal interference effects that may sometimes have significant and even dominant implications for macroscopic chemical dynamics.

It is not possible for a book like this to present a comprehensive overview of all activity in the fields covered. But it does, I believe, provide a snapshot of some of the dominant strains of current research in these rapidly growing fields.

I am grateful to James Kinsey, Al Kwiram, Sue Roethel, Robin Giroux, Brenda Ford, and the chapter referees for their assistance with the Symposium and the volume.

DONALD G. TRUHLAR
University of Minnesota
Minneapolis, Minnesota

June 1984

GENERAL DISCUSSIONS

1

Roles Played by Metastable States in Chemistry

JACK SIMONS

Department of Chemistry, University of Utah, Salt Lake City, UT 84112

Metastable states are important in chemistry for reasons which relate to the fact that such states have finite lifetimes and finite Heisenberg energy widths. They are observed in spectroscopy as peaks or resonances superimposed on the continua in which they are buried. Their fleeting existence provides time for energy transfer to occur between constituent species which eventually become separated fragments. It is often the rate of such intrafragment energy transfer which determines the lifetimes of resonances. The theoretical exploration of metastable states presents special difficulties because they are not discrete bound states. However, much of the machinery which has proven so useful for stationary electronic and vibrational-rotational states of molecules has been extended to permit resonance energies and lifetimes to be evaluated. In this contribution, examples of electronic shape and Feshbach, rotational and vibrational predissociation, and unimolecular dissociation resonances will be examined. Finally, a novel situation will be treated in which the energy transfer dictating the decay rate of the metastable species involves vibration-to-electronic energy flow followed by electron ejection.

The purposes of this chapter are to provide overview and perspective concerning the various kinds of metastable species found in chemical systems as well as to focus attention on an interesting class of temporary anions (1-4) whose lifetimes are governed by vibration-electronic coupling strengths. To emphasize the importance of metastable states in experimental chemistry, it is useful to first analyze how they are created via collisional or photon absorption processes. This provides a basis for discussing the signatures which metastable states leave in the instrumental responses seen in the laboratory. Having introduced metastable states in relation to the experimental situations in which they arise, it is useful to

0097-6156/84/0263-0003$06.00/0

distinguish between two primary categories of such species and to classify many of the systems treated in this symposium. This overview and categorizing focuses attention on the properties (e.g., mass, angular momentum, energy, potential energy surfaces, internal energy distribution) which determine the decay rates or lifetimes of particular species.

After giving an overview and interpretative treatment of a wide variety of metastable systems, this paper treats in somewhat more detail a class of highly vibrationally excited molecular anions which undergo electron ejection at rates determined by the strength of vibration-electronic coupling present in the anion. The findings of theoretical simulations of the electron ejection process as well as the experimental relevance of such temporary anions are discussed.

Metastable states as they occur in collisions and half collisions.

One can form metastable states (denoted here by AB*) either by bringing together two fragments (A and B) in a collision experiment or by exciting a bound state of the AB system using, for example, photon absorption or electron impact excitation. In situations where bound-state excitation is employed, the subsequent decay of the metastable to produce fragments (A and B) is termed a "half collision". The decay rate or lifetime (τ) of the excited state can, in some cases, be inferred from the Heisenberg component to the width (Γ) of the measured resonance feature in the bound-state absorption or excitation spectrum. It is often very difficult to extract from the total observed linewidth the component due to the decay of the corresponding state. Unresolved rotational structure and Doppler broadening often dominate the linewidth. Only for lifetimes shorter than 10^{-9} s (or $\Gamma \sim 0.03\ cm^{-1}$) is it likely that the Heisenberg width will be a major component. For very long lived states, the lifetime may be measured by monitoring the time evolution of the product fragment species, by, for example, laser induced fluorescence or the absorption spectrum of one of the fragments produced. If one or both of the fragments are ionic, ion detection methods can be used. The appearance of structure in the absorption spectrum superimposed upon a background continuum is a result of the strong-interaction region component of the resonance-state wavefunction. This is the component of the metastable specie's wavefunction in which the fragments A and B reside within the region where their interfragment potential energy is significant. For fragment separations outside this region, the term "asymptotic" is employed. It is this "in close" part of the wavefunction that characterizes resonances and that may carry strong oscillator strength from the underlying bound state. Since in the lower AB state the A--B relative motion is bound and hence localized, it is only the localized part of the excited-state wavefunction which will appreciably overlap the ground-state wavefunction. Hence for excited metastable states, which possess large valence-region components, the electric dipole transition matrix element can be substantial. In contrast, excitation from the lower bound state to nonresonant dissociative excited states gives rise to smaller transition dipoles, because such excited states have small valence-region components and hence weaker absorption intensities.

For experiments in which two fragments collide to produce metastable species which subsequently undergo decay, the lifetime of the metastable state is reflected in the kinetic energy dependence of the elastic, inelastic, and reactive cross-sections characterizing the collision events. Such experiments measure cross-sections as the initial relative kinetic energy of the fragments is scanned. They may also monitor the kinetic energy of one or both of the fragments ejected from the metastable species. Such cross-sections when plotted as functions of the fragment relative kinetic energy display resonance features in the vicinity of each metastable-state energy. Collisionally formed states are called metastable or resonances if their lifetimes are substantially longer than the "transit times" which one would calculate for the fragments based upon knowledge of their asymptotic relative kinetic energies. The simple fact that the fragments remain in close contact longer than expected based on their transit time is what allows interfragment energy transfer to be so efficient in metastable species. Given interaction times of the order of $10-10^{13}$ times the transit time, [a typical transit time for molecules colliding at room temperature is $\sim 10^{-13}$ s], even small rates of energy transfer per interfragment vibration can result in large net energy transfers during the state's lifetime. Such enhanced efficiency for energy transfer results in higher probabilities that the product fragments exit in different internal states than were occupied in the initial collision pair. This is one of the most chemically important characteristics of metastable states.

Two Principle Categories of Resonance States. Denoting the internal energies of the separated fragments (A and B) by $E_{Ai} + E_{Bj}$, all metastable or resonance states can be subdivided into those whose total energy E_r lies either below or above each such asymptotic energy. As is shown below, this division is chemically meaningful because the physical features which determine the energies and lifetimes of the two types of resonances defined in this manner are different.

At energies above each $E_{Ai} + E_{Bj}$ one can expect to find orbiting resonances (also known as shape resonances) if the fragments experience an attractive potential (V) varying more strongly than r^{-2} and have non-zero relative angular momentum (ℓ). Here r represents the coordinate which asymptotically is the distance between the centers of mass of the two fragments. In elastic or inelastic collisions, the entering and exiting fragments have identical chemical composition but their internal (e.g., vibrational, rotational, electronic) states may differ. For reactive processes which proceed through a metastable state, the entering and exiting species are chemically different. The effective radial potentials $V + \hbar^2\ell(\ell+1)/2r^2\mu$ depicted in Figure (1) for various ℓ values may then support states whose total energy exceeds $E_{Ai} + E_{Bj}$ but which can decay only by the two fragments of relative reduced mass μ tunneling through the "angular momentum barrier" arising from the $\hbar^2\ell(\ell+1)/2\mu r^2$ factor in interfragment angular kinetic energy (i.e., the interfragment rotation). The tunneling lifetimes of these resonances are determined by the height and thickness of the angular barrier and the fragment reduced mass μ.

At energies below each fragment energy $E_{Ai} + E_{Bj}$, target-excited or Feshbach resonances are to be expected. Although these metastable states cannot decay to the $E_{Ai} + E_{Bj}$ asymptotic state, they possess enough energy to couple to the continua of lower fragment states (e.g., to $E_{Ai-1} + E_{Bj}$ or $E_{Ai} + E_{Bj-1}$; see Figure (2)). What makes these states metastable is the fact that not enough of their total energy is initially (i.e., as they are created) contained in the dissociative coordinate r. They have ample energy in the internal intrafragment degrees of freedom; however, some of this energy must be redistributed to the asymptotically dissociative motion before the resonance state can decay. The lifetimes of such target-excited states are determined by the strength of the coupling between the fragment internal degrees of freedom and the interfragment motion which, in turn, depends upon the curvature and anharmonicity of the system's potential energy surface. The position or energy E_r of the resonance state relative to its asymptotically closed reference energy $E_{Ai} + E_{Bj}$ is determined by the strength of the attractive potential between the A and B fragments in the E_{Ai} and E_{Bj} states; the stronger this attraction, the further below $E_{Ai} + E_{Bj}$ the Feshbach resonance should lie. A decay channel is said to be closed if the total energy of the system lies below the asymptotic energy of this channel. Thus metastable species can only decay to produce open-channel products. It is also possible to have target-excited resonances that lie above the $E_{Ai} + E_{Bj}$ asymptotic; these are discussed elsewhere (5).

Examples of Orbiting and Target-Excited Resonances

To better appreciate the chemical significance of both classes of metastable species it is helpful to examine several examples of such species. This overview is by no means intended to be exhaustive either with respect to the systems it covers or concerning references to workers in the area. Its purpose is to illustrate how metastable states can have important effects in chemical experiments. Therefore, the treatment given here is rather qualitative and directed toward the interested but non-expert reader.

Because electrons have much smaller mass than atomic nuclei, it is natural to distinguish between situations in which an electron is one of the fragment particles and those in which both fragments are so called heavy particles (i.e., atoms, molecules or ions). In the former case, the asymptotic fragment internal energies $E_{Ai} + E_{Bj}$ reduce to the energies of the atomic or molecular fragment since the electron fragment has no internal energy levels of its own (external magnetic field effects are ignored).

Electronic Shape Resonances. In the first three examples given below in Table I, the orbiting resonance can be viewed as forming when an electron of kinetic energy E_r enters an empty low-energy orbital of the atomic or molecular target fragment. As indicated in Table I, the active orbital has, in every case, nonzero angular momentum components which give rise to the barrier through which the electron must tunnel to escape. The attractive part of the

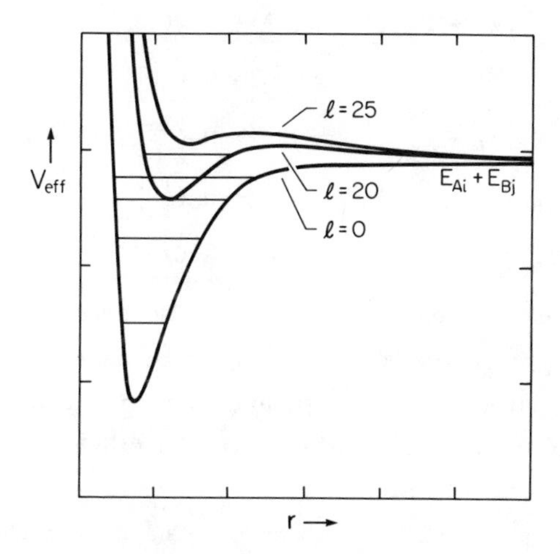

Figure 1: Effective potentials V + $\hbar^2 J(J(1)/2\mu r^2$ for ℓ = 0, 20, 25.

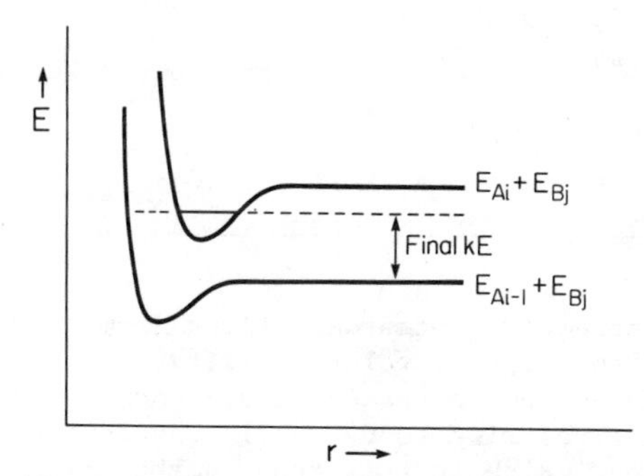

Figure 2: A Feshbach resonance exists below E_{Ai} + E_{Bj} but above dissociation to E_{Ai-1} + E_{Bj}. The kinetic energy (KE) of the ejected A and B fragments is also shown.

Table I. Electronic Shape Resonance Examples

Formation Process	Decay Products
$e(3.7\ eV) + H_2(1\sigma_g^{\ 2},v) \rightarrow H_2^-(1\sigma_g^{\ 2}1\sigma_u) \rightarrow$	$H_2(1\sigma_g^{\ 2},v') + e(p\ wave)$[a]
	$\rightarrow H + H^-$, at 3.73 eV (Ref.10)
$e(2.3\ eV) + N_2\ (^1\Sigma_g^+,v) \rightarrow N_2^-(^2\pi_g,v'')$	$\rightarrow N_2(^1\Sigma_g^+,v') + e(d\text{-}wave)$[b]
$e(0.3\ eV) + Mg(^1S) \rightarrow Mg^-(^2P^o)$	$\rightarrow Mg(^1S) + e(p\text{-}wave)$[c]
$NaCl^-(^2\Sigma^+) + h\nu(2.1\ eV) \rightarrow NaCl^-(^2\pi)$	$\rightarrow NaCl\ (^1\Sigma^+) + e(p\text{-}wave)$[d]

a) Vibrational excitation of H_2 occurs. The ejected electron comes off primarily in a p-wave distribution because the $1\sigma_u$ orbital of H_2 is dominated by the p symmetry when it is expanded about the center of mass of the molecule. The width of this state is greater than 2 eV (Refs. 6-9).

b) Again vibrational excitation occurs. The d-wave character is dictated by the d-like symmetry of the active $1\pi_g$ antibonding orbital of N_2. The width $\Gamma \sim 0.6$ eV corresponds to a lifetime of $\sim 10^{-14}$ s (Refs. 11-14).

c) The width $\Gamma \sim 0.1\text{-}0.2$ eV corresponds to a lifetime of $2 - 4 \times 10^{-14}$ s (Refs. 15-18).

d) The lifetime is $\sim 10^{-12}$ s (Refs. 19-20).

electron-target effective potential arises from charge-dipole, charge-induced dipole (i.e., polarizability) and valence-level interactions. The shape resonance occurs when the energy of the ejected electron gives rise to deBroglie wavelengths in the strong-interaction region, which in this case is the valence region, that permit the electron's radial wavefunction to establish a standing wave pattern in the region between the two inner turning points of the effective potential (see Figure 1). In all of the examples illustrated here, only the ground electronic-state fragment is involved. Such does not have to be the case, however; orbiting resonances can arise from the binding of an electron with $\ell \neq 0$ to an excited electronic state of the fragment.

The relevance of the above kind of electronic shape resonance to chemistry is twofold. First, in environments such as plasmas, electrochemical cells, and the ionosphere, where free electrons are prevalent, the formation of such temporary anions can provide avenues for the free electrons to "cool down" by transferring kinetic energy to the internal (vibrational and/or electronic) degrees of freedom of the fragment. (6-14) Second, metastable states may play important roles in quenching excited electronic states of atoms and molecules by providing a mechanism through which electronic energy can be transformed to fragment internal energy. For example, H_2^- ($1\sigma_g^2 1\sigma_u$) is invoked as an intermediate to explain the quenching of electronically excited (21-21) Na(3p;$^2P^o$) and (23 Mg(^1P) by H_2 ($1\sigma_g^2$,v) to yield vibrationally hot H_2:

$$Na(^2P) + H_2(v) \rightarrow Na^+H_2^- \rightarrow Na(^1S) + H_2(1\sigma_g^2, v'),$$

$$Mg(^1P^o) + H_2(v) \rightarrow Mg^+(^2S)H_2^- \rightarrow Mg(^1S) + H_2(v').$$

In these examples, the metastable H_2^- does not actually undergo electron loss because the Na^+ or Mg^+ ion "retrieves" the electron as the complex dissociates leaving the H_2 vibrationally excited and the Na or Mg atom in its ground state. Recent work on electron transmission spectroscopy studies of unsaturated hydrocarbons (24) demonstrates that electronic shape resonances may be essentially ubiquitous in chemical systems which possess low-energy vacant orbitals and the availability of electron density to enter such orbitals.

Electronic Feshbach Resonances. In Table II are a few examples of target-excited or Feshbach resonances in which the ejected electron is initially attached to an electronically excited state of the fragment. Note that the angular properties (i.e., p-wave, s-wave, etc.) of the ejected electron are again constrained by the symmetries of the metastable state and the lower lying target state to which it decays. In the case of $H^-(2p^2, {}^3P^e)$, parity constraints even forbid direct ejection of a p-wave electron (odd parity) to leave the H atom in the H(1s,^2S) (even parity) ground state. $H^-(2p^2, {}^3P^e)$ must first radiate to produce the $H^-(2s2p, {}^3P^o)$ metastable state which can then undergo Feshbach decay to produce H(1s,^2S) and a p-wave electron. In contrast, the seemingly similar metastable $Na^-(3p^2, {}^3P^e)$ state of Na^- can undergo radiative decay to $Na^-(3s3p, {}^3P^o)$ and subsequent tunneling (not Feshbach) decay to Na (3s,^2S) and e(p-wave); direct Feshbach decay of $Na^-(3p^2, {}^3P^e)$ to Na(3s,^2S) and e(p-wave) is parity forbidden as was the case in H^-.

Electronic Feshbach resonances are often very long lived and hence have narrow (often $\leq$ 0.01 eV) widths. Their lifetimes are determined by the coupling between the quasibound and asymptotic components of their electronic wavefunctions. Because the Feshbach decay process involves ejection of one electron and deexcitation of a second, it proceeds via the two-electron terms e^2/r_{ij} in the Hamiltonian. For example, the rate of electron loss in $H^-(2s2p, {}^3P^o)$ is proportional to the square of the two-electron integral $\langle 2s2p | e^2/r_{12} | 1s\ kp \rangle$, where kp represents the continuum p-wave orbital. This integral, and hence the decay rate, is often quite small because of the size difference between the 2s or 2p and 1s orbitals and because of the oscillatory nature of the kp orbital. In fact, series of Feshbach resonances involving, for example, nsnp $\rightarrow$ n's kp decay often show (29) lengthening lifetimes as functions of increasing n. This trend can be explained in terms of both the greater radial size difference and the increasing oscillatory character (due to increased kinetic energy of the ejected electron) of the kp orbital as n increases. Both trends tend to make the coupling integral $\langle nsnp | e^2/r_{12} | n'skp \rangle$ smaller.

Table II. Electronic Feshbach Resonance Examples

Metastable Species		Decay Product
$He^-(1s2s^2, {}^2S)$	→	$He(1s^2, {}^1S)$ + e(s-wave)[a]
$H^-(2p^2, {}^3P^e)$[b]	→	$H^-(2s2p, {}^3P^o)$ + hν ↓ $H(1s, {}^2S)$ + e(p-wave)
$O_2^-(1\pi_g^1\ 3s\ \sigma_g^2, {}^2\pi_g)$[c]	→	$O_2(1\pi_g^2, {}^3\Sigma_g^+)$ + e (p wave)
$Na^-\ (3p^2, {}^3P^e)$ → $Na^-(3s3p, {}^3P^o)$ + hν	→	$Na(3s, {}^2S)$ + e(p-wave)[d]

a) The electron is ejected with 19.4 eV of kinetic energy. The width of the resonance is 0.008 eV ($\tau \sim 5 \times 10^{-13}$ s) (Ref. 25).

b) This resonance lies ~ 0.01 eV below the 2p state of H and has a radiative decay rate to the ${}^3P^o$ state of $2.5 \times 10^6\ s^{-1}$. The ${}^3P^o$ state then decays via a Feshbach mechanism to H(1s) and e(p) at a rate of $6 \times 10^8\ s^{-1}$ (Ref. 26).

c) At 8.04 eV above the ground state of O_2, this resonance is thought to involve two electrons in a Rydberg-like $3s\sigma_g$ orbital bound to essentially an O_2^+ core (Ref. 27).

d) As explained in the text, the decay of $Na^-(3s3p, {}^3P^o)$ is not a Feshbach decay. This example is listed in this Table only to demonstrate that its seeming similarity to the above H^- case is deceiving. The ${}^3P^e$ state of Na^- lies 0.06 eV below the ${}^3P$ state of Na (Ref. 28).

The chemical importance of electronic Feshbach resonances derives from essentially the same effects as were given earlier for shape resonances. They allow either free electrons or electron density from a collision partner to give energy to the internal (vibrational, rotational, or electronic) degrees of freedom of the target.

Heavy Particle Shape Resonances. The van der Waals attraction between two Ar atoms is shown in Figure 3. It supports eight bound vibrational states if the diatom's rotational quantum number j is zero. For j = 30, only two vibrational states are bound. The angular momentum barrier arising from the $j(j+1)\hbar^2/2\mu r^2$ factor in the Ar_2 rotational kinetic energy gives rise to an effective radial potential (see Figure 3) which for j = 30 supports only two bound states and one metastable orbiting resonance state. Lower j values give rise to more bound states and more shape resonance states. The lifetimes of these states depend upon the height and thickness of the barrier, which depend upon j and the reduced mass and actually range (30) from 10^{-11} s to 10^{10} s.

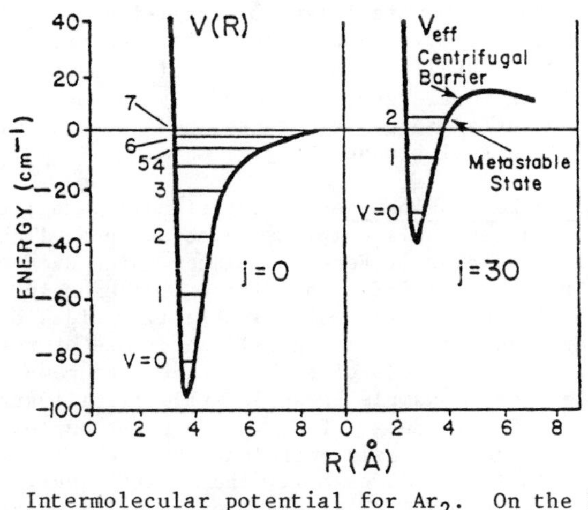

Figure 3: Intermolecular potential for Ar_2. On the left is the potential curve for non-rotating (j=0)Ar_2. On the right is the effective potential for rotating (j=30)Ar_2.

One can, of course, have orbiting resonances for larger molecules and for molecules in which chemical bonds (rather than van der Waals interactions) are operative. For example, H_2 (v = 0, j = 38) undergoes rotational predissociation to produce two H atoms. This metastable shape resonance state has a Heisenberg width (31) of 90 cm^{-1}; the v = 0, j = 37 state of H_2 decays with a width of 6 cm^{-1} and the v = 14, j = 4 state does so with $\Gamma \sim 0.007$ cm^{-1} (even though the v = 14, j = 4 state has far more total energy than the v = 0, j = 38 one).

Orbiting resonances are very important in chemistry. They have observable effects on the transport properties of dense gases and liquids (32), and they are thought to provide a mechanism for atom-atom recombination to occur via metastable shape resonances which live long enough to become stabilized by collision (33) with another molecule or with a surface. Shape resonances also give rise to broadening in experimentally observed absorption spectra; transitions into high rotational levels often result in populating rotationally predissociating shape resonances whose natural lifetime is reflected in the spectral broadening.

Heavy-Particle Feshbach Resonances. Vibrationally and rotationally predissociating van der Waals complexes such as those listed in Table III provide examples of Feshbach states which decay via intramolecular energy transfer. The first three examples involve transfer of vibrational energy, which initially resides essentially in one molecular fragment, to the radial coordinate of the weak van der Waals bond. The lifetimes of such species span many orders of magnitudes. The fourth example shown in Table III illustrates a case in which rotational energy of H_2 in the H_2Ar complex is transferred to the van der Waals coordinate upon which the system can dissociate. In the final example, the electronically excited HCN can, if it is prepared with excess energy in its bending vibrational mode (ν_2), undergo CH bond rupture if the excess bending energy transfers to the CH stretching mode (ν_1). The coupling between ν_2 and ν_1 is caused by strong off-diagonal curvature of the HCN molecule's potential energy surface between the bending and CH stretching coordinates, so the strength of this strong off-diagonal curvature determines the decay rate of C^1A' HCN.

The chemical importance of heavy-particle Feshbach resonances cannot be overstated. They are present in unimolecular rearrangements, both thermal ones and those which occur in organic photochemical reactions, in mass spectroscopic ion fragmentations, and in bimolecular collisions which proceed through long-lived intermediates.

A Novel Class of Target-Excited Resonance

Experiments have recently been carried out (1,3) in which polyatomic molecular anions trapped in an essentially collisionless ion cyclotron resonance cell are vibrationally excited using an infrared laser laser of 0.1-6 J/cm^2 fluence and ~ 1000 cm^{-1} energy. Electron ejection from the anions is observed to occur at rates which are fluence dependent. The mechanism of this ejection is the subject of these remarks.

Table III. Heavy-Particle Feshbach Resonance Examples

Metastable Species		Decay Products
$I_2(B\ ^3\Pi, v)He$	→	$I_2(B\ ^3\Pi, v') + He$ [a]
$(C\ell_2)_2$	→	$2C\ell_2$ [b]
$NO_2(^2B)He$	→	$NO_2(^2B) + He$ [c]
$H_2(j,v)Ar$	→	$H_2(j',v) + Ar$ [d]
$HCN(C\ ^1A', \nu_1 \nu_2 \nu_3)$	→	$CN(B\ ^2\Sigma) + H$ [e]

a) A very strong propensity is observed for the $v' = v - 1$ channel. The rate of decay is $4.5 \times 10^9\ s^{-1}$ for $v = 12$ and increases to $2.6 \times 10^{10}\ s^{-1}$ for $v = 26$ (Refs. 35,36).

b) The vibrationally excited $(C\ell_2)_2$ requires $\sim 10^{-4}$ s to decay. This time is 10^8-10^9 times the vibrational period of the $C\ell_2$ moiety (Ref. 37).

c) The observed lifetime for the vibrationally hot ^2B state of NO_2 to eject the van der Waals bound He atom is 10^{-11} s (Ref. 38).

d) The rotationally hot H_2 moiety in H_2Ar decays via transfer of rotational energy to $H_2 \cdots Ar$ relative motion with a width of $\sim 10^{-3}\ cm^{-1}$ for $j = 2$ and $4 \times 10^{-4}\ cm^{-1}$ for $j = 4$ (Refs. 39,40).

e) The bent excited state of HCN, when photochemically prepared from the $X\ ^1\Sigma^+$ ground state, requires transfer from the bending (ν_2) motion to the CH stretching motion (ν_1) before dissociation can occur (Refs. 41,42).

Laser pulses of 3 μs duration and a typical fluence of 1.0 J/cm^2 and infrared absorption cross sections of 10^{-18}-10^{-20} cm^2 yield photon absorption rates of 10^5-10^7 photons/s, which are much slower than the rate of intramolecular vibrational energy redistribution in systems such as benzyl anion. Hence, the anions are excited in a sequential process in which the vibrational energy is redistributed before the next photon is absorbed. This means that such experiments cannot determine in which vibrational mode(s) the energy resides.

It has also been demonstrated (3), by studying the competition between electron loss and a unimolecular decomposition of known activation energy, that sequential infrared absorption continues to occur even after the anion has achieved enough total internal energy to reach its electron detachment threshold. This implies that, near threshold, electron ejection must not be occuring faster than the 10^5-10^7 s^{-1} photon absorption rate. These experiments do not, however, allow one to conclude with much certainty how far above threshold photon absorption continues to occur; clearly, once the electron loss rate exceeds the absorption rate the anion will no longer absorb.

The above experiments are hampered by lack of knowledge of the total energy content and internal energy distribution of the anions. This interpretation would be helped by order of magnitude estimates of the rate of electron ejection as a function of energy above threshold. The author's group recently undertook (4) an *ab initio* simulation of such ejection rates for two prototype anions (OH^- and LiH^-) which are viewed as limiting cases of slow (OH^-) and rapid (LiH^-) vibration induced electron ejection. Diatomic anions were chosen to obviate questions about where (i.e., in what vibrational mode) the internal energy is residing. OH^- is an ideal candidate for slow electron loss because the energy of its 1π active orbital is only weakly dependent upon bond length and because its detachment threshold is large (1.82 eV). In contrast, the 3σ active orbital of LiH^-, which consists primarily of a non-bonding $2s$-$2p\sigma$ hybrid localized on the Li center and directed away from the H center, is quite strongly affected by movement of the very polar (Li^+H^-) bond. As a result, the detachment energy of LiH^- varies substantially with bond length; even at the equilibrium bond length it is only 0.3 eV.

For neither LiH^- nor OH^- do the anion and neutral Born-Oppenheimer potential energy curves cross. Such is also thought to be the case for the energy surfaces of the polyatomic anions studied in the experiments of Refs. (1) and (3) and for the curves of (B $^2\Sigma_u^+$) C_2^- and (X $^1\Sigma_g^+$ or a $^3\Pi_u$) C_2 where Lineberger (43) has observed (B $^2\Sigma_u^+$) C_2^- to be metastable with respect to electron ejection. Hence the electron loss mechanism does not involve the anion sampling, through its vibrational motion, geometries where electronic shape or Feshbach resonances or direct detachment occurs. It requires coupling between electronic and vibrational motion and can be viewed as a radiationless transition between the two components of the metastable state (the vibrationally excited anion and the vibrationally cooler neutral with an outgoing electron) induced by the nonadiabatic coupling terms which the Born-Oppenheimer Hamiltonian neglects. In this sense, such metastable states can be viewed as target-excited states. They are analogous to vibrationally excited Rydberg states which undergo autoionization (44).

It was found in Ref. (4) that the lifetimes for electron ejection ranged from 10^{-9} to 10^{-10} s for LiH^- in vibrational levels between v = 3 and v = 10 and from 10^{-5} to 10^{-6} s for OH^- from v = 5 to v = 11. The ejection rates did not increase strongly with increasing vibrational energy above threshold. In all cases, the anions were found to decay preferentially to the energetically closest vibrational state of the neutral; the branching ratios for decay into various neutral vibrational levels correlated well with ion-neutral Franck-Condon-like factors.

This range of lifetimes (10^{-9}-10^{-10} to 10^{-5}-10^{-6} s) for prototype fast and slow electron ejectors is entirely consistent with the estimates made in Refs. (1), (3) and (43). The weak dependence of these rates on the energy above threshold indicates that many polyatomic anions (i.e., those for which the ejection rate near threshold is not faster than the photon absorption rate) continue to absorb infrared photons far above their detachment thresholds. This information, when combined with the observation of a propensity to decay to the closest vibrational level of the neutral, suggests that vibrationally hot neutrals can be expected to be produced in such infrared pumping experiments. When considering the reverse reaction, the attachment of an electron to a molecular target to produce a vibrationally hot anion, the rate of electron ejection as studied here can be combined with estimates of the equilibrium constant for

$$e + A(v) \rightleftarrows A^-(v')$$

to estimate the rate constant for the attachment step. Benson has actually done so (45) in a recent study of SF_6^-.

Acknowledgments

Acknowledgment is made to the National Science Foundation (Grant #8206845), as well as to the Donors of the Petroleum Research Fund, administered by the American Chemical Society (PRF 14446-AC6), for their support.

Literature Cited

1. Meyer, F. K.; Jasinski, J. M.; Rosenfeld, R. N.; Brauman, J. I. J. Am. Chem. Soc. 1982, 104, 663; Rosenfeld, R. N.; Jasinski, J. M.; Brauman, J. I. J. Chem. Phys. 1979, 71, 1030; Wight, C. A.; Beauchamp, J. L. J. Am. Chem. Soc. 1982, 103, 6501.
2. Simons, J. J. Am. Chem. Soc. 1981, 103, 3971.
3. Foster, R. F.; Tumas, W.; Brauman, J. I. J. Chem. Phys. 1983, 79, 4644.
4. Acharya, P. K.; Kendall, R. A.; Simons, J. J. Am. Chem. Soc. 1984, 00, 0000.
5. Taylor, H. S.; Nazaroff, G. V.; Golebiewski, A. J. Chem. Phys. 1966 45, 2872.
6. Bardsley, J.N.: Herzenberg, A.; Mandl, F. Proc. Phys. Soc. London 1966, 89, 321.
7. Eliezer, I.; Taylor, H.S.; Williams, J. K. J. Chem. Phys. 1967, 47, 2165.
8. Schulz, G. J. Phys. Rev. 1964, 135, A988; 1964, 136, A650.
9. Trajmar, S.; Truhlar, D. G.; Rice, J. K.; Kuppermann, A. J. Chem. Phys. 1970, 52, 4516.
10. Schulz, G. J.; Asundi, R. K. Phys. Rev. 1967, 158, 25.
11. Herzenberg, A.; Mandl, F. Proc. Roy. Soc. 1962, A270, 48.
12. Schulz, G. J.; Koons, H. C. J. Chem. Phys. 1966, 44, 1297.
13. Herzenberg, A. J. Phys. 1968, B1, 548.
14. Birtwistle, D. T.; Herzenberg, A. J. Phys. 1971, B4, 53.
15. Burrow, P. D.; Michejda, J. A.; and Comer, J. J. Phys. 1976, B9, 3255.
16. Hazi, A. V. J. Phys. 1978, B11, L259.
17. Donnelly, R. A. J. Chem. Phys. 1982, 76, 5914.
18. Mishra, M.; Kurtz, H.; Goscinski, O.; Ohrn, Y. J. Chem. Phys. 1983, 79, 1896.

19. Novick, S. E.; Jones, P. L.; Mulloney, T. J.; Lineberger, W. C. J. Chem. Phys. 1979, 70, 2210.
20. Collins, L.A.; Norcross, D. W. Phys. Rev. Lett. 1977, 38, 1208.
21. Blais, N. C.; Truhlar D. G. J. Chem. Phys. 1983, 79, 1334; McGuire, P.; and Bellum, J. C. J. Chem. Phys. 1979, 71, 1975.
22. Hertel, I. V.; Hofmann, H.; Rost, K. A. Chem. Phys. Lett. 1977, 47,163.
23. Breckenridge, W. H.; Umemoto, H. J. Chem. Phys. 1981, 75, 698.
24. Jordan, K. D.; Burrow, P. D. Acc. Chem. Res. 1978, 11, 341.
25. Sanche, L.; Schulz, G. J. Phys. Rev. 1972, A5, 1672.
26. Drake, G. W. F. Astrophys. J. 1973, 184, 145.
27. Sanche, L.; Schulz, G. J. Phys. Rev. 1972, A6, 69.
28. Norcross, D. W. Phys. Rev. Lett. 1974, 32, 192.
29. Moiseyev, N.; Weinhold, F. Phys. Rev. 1979, A20, 27.
30. Ewing, G. E. Can. J. Phys. 1976, 54, 487; Acc. Chem. Res. 1975, 8, 185.
31. Leroy, R. J. Ph.D. Thesis, University of Wisconsin, 123, 1971.
32. Hirschfelder, J. O.; Curtiss, F. F.; Bird, R. B. Molecular Theory of Gases and Liquids 1967, J. Wiley and Sons, N.Y. See p. 555.
33. Roberts, R. E. Ph.D. Thesis, University of Wisconsin, 1968.
34. Herzberg, G. Molecular Spectra and Molecular Structure I. Spectra of Diatomic Molecules 1950, Van Nostrand Reinhold, N.Y. See pp. 425-430.
35. Johnson, K. E.; Wharton, L.; Levy, D. H. J. Chem. Phys. 1978, 69, 2719.
36. Beswick, J. A.; Jortner, J. Adv. Chem. Phys. 1981, 47, 363.
37. Dixon, D. A.; Herschbach, D. R. Ber. Bunsen Ges. Phys. Chem. 1977, 81, 145.
38. Smalley, R.; Wharton, L.; Levy, D. H. J. Chem. Phys. 1977, 66, 2750.
39. Leroy, R. J.; Carley, J. S. Adv. Chem. Phys. 1980, 42, 353.
40. Beswick, J. A.; Requena, A. J. Chem. Phys. 1980, 72, 3018.
41. Chuljian, D. T.; Ozment, J.; Simons, J. J. Chem Phys. 1984, 80, 176; Inter. J. Quantum Chem. 1982, S16, 435.
42. Macpherson, M. T.; Simons, J. P. J. Chem. Soc. Faraday Trans. II 1978, 74, 1965.
43. Jones, P. L.; Mead, R. D.; Kohler, B. E.; Rosner, S. D.; Lineberger, W. C. J. Chem. Phys. 1980, 73, 4419.
44. Berry, R. S. J. Chem. Phys. 1966, 45, 1228.
45. Heneghan, S. P.; Benson, S. W. Inter. J. Chem. Kin. 1983, 15, 109.

RECEIVED June 11, 1984

2

Direct Variational Methods for Complex Resonance Energies

C. WILLIAM MCCURDY

Department of Chemistry, Ohio State University, Columbus, OH 43210

In contrast to earlier complex coordinate methods which require a specific analytic continuation of the Hamiltonian, complex basis function methods for resonances rely on the existence of a complex variational principle for complex resonance energies, $E_r(R) - i\Gamma(R)/2$, and are thus considerably more general, flexible and successful than their antecedents. A summary is presented of several methods which have appeared in the literature in this context with a view to displaying their similarities and common conceptual grounding. Those methods include: (1) complex SCF, (2) complex stabilization (CI), (3) the saddle-point coordinate rotation methods, and (4) analytic continuation of stabilization graphs. The last of these approaches requires only real-valued eigenvalue calculations, but nonetheless yields complex resonance energies directly.

Recent years have seen the development of a number of methods for direct calculation of complex resonance energies, both in the context of electron scattering from atoms and molecules and in the context of heavy-particle scattering. This article is not intended as a review of that literature. Rather, I present here a brief summary of some of our own work together with a description of a few other approaches with the intention of exhibiting the common theme they employ. That theme is a generalized complex variational principle for resonances, and the various methods discussed here differ only in the forms of the trial wavefunctions they prescribe. Before beginning the specific description of these methods I will describe the variational principle and, to some extent, its origins.

0097-6156/84/0263-0017$06.00/0

Complex Variational Principles for Resonances

The method of complex scaling of coordinates, known variously as rotated coordinates, complex coordinates and dilatation analyticity, is the predecessor of all of the complex variational methods we will consider. Although complex coordinates have long been a textbook device for establishing the analytic structure of the S-matrix in potential scattering (1), it was the theorems of Aguilar, Balslev and Combes (2,3) on many-particle systems which formed the basis of the first atomic resonance calculations by this approach. Their now familiar results state that if all the coordinates in an atomic Hamiltonian are scaled according to

$$\vec{r} \rightarrow \vec{r}e^{i\theta} \tag{1}$$

the spectrum of the resulting nonhermitian Hamiltonian consists of discrete eigenvalues at the bound-state energies and at the complex energies of resonances, $E_r - i\Gamma/2$, which have been exposed by the rotation of the continuous spectra associated with each scattering threshold into the complex plane by an angle equal to -2θ. Moreover the eigenfunctions associated with the complex resonance eigenvalues are square integrable. Figures showing the complex spectrum of the complex scaled Hamiltonian, $H_\theta = H(\{\vec{r}_i e^{i\theta}\})$, can be found in several reviews on the subject (4-6).

The first computational implementations of these theorems were made by Doolen, Hidalgo, Nuttall and Stagat (7-9) for two-electron atoms. These were configuration interaction (CI) calculations using real valued radial basis functions and the complex Hamiltonian, H_θ. They made use of a complex variational principle of the form

$$\delta E_\theta[\Psi]/\delta\Psi = 0 \tag{2}$$

with

$$E_\theta[\Psi] = \frac{\int (\Psi)^{(*)} H_\theta \, \Psi \, d\tau}{\int (\Psi)^{(*)} \Psi \, d\tau} \tag{3}$$

where the notation $(\Psi)^{(*)}$ means that only the angular factors (spherical harmonics) in the wavefunctions are complex conjugated. The variations in Equation 2 in these calculations (7-9) were variations in complex coefficients in an otherwise ordinary CI expansion in Hylleraas functions. The validity of this approach rests on the fact that the eigenfunctions of H_θ in question are square integrable.

Equations 2 and 3 state a complex variational theorem, which has been used and discussed extensively in the literature (4-6,10), but it is not the variational theorem on which the most successful methods are based and which concerns us here. The more general complex variational principle which does concern us makes use of a real-valued Hamiltonian and can be stated simply as

$$\delta E[\Psi]/\delta\Psi = 0 \tag{4}$$

with

$$E[\Psi] = \frac{\int (\Psi)^{(*)} H \, \Psi \, d\tau}{\int (\Psi)^{(*)} \, \Psi \, d\tau} \tag{5}$$

where arbitrary complex variations in the square-integrable wavefunction Ψ are allowed. It should be emphasized that this is a stationary principle only and does not directly provide variational bounds for resonance energies.

Equation 5 encompasses the complex scaling variational principle of Equation 3 as a special case, and this can be seen, for example, in a one-particle problem from the following identity for radial matrix elements with respect to square-integrable basis functions.

$$\int_0^\infty \phi_i(r) \, H(re^{i\theta}) \, \phi_j(r) \, dr \tag{6}$$

$$= \int_0^\infty e^{-i\theta/2} \phi_i(re^{-i\theta}) H(r) e^{-i\theta/2} \phi_j(re^{-i\theta}) dr$$

With this identity it becomes clear that Equation 5 with complex basis functions of the form $e^{-i\theta/2}\phi(re^{-i\theta})$ is identical to Equation 3 with real basis functions of the form $\phi(r)$. Thus this more general variational principle can be used to reinterpret calculations which used complex scaling in the Hamiltonian but various kinds of complex basis functions in Equation 3. Such calculations (11,12) were the first successful applications of these ideas to systems of more than two electrons.

The generality of the variational principle expressed by Equations 4 and 5 is seen primarily in calculations which cannot be reformulated in terms of complex coordinates, and there are now many such examples, several of which are described below. The formal footing of this generalized variational principle is somewhat ambiguous at this point because attempts to justify it have, to my knowledge, involved appeals to analytic continuation of the Hamiltonian along general contours in r such as Simon's exterior scaling contour (13), and there is no such unique contour implicit in most applications of the idea (14). This ambiguity, however, did not prevent the implicit use of this principle in early studies by Herrick and coworkers (15-17) or its explicit use in the first such calculation on a molecular problem by McCurdy and Rescigno (18). Nor did the lack of formal footing prevent Junker (19,20) from giving a clear statement of this idea, which he called "complex stabilization," in a context completely divorced from any consideration of complex coordinates. Leaving formal arguments on this point (21) aside, the following sections present brief summaries of several approaches as direct applications of Equations 4 and 5.

Specific Approaches in Direct Variational Resonance Calculations

The methods we discuss here are distinguished primarily by the forms of the trial wavefunctions they employ. In each case we will begin with a brief description of the trial wavefunction to be used with the generalized variational principle in Equations 4 and 5 in that method, and also quote the results of representative applications. For the (often critical) details of each calculation the original references should be consulted.

Complex Self-Consistent-Field (CSCF). Consider a resonance problem in electron-atom or electron-molecule scattering so that the Hamiltonian in Equation 4 is the real valued (Born-Oppenheimer) electronic Hamiltonian. Then if a trial wavefunction consisting of a single electronic configuration is used in the generalized complex variational principle, a set of complex SCF equations is obtained which differs from the ordinary SCF equations of electronic structure theory only by the absence of the usual complex conjugation in the matrix elements involved. In the molecular calculations presented here, and in those described in references 22 and 24-29, cartesian Gaussian basis functions were used and no complex conjugates appear at all. It should be noted that equivalent calculations could have been performed on diatomic molecules in prolate spheroidal coordinates. In that case the complex variational principle in equation 5 would have involved complex conjugation of the eigenfunctions of the projection of the orbital angular momentum along the internuclear axis. Reinhardt (4) comments briefly on the equivalence of these two approaches and the ambiguities in the present understanding of the complex variational principle.

Electronic resonances are generally open shell states. The simplest case is that of a single electron outside of a closed shell for which a single configuration is also a single determinant,

$$\Psi_{trial} = A[\phi_1\bar{\phi}_1\phi_2\bar{\phi}_2\cdots\phi_n\bar{\phi}_n\phi_{n+1}] \qquad (7)$$

where A is the antisymmetrizer and ϕ_i and $\bar{\phi}_i$ denote spin orbitals of opposite spin. However more general open-shell CSCF calculations are also possible.

The first analysis and calculations by this method were presented by McCurdy, Rescigno, Davidson and Lauderdale (22) although the idea was developed concurrently by Mishra, Ohrn and Froelich (23). A number of calculations have been performed on atomic and molecular electronic resonances using this approach (22-27) but we will mention only two, both of which made use of Gaussian basis functions. In each of these calculations real Gaussian basis functions were used to span the region of space occupied (approximately) by the electrons of the target atom or molecule and this real basis was augmented by a complex basis of diffuse Gaussians of the form

$$\chi_j(\vec{r}) = N_j(\alpha_j e^{-2i\theta})(x-A_x)^{\ell_j}(y-A_y)^{m_j}(z-A_z)^{n_j}\, e^{-\alpha_j e^{-2i\theta}(\vec{r}-\vec{A})^2} \qquad (8)$$

in which the exponent, α, has been scaled by the complex factor $e^{-2i\theta}$. Writing this factor as $e^{-2i\theta}$ rather than $e^{-i\theta}$ is a vestige of attempts to make formal connections with complex scaling of coordinates (14,18).

In a calculation on the lowest ^2D shape resonance state of Ca^-, McCurdy, Lauderdale and Mowrey (25) provided a dramatic demonstration that this method is based on the generalized complex variational principle and not on any direct connection with complex coordinates. They converged CSCF calculations on this resonance for a variety of values of θ in an extensive basis of complex functions of the form of Equation 8 (which, as stated above, were used together with real basis functions). Figure 1 shows a trace in the complex plane of the converged CSCF energy as a function of θ. What is remarkable about this calculation is the appearance of the point $\theta = 0$ in Figure 1. That point represents a CSCF calculation in which all the basis functions are real valued as is the Hamiltonian. Only the coefficients in the basis expansion of the orbitals are complex and are determined by the generalized complex variational principle. For this calculation, it is certainly the case that no connection can be made to any particular complex scaling of the electronic coordinates.

Figure 2 shows the results of CSCF calculations in a molecular case, namely the well studied $^2\Pi_g$ shape resonance state of N_2^- (27). This is a case for which the SCF approximation evidently describes the dominant contributions to the width over the range of internuclear distances which are involved in vibrational excitation of N_2 through this resonance, as indicated by comparisons of the CSCF width with other calculations as given in Figure 2b.

However a shortcoming of the CSCF approximation for resonances is also in evidence in Figure 2. The small arrow in Figure 2a indicates that the point at which the CSCF energy for N_2^- becomes complex is not the same point at which the real part of the CSCF energy for N_2^- crosses the SCF energy for the ground state of N_2. This is a general phenomenon in SCF calculations due to orbital relaxation in the anion which causes the SCF descriptions of the neutral and anion to be inconsistent with one another to this extent. The fact that both neutral and anion potential curves are required in a calculation of vibrational excitation cross sections is an inducement for finding an approximate remedy for this problem. As discussed in Reference 26, Koopmans' Theorem evidently does not provide that remedy.

Finally we note that in the case of a Feshbach resonance a single configuration is insufficient to account for both the resonance and decay channels, as shown by McCurdy, Rescigno, Davidson and Lauderdale (22), so that multiconfiguration complex SCF (or some other configuration interaction approach) is required. This point brings us to variational methods for electronic resonances which do incorporate the effects of electronic correlation.

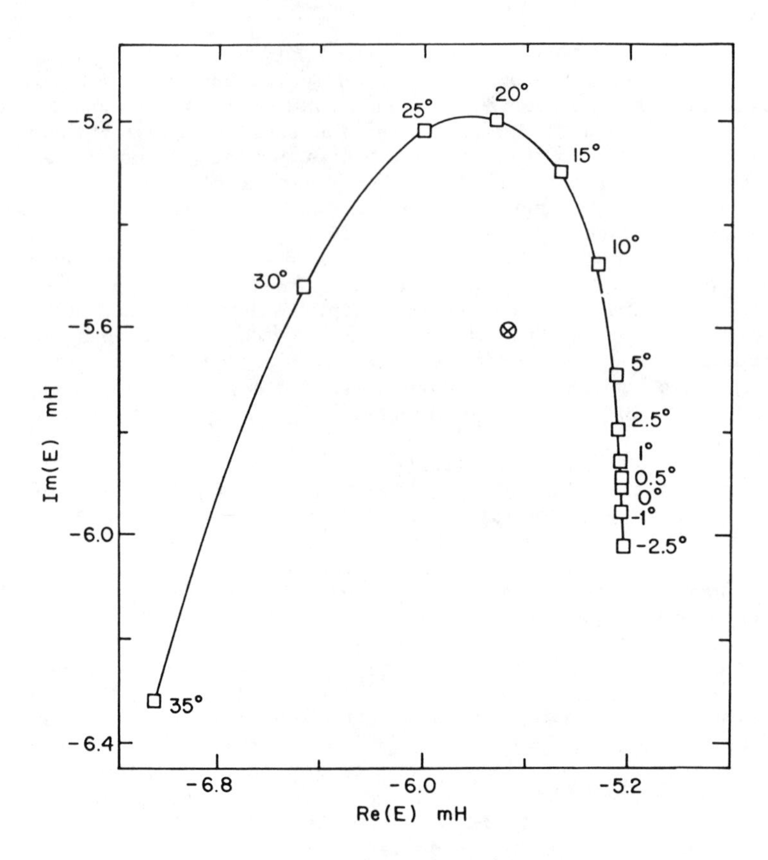

Figure 1. Trace of CSCF energy + 676.66 E_h (where E_h = 1 hartree = 4.3598 × 10^{-18} J) for the Ca^- ^2D resonance as a function of θ. mH denotes 10^{-3} E_h. Reproduced with permission from Ref. 25. Copyright 1981, American Physical Society.

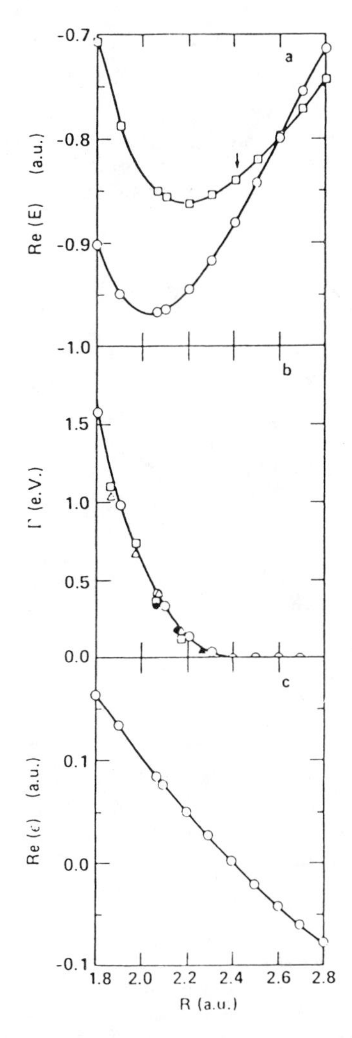

Figure 2. (a) SCF energy for N_2 (o) and real part of CSCF energy for N_2^- (□) relative to -108 E_h. To the left of the arrow the real part of the N_2^- $1\pi_g$ orbital energy is positive. (b) Width as a function of internuclear distance for N_2^- from CSCF calculations (o), R-matrix calculations from Ref. 30 (•), Stieltjes calculations (set III from Ref. 31) (Δ), and T-matrix calculation from Ref. 32 (□). (c) Real part of the $1\pi_g$ orbital energy of N_2^- from CSCF calculations. Reproduced with permission from Ref. 27. Copyright 1983, American Physical Society.

Complex Configuration Interaction (CCI). A logical extension of the CSCF idea is complex CI with a CSCF reference configuration. The trial function is now

$$\psi_{trial} = c_1\Phi_{CSCF} + \sum_{i=2}^{N} c_i\Phi_i \quad (9)$$

Such a calculation was performed by McNutt and McCurdy (28) on the lowest ^2P resonance state of Be^-, for which the CSCF configuration is $1s^2 2s^2 2p$. Table I gives their results on this system and illustrates the extent to which correlation can be important in the case of atomic shape resonances.

Table I. Calculations on the Be^- ^2P Shape Resonance

Reference	Method	E_r(eV)	Γ(eV)
29	Static Exchange Approximation (Complex basis functions for scattering electron only)	0.76	1.11
22	CSCF	0.70	0.51
28	CCI, singles and doubles from valence subshells	0.58	0.38
28	CCI, singles, doubles and triples from valence subshells	0.323	0.296

No experimental results are available for this system and these results should be compared with other theoretical calculations as is done in reference 29. Suffice it to say that other *ab initio* calculations have not given positions and widths smaller than those in the last line of Table 1 while model potential calculations (*i.e.*, calculations employing polarizability-based effective potentials of the $-\alpha/r^4$ type) give values about half as large (29). The point to be made here is that even though the CSCF approximation may be reliable for many cases of molecular shape resonances, it generally appears that atomic shape resonances are more sensitive to correlation effects, because there are typically more low lying excited states of the target to act as correlating configurations for the resonance anion in the atomic case.

Complex Stabilization. This idea is due to Junker (19,20), and although it is logically a form of CCI it predates the more straightforward CCI calculations of the previous section. It was also in these papers (19,20) that Junker discussed the generalized complex variational principle in a form completely unrelated to complex scaling. That this idea is not just a variant of complex scaling can be seen in the complex stabilization trial wavefunction which Junker employed in a calculation on the He^- $(1s,2s^2)$ ^2S Feshbach resonance:

$$\Psi_{trial} = \Psi_Q + \Psi_P \tag{10}$$

with

$$\Psi_P = A\phi_T(1,2) \{ \sum_{i=1}^{15} c_i \chi_i(r_3) + e^{ikr_3}(c_{16}\, e^{-\alpha r_3} + c_{17}\, e^{-(\alpha+\varepsilon)r_3} \} \tag{11}$$

$$\Psi_Q = \sum_{n=18}^{M} c_n\, \Phi_n(1,2,3) \tag{12}$$

In Equations 10-12 A is the antisymmetrizer; $\chi_i(r_3)$ is a real-valued Slater type basis function; $\phi_T(1,2)$ is an approximate target ground state (real-valued) wavefunction; and $\Phi_n(1,2,3)$ is a real-valued correlating configuration. The three parameters k, α, and ε were taken as nonlinear variational parameters and were varied to (approximately) satisfy the conditions

$$(\frac{\partial E}{\partial k}) = 0,\ (\frac{\partial E}{\partial \alpha}) = 0,\ (\frac{\partial E}{\partial \varepsilon}) = 0 \tag{13}$$

while the linear parameters, c_i, were determined from the CCI secular equation. What is remarkable about this calculation is Junker's observation that, for small ε in Equation 11, $c_{16} \simeq - c_{17}$. Therefore the "spherical wave" like function, $\exp(ikr_3)$, contributes predominantly for large r_3, and this is the reason for the numerical stability of these calculations.

The results for this resonance are (20) E_r = 19.3875 eV and Γ = 12.1 meV, in good agreement with the experimental values (42) of 19.35±0.02 eV and Γ = 13 meV. While this method can be viewed simply as CCI with variation of some nonlinear parameters, the specific form of the trial function in Equations 10-12 is critical and represents considerable insight into the properties of correlated wavefunctions for resonances (19,20).

Saddle Point Coordinate-Rotation. This approach, which was developed by Chung and Davis (33), can also be viewed as a variant of CCI, but again with a clever and physically motivated choice of trial function. Chung and Davis described their approach in terms of the variational principle in Equations 2 and 3 and therefore in terms of the complex scaling transformation of Equation 1. However, using identities like the one in Equation 6 for the matrix elements involved in this calculation, this approach can be seen to be identical to one using the generalized complex variational principle with a real Hamiltonian. In this picture the trial function of the saddle point coordinate-rotation method is, for the case of the $(2s^2)\ ^1S$ autoionizing state of He for example,

$$\Psi_{trial} = \sum_j c_j \Phi_j(1,2) + A\phi_{1s}(1) \sum_k d_k r_2^{\,k} e^{-\beta e^{-i\theta} r_2} \mathcal{S}(1,2) \tag{14}$$

In Equation 14 Φ_j denotes a real valued two electron configuration, ϕ_{1s} is the spatial part of the 1s state of He^+ to which this resonance decays, $\mathcal{S}(1,2)$ is a spin two-electron function, A is the antisymmetrizer, and β and θ are real parameters.

In this trial function the configuration functions $\Phi_j(1,2)$ are first optimized in a real-valued projected CI calculation called the "saddle point" method. In this calculation a vacancy 1s orbital is projected out of the $\Phi_j(1,2)$ configurations. Then in a calculation including only the first sum in equation 14 the CI energy is minimized with respect to nonlinear parameters in Φ_j and maximized with respect to the parameters of the vacancy orbital. Then these optimized configurations are used in a CCI calculation which determines the linear parameters c_j and d_k and in which the nonlinear parameters β and θ are also varied. Chung and Davis varied β and θ separately, but in view of our discussion of the complex variational principle this amounts to varying a single complex variational parameter, $\beta e^{-i\theta}$.

Calculations by this method have shown remarkable insensitivity to the nonlinear parameters of the complex part of the trial function. For the $2s^2\ ^1S$ autoionizing state, for which Equation 14 is the trial function, Chung and Davis (33) obtained $E_r = 57.8483$ eV and $\Gamma = 0.12468$ eV which compare well with the experimental (34) values of $E_r = 57.82\pm0.04$ eV and $\Gamma = 0.138\pm0.15$ eV. The saddle point complex-rotation method is strictly speaking another variant of CCI, but it demonstrates the premium in accuracy and efficiency to be gained from a well chosen trial function, in this case one in which the Feshbach Q-space (resonance) part of the trial wavefunction is optimized.

Analytic Continuation of Stabilization Graphs. Once the notion of the generalized complex variational principle is established, an obvious question is: can this principle be used by extrapolating (analytically continuing) the results of real-valued calculations from real to complex values of variational parameters? This question is strongly suggested by a "stabilization graph" of the eigenvalues of a matrix representation of the Hamiltonian in a real basis as a function of a parameter of the basis (35). For example consider the simple potential scattering problem of an electron scattering from the potential (in atomic units)

$$V(r) = \frac{15}{2} r^2 e^{-r} \tag{15}$$

Figure 3 shows the eigenvalues for this (s-wave) problem in a basis of 50 Laguerre functions of the form

$$\phi_n(r) = \eta^{3/2}[(n+1)(n+2)]^{-1/2} r\, L_n^{(2)}(\eta r) e^{-\eta r/2} \tag{16}$$

as a function of η. The stabilization phenomenon is the approach of $(\partial E/\partial \eta)$ to a small (but nonzero) value for each of the eigenvalues in turn. So why not just analytically continue one of the eigenvalues, $E_n(\eta)$, to complex η and look for its stationary points?

Thompson and Truhlar (36) did just that, using polynomial analytic continuation, but found that they could only obtain results which were at best reliable to only one or two significant figures in the width. In other words the analytic continuation of an eigenvalue, $E_n(\eta)$, to complex η is unstable. Why?

The answer, and a way around the instability, was provided by McCurdy and McNutt (37). The $E_n(\eta)$ are not separate functions; they are the branches or Riemann sheets of a single analytic function because they are the roots of the polynomial equivalent to the secular equation

$$\det|\underline{H} - E\underline{I}| = 0 \tag{17}$$

Every avoided crossing on the real η axis indicates that there is an actual crossing at a nearby complex value of η (and its complex conjugate). These crossings, where two branches of the multibranched eigenvalue function are equal, are branch points of $E_n(\eta)$. McCurdy and McNutt (37) showed by considering a simple model that the stationary points of $E_n(\eta)$ are typically just beyond its branch points in the complex η plane. Figure 4 shows a sketch of the situation in the complex η plane. The fact that the stationary points are beyond the branch points is the reason for the instability in analytic continuation of a single $E_n(\eta)$ eigenvalue.

The solution to this problem is to analytically continue something else instead of $E_n(\eta)$. The something else is the set of coefficients, $P_n(\eta)$, of the characteristic polynomial of the Hamiltonian matrix. The polynomial equation equivalent to Equation 17 is

$$(-1)^N[E^N - P_{N-1}(\eta)E^{N-1} - P_{N-2}(\eta)E^{N-2} - \cdots - P_0(\eta)] = 0 \tag{18}$$

which can be constructed for real η from the eigenvalues because it is the same as

$$\prod_{n=1}^{N} (E_n(\eta) - E) = 0 \tag{19}$$

McCurdy and McNutt showed how to construct a truncated version of Equation 18 using two or more eigenvalue curves, $E_n(\eta)$, near an avoided crossing. The coefficients $P_n(\eta)$ thus constructed are analytic, single-valued functions of η which can be analytically continued in a stable way. Furthermore, if the continued fraction representation of $P_n(\eta)$ suggested in reference 37 is used, several simplifications are possible in locating the stationary point(s)

$$\left(\frac{\partial E}{\partial \eta}\right) = 0 \tag{20}$$

and in the calculation of $E(\eta)$ at those points.

We will give two examples of applications of this idea in both of which a "two root" continuation is used. Specializing to the quadratic case of Equation 18 is instructive and also allows us to

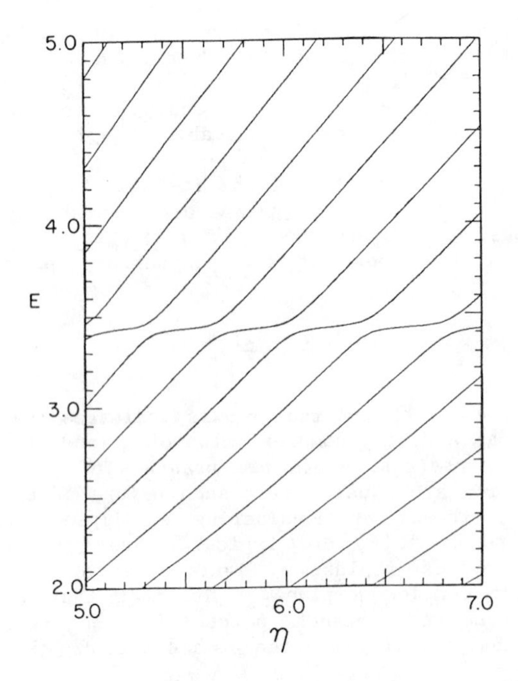

Figure 3. Stabilization graph for model problem of Equation 15 with the basis of Equation 16. Reproduced with permission from Ref. 37, copyright 1983, North Holland.

η plane

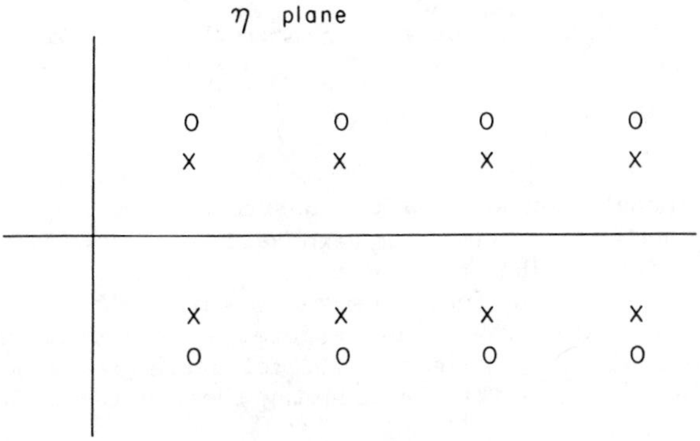

Figure 4. Qualitative sketch of locations of branch points (x) and stationary points (o) in the η plane associated with avoided crossings in the stabilization graph. Reproduced with permission from Ref. 37, copyright 1983, North Holland.

recast the procedure to gain slightly more numerical stability. In this case we choose two eigenvalues $E_1(\eta)$ and $E_2(\eta)$ which show an avoided crossing with each other. Then we write the apparently circular relations

$$E_{\pm}(\eta) = \frac{1}{2}[f_1(\eta) \pm (f_2(\eta))^{1/2}] \tag{21}$$

with

$$f_1(\eta) = E_1(\eta) + E_2(\eta) \tag{22}$$

$$f_2(\eta) = (E_1(\eta) - E_2(\eta))^2 \tag{23}$$

The functions $f_1(\eta)$ and $f_2(\eta)$ are obviously single valued in the region of η for which $E_1(\eta)$ and $E_2(\eta)$ have a branch point connecting them (interchanging E_1 and E_2 leaves f_1 and f_2 unchanged). Given values of $E_1(\eta)$ and $E_2(\eta)$ at M discrete values of η we can form a continued fraction representation of $f_1(\eta)$ and $f_2(\eta)$, which can be used to evaluate Equation 21 (and its derivative with respect to η) at complex values of η. Finding the stationary points at which Equation 20 is satisfied is then a simple root search problem. Equations 21-23 also make it evident that a prescription given earlier by Simons (38) for order-of-magnitude estimates of resonance widths is a linearized (in η) version of these equations. An application of this procedure to the potential scattering problem of Equation 15 and Figure 3 is given in Table II.

Table II. Two Root Analytic Continuation in a Model Problem (39).

Number of real valued input points, M	E_r(a.u.)	Γ(a.u.)
10	3.42631	0.02594
20	3.42641	0.02556
30	3.42641	0.02555
40	3.42640	0.02556
Exact value	3.42639	0.02555

This example indicates the kind of numerical stability possible with this approach. Another application of the method, this time to a "heavy particle" problem was given in a calculation by Bai, Hose, McCurdy and Taylor (40) on the Hénon-Heiles system

$$H = -\frac{1}{2}\frac{\partial^2}{\partial x^2} - \frac{1}{2}\frac{\partial^2}{\partial y^2} + \frac{1}{2}x^2 + \frac{1}{2}y^2 - \frac{1}{3}\lambda x^3 + \lambda xy^2 \tag{24}$$

where the analytic continuation was performed in a variable appearing only in basis functions for the x degree of freedom.

Figure 5 shows the results of this calculation for the widths of the states of E symmetry for which matrices of up to order 250 were required. In this example the widths varied over 4 orders of magnitude, and the fact that the potential is unbounded as x goes to infinity raises formal as well as numerical problems. Nonetheless the analytic continuation procedure was able to produce useful results.

The method expressed in Equations 18-23 has seen only a few numerical applications, but it represents the ultimate application of the generalized complex variational principle in that only real valued eigenvalue calculations are required. Doubtless the use of this idea coupled with more cleverly chosen trial functions, such as those of Junker (19,20) and Chung and Davis (33) described above will also be possible.

Some Open Questions, both Formal and Practical

Two questions obvious from our discussion here are: what is the basis for the generalized variational principle of Equations 4 and 5, and what are its limitations? For example, it is not known to what extent this principle is applicable for problems in which the potentials do not approach a constant at large distances as in the case of the Stark effect. The Heinon-Heiles potential is another such problem, but the calculation in reference 40, which blithely ignored this question, apparently gave meaningful results. The state of affairs is that we have a number of successful and extremely suggestive numerical experiments, but, except for calculations which can be related to a specific analytic continuation of the Hamiltonian as a function of its coordinates, they have taken Equations 4 and 5 as an ansatz and we have no fundamental proof that it is correct to do so.

The problem in constructing such a proof can be seen by asking another question: what do the wavefunctions from an application of Equations 4 and 5 mean? We have argued using equation 6 that, in the simplest case with all basis functions complex, particular complex basis calculations are equivalent to rotated coordinate calculations for the resonance wavefunction, $\Psi_{res}^{(\theta)}(r)$ for particular values of θ. These wavefunctions for various values of θ are equivalent analytic continuations of each other. However, if we start with an arbitrary mixed set of real and complex basis functions in Equations 4 and 5, which one of these equivalent wavefunctions (if any) are we approximating? Since we do not know the answer to this question we cannot use our wavefunctions to compute dipole matrix elements, for example,

$$\int \Psi_o(\vec{r})\mu\, \Psi_{res}(\vec{r})\, d^3r$$

involving the complex resonance wavefunction, $\Psi_{res}(\vec{r})$, from a general complex variational calculation of the resonance energy.

On the more practical side, it has not yet been shown that a version of Junker's (19,20) trial function can be fabricated from Gaussian basis functions so that it can be used in a calculation on

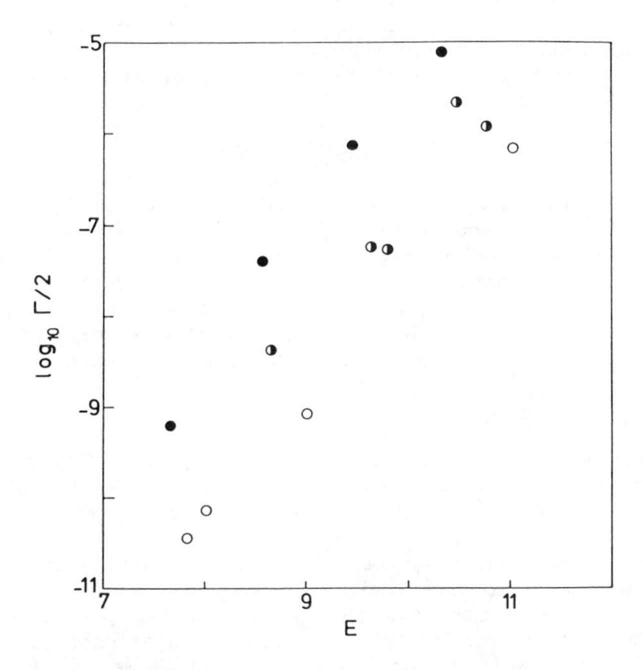

Figure 5. Logarithm of half the width ($\Gamma/2$) for metastable E states of the Hénon-Heiles Hamiltonian (λ^2 = 0.0125). Solid, semi-solid, and empty circles denote Q^{II}, N, and Q^{I} type states in the notation of Ref. 40. Reproduced by permission, copyright 1983, North Holland.

a molecular resonance for which Gaussian, as opposed to Slater, basis functions are far more convenient. Molecular applications of these ideas are currently stalled for the lack of practical basis sets for nonlinear polyatomic molecules.

Another practical question is applicability of a complex version of multiconfiguration SCF to Feshbach resonances. In reference 22 it was pointed out that for a Feshbach resonance at least two configurations must be included in an SCF calculation, the resonance configuration and one representing the decay channel. For the (2s2p) autoionizing states of He, for example, we would need (at least) the trial wavefunction

$$\Psi_{trial} = a(2s2p) + b(1skp)$$

where k_p denotes the square-integrable complex "continuum" orbital. Such a calculation might suffer even more severe convergence problems than a typical multiconfiguration SCF calculation on an excited state. On the other hand, it would be easy to build trial functions in this approach which should provide excellent approximations to the complex resonance energy. To this author's knowledge, no such calculation has yet been attempted.

Conclusion

In this discussion we have attempted to display the unifying theme of direct variational calculations on resonances. Since this was not intended to be a review article, we have omitted several important contributions to the subject and apologies are due to those authors whose work is neglected here. In particular work on variational bounds on complex energies, (41,43) is a very promising development. Our hope has been that by presenting the subject in this simplified way we could stimulate further applications of the basic idea, particularly in heavy-particle resonance problems for which even more successful versions of the analytic continuation of stabilization graphs can surely be developed. Because they have practical implications we have also emphasized some formal problems which have gone largely neglected in the literature. In particular we cannot address the problem of line shapes in photoionization by adding a background contribution to the resonance contribution to the dipole oscillator strength until we know for certain how to use our complex variational wavefunctions to do dipole matrix elements.

Acknowledgments

Our work in this area has been supported by the National Science Foundation. I would also like to thank Tom Rescigno, Bobby Junker, Bill Reinhardt, Barry Simon, and others working in this area for stimulating conversations which have often set straight my ideas on the subject.

Literature Cited

1. See for example: Taylor, J. R. "Scattering Theory"; John Wiley and Sons: New York, 1972; p. 222.

2. Aguilar, J.; Combes, J. Commun. Math. Phys. 1971, 22, 269.
3. Balslev, E.; Combes, J. Commun. Math. Phys. 1971, 22, 280.
4. Reinhardt, W. P. Ann. Rev. Phys. Chem. 1982, 33, 223.
5. McCurdy, C. W. In "Autoionization II"; Temkin, A., Ed.; Plenum: New York, 1984 (to appear).
6. Simon, B. Int. J. Quantum Chem. 1978, 14, 529. The issue in which this article appears is devoted entirely to complex scaling.
7. Doolen, G. D.; Hidalgo, M.; Nuttall, J.; Stagat, R. W. In "Atomic Physics"; Smith, S. J.; Walters, G. K., Eds.; Plenum: New York, 1973, p. 257.
8. Doolen, G. D.; Nuttall, J.; Stagat, R. W. Phys. Rev. A 1974, 10, 1612.
9. Doolen, G. D.. J. Phys. B 1975, 8, 525.
10. Moiseyev, N. Mol. Phys. 1982, 47, 585.
11. Rescigno, T. N.; McCurdy, C. W.; Orel, A. E.. Phys. Rev. A 1978, 17, 1931.
12. Junker, B. R.; Huang, C. L. Phys. Rev. A 1978, 18, 313.
13. Simon, B. Phys. Lett. 1979, 71A, 211.
14. McCurdy, C. W. Phys. Rev. A 1980, 21, 464.
15. Herrick, D. R.; Stillinger, F. H. J. Chem. Phys. 1975, 62, 4360.
16. Herrick, D. R. J. Chem. Phys. 1976, 65, 3529.
17. Sherman, P. R.; Herrick, D. R. Phys. Rev. A 1981, 23, 2790.
18. McCurdy, C. W.; Rescigno, T. N. Phys. Rev. Letts. 1978, 41, 1364.
19. Junker, B. R. Phys. Rev. Letts. 1980, 44, 1847.
20. Junker, B. R. Int. J. Quantum Chem. 1980, 14S, 53.
21. Brändas, E.; Froelich, P.; Obcema, C. H.; Elander, N.; Rittby, M. Phys. Rev. A 1982, 26, 3656.
22. McCurdy, C. W.; Rescigno, T. N.; Davidson, E. R.; Lauderdale, J. G. J. Chem. Phys. 1980, 73, 3268.
23. Mishra, M.; Ohrn, Y.; Froelich, P. Phys. Lett. A 1981, 84, 4.
24. Rescigno, T. N.; Orel, A. E.; McCurdy, C. W. J. Chem. Phys. 1980, 73, 6347.
25. McCurdy, C. W.; Lauderdale, J. G.; Mowrey, R. C. J. Chem. Phys. 1981, 75, 1835.
26. McCurdy, C. W.; Mowrey, R. C. Phys. Rev. A 1982, 25, 2529.
27. Lauderdale, J. G.; McCurdy, C. W.; Hazi, A. U. J. Chem. Phys. 1983, 79, 2200.
28. McNutt, J. F.; McCurdy, C. W. Phys. Rev. A 1983, 27, 132.
29. Rescigno, T. N.; McCurdy, C. W.; Orel, A. E. Phys. Rev. A 1978, 17, 1931.
30. Schneider, B. I.; Le Dourneuf, M.; Vo Ky Lan Phys. Rev. Lett. 1979, 43, 1926.
31. Hazi, A. U.; Rescigno, T. N.; Kurilla, M. Phys. Rev. A 1981, 23, 1089.
32. Levin, D. A.; McKoy, B. V. (unpublished results).
33. Chung, K. J.; Davis, B. F. Phys. Rev. A, 1982, 26, 3278.
34. Hicks, P. J.; Comer, J. J. Phys. B 1975, 8, 1866.
35. Hazi, A. U.; Taylor, H. S. Phys. Rev. A 1970, 1, 1109 gives a detailed treatment of the stabilization phenomenon.
36. Thompson, T. C.; and Truhlar, D. G. Chem. Phys. Letts. 1982, 92, 71.
37. McCurdy, C. W.; McNutt, J. F. Chem. Phys. Letts. 1983, 94, 306.

38. Simons, J. J. Chem. Phys. 1981, 75, 2465.
39. Mowrey, R. C.; McCurdy, C. W. (unpublished results).
40. Bai, Y. Y.; Hose, G.; McCurdy, C. W.; Taylor, H. S. Chem. Phys. Letts. 1983, 99, 342.
41. Froelich, R.; Davidson, E.R.; Brändas, E. Phys. Rev. A 1983, 28, 2641.
42. Golden, D.E.; Schowengerdt, F.D.; Macek, J. J. Phys. B 1974, 7, 478. The resonance energy of reference 20 is compared with this measurement using a ground state energy of -79.0016 eV and a conversion factor of 27.211652 eV/E_h.
43. Moiseyev, N.; Froelich, P.; Watkins, E. J. Chem. Phys. 1984, 80, 3623.

RECEIVED June 11, 1984

3

Complex Energy Quantization for Molecular Systems

R. LEFEBVRE

Laboratoire de Photophysique Moléculaire, Université Paris-Sud, 91405 Orsay, France

The method of deriving resonance energies from propagations and a matching of the solutions of a set of coupled equations with appropriate boundary conditions is briefly presented. Complete or incomplete complex rotation of the interfragment coordinate gives the benefit of allowing the use of the same boundary conditions as for a bound state. The extension of propagation and matching to arbitrary paths in the complex plane is also described and an application is made to the treatment of electronic non adiabatic transitions.

Of the many approaches developped for the treatment of resonances, none has been in recent years so extensively used as the method of complex rotation or complex scaling (for reviews see 1-3). The appealing aspects of the method are :

a) the definition of a resonance energy as being a rigorous complex eigenvalue of the system wave equation solved under the so-called Siegert boundary conditions (4).
b) the transformation under complex scaling of the associated eigenfunction into an integrable function.

For a physical understanding of the mathematical background (5-6) but without delving too much into it, one may consider (7) the case of a single channel with an asymptotic solution of the form

$$\psi(k,r) \sim -e^{-ikr} + S(k)\, e^{ikr} \qquad (1)$$

with k being the wave number and r an interfragment coordinate. The Siegert wave corresponds to choosing k to be a pole of the scattering amplitude S(k) (so that only the outgoing term survives in (1). Complex scaling amounts to replacing r by $\rho\, \exp(i\theta)$ (ρ real). If k is complex and of the form $k = K \exp(-i\beta)$ ($\beta > 0$), the Siegert wave vanishes asymptotically if the condition

$$0 < \theta - \beta < \pi$$

0097-6156/84/0263-0035$06.00/0

is fulfilled. Such a value of k gives a complex energy with a negative imaginary part which is often simply related to a decay rate. These considerations can be easily extended to a multichannel situation.

Although the previous argument is based on the form taken by the scattering function, most implementations of the complex scaling method are made by expanding the wave function in terms of integrable functions. In this way advantage is taken of the vast body of efficient methods developped for treating the bound states of atomic or molecular systems. However there exists also a number of techniques for treating bound states from a coupled-channel point of view (8-10), and it is therefore worthwhile to investigate complex scaling in such a context. This was done by Atabek and the author and applied to various processes such as photodissociation of linear triatomics (11, 12) or rotational predissociation of van der Waals complexes (13, 14). The next section describes this algorithm and gives two examples of the high accuracy which can be reached in the determination of resonance energies. The other two sections describe a recent development of this method which allows the consideration of paths in the complex plane which generalize those considered in the complex scaling procedure (either in its original form or in the so-called exterior scaling version (15)). An application is made to a problem involving electronic non-adiabatic transitions.

Coupled Channel Approach to the Determination of Resonances

Consider a system whose dissociation is described by letting an interfragment coordinate r go to infinity. We assume that the description of the system has been reduced to a set of coupled equations determining a n-dimensional column vector U(r). If this vector is to obey prescribed boundary conditions for either small r (say r_1) or large r (say r_2), there are in fact n independent such vectors which obey these conditions. They can be combined in a matrix $\tilde{U}(r)$ which, when propagated outward with the help of some algorithm starting from its initial value $\tilde{U}(r_1)$, leads at some intermediate point r_m to $\tilde{U}^o(r_m)$. When propagated inward from its initial value $\tilde{U}(r_2)$ we reach at r_m a matrix $\tilde{U}^i(r_m)$. It is then possible to take advantage of the fact that we can freely combine the outward or inward solutions without affecting the boundary conditions to set up a relation to be fulfilled so that the same column vector $U(r_m)$ will be produced. This condition can be met only for particular values of the energy, which is then quantized. This is the problem which was solved for bound states by Gordon (8) and Johnson (9). In this case the boundary conditions are usually $\tilde{U}(r_1)=\tilde{U}(r_2)=0$. We make now several remarks :

a) in practice one does not propagate the wave function because of instabilities which may set in. In our own approach (10) we have chosen to propagate the product (see (16)) :

$$\tilde{P}(r) = \tilde{U}(r-h)\ \tilde{U}^{-1}(r) \qquad (2)$$

h being a propagation step. The matching condition then reads :

$$\det |\tilde{P}^i(r_m) - \tilde{P}^o(r_{m+1})^{-1}| = 0 \qquad (3)$$

In the one channel case this reduces to the trivial condition

$$U^i(r_m+h) \ / \ U^i(r_m) = U^o(r_m+h) \ / \ U^o(r_m) \tag{4}$$

b) the matching condition is <u>independent</u> of the choice of the boundary conditions. Thus we may for instance choose $\underset{\sim}{P}^i(r_2)$ to have a form suggested by the Siegert condition for a resonance, that is outgoing waves only in all open channels. In view of the complicated dependence of (3) on energy, an iterative procedure has to be used to reach an eigenenergy. It has been shown in several multichannel examples (13) that this procedure leads indeed to the correct resonance energies.

c) we may now substitute by ρ exp $(i\theta)$ (ρ real), and we substitute h by h exp $(i\theta)$ in the propagation algorithm and calculate the potential matrix with the new variable. Since this rotation has the effect (for θ large enough) of changing the diverging Siegert wave function into an asymptotically vanishing one, the matching method now involves boundary conditions that are exactly those of bound states.

We give now two examples showing, either in a one-channel case or in a multichannel case, that very high accuracy can be achieved in this way. The propagation technique itself has not been commented upon. We leave this for the next sections where more general paths than those implied by ordinary complex scaling are considered.

- A one channel example : the resonances associated with a repulsive exponential potential.

It is not always obvious whether a given potential supports or does not support resonance states. The exponential well for instance (i.e. a potential of the form $-A \exp(-\alpha r)$ with A, $\alpha>0$) supports only bound and virtual states (17). The exponential repulsive potential on the other hand (i.e. of the form $A \exp(-\alpha r)$ with A, $\alpha>0$) supports resonances and virtual states (18). The potential 9 exp (- 2r) (with the kinetic energy operator written as $-d^2/dr^2$) supports two (and only two) resonance states with the energies calculated (19) from an analytical treatment to be

$$E_1 = -\ 2.421833 - i\ 7.280682$$
$$E_2 = -14.627351 - i\ 3.915473$$

and calculated from the matching technique combined with complex rotation to be

$$E_1 = -\ 2.421830 - i\ 7.280682$$
$$E_2 = -14.627353 - i\ 3.915471.$$

- A multichannel example : Stark ionization of the hydrogen atom.

In a recent paper Maquet, Chu, and Reinhardt (20) have extended their previous work on the Stark ionization of the hydrogen atom and obtained very accurate resonance energies from matrix techniques using bases of integrable functions. For the ground state in a field of strength 0.1 a.u., the following stabilized energy (in a.u.) is obtained with 10 ℓ-blocks (where ℓ denotes angular momentum) :

$$E_o = -\ 0.527418173 - i\ 0.007269057$$

This problem has also been considered in a coupled-channel approach (21). A recent extension of this calculation to meet this challenging result has been made, with 10 channels, propagation step h=0.01, and rotation angle θ=0.5. The resonance energy is found to be :

$$E_o = - 0.52741817 -i\ 0.0072690633$$

An interesting aspect of this case is that Siegert boundary conditions with a real coordinate are difficult to formulate in spherical coordinates (the diagonal elements of the potential matrix go to zero and the off-diagonal ones to infinity). Complex rotation obviates these difficulties. The initial boundary condition for inward propagation can be taken in all cases to be $\underset{\sim}{P}(r_2)=0$.

Propagation along a Mixed Path

The initial value given for inward propagation to $\underset{\sim}{P}(r_2)$ deserves a comment since in a one channel case we expect P(r) (Equation (2)) in the asymptotic region to be exp (ikh exp (iθ)) if use is made of the complex step h exp (iθ). We will show by an example that the justification for choosing $\underset{\sim}{P}(r_2)$ equal to zero is not that this is a sensible guess, but that the form given to $\underset{\sim}{P}(r_2)$ really does not matter. In view of this, $\underset{\sim}{P}(r_2)=0$ amounts to attempting no guess at all. This will be done by integrating inward along the mixed path (complex and then real) which is to be used in the exterior scaling procedure (15). This method was originally devised by Simon to avoid the singularities arising from complex scaling the electronic coordinates in the interior region of a molecule but another use for it is to allow (14, 24) for an application of complex rotation in its propagation-matching version to piecewise analytic or numerical potentials. If the potential matrix is given an analytical asymptotic form beyond a certain radius r_o, it is possible to make, for $r > r_o$, the coordinate transformation.

$$r \rightarrow r_o + (\rho - r_o) \exp (i\theta) \tag{5}$$

while maintaining inside $r=r_o$ the original form for the matrix. For the determination of $P^i(r_m)$ the propagation is first along a rotated axis until point r_o is reached. The propagation has then to continue along the real axis (we assume $r_m < r_o$). The observation has been made (14) that the change of path from complex to real can be performed very accurately by using the Numerov algorithm with variable steps h_1 and h_2 (in which case the step is $h_1=-he^{i\theta}$ just before r_o and $h_2=-h$ just after it). This algorithm consists in building the three matrices (16, 22) :

$$\begin{aligned}
\underset{\sim}{\alpha}(r) &= h_2\ \{1+ \frac{1}{12}(h_1{}^2+h_1h_2-h_2{}^2)[E\underset{\sim}{1}-\underset{\sim}{V}(r-h_1)]\} \\
\underset{\sim}{\beta}(r) &= (h_1+h_2)\{1- \frac{1}{12}(h_1{}^2+3h_1h_2+h_2{}^2)[E\underset{\sim}{1}-\underset{\sim}{V}(r)]\} \\
\underset{\sim}{\gamma}(r) &= h_1\{1+ \frac{1}{12}(-h_1{}^2+h_1h_2+h_2{}^2)[E\underset{\sim}{1}-\underset{\sim}{V}(r+h_2)]\}
\end{aligned} \tag{6}$$

and using the propagation relation

$$\underset{\sim}{U}(r)\,\underset{\sim}{U}^{-1}(r+h_2) = [\underset{\sim}{\beta}(r)-\underset{\sim}{\alpha}(r)\underset{\sim}{U}(r-h_1)U^{-1}(r)]^{-1}\ \gamma(r) \qquad (7)$$

In (6) E is the energy, $\underset{\sim}{1}$ the unit matrix, and $\underset{\sim}{V}(r)$ the potential matrix.

Our example will be the resonance of the potential $15\ r^2 e^{-r}$ (with $-d^2/dr^2$ for the kinetic energy operator) first studied by Bain, Bardsley, and Sukumar (7). The best available determination of the resonance energy appears to be that performed by Meyer (23) who gives :

$$E = 6.85278062 -i\ 0.02554976$$

Table I shows how starting from an initial ratio equal to zero, the correct Siegert ratio $P_1 = \exp\ [ikh\ \exp\ (i\theta)]$, is progressively built when propagating inward in the asymptotic region along a complex axis making an angle θ with the real axis. The value given to k is obviously not critical in this calculation, since quantization results from a comparison of outward and inward ratios. Shown also is the ratio P_2' obtained in one further real step and the ratio $P_2 = \exp\ (ikh)$. The role of complex scaling is therefore to build the correct Siegert ratios. In the exterior scaling version, we can finally obtained the ratio which should be used in a calculation with no scaling at all.

Table I

Illustration of the building of the Siegert ratios as a result of an inward propagation along a complex path, followed by a single further real step. Underlined are the figures showing the errors in the numerical results. Step size $h = 0.5\ 10^{-2}$, rotation angle $\theta = 0.5$ rad., $k = \sqrt{E}$, E resonance energy of reference (7). Column (a) : number of steps. Column (b) ratio of the wave functions at two adjacent points.

(a)	(b)		
250	0.993741202828	+i	0.0126036189460
500	0.993726019270	+i	0.0114688444063
1 000	0.993700162193	+i	0.0114263762517
2 000	0.993700066665	+i	0.0114263804773
P_1	0.993700066666	+i	0.0114263804783
P_2'	0.999938739538	+i	0.0130888764967
P_2	0.999938739535	+i	0.0130888764899

Table II illustrates another aspect of the inward propagation, with P_1^{-1} as the initial ratio. A longer path is needed to recover P_1. Thus the initial value does not really matter. This is to be understood as due to the different behaviour of the ingoing and outgoing components of the scattering wave function. Even with an initial ratio contaminated by the ingoing component, the outgoing component which increases in an inward propagation will always dominate. For virtual states, with or without rotation, the reverse is true (19) and the inward propagation fails to produce the correct ratio.

Table II

Building of a correct Siegert ratio when starting from the inverse of the Siegert ratio. Same parameters as in Table I. Column (a) : number of steps. Column (b) : ratios of the wave function at two adjacent points.

(a)	(b)		
1 000	0.100620705675	-i	0.0115702161521
2 000	0.997591288035	-i	0.00464400289036
3 000	0.993700025456	-i	0.0114263819099
4 000	0.993700066666	-i	0.0114263804777
P_1	0.993700066666	-i	0.0114263804783

The value obtained for the resonance of this example with matching and exterior scaling is :

$$E = 6.85278062 - i\ 0.02554896.$$

Complex Adiabatic Paths

The fact that the Numerov algorithm with variable steps allows for a change from a complex to a real coordinate along the integration path opens the way to paths of a more general type. A code has been written to integrate along a path made of segments of arbitrary lengths in the complex plane. Provision is being made for different potential matrices for the different segments (to perform, for instance, a "diabatic by sectors" calculation (25, 26)). We will illustrate the use of such a program for determining predissociation rates by quantization of the energy along a path suggested by the semi-classical theory of electronically nonadiabatic transitions (27, 28, 29). We consider a situation with two adiabatic potentials which arise from a configuration interaction treatment of a diabatic curve crossing situation. It is not possible to go from the left turning point of the lower potential to the right turning point of the upper one by following one of the potentials. However there exists the possibility of switching from one of the adiabatic potential to the other by following a complex path going across one of the complex intersection points. Quantization can be obtained by matching the two solutions obtained from outward or inward propagation starting from the non-classical regions. The quantized energies reported here correspond to a model of a Morse potential and an exponential one studied previously by Child and the author (30).

Table III gives some of the quantized energies obtained from a traditional 2-channel phase shift analysis (30) or from the present procedure. Levels with $v < 16$ have no sizeable width. The shifts for such levels are in remarkable agreement. There is an obvious parallelism between both sets of energies up to $v = 20$.

Table III

Complex adiabatic energies (a) compared to resonance energies (b) for a model of a closed and an open channel described in reference (30). The zeroth order energy has been subtracted out. Unit : cm^{-1}. Interchannel coupling 200 cm^{-1}. Crossing diabatic energy equal to the energy of the level with v = 18.

v	(a)	(b)
14	-1.53 -i 0.00	-1.56 -i 0.00
15	-1.97 -i 0.02	-1.97 -i 0.01
16	-2.74 -i 0.17	-2.76 -i 0.10
17	-4.03 -i 1.03	-4.27 -i 0.86
18	-4.94 -i 3.76	-6.08 -i 4.11
19	-2.33 -i 7.61	-2.18 -i 8.66
20	+3.76 -i 6.54	+4.81 -i 5.16
21	+3.07 -i 0.31	+0.50 -i 0.22

The conclusion from this study and from some others made on different models (31), are that with an energy close to the crossing diabatic energy (both above and below it) there is good general agreement between the exact resonance energies and the energies of this adiabatic quantization procedure. The reasons for the successes and failures of this approach are not yet completely clear, although a semi-classical argument (31) shows that in any case the right order of magnitude for the shifts and widths are to be expected.

Acknowledgments

This work has been supported by a grant of computing time on the CCVR CRAY.

Literature cited

1. Reinhardt, W.P., Ann. Rev. Phys. Chem. 1982, 33, 223.
2. Junker, B.R., Adv. Atom. and Mol. Phys. 1982, 18, 208.
3. Ho., Y.K., Phys. Rev. 1983, 99, 1.
4. Siegert, A.F., Phys. Rev. 1939, 56, 750.
5. Aguilar, J., Combes, J.M., Comm. Math. Phys. 1971, 22, 269.
6. Balslev, E., Combes, J.M., Comm. Math. Phys. 1971, 22, 280.
7. Bain, R.A., Bardsley, J.N., Sukumar, C.V. J. Phys. B. 1974, 7, 2189.
8. Gordon, R.G., J. Chem. Phys. 1969, 14, 51.
9. Johnson, B.R., J. Chem. Phys. 1978, 69, 4678.
10. Atabek, O., Lefebvre, R., Chem. Phys. 1980, 52, 199.
11. Atabek, O., Lefebvre, R., Phys. Rev. 1980, A22, 1817.
12. Atabek, O., Lefebvre, R., Chem. Phys. 1981, 56, 195.
13. Atabek, O., Lefebvre, R., Chem. Phys. 1981, 55, 395.
14. Lefebvre, R., J. Phys. Chem. (in press).
15. Simon, B., Phys. Lett., 1979, 71A, 211.
16. Norcross, D.W., Seaton, M.J., J. Phys. B. 1973, 6, 614.

17. De Alfaro, V., Regge, T., "Potential scattering" ; 1965 North-Holland, Amsterdam.
18. Jost, R., Helv. Phys. Acta, 1947, 20, 256.
19. Atabek, O. Lefebvre, R., Jacon, M., J. Phys. B., 1982, 15, 2689.
20. Maquet, A., Chu, Sh-I, Reinhardt, W.P., Phys. Rev. A., 1983, 27, 2946.
21. Atabek, O. Lefebvre, R., Int. J. Quant. Chem., 1981, 19, 901.
22. Bergeron, G., Chapuisat, X., Launay, J.M., Chem. Phys. Lett., 1976, 38, 349.
23. Meyer, H.D., Walter, O., J. Phys. B, 1982, 15, 3647.
24. Turner, J. McCurdy, C.W., Chem. Phys., 1982, 71, 127.
25. Kuppermann, A., Kaye, J.A., Dwyer, J.P., Chem. Phys. Lett., 1980, 74, 257.
26. Launay, J.M., Le Dourneuf, M., J. Phys. B, 1982, 15, L455.
27. Dykhne, A.M., Sov. Phys. JETP, 1962, 14, 941.
28. Miller, W.H., George T.F., J. Chem. Phys., 1972, 56, 5637.
29. Davis, J.P., Pechukas, P., J. Chem. Phys., 1976, 64, 3129.
30. Child, M.S., Lefebvre, R., Mol. Phys., 1977, 34, 979.
31. Lefebvre, R., Chem. Phys. Lett., in press.

RECEIVED June 29, 1984

4

Approximate Quantum Approaches to the Calculation of Resonances in Reactive and Nonreactive Scattering

JOEL M. BOWMAN[1], KI TUNG LEE[1,3], HUBERT ROMANOWSKI[1,4], and LAWRENCE B. HARDING[2]

[1]Department of Chemistry, Illinois Institute of Technology, Chicago, IL 60616
[2]Chemistry Division, Argonne National Laboratory, Argonne, IL 60439

Reduced-dimensionality, quantum differential cross sections and partial wave cumulative reaction probabilities are presented for resonant and non-resonant scattering in the $H+H_2(v=0)$ reaction. The distorted wave Born approximation is tested against previous complex coordinate calculations of resonance energies and widths for a model van der Waals system. This approximation is subsequently used to obtain resonance energies and widths for the HCO radical using an approximate scattering path hamiltonian based on an **ab initio** potential energy surface. Several advantages of this hamiltonian for addition reactions are discussed.

It is by now well established theoretically and experimentally that resonances are prominent features in non-reactive molecular systems. Although less evidence is available for reactive systems it is clear that resonances are also present for them. We have been investigating resonances in both kinds of systems and we present some of our recent results in this paper. Depending on the approach, resonances

[3]Current address: Department of Chemistry, University of Rochester, Rochester, NY.
[4]Permanent address: Institute of Chemistry, The Wroclaw University, 50-383 Wroclaw, Poland.

0097-6156/84/0263-0043$06.00/0

can be difficult or easy to obtain theoretically. They are difficult to locate in scattering calculations because their location is effectively unknown and must be uncovered by a search method. Direct, easy methods are those L^2 ones which locate the resonances and only the resonances. Of course the scattering methods give the most detailed information possible about the scattering at a resonance and indeed the two methods can be used in a complementary fashion. Both types of calculations scale exponentially with the number of coupled degrees of freedom and therefore they can quickly become computationally intractable. Thus, for both the scattering and direct methods we have been interested in reduced-dimensionality strategies to reduce the number of coupled degrees of freedom.

In the reduced-dimensionality space the dynamics is treated exactly, e.g., by the quantum coupled-channel approach. The remaining degrees of freedom are described in one of several approximate ways which will be reviewed below. The advantage of this approach is that it is feasible for systems of arbitrary complexity. In addition, it enables one to calculate cross sections, rate constants, etc. that are implicitly averaged over those degrees of freedom not explicitly treated dynamically, thus enabling a direct comparison to experiments which in most cases are not fully state-resolved. The degrees of freedom which are neither state-resolved experimentally nor treated dynamically will often be the same because they are usually the low-frequency motions such as rotation which are widely populated initially and finally in a collision. In the next section we shall review the elements of this theory for reactive systems with particular emphasis on resonances.

Following that we present calculations on resonances in a two-mathematical-dimensional (2MD) model for van der Waals systems. We will compare the complex eigenvalues obtained previously(1) by the complex coordinate method with those obtained from the distorted wave Born approximation (DWBA). Part of the motivation for making this comparison is to assess the accuracy of the DWBA before applying it to more realistic problems.

That is done in the penultimate section where we present some preliminary DWBA calculations of the resonances in the H+CO -> HCO addition reaction using a fit to an **ab initio** potential energy surface. A reduced-dimensionalty scattering space is derived based on a novel scattering path hamiltonian.

A summary and prognosis for further work are given in the final section.

Model differential cross sections for $H+H_2(v=0)$

For the reactive systems we have studied the choice of a reduced-dimensionality space is obvious because the lowest-energy configuration in all of these systems is the collinear one. This allows us to easily describe the reaction with standard Jacobi scattering coordinates, for which the kinetic energy operator is quite simple. With this approximation we have calculated quite accurate rotationally averaged and summed but vibrational state-to-state cross sections and rate constants for $H+H_2(v=0,1)$(2-4), $D+H_2(v=0,1)$(5), $O(^3P)+H_2(v=0,1)$ and $D_2(v=0,1)$(6,7). We have also given an expression for the differential cross section which is intended to be a simple model to predict the effect of resonances which are present in the reduced-dimensionality space on differential cross sections. This model has been applied to the $F+H_2$(8) and F+HD(9) reactions. The differential cross section for the vibrational state-to-state transition v to v' averaged over initial rotational states and summed over final ones is given by(8-10)

$$d\bar{\sigma}_{v->v'}(\Theta)/d\Omega = 1/(4\bar{k}_v^2)\,|\sum_{J=0} (2J+1)S^{Jn\Omega}_{v->v'}d^J_{\Omega\Omega}(\Theta)|^2 \tag{1}$$

where

$$\bar{k}_v^2 = \sum_{j=0} (2j+1)k^2_{vj} \quad \text{and} \quad k^2_{vj} = 2\mu(E-E_v-E_j)/\hbar^2 \tag{2}$$

E is the total energy and E_v and E_j are the vibrational and rotational energies of the reactant diatomic molecule. $d^J_{\Omega\Omega}(\Theta)$ equals $D^J_{\Omega\Omega}(0,\Theta,0)$, the rotation matrix and Θ is the angle between the reactant and product arrangement channel body-fixed z-axes(11). $S^{Jn\Omega}_{v->v'}$ is the reduced-dimensionality partial wave scattering matrix which is obtained from approximate solutions to the three-dimensional Schroedinger equation, expressed in body-fixed coordinates(11-13). These solutions are based on the centrifugal sudden(14) and adiabatic approximation applied to the three-atom bending motion(10). For collinearly-favored reactions the adiabatic bending state, $|n\Omega\rangle$, is labeled by two quantum numbers, n and Ω, where Ω is the z-component of the molecular angular momentum in the body-fixed frame (which also equals the z-component of the total angular momentum J in the body-fixed frame)(11). In the centrifugal sudden approximation Ω is a good quantum number, and for the adiabatic bending wavefunction n is restricted to the range $-\Omega \leq n \leq \Omega$ in steps of two(15). In terms of $S^{Jn\Omega}_{v->v'}$, the partial wave rotationally cumulative reaction probability is given by(10)

$$\bar{P}^J_{v->v'} = \sum_{n\Omega} |S^{Jn\Omega}_{v->v'}|^2 \tag{3}$$

where the summation is over the bending states. This probability is related to the physical ones by

$$\bar{P}^J_{v->v'} = \sum_{j\Omega}\sum_{j'\Omega'} P^J_{vj\Omega->v'j'\Omega'} \tag{4}$$

where $P^J_{vj\Omega->v'j'\Omega'}$ is the complete vibrational-rotational, state-to-state reaction probability.

The scattering matrix $S^{Jn\Omega}_{v->v'}$ is obtained from the asymptotic behavior of the scattering solutions to the following body-fixed Schroedinger equation(10)

$$\{-(\hbar^2/2\mu)(\partial^2/\partial R^2 + \partial^2/\partial r^2) + [J(J+1)-2\Omega^2]\hbar^2/(2\mu R^2)$$

$$+ V(r,R,\gamma=0) + \varepsilon_{n\Omega}(r,R) - E\} U^J_{n\Omega}(r,R) = 0 \quad (5)$$

where R is the mass-scaled Delves radial distance of the atom A with respect to the center-of-mass of the reactant diatom BC and r is the mass-scaled diatom separation, $V(r,R,\gamma=0)$ is the collinear potential energy surface, $\varepsilon_{n\Omega}(r,R)$ is the ABC adiabatic bending eigenvalue which depends on r and R(10). The reduced mass μ is given by(11)

$$\mu = m_A m_B m_C/(m_A+m_B+m_C) \quad (6)$$

Consider the particular case of the ground state bend, i.e., $n=\Omega=0$. We have introduced two approximations to Equation 5. within the spirit of transition state theory(2,3,10). The first one consists of replacing the centrifugal potential $J(J+1)\hbar^2/(2\mu R^2)$ by its value at the transition state, $E^{\ddagger}_J$, i.e., $J(J+1)\hbar^2/(2\mu R^{\ddagger 2})$, where $R^{\ddagger}$ is the value of R at the transition state. In this approximation $S^{J00}_{v->v'}$ is related to the scattering matrix corresponding to Equation 5. with no centrifugal potential, $S^{CEQB}_{v->v'}(E)$, by (10)

$$S^{J00}_{v->v'}(E) = (-1)^J S^{CEQB}_{v->v'}(E-E^{\ddagger}_J|n=\Omega=0) \quad (7)$$

The corresponding CEQB probability is

$$\bar{P}^J_{v->v'} = |S^{CEQB}_{v->v'}(E-E^{\ddagger}_J|n=\Omega=0)|^2 = P^{CEQB}_{v->v'}(E-E^{\ddagger}_J) \quad (7a)$$

The notation CEQB stands for collinear exact quantum with an adiabatic treatment of the bending motion. The factor $(-1)^J$ accounts for a trivial difference in the asymptotic behavior of $U^J_{00}(r,R)$ and $U^{CEQB}_{00}(r,R)$, the CEQB wavefunction for the ground state bend, from which $S^{CEQB}_{v->v'}$ is obtained(10). The replacement of the centrifugal potential by an energy shift $E^{\ddagger}_J$ reduces the computational effort considerably.

Another and further approximation we have considered is to replace $\varepsilon_{00}(r,R)$ by $\varepsilon^{\ddagger}_{00}$, the value of the ground state bending energy at the transition state. In this further approximation $S^{J00}_{v->v'}$ is related to CEQ (collinear exact quantum) scattering matrix by

$$S^{J00}_{v->v'}(E) = (-1)^J S^{CEQ}_{v->v'}(E-E^{\ddagger}_J-\varepsilon^{\ddagger}_{00}) \quad (8)$$

where $S^{CEQ}_{v->v'}$ is the scattering matrix corresponding to Equation 5. with no centrifugal potential and no adiabatic bending eigenvalue. In this case that equation looks exactly like the usual collinear Schroedinger equation.

We have focused on the ground state bending case for the following reason. As noted, Equation 1. represents a rotationally averaged and summed differential cross section. Specifically,

$$d\bar{\sigma}_{v->v'}(\Theta)/d\Omega = (1/\bar{k}^2_v) \sum_j (2j+1)k^2_{vj} d\sigma_{vj->v'j'}(\Theta)/d\Omega \quad (9)$$

where $d\sigma_{vj->v'j'}(\Theta)/d\Omega$ is the usual degeneracy-averaged differential cross section. While we do not assume an adiabatic correlation between bending and asymptotic free-rotor states, we do expect (assume) that the single term, $n=\Omega=0$, in Equation 1. contains information mainly about the low-lying j and j' states in Equation 9. Based on this expectation we have restricted our previous calculations of differential cross sections (using the CEQ approximation to $S^{J00}_{v->v'}$) to the ground state bend(8,9). In this case Equation 1. can be written explicitly as

$$d\bar{\sigma}_{v->v'}(\theta)/d\Omega = 1/(4\bar{k}_v^2)\,|\sum_{J=0}(2J+1)S^{J00}_{v->v'}P_J(\cos\theta)|^2 \qquad (10)$$

where $P_J(\cos\theta)$ is the Legendre polynomial of order J.

We have not yet implemented the fully adiabatic theory represented by Equation 5. That theory bears some resemblance to the bending-corrected rotating linear model (BCRLM)(16-18). In this model a partial wave hamiltonian is given by

$$H^{\ell} = -(\hbar^2/2\mu)(\partial^2/\partial u^2 + \partial^2/\partial w^2) + [\ell(\ell+1)+1]\hbar^2/2\mu Q^2$$
$$+ V(u,w) + E_0^b(u) \qquad (11)$$

where ℓ is the orbital angular momentum, u is a reference reaction path coordinate, w is the coordinate transverse to u, and $E_0^b(u)$ is the ground state adiabatic bending energy evaluated on u, i.e., for w=0, and

$$Q^2 = R^2 + r^2 \qquad (12)$$

Scattering solutions to $(H^{\ell} - E)\psi^{\ell}(u,w)$ are obtained and the differential cross section is given by

$$\sigma^{BCRLM}_{v->v'}(\theta) = (1/4k^2_{vj=0})\,|\sum_{\ell=0}(2\ell+1)S^{\ell}_{v->v'}P_{\ell}(\cos\theta)|^2 \qquad (13)$$

where $k^2_{vj=0}$ is given by Equation 2.

There are a number of differences between the BCRLM and the present adiabtic bending model. First, the total angular momentum J is used in Equation 5. whereas the orbital angular momentum ℓ appears in the BCRLM equations. Second, the centrifugal potentials containing these angular momenta are different, the one given in BCRLM does not vanish asymptotically, whereas the one in Equation 5. does. Another difference is the use of $k^2_{vj=0}$ in Equation 13. instead of the average squared wavector $\bar{k}_v^2$. This last difference arises because the differential cross section we have given is explicitly a rotationally summed and averaged quantity whereas the exact identity of the BCRLM differential cross section is still ambiguous(17).

We applied Equations 3., 7a. and 10. to a study of cumulative reaction probabilities and differential cross sections for $H+H_2(v=0) -> H_2(v'=0)+H$ using the PK2 potential energy surface(19) in the total energy range 0.6 to 0.95 eV. The upper end of this range contains part of a broad resonance which is centered at 0.965 eV, in good agreement with the accurate quantum calculations for the J=0 partial wave of Schatz and Kuppermann(20), who found the resonance centered at 0.973 eV.

In Figure 1 we have plotted the partial wave cumulative reaction probabilities $\bar{P}^J_{0->0}$ versus J and the corresponding differential cross sections for total energies of 0.6 and 0.7 eV. Both sets of probabilities display a monotonic decrease with J and the differential cross sections both decline smoothly from Θ equal to 180°, in qualitative agreement with the accurate quantum results of Schatz and Kuppermann(11). Above 0.7 eV there are no accurate quantum differential cross sections for this reaction with which to compare the present calculations and so the results in Figure 2 should be regarded cautiously. There the partial wave cumulative reaction probabilities and differential cross sections are given for total energies of 0.8, 0.85, and 0.95 eV. As seen the $\bar{P}^J_{0->0}$ show a flatter dependence on J for E equal to 0.8 and 0.85 eV and distinct structure at 0.95 eV. This latter behavior is quite interesting in that it shows a maximum at non-zero J. This is simply due to the resonance in the CEQB reaction probability $P^{CEQB}_{0->0}$ manifesting itself in the partial wave reaction probability according to Equation 7a. The minimum in the CEQB resonance occurs at a slightly higher energy than 0.95 eV, as already noted, however, the behavior of $P^{CEQB}_{0->0}$ is such that as E decreases somewhat from 0.95 eV the probability increases until E decreases sufficiently so that $P^{CEQB}_{0->0}$ declines to zero. It is interesting that the differential cross sections at 0.8, 0.85, and 0.95 eV all show structure, even though the partial wave cumulative reaction probabilities do not show obvious structure at the two lower energies. This suggests that there is no simple correlation between structure in the differential cross section and structure in the partial wave reaction probabilities. However, it does seem reasonable that if $P^{CEQB}_{0->0}$ does show structure then so will the corresponding differential cross section.

The results presented here are in support of earlier studies which have correlated structure in the partial wave reaction probabilities and structure in the differential cross sections. Especially noteworthy among these earlier studies are the ones of Redmon and Wyatt(21) who first suggested this correlation based on their j_z-conserving quantum calculations of the $F+H_2$ -> HF+H reaction. We applied the CEQ version of the reduced dimensionality method to a study of that reaction(8) as well as the F+HD -> HF+D, DF+H reactions(9) and found similar correlations. More recently the BCRLM was used in an extensive study of differential cross sections in the $F+H_2$, D_2, and HD reactions(17). (Also, see the paper by Hayes and Walker in this volume.)

A scattering calculation gives the most complete description of resonances in reactive and non-reactive systems. However, if the resonances are the major features of interest a more direct approach to obtain them is desirable. A number of such aproaches exist and are reviewed in this volume and in the following sections of this paper. These have been applied mainly to non-reactive systems, however, they are beginning to be used in reactive systems. Thus far they have been applied to collinear reactive systems where their accuracy is being tested. The status of these calculations is discussed in a paper by Garrett et al. in this volume.

Of these direct approaches the complex coordinate method is the most rigorous one. In principle it yields the exact energies of the poles of the scattering matrix, which, ignoring the background contribution to the scattering, gives the resonance position and width.

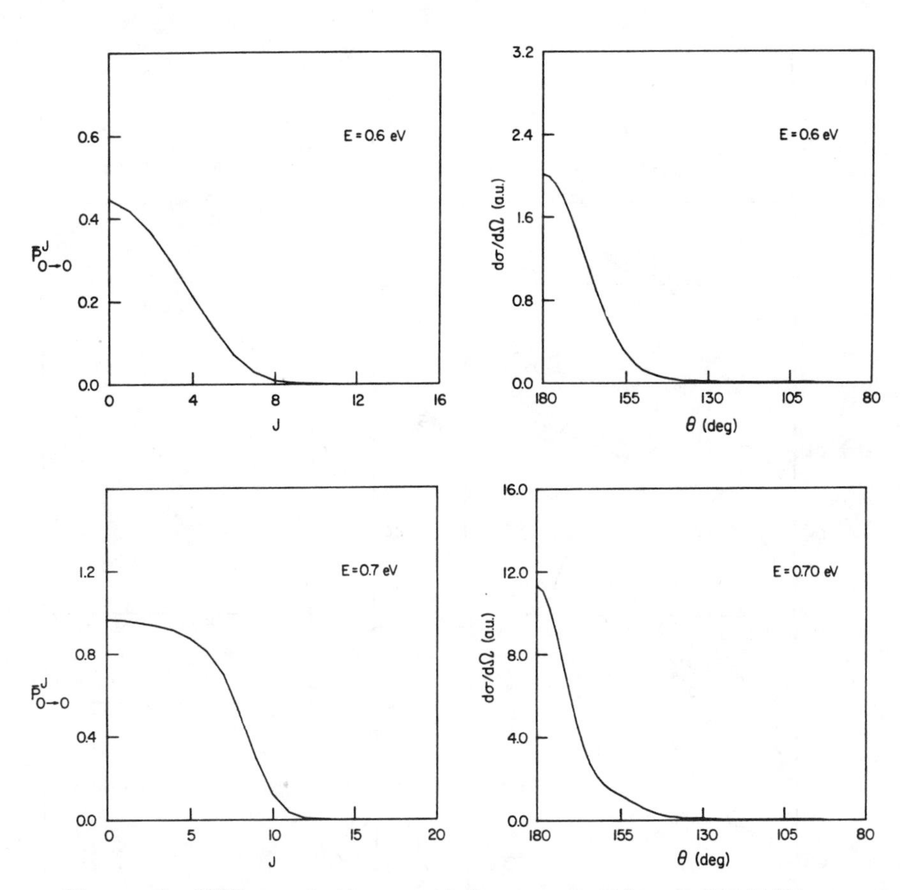

Figure. 1 CEQB cumulative partial wave reaction probabilities versus total angular momentum for $H+H_2(v=0) \rightarrow H_2(v'=0)+H$ and corresponding differential cross section for total energies of 0.6 and 0.7 eV.

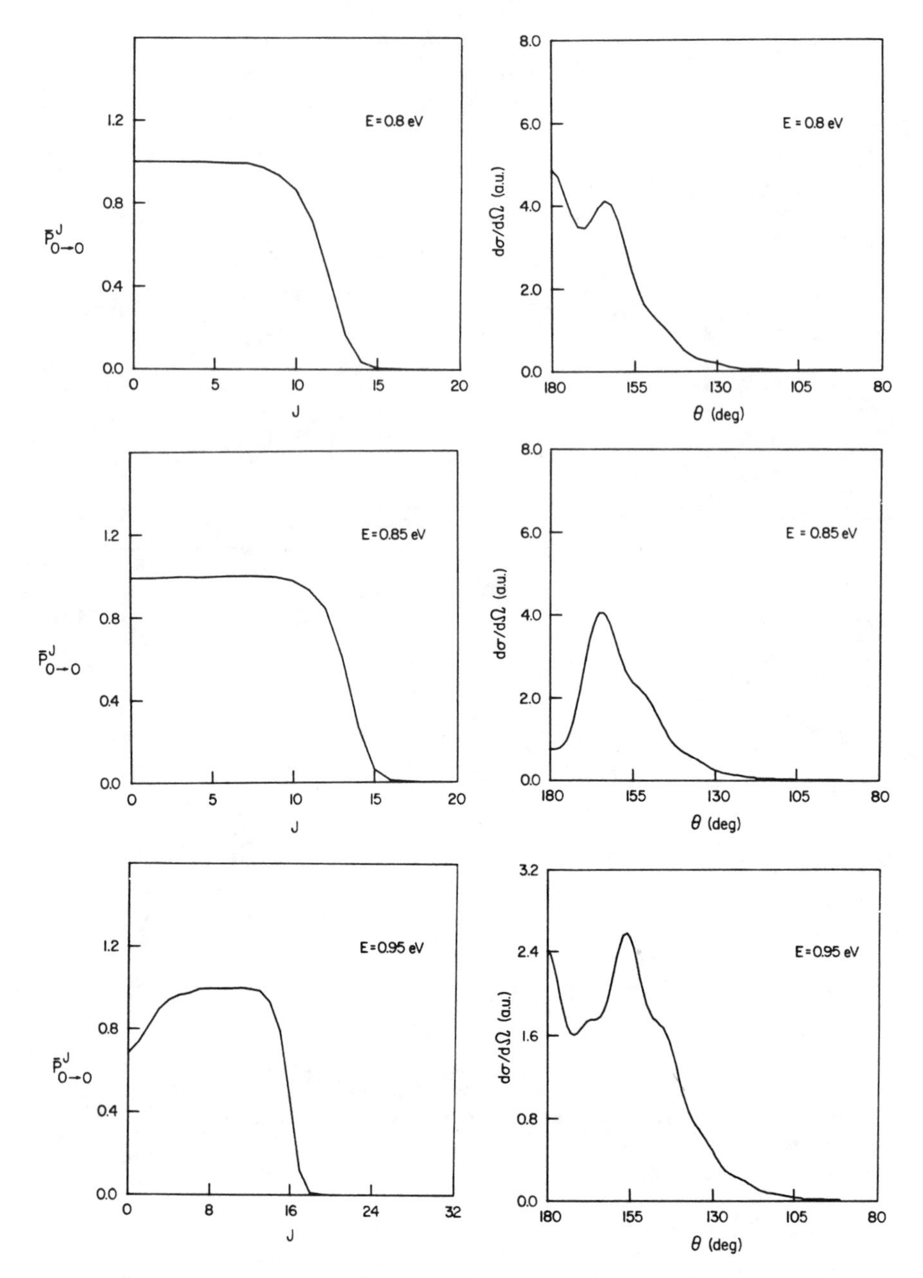

Figure. 2 Same as Fig. 1 but for total energies just below the center energy of the resonance, 0.965 eV.

There are many other approaches to obtain resonance energies and widths, many are reviewed in this volume. One that we consider in the next two sections is the distorted wave Born approximation (DWBA). In the following section the DWBA is tested against accurate complex coordinate calculations reported previously for a collinear model van der Waals system(1). The DWBA is then used to obtain the resonance energies and widths for the HCO radical. A scattering path hamiltonian is developed for that system and a 2MD approximation to it is given for the J=0 state.

Distorted wave Born and complex coordinate resonances for a model van der Waals system.

The complex coordinate method was applied recently by Christoffel and Bowman(1) to a model collinear hamiltonian used earlier by Eastes and Marcus(22) in quantum and semiclassical studies of the resonances in that system. The hamiltonian is ($\hbar$=1)

$$H = -1/m(\partial^2/\partial x^2 + \partial^2/\partial y^2) + y^2$$

$$+ D\{\exp[-2\alpha(x-y)]-2\exp[-\alpha(x-y)]\} \qquad (14)$$

where the energy is in units of the zero-point energy of the harmonic oscillator (y-mode), m = 0.2, D = 1.5, and α = 0.1. This hamiltonian describes a harmonic oscillator the end atom of which interacts with a third atom by a Morse potential. This model has served as a paradigm in many quantum scattering studies(22-27). It is also a suitable model for linear atom-diatom van der Waals complexes and has been adopted for that purpose in a series of papers by Beswick and Jortner(28). For this hamiltonian the DWBA provides analytical expressions for the resonance widths, as shown by Beswick and Jortner(29) and much earlier by Rosen(30). A very simple picture of the (Feshbach) resonances is given by partitioning H as

$$H = H_o + V_c(x,y) \qquad (15)$$

where H_o is given by Equation 14. with y equal to zero in the Morse potential and therefore

$$V_c(x,y) = D\{1-\exp[\alpha(x-y)]\}^2 - D\{1-\exp(-\alpha x)\}^2 \qquad (16)$$

Those zero-order bound state eigenfunctions $\phi_v(y)\chi_m(x)$ which are energetically degenerate with continuum ones $\phi_{v'}(y)\chi_\varepsilon(x)$ are the resonance states, where $\phi_v(y)$ is a harmonic oscillator eigenfunction and $\chi_m(x)$ and $\chi_\varepsilon(x)$ are bound and energy-normalized continuum Morse oscillator eigenfunctions, respectively. For the system parameters chosen all zero-order bound states with v>0 are resonance states. This is because the harmonic oscillator quantized energy, 2v+1, is greater than D for v>0. In the DWBA the width of a resonance is given by(30)

$$\Gamma_{vm->v'\varepsilon} = 2\pi|\langle v'\varepsilon|V_c|vm\rangle|^2 \qquad (17)$$

The matrix element in Equation 17. can be evaluated analytically(29) and of course numerically. We did both to test the numerical method, based on a Cooley-Numerov integrator(31,32) which was used in later calculations.

The widths $\Gamma_{vm->v'\varepsilon}$ were calculated for v=1,2, and 3 and numerous values of m (there are ten bound states in the Morse potential for m=0.2) and for v'=v-1. The widths for v' less than v-1 are much smaller than the single quantum changes ones. The widths $F_{vm->v'\varepsilon}$ are related to the imaginary part of the complex energy pole of the scattering matrix by

$$\Gamma_{vm->v'\varepsilon} = -2\mathrm{Im}(E_{m,v}) \quad (18)$$

The DWBA and available complex coordinate results for $-\mathrm{Im}(E_{m,v})$ are given in Figure 3. The zero-order resonance energies are given by

$$\mathrm{Re}(E_{m,v}) = 2v+1 + E_m \quad (19)$$

where E_m are the bound state Morse oscillator energies. The results naturally group into series for a given v in which m increases. Each group is bounded above in energy by 2v+1 because the last bound Morse level is close to but less than zero. There is good agreement between the complex coordinate and DWBA results, the latter are roughly 30% smaller than the accurate results. A striking mode specificity is seen in these results; clearly energy in the dissociative Morse degrees of freedom is more effective in the dissociation than energy in the harmonic degree of freedom. This is not a surprising result for such a weakly coupled system.

These results confirm and complement the earlier work of Beswick and Jortner who compared DWBA widths with those from collinear coupled-channel scattering calculations(33) where, as here, good agreement was observed. In the next section the DWBA widths are calculated for the chemically bonded HCO radical to give H+CO. Based on the present comparisons with exact results we are optimistic that the DWBA will provide realistic widths for this system.

Resonances in the HCO system using an **ab initio** potential

The HCO potential surface was generated with Hartree-Fock plus all single and double excitation, configuration interaction calculations to which the Davidson correction(34,35) for quadruple excitations was added. The basis set employed was the standard polarized valence double zeta basis set of Dunning and Hay(36). These calculations lead to structures and energies, shown in Table I., similar to those obtained by Dunning(37). For the dynamical calculations presented here it is necessary to have global information on the surface in addition to the properties of the stationary points. This was obtained by carrying out **ab initio** calculations at approximately 2000 geometries. The energies from these calculations were fit, in localized regions, to Taylor series polynomials of the Simons-Parr-Finlan form(38,39). These analytic representations of local regions of the potential surface were then connected by hyperbolic tangent switching functions to give an analytic representation of the global surface having the same stationary point properties as the **ab initio** surface (Table I.).

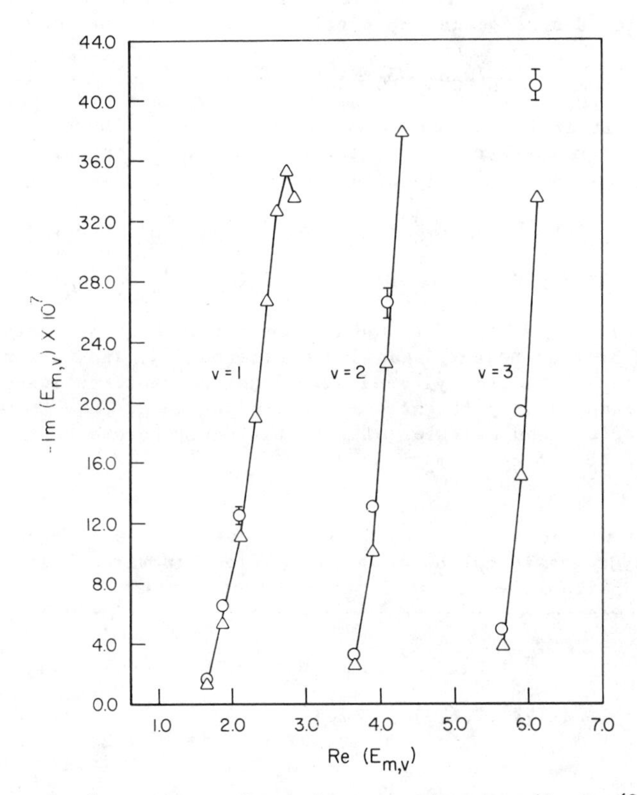

Figure. 3 Comparison of previous complex-coordinate (O) (ref. 1) and present distorted wave Born (Δ) resonance half-widths and energies for a model van der Waals system. The half-widths are given by $-\mathrm{Im}(E_{m,v})$ and the energies by $\mathrm{Re}(E_{m,v})$.

The **ab initio** calculations used here lead to a barrier height of 5.8 kcal/mol, while the observed activation energy for this reaction is reported to be approximately 2 kcal/mol(40), implying an error in the calculated barrier height of approximately 4 kcal/mol. Similarly, the **ab initio** calculations predict the addition of hydrogen to carbon monoxide to be 18.1 kcal/mol exothermic, while the best experimental estimate(41) of this energy difference is approximately 1 kcal/mol higher. To correct for these deficiencies in the calculated surface an empirical correction of the form,

$$E_{CORR} = (E_C/R_CH + E_O/R_{OH})/(1/R_{CH} + 1/R_{OH}) \quad (20)$$

was added to the **ab initio** potential energy surface, where R_{CH} is the distance from the hydrogen to the carbon, R_{OH} is the distance from the hydrogen to the oxygen, and where

$$E_C = A_C/\{\exp[B_C(R_{CH}-R^o_{CH})] + \exp[-C_C(R_{CH}-R^o_{CH})] + D_C\} \quad (21)$$

$$E_O = A_O/\{\exp[B_O(R_{OH}-R^o_{OH})] + \exp[-C_O(R_{OH}-R^o_{OH})] + D_O\}. \quad (22)$$

The parameters, A, B, C, D, R^o_{CH} and R^o_{OH} were chosen to match the experimental barrier height, reaction exothermicity, and CH stretching frequency of the formyl radical. Other characteristics of the surface are changed only slightly by this adjustment. The geometries, frequencies, and energies of the stationary points on this adjusted surface are also shown in Table I.

Table I. Calculated SDQ-CI structures and energies for the addition of atomic hydrogen to carbon monoxide. Values in parenthesis are from the adjusted surface (see text).

	H + CO	H - CO	HCO
Geometries			
R_{CO}	2.173 (2.173)	2.193 (2.181)	2.259 (2.259)
R_{CH}	-	3.431 (3.495)	2.116 (2.124)
<HCO(deg)	-	119.0 (117.2)	124.5 (124.2)
Harmonic frequencies (cm^{-1})			
ω_1	2173 (2173)	1963 (2120)	2815 (2748)
ω_2	-	400 (400)	1903 (1905)
ω_3	-	1095i (589i)	1156 (1145)
Total energies (hartree)	-113.54392	-113.53472	-113.57284
Relative energies	0.0 (0.0)	5.8 (1.6)	-18.1 (-19.4)

An equipotential contour plot of the potential surface is given in Figure 4 for a fixed R_{CO} distance of 2.25 a_o, its value at the HCO minimum. The steepest descent paths (in mass-weighted cartesian coordinates) to the HCO minimum are shown from the primary, low-energy HCO saddle point and from a secondary, higher-energy saddle point for the collinear HCO configuration. Note that R_{CO} does vary along these paths, although its value was fixed for the potential surface contour, as noted. Several features of this reaction path are noteworthy. First, R_{CO} does not change very much along them. For example R_{CO} changes by only +0.078 a_o from its value at the primary energy saddle point to the minimum. Second, that branch of the path is nearly linear, except in the vicinity of the minimum. These facts suggest the following scattering scenario. The reactants approach along the reaction path initially, however, due to the presence of the HCO minimum and the rather weak coupling the H atom picks up kinetic energy rapidly and tends to follow a straight-line (diabatic) path into the hard-wall region of the potential, where it may undergo a resonant transition down into a bound state of the well while temporarily exciting the CO-stretch. For these reasons, we have adopted another path which we term a scattering path. This is a straight-line path connecting the low energy saddle point with the minimum and for a constant value of R_{CO} given by its value at that saddle point, $R^{\ddagger}_{CO}$. The distance along this path is denoted t, the orthogonal transverse coordinate is u, and R_{CO} is unchanged. For simplicity we shall make the following changes in notation, $R=R_{H,CO}$ and $r=R_{CO}$. Also, we shall replace γ by the distance $r^{\ddagger}\gamma$. Thus, the transformation between the coordinates $(r^{\ddagger}\gamma,R)$ and (u,t) is given by

$$\begin{bmatrix} R-R^{\ddagger} \\ r^{\ddagger}(\gamma - \gamma^{\ddagger}) \end{bmatrix} = \begin{bmatrix} \cos\theta & -\sin\theta \\ \sin\theta & \cos\theta \end{bmatrix} \begin{bmatrix} t \\ u \end{bmatrix} \tag{23}$$

where

$$\tan\theta = r^{\ddagger}(\gamma^{\ddagger} - \gamma_o)/(R^{\ddagger} - R_o) \tag{24}$$

where γ_o and R_o are the values of γ and R at the HCO minimum.

To obtain the scattering path hamiltonian we would begin with the general body-fixed hamiltonian in the variables (R,γ,r)(<u>13</u>) and simply re-express it in terms of the variables (t,u,r) using the above transformation. For simplicity we consider the zero partial wave, J=0, and we obtain

$$H = \frac{-\hbar^2}{2\mu_{H,CO}} \frac{\partial^2}{\partial R^2} \quad \frac{-\hbar^2}{2\mu_{CO}} \frac{\partial^2}{\partial r^2} \quad \frac{-\hbar^2}{2I(r,R)} \left(\frac{\partial^2}{\partial\gamma^2} + \cot\gamma\frac{\partial}{\partial\gamma}\right) + V(R,r,\gamma), \tag{25}$$

where

$$I^{-1}(R,r) = (\mu_{H,CO}R^2)^{-1} + (\mu_{CO}r^2)^{-1} \tag{26}$$

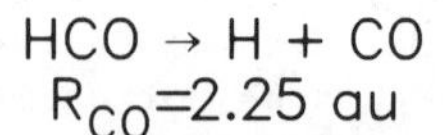

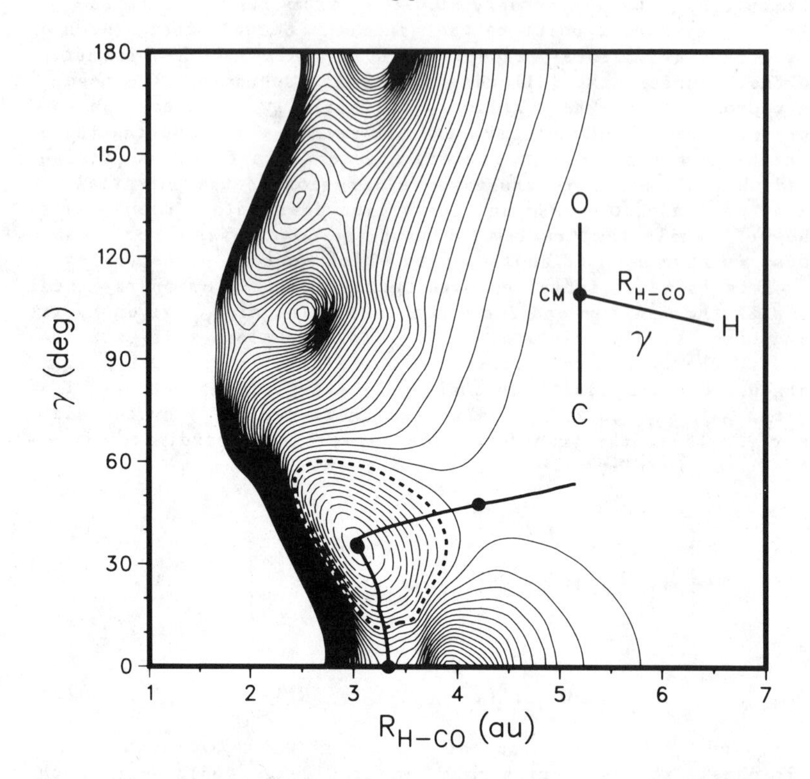

Figure. 4 Equipotential energy contours obtained from **ab initio** calculations for the HCO system for R_{CO} fixed at its value in HCO. The zero-energy contour is given by the thick dashed line, negative and positive energy contours are given by thin dashed and solid lines, respectively. The energy difference between adjacent contour lines is 2 kcal/mole. The paths shown are the paths of steepest descent in mass-weighted Cartesian coordinates (along which R_{CO} varies) from the primary low energy saddle point to the HCO minimum and then to the secondary high energy collinear saddle point.

The Schroedinger equation is

$$(H-E)\,\psi(R,r,\gamma) = 0 \tag{27}$$

where

$$\psi(R,r,\gamma) = (rR)\,\Psi(R,r,\gamma) \tag{28}$$

and $\Psi(R,r,\gamma)$ is the eigenfunction of the body-fixed hamiltonian containing first derivative terms in R and r. Before rewriting Equation 27. in terms of the new variable (t,u,r) we approximate $I^{-1}(R,r)$ by replacing r by $r^{\ddagger}$. This is justified in the present case because as noted r does not change very much from $r^{\ddagger}$ along the reaction path and is assumed to be constant along the scattering path. Thus, we replace I^{-1} by the approximation

$$I^{-1} = [(1/\mu_{H,CO})(r^{\ddagger}/R)^2 + 1/\mu_{CO}]/r^{\ddagger 2} \tag{29}$$

and rewrite Equation 25. as

$$H = \frac{-\hbar^2}{2\mu_{H,CO}}\frac{\partial^2}{\partial R^2} \quad \frac{-\hbar^2}{2\mu_{CO}}\frac{\partial^2}{\partial r^2} \quad \frac{-\hbar^2}{2\mu(R)}\left(\frac{1}{r^{\ddagger 2}}\frac{\partial^2}{\partial \gamma^2} + \cot\gamma\frac{\partial}{\partial \gamma}\right) + V(R,r,\gamma) \tag{30}$$

where

$$1/\mu(R) = 1/\mu_{CO} + 1/[\mu_{H,CO}(R/r^{\ddagger 2})] \tag{31}$$

The transformation of Equation 30. to the (t,u,r) coordinate system is straightforward, with the following result.

$$H = -\frac{\hbar^2}{2}\left(\frac{\cos^2\theta}{\mu_{H,CO}} + \frac{\sin^2\theta}{\mu(R)}\right)\frac{\partial^2}{\partial t^2} \quad \frac{-\hbar^2}{2}\left(\frac{\cos^2\theta}{\mu(R)} + \frac{\sin^2\theta}{\mu_{H,CO}}\right)\frac{\partial^2}{\partial u^2} \quad \frac{-\hbar^2}{2\mu_{CO}}\frac{\partial^2}{\partial r^2}$$

$$+ T(u,\partial/\partial u,t,\partial/\partial t) + V(t,r,u) \tag{32}$$

where

$$T(u,\partial/\partial u,t,\partial/\partial t) = -[\hbar^2/2\mu(R)](\cot\gamma/r^{\ddagger})(\sin\theta\partial/\partial t + \cos\theta\partial/\partial u) \tag{33}$$

One advantage of this scattering path hamiltonian is that the potential can be written as

$$V(t,r,u) = V_o(t) + V'(r,u,t) \tag{34}$$

where $V_o(t)$ is the potential along the scattering path. This will be shown and discussed below. Another advantage of Equation 32. is that reduced dimensionality quantum approaches to the calculation of the inelastic scattering can be easily implemented. Our focus here is on the resonance energies and widths of the HCO complex and in particular those which are due to temporary excitation of the CO stretch. Thus, we shall simply ignore the u-motion hereafter. (Other approximate, e.g. adiabatic, treatments of it could be considered.) In addition, $\cos^2\theta = 0.87$, $\sin^2\theta = 0.13$ and $\mu(R)$ is roughly several times $\mu_{H,CO}$ for values of R in the vicinity of the HCO minimum. Thus, the term $\sin^2\theta/\mu(R)$ is roughly an order of magnitude smaller than $\cos^2\theta/\mu_{H,CO}$, and so we shall ignore it.

This gives

$$h = \frac{-\hbar^2}{2\mu_{H,CO}/\cos^2\theta} \frac{\partial^2}{\partial t^2} \frac{-\hbar^2}{2\mu_{CO}} \frac{\partial^2}{\partial r^2} + V_o(t) + V'(r,t) \quad (35)$$

for the reduced-dimensionality hamiltonian. This hamiltonian was used in distorted wave Born approximation calculations of resonances, as described in the previous section and the results are given below.

DWBA calculation of resonances. The variation of R_{CO}, $R_{H,CO}$, γ, and the potential with t is shown in Figure 5 for the straight-line scattering path and the portion of the reaction path shown in Figure 4 from the primary saddle point to the HCO minimum. There is good agreement between the paths, however, when considering the potential $V'(r,t)$ (recall $r=R_{CO}$) there are important differences. For the scattering path

$$V'(r,t) = a(t)(r-r^{\ddagger}) + b(t)(r-r^{\ddagger})^2 + \ldots \quad (36)$$

whereas for the reaction path the normal modes transverse to it (one of which is close to being the CO stretch) do not contain a linear term in a Taylor series representation of the potential(42,43). This difference between the form of the potential for the mode transverse to the reference scattering or reaction path has important consequences for the nature of resonance states, as discussed below.

Distorted wave Born approximation resonance energies and widths were calculated numerically using Equation 17. for the reduced-dimensionality hamiltonian given by Equation 35. and employing Equation 36. for $V'(r,t)$ up to second order. The coefficients $a(t)$ and $b(t)$ were determined numerically from the **ab initio** potential surface. Zero-order wavefunctions $\phi_v(r)$, $\chi_m(t)$ and $\chi_\varepsilon(t)$ were determined as follows. The $\phi_v(r)$ are harmonic oscillator functions for CO with a frequency of 2172.6 cm^{-1}, as determined from the **ab initio** surface and $\chi_m(t)$ and $\chi_\varepsilon(t)$ are the bound and continuum eigenfunctions of the zero-order hamiltonian

$$h_m = \frac{-\hbar^2}{2\mu_{H,CO}/\cos^2\theta} \frac{d^2}{dt^2} + V_o(t) \quad (37)$$

They were determined numerically by the Cooley-Numerov method(31,32), with a spline representation of $V_o(t)$. The $V_o(t)$ potential supports three bound states.

The general expression for the DWBA width (cf. Equation 17.) can be written as

$$\Gamma_{vm\rightarrow v'\varepsilon} = 2\pi|\langle v'|r-r^{\ddagger}|v\rangle\langle\varepsilon|a(t)|m\rangle+\langle v'|(r-r^{\ddagger})^2|v\rangle\langle\varepsilon|b(t)|m\rangle|^2 \quad (38)$$

Thus, only $\Delta v = -1$ and $\Delta v = -2$ transitions can occur, the first term is responsible for the former and the second term is responsible for the latter. (Strictly, because $r^{\ddagger}$ is not quite equal to the equilibrium CO distance, the second term can also give rise to $\Delta v = -1$ transitions, however, those are negligible compared to those due to

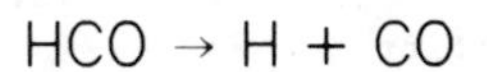

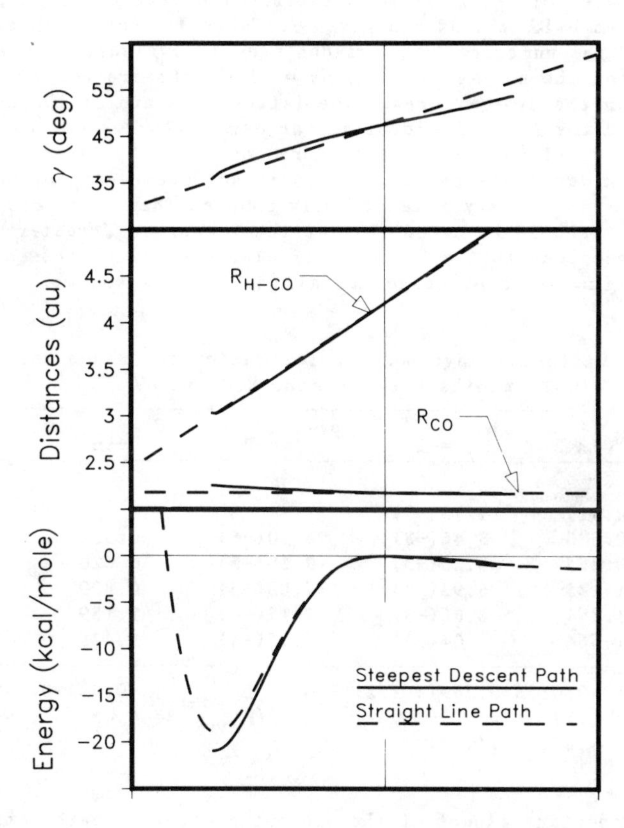

Figure. 5 The variation of γ, R_{H-CO}, R_{CO} and the potential energy along the steepest descent (solid line) and straight line (dashed line) paths described in the text. Only the portion of the former path from the primary saddle point to the HCO minimum is shown.

the first term because the above two distances differ by only 0.008 a_o.)

Because only three zero-order bound states are supported by $V_o(t)$ it is possible to classify the widths into three categories. For the $\Delta v = -1$ transitions only the highest bound state of $V_o(t)$ is metastable. That is, the energy released in the $\Delta v = -1$ transition is only sufficient to cause this highest bound state to make the transition to the continuum. For $\Delta v = -2$ transitions the two excited states of $V_o(t)$ are metastable. The zero-order resonance energies and DWBA widths are given in Table II. for a number of states and the numerical expressions used to evaluate them are given as footnotes there. As seen the $\Delta v = -1$ widths are considerably larger than the $\Delta v = -2$ ones. The latter ones are on the order of those found for $\Delta v = -1$ widths in van der Waals systems(28). The present $\Delta v = -1$ widths are therefore considerably larger than those seen in van der Waals systems. This is not unexpected because the HCO complex is clearly more strongly coupled than a van der Waals system. It should be noted that at total energies greater than 1.5 eV resonances due to the HOC complex also occur and these are discussed in the paper by Geiger et al.(44) in this volume.

Table II. Distorted wave Born approximation resonance energies and widths for two-mode HCO (in eV).

v	a) $E_{v,m=2}$	b) $\Gamma_{\Delta v=-1}$	c) $\Gamma_{\Delta v=-2}$	d) $E_{v,m=1}$	e) $\Gamma_{\Delta v=-2}$
1	0.317	1.73(-3)	--	--	--
2	0.586	3.46(-3)	8.90(-6)	0.351	1.17(-5)
3	0.855	5.20(-3)	3.50(-5)	0.620	2.67(-5)
4	1.125	6.93(-3)	7.00(-5)	0.890	5.34(-5)
5	1.394	8.66(-3)	1.17(-4)	1.159	8.90(-5)
6	1.664	1.04(-2)	1.75(-4)	1.428	1.33(-4)

a) $E_{v,m=2} = -0.0873+0.269(v+1/2)$
b) $\Gamma_{\Delta v=-1} = 1.73(-3)v$
c) $\Gamma_{\Delta v=-2} = 5.83(-6)v(v-1)$
d) $E_{v,m=1} = -0.322+0.269(v+1/2)$
e) $\Gamma_{\Delta v=-2} = 4.45(-6)v(v-1)$

An important aspect of the present scattering path hamiltonian is the presence of the linear term in the potential $V'(r,t)$ (cf. Equation 36.). This gave rise to the important $\Delta v = -1$ resonances. In contrast the reaction path hamiltonian does not contain linear terms in the potential and so $\Delta v = -1$ transitions (within the DWBA) would arise only from curvature terms in the kinetic energy operator(42,43). (Thus, a zero-curvature reaction path hamiltonian would not produce $\Delta v = -1$ resonances.) For this reason and for its simplicity, the use of a scattering path hamiltonian should be considered along with the reaction path hamiltonian for addition reactions. Skodje et al. recently reported DWBA calculations of resonance energies and widths for several collinear reactive scattering systems using a vibrationally adiabatic model based on a reaction path hamiltonian(45).

Summary and prognosis

We have presented a sample of resonance phenomena and calculations in reactive and non-reactive three-body systems. In all cases a two-mathematical dimensional dynamical space was considered, leading to a great simplification in the computational effort. For the H+CO system, low-energy coupled-channel calculations are planned in the future to test the reliablity of the approximations used here, i.e., the scattering path hamiltonian as well as the distorted wave Born approximation. Hopefully these approximations will prove useful in larger systems where coupled-channel calculations would be prohibitively difficult to do. Such approximations will be necessary as resonance phenomena will continue to attract the attention of experimentalists and theorists for many years.

Acknowledgments: The work at IIT was supported in part by the National Science Foundation (CHE-811784) (the complex coordinate calculations) and the U. S. Department of Energy, Office of Basic Energy Sciences (DE-AC02-81ER10900) (the $H+H_2$ resonances). The work at Argonne National Laboratory was supported by the U. S. Department of Energy, Division of Chemical Sciences under contract W-31-109-Eng-38.

Literature Cited

1. Christoffel, K. M.; Bowman, J. M. J. Chem. Phys. 1983, 78, 3952.
2. Bowman, J. M.; Ju, G.-Z.; Lee, K.-T. J. Chem. Phys. 1981,75, 5199.
3. Bowman, J. M.; Ju, G.-Z.; Lee, K.-T. J. Phys. Chem. 1982, 86, 2232.
4. Bowman, J. M.; Lee, K.-T. Chem. Phys. Lett. 1983, 74, 363.
5. Bowman, J. M.; Lee, K.-T.; Walker, R. B. J. Chem. Phys. 1983, 79, 3742.
6. Lee, K.-T.; Bowman, J. M.; Wagner, A. F.; Schatz, G. C. J. Chem. Phys. 1982, 76, 3583.
7. Bowman, J. M.; Wagner, A. F.; Walch, S. P.; Dunning, Jr., T. H. "Reaction dynamics for $O(^3P)+H_2$ D_2. IV. Reduced dimensionality quantum and quasiclassical rate constants with an adiabatic incorporation of the bending motion", J. Chem. Phys., accepted for publication.
8. Bowman, J. M.; Lee, K.-T.; Ju, G.-Z. Chem. Phys. Lett. 1982, 86, 384.
9. Lee, K.-T.; Bowman, J. M. J. Phys. Chem. 1982, 86, 2289.
10. Bowman, J. M. "Reduced dimensionality quantum theories of reactive scattering", Adv. Chem. Phys., to be published.
11. Schatz, G. C.; Kuppermann, A. J. Chem. Phys. 1976, 65, 4642, 4668.
12. Curtiss, C. F.; Hirschfelder, J. O.; Adler, F. T. J. Chem. Phys. 1950, 18, 1638.
13. Pack, R. T J. Chem. Phys. 1974, 60, 633.
14. Kuppermann, A.; Schatz, G. C.; Dwyer, J. P. Chem. Phys. Lett. 1977, 45, 71.
15. See, for example, Townes, C. H.; Schalow, A. L. "Microwave Spectroscopy", McGraw-Hill: New York, 1955, Chapt. 3.

16. Walker, R. B.; Hayes, E. F. J. Phys. Chem. 1983, 87, 1255.
17. Hayes, E. F.; Walker, R. B. "Reactive differential cross sections in the rotating linear model: reactions of fluorine atoms with hydrogen molecules and their isotopic variants", J. Phys. Chem., submitted
18. Walker, R. B.; Blais, N. C.; Truhlar, D. G. J. Chem. Phys. 1984, 84, 246.
19. Porter, R. N.; Karplus, M. J. Chem. Phys. 1964, 40, 1105.
20. Schatz, G. C.; Kuppermann, A. Phys. Rev. Lett. 1975, 35, 1266.
21. Redmon, M. J.; Wyatt, R. E. Chem. Phys. Lett. 1979, 63, 209.
22. Eastes, W.; Marcus, R. A. J. Chem. Phys., 1973, 59, 4757.
23. Secrest, D.; Johnson, B. R. J. Chem. Phys., 1966, 45, 4556.
24. Secrest, D.; Eastes, W. J. Chem. Phys. 1972, 56, 2502.
25. Heller, E. J. Chem. Phys. Lett. 1973, 23, 102.
26. Numrich, R. W.; Kay, K. G. J. Chem. Phys. 1979, 70, 4343.
27. Truhlar, D. G.; Schwenke, D. W. Chem. Phys. Lett. 1983, 95, 83.
28. For a review, see Beswick, J. A.; Jortner, J. Adv. Chem. Phys. 1981, 47, part 1, 363.
29. Beswick, J. A.; Jortner, J. J. Chem. Phys. 1978, 68, 2277.
30. Rosen, N. J. Chem. Phys. 1933, 1, 319.
31. Cooley, J. W. Math. Comput. 1961, 15, 363.
32. We thank Professor C. W. McCurdy for a copy of his Cooley-Numerov code.
33. Beswick, J. A.; Jortner, J. J. Chem. Phys. 1978, 69, 512.
34. Langhoff, S. R.; Davidson, E. R. Int. J. Quant. Chem. 1974, 8, 61.
35. Davidson, E. R.; Silver, D. W. Chem. Phys. Lett. 1978, 52, 403.
36. Dunning, Jr.; T. H., Hay, P. J. "Methods of Electronic Structure Theory", Schaefer, III, H. F., Ed.; Plenum: New York, 1971, Chapt. 1.
37. Dunning, Jr., T. H. J. Chem. Phys. 1980, 73, 2304.
38. Simons, G.; Parr, R. G.; Finlan, J. M. J. Chem. Phys. 1973, 59, 3229.
39. Simons, G. J. Chem. Phys. 1974, 61, 369.
40. Wang, H. Y.; Eyre, J. A.; Dorfman, L. M. J. Chem. Phys. 1973, 59, 5199.
41. Warneck, P. Z. Naturforsch. Teil A 1974, 29, 350.
42. Marcus, R. A. J. Chem. Phys. 1968, 49, 2610 and references therein.
43. Miller, W. H.; Handy, N. C.; Adams, J. E. J. Chem. Phys. 1980, 72, 99.
44. Geiger, L. C.; Schatz, G. C.; Garrett, B. C. "Resonances in the collisional excitation of CO by fast H atoms", this volume.

RECEIVED June 11, 1984

ELECTRON–MOLECULE SCATTERING AND PHOTOIONIZATION

5

Resonances in Electron–Molecule Scattering and Photoionization

B. I. SCHNEIDER and L. A. COLLINS

Los Alamos National Laboratory, Los Alamos, NM 87545

The development of reliable theoretical models for calculating the decay of quasi-stationary states of molecular systems has become an important endeavor for theoretical chemists. The understanding and analysis of a wide variety of physical and chemical phenomena depend on a knowledge of the behavior of these states in both collisional and photoionization problems. In this article we describe the theory and calculation of these cross sections using our Linear Algebraic/Optical Potential method. The theory makes optimal use of the numeriical methods developed to solve large sets of coupled integral equations and the bound state techniques used by quantum chemists. Calculations are presented for a representative class of diatomic and triatomic molecules at varying levels of sophistication and for collisional and photoionization cross sections.

The formation and subsequent decay of highly excited neutral molecules and molecular negative ions plays an important role in many physical and chemical problems (1,2,3). These metastable states are formed when electrons collide with or are ejected from molecular systems. Consequently they are seen in photoionization cross sections (4,5) as well as collisional excitation problems involving electronic, (6) vibrational, (7) and dissociative channels (8). The quasi-stationary nature of these temporary states has interested theoretical chemists for many years. In fact a large number of the early predictions and calculations of the energies of these states were based on variants of bound state techniques used widely in quantum chemistry. These methods exploited the localized nature of the resonant scattering wavefunction. Later methods such as the Stieltjes imaging (5,6) and complex co-ordinate techniques (9,10) went further and calculated the lifetime (width) of these metastable states. However it should be recognized that these resonant states are a subset of those treated by more standard collisional techniques. Methods such as the close-coupling (11),

0097–6156/84/0263–0065$07.00/0

Kohn-variational (12), Schwinger-variational, (13) R-Matrix (14), and linear algebraic techniques (15,16) have been quite successful in calculating collisional and photoionization cross sections in both resonant and nonresonant processes. These approaches have the advantage of generality at the cost of an explicit treatment of the continuous spectrum of the Hamiltonian and the requisite boundary conditions. In the early molecular applications of these scattering methods, a rather direct approach based on the atomic collision problem was utilized which lacked in efficiency. However in recent years important conceptual and numerical advances in the solution of the molecular continuum equations have been discovered which have made these approaches far more powerful than those of a decade ago (13,15,16). These new methods make extensive use of the ideas of bound state quantum chemistry to treat electron exchange, polarization and correlation (17,18). On the other hand they treat the molecular scattering function either numerically or in a basis set of numerical continuum functions (19). This has the advantage of providing an accurate representation of the molecular continuum function without causing undue strain on the numerical representation. In the next section we develop one of these approaches, the linear algebraic/optical potential method (LAMOPT) (15), in some detail. The last section is devoted to discussing the numerous applications to photoionization and electron scattering which have been made with the method as well as some earlier work on $e+N_2$ vibrational excitation (7) using the R-matrix technique.

Theoretical Methods

One of the more important features underlying all of the theoretical approaches used in the molecular continuum problem is the division of space into a strongly and weakly intereacting part. This division may be performed in function space as in the early "stabilization" method (3) or in co-ordinate space as in the R-matrix method (14). In the strongly interacting subspace, it is necessary to deal with all the complications of the full many-body problem. However, the strength of the interaction in this region is such that the difference between bound and continuum states is not substantial. These facts are what led to the early successes of the stabilization, Stieltjes and complex co-ordinate methods in treating resonances. The difficulties of applying these approaches to the entire molecular continuum led to the development of more powerful techniques. The R-matrix method, which had some spectacular successes for atomic collisions (20), seemed an excellent choice to fill the void. One of the most attractive features of the method is its use of a discrete basis set to expand the continuum orbital in the inner R-matrix region. The application of the standard Gaussian and Slater basis sets in the R-matrix formalism was made practical by the use of the Bloch $\mathcal{L}$-operator formalism (21,22). This was successful in a number of cases (23,24,25) most notably $e-N_2$ scattering. However certain difficulties arose which pointed to the need of more general and flexible basis sets for rapid (practical) convergence. In spite of these difficulties the physical division of space into an internal and external region remains viable. The R-matrix formalism allows the flexibility of exploiting the different physics of the two regions optimally and adapting the mathematics to accomplish the purpose in the most efficient manner. We proceed by defining

$$(\mathscr{H} + \mathscr{L}_b - E)|\psi) = \mathscr{L}_b|\Psi) \tag{1a}$$

in the internal region, where

$$\mathscr{H} = \mathscr{H}_T + T_e + V_{eT} + T_R \tag{1b}$$

$$\mathscr{H}_T = \text{Target electronic Hamiltonian} \tag{1c}$$

$$T_e = \text{Kinetic energy of the scattering electron} \tag{1d}$$

$$V_{eT} = \text{Electronic potential} = -\sum_q^{n_q} \frac{Z_q}{|\vec{R}_q - \vec{r}_e|} + \sum_{i=1}^{n} \frac{1}{|\vec{r}_e - \vec{r}_i|} \tag{1e}$$

$$T_R = \text{Nuclear kinetic Energy} = -\frac{1}{2} \sum_q \nabla^2_{R_q} \tag{1f}$$

$$\mathscr{L}_b = \frac{1}{2} \sum_c |c)\, \delta(r_e - a) \left(\frac{\partial}{\partial r_e} - b\right)(c| \tag{1g}$$

$$a = \text{R-matrix radius} \tag{1h}$$

$$b = \text{Surface log derivative} \tag{1i}$$

and $|c)$ is a target eigenfunction of $\mathscr{H}_T$.

Aside from the Bloch $\mathscr{L}$-operator, $\mathscr{L}_b$, which is added and subtracted to the Hamiltonian to ensure Hermiticity, the division into target and incident particle Hamiltonian is standard. In equation (1g) we have only allowed for open electronic channels. A Bloch operator for the nuclear coordinate would need to be added if dissociation were included. A formal solution to the problem may be written as

$$|\Psi) = g\, \mathscr{L}_b |\psi) \tag{2a}$$

where

$$g = (\mathscr{H} + \mathscr{L}_b - E)^{-1} \tag{2b}$$

By projecting equation (2a) onto the channels, $|c)$, we obtain

$$F_c(r) = \sum_{c'} g_{cc'}(r|a) \left(\frac{\partial F_{c'}}{\partial r} - bF_{c'}\right)_a \tag{3}$$

and setting r=a, we get

$$F_c(a) = \sum_{c'} \mathscr{R}_{cc'}\left(\frac{\partial F_{c'}}{\partial r} - bF_{c'}\right)_a \tag{4a}$$

$$\mathscr{R}_{cc'} = g_{cc'}(a|a) = \text{R-matrix} \tag{4b}$$

If one knows the functional form of F_c at r=a it is possible by a simple matching procedure to extract the scattering information. Alternatively, it is possible to devise numerical procedures to propagate the R-matrix from r=a to very large values of the radial coordinate. At these values of r, a matching to free waves is possible. In order to do this it is necessary that the coupling potential be local beyond r=a. In addition, if it is weak and multipolar in form, the R-matrix propagation method (26,27) can be made very efficient. In essense then the difficult part of the calculation is the construction of the Green's function inside the spherical surface r<a. As is true of all boundary value problems there are essentially two methods for the construction of the Green's function. The first and perhaps most sraightforward is to construct solutions of the problem.

$$(\mathscr{H} + \mathscr{L}_b - E_i)|\Psi_i) = 0 \tag{5}$$

which enables us to write

$$g_{cc'}(r|r') = \sum_i \frac{\gamma_{ci}(r)\gamma_{c'i}(r')}{E_i - E} \tag{6}$$

This spectral form has the advantage that a single diagonalization of the Hamiltonian allows one to construct easily the R-matrix at all energies. In order to accomplish this it is necessary to introduce a basis set of many-electron functions and solve equation (5) variationally (23,24,25). The many-electron functions themselves are constructed as products of one-electron orbitals, expanded in some primitive basis set. For molecular systems with more than one nuclear center the use of multicenter Gaussian or Slater functions does much in representing the continuum function near the nuclei. However these functions are not particularly good at representing the scattering function away from the nuclei where they oscillate rather than decay as true bound states. Thus one is faced with the following dilemma: use large traditional basis sets of Slater and/or Gaussian functions for which it is possible to do the one and two electron integrals efficiently or look for a better representation. By a better representation we mean either a more efficient one-electron basis to expand the molecular continuum function or an alternative procedure for the solution of the equation for the Green's function. In searching for a better representation it should be realized that some of the numerical procedures might have to be replaced by others which are not quite so "simple" or efficient. Thus, for example, the use of numerical continuum orbitals within the R-matrix formalism requires the use of single-center expansion techniques and numerical quadratures for certain classes of one and two-electron matrix elements. For diatomics the procedure can be made reasonably efficient but for polyatomics the jury is still out. The approach we have pursued is rooted in the second method for the construction of the Green's function. If we return to the integral equation (2) and divide the Hamiltonian into an unperturbed ($\mathscr{H}_0$) and perturbed (V) part, we may write

$$|\Psi) = (G_0 + G_0 V g)\mathscr{L}_b|\Psi) \tag{7a}$$

$$= G_0\mathscr{L}_b|\Psi) + G_0 V|\Psi)$$

where

$$G_0 = (\mathscr{H}_0 + \mathscr{L}_b - E)^{-1} \tag{7b}$$

The division of the Hamiltonian into an unperturbed and perturbed part is of course arbitrary. However in most cases $\mathscr{H}_0$ is chosen to make the scattering particle Green's function simple to calculate. Typical choices would be the target plus a free particle or in the case of positive ions, target plus Coulomb wave. Equation (7) has a particularly simple structure for the scattering of an electron from a static potential. The basic starting point is the expansion of the wavefunction, Green's function and potential in spherical harmonics. In contrast to atomic scattering problems, the potential is not diagonal in the angular momentum quantum number of the scattered electron ℓ. This leads to the following set of coupled integral equations,

$$\Psi_\ell(r) = G_\ell(r|a)\left(\frac{\partial\Psi_\ell}{\partial r} - b\Psi_\ell\right)_{r=a} + \sum_{\ell'} \int G_\ell(r|r')V_{\ell\ell'}(r')\Psi_{\ell'}(r')dr' \tag{8a}$$

where

$$G_\ell(r|r') = R_\ell(r_<)I_\ell(r_>) \tag{8b}$$

The Green's function is chosen to satisfy the boundary condition demanded by the R-matrix method. This is easily accomplished by choosing $R_\ell(r)$ to be regular at the origin and requiring

$$\frac{dI_\ell}{dr} = bI_\ell \tag{8c}$$

at r=a. The linear algebraic method proceeds by introducing a quadrature scheme into the set of equations (8a) to get

$$\Psi_\ell(r_i) = G_\ell(r_i|a)\left(\frac{\partial\Psi_\ell}{\partial r} - b\Psi_\ell\right)_a + \sum_{\ell' j} G_\ell(r_i|r_j)W_j V_{\ell\ell'}(r_j)\Psi_{\ell'}(r_j) \quad (9)$$

By defining

$$\mathcal{M}_{\ell i,\ell' j} = \delta_{\ell\ell'}\delta_{ij} - G_\ell(r_i|r_j)W_j V_{\ell\ell'}(r_j) \quad (10)$$

one obtains

$$\sum_{\ell' j} \mathcal{M}_{\ell i,\ell' j}\Psi_{\ell'}(r_j) = G_\ell(r_i|a)\left(\frac{\partial\Psi_\ell}{\partial r} - b\Psi_\ell\right)_a \quad (11)$$

The solution may be written as

$$\Psi_\ell(r_i) = \sum_{\ell' j} \mathcal{M}^{-1}_{\ell i,\ell' j} G_{\ell'}(r_j|a)\left(\frac{\partial\Psi_{\ell'}}{\partial r} - b\Psi_{\ell'}\right)_a$$

$$= \sum_{\ell'} \gamma_{\ell i,\ell' a}\left(\frac{\partial\Psi_{\ell'}}{\partial r} - b\Psi_{\ell'}\right)_a \quad (12)$$

Setting r_i=a gives

$$\Psi_\ell(a) = \sum_{\ell'} \gamma_{\ell a,\ell' a}\left(\frac{\partial\Psi_{\ell'}}{\partial r} - b\Psi_{\ell'}\right) = \sum_{\ell'} \mathcal{R}_{\ell\ell'}\left(\frac{\partial\Psi_{\ell'}}{\partial r} - b\Psi_{\ell'}\right)_a \quad (13)$$

The R-matrix is thus seen to be the value of the full Green's function on the surface of the sphere enclosing the internal region. The solution of equation (11) may be accomplished using standard techniques of linear algebra (28). The reduction to a matrix equation has the advantage that vector processors such as the CRAY I can solve such equations 15-20 times faster than scalar computers. If the matrices become too large for central memory, partitioning techniques and/or iterative methods may be used to solve the equations. These approaches may slow down the calculation somewhat but experience has shown that the methodology is still quite efficient. Perhaps the greatest difficulty with the LAM as we have described it is the use of the single-center expansion method to obtain equation (8a). It has been known for many years that such single-center techniques are very slowly convergent for molecular systems. The situation for the calculation of bound states is much worse than for low energy electron scattering due to the strong dependence of the energy on the region near the atomic nuclei. For continuum electrons, which do not penetrate too deeply into the electron cloud, the expansion is slowly convergent but practical techniques can be developed to aid the convergence. The use of the spherical harmonic expansion at the level of equation (8) may in fact be superior to using numerical continuum functions and multicenter functions in the standard R-matrix formalism. The latter approach requires the single center decomposition of the multicenter orbitals and a numerical integration to form the required two-electron matrix elements. The LAM "matrix elements" are very simple functions requiring little computational effort for their formation. The major effort is placed on the solution of the linear equations, which are well suited to vector prescriptions. In addition schemes can be devised which utilize different quadrature meshes for each partial wave. Since the higher partial wavefunctions are strongly peaked near the nuclear singularity and die off quite rapidly thereafter, it is possible to get accurate representations with very few points. This is very similar in philosophy to the use of multicenter basis sets in conventional approaches. Another approach would be to combine a multicenter basis set expansion with the numerical wavefunction for low partial waves,

$$\Psi(\vec{r}) = \sum_{\ell=0}^{L} \sum_{m=-\ell}^{\ell} F_{\ell m}(r) Y_{\ell m}(\Omega) + \sum_{q} C_q \Phi_q(\vec{r}) \tag{14}$$

By substituting Equation (14) into Equation (2a) we can derive a set of coupled equations for $F_{\ell m}(r)$ and C_q. These may in turn be reduced to linear algebraic equations by introducing quadratures. The advantage of this latter approach is the possibility of representing the large number of high angular momentum terms by a few $\Phi_q(\vec{r})$. In addition these functions could be chosen to be bound-state Cartesian Gaussians for which much intuition has been developed over the past few decades. However, like the standard R-matrix method it is necessary to perform numerical integrations to calculate the required matrix elements. The efficacy of this can only be ascertained by experimentation. Now that we have outlined the basic theory and numerical technique of the LAM let us turn to the calculation of the exchange and correlation terms which provide the major difficulties of the full many-body problem.

Electron Exchange. The need for an anti-symmetric wavefunction for incident and bound electrons gives rise to nonlocal interactions which greatly complicate the solution of the scattering equations. These exchange interactions have the form,

$$\int K(\vec{r}|\vec{r}\,')F(\vec{r}\,')\,d\vec{r}\,' = \left[\int \Phi_B(\vec{r}\,') \frac{1}{|\vec{r}-\vec{r}\,'|} F(\vec{r}\,')d\vec{r}\,'\right]\Phi_B(r) \tag{15}$$

where $\Phi_B(\vec{r})$ is a bound-state molecular orbital and $K(r|r')$ is the exchange kernel. The difficulty with these interactions is not so much their nonlocality as their nonseparability. The nonseparability arises because the interaction $\frac{1}{r_{12}}$ does not decompose into a product of functions of $\vec{r}_1$ and $\vec{r}_2$. If one examines these exchange operators more closely one notices that they are rather short range functions. The reason for this is the quite physical fact that the incident electron can only exchange with a bound state electron when the two are close together. Since the electron cloud of the atom or molecule is spatially localized these interactions fall off quite rapidly away from the target.

These considerations suggest that it should be possible to expand these integral kernels as a sum of separable terms (15,16) using bound state functions as the expansion set,

$$K(\vec{r}|\vec{r}\,') = \sum_{i,j} \Phi_i(\vec{r}) K_{ij} \Phi_j(\vec{r}\,') \tag{16}$$

The advantage of equation (16) is twofold. First, the nonseparability is avoided by using a basis set (separable) expansion of the operator. Second, the matrix elements K_{ij} may be extracted from standard bound-state programs available from a wide variety of sources. In addition the matrix elements are independent of energy and need be computed only once even if scattering calculations are to be performed for a range of energies. A practical question which must be answered is the rate of convergence of the separable expansion. Numerous calculations at the static-exchange or Hartree-Fock level on a wide variety of diatomic and triatomic molecules have shown quite rapid convergence. In fact, in many cases, the use of standard SCF basis sets has given results of 10-20% accuracy. By augmenting these sets slightly, we can reach the 1-5% level of accuracy with little difficulty. Another feature of the use of separable expansions may be illustrated by considering the following equation,

$$(\mathscr{L} - \mathscr{E})\,|F\rangle = |\chi\rangle\langle\phi|F\rangle \tag{17}$$

where $\mathscr{L}$ is a local operator. The solution of equation (17) may be written as

$$|F\rangle = |F^0\rangle + |F^1\rangle\langle\phi|F\rangle \tag{18a}$$

where

$$(\mathscr{L} - \mathscr{E})|F^0\rangle = 0 \tag{18b}$$

$$(\mathscr{L} - \mathscr{E})|F^1\rangle = |\chi\rangle \tag{18c}$$

The unknown constant, $\langle\phi|F\rangle$, may be determined after the solution of equations (18b,c) by quadrature. The generalization to an n-term separable expansion is straightforward requiring n inhomogeneous equations to be solved and the inversion of an (n*n) matrix for the unknown constants. The procedure is quite similar to the treatment of LaGrange undetermined multipliers in standard scattering formalisms. The method is particularly convenient for the LAM since the most difficult computational step in the scattering involves the reduction of the algebraic matrix to $\underset{\sim}{L}\underset{\sim}{U}$ form where $\underset{\sim}{L}(\underset{\sim}{U})$ are lower (upper) triangular matrices. The work required for additional right-hand sides (inhomogeneities) is usually quite small. The reduction in computational time which is achieved by the use of separable exchange varies from factors of about 3 to 10 over standard approaches. This savings in time increases dramatically with the number of incident energies since as mentioned above the difficult step in the calculation, the formation of the K_{ij} matrix elements, is energy independent. The success of the separable expansion for exchange led us to ask if it would be possible to extend this kind of approach to the treatment of polarization and correlation and ultimately to the treatment of inelastic scattering. The results of that inquiry appear in the next section.

Polarization and Correlation. In order to satisfactorily explain the details of low-energy electron-molecule collisions it is necessary to go beyond the static-exchange level and include correlation effects. It has been known for many years that a straightforward close-coupling expansion is very slowly convergent for elastic scattering for many systems. The basic difficulty is that the physical closed channels are too delocalized in space to adequately describe what is happening near the bound electrons. Pseudostates, (11,12), which are usually derived from a perturbation treatment of the distortion of the molecular charge cloud by an electric field, are far better functions for treating polarization and correlation. These pseudostates may be used directly in the close-coupled equations or included in the open-channel space as a nonlocal energy-dependent optical potential. The latter approach has the advantage of being able to include more functions by using bound-state matrix methods. In addition, the dimensionality of the scattering equations does not increase beyond the static-exchange approximation. However in order to efficiently use the optical potential formalism it is necessary to be able to calculate and manipulate the required Hamiltonian matrix elements rapidly. Since bound state configuration interaction (CI) programs were developed by quantum chemists for just this purpose we began to examine the possibility of using them in the scattering problem. In order to proceed, it is essential for the purpose of the formalism as well as the numerics to introduce a basis set expansion of the continuum. This expansion must be complete enough to represent the important physical effects in the problem. In contrast to the exchange kernel, the optical potential has some long-range character which suggests convergence may be somewhat more difficult than for the static-exchange case. Again, only numerical experimentation would allow us to decide on the efficacy of the approach. From a purely formal standpoint the partitioning of function space into an open (P) and closed (Q) channel part results in the following equation,

$$[\mathcal{H}_{PP} + \mathcal{L}_b - E + \mathcal{H}_{PQ}(E - \mathcal{H}_{QQ})^{-1}\mathcal{H}_{QP}]P|\Psi\rangle = \mathcal{L}_b P|\Psi\rangle$$

for the scattering electron, where

$$P = \sum_{\alpha} |A(\phi_0(1\text{-}N)F_{\alpha}(N+1))\rangle\langle A(\phi_0(1\text{-}N)F_{\alpha}(N+1))| \qquad (20a)$$

$$P + Q = I \qquad (20b)$$

The Q space configurations, which contain single, double etc. excitations away from the reference set, account for the polarization and correlation. The use of the discrete expansion enables us to write,

$$V_{opt} = \mathcal{H}_{PQ}(E-\mathcal{H}_{QQ})^{-1}\mathcal{H}_{QP} = \sum_{\alpha,\beta} |F_\alpha\rangle V_{\alpha\beta}(E)\langle F_\beta| \tag{21}$$

which is, of course, a separable form (16,17,18). Thus the formation of the optical potential requires a standard CI program to form the matrix elements and the solution of the linear equations,

$$(E-\mathcal{H}_{QQ})X_{QP} = \mathcal{H}_{QP} \tag{22a}$$

$$\mathcal{H}_{PQ}X_{QP} = V_{opt} \tag{22b}$$

to get the matrix optical potential. The essential difference between the problem with and without correlation is the need to construct V_{opt} at each incident energy. This in turn requires that equation (22) be solved numerous times. However, the Hamiltonian matrix need be computed only one time. Once the optical potential is formed the solution of the scattering equations is identical to that of the static-exchange case. Thus computer codes which were developed for the static-exchange problem may be used without any modifications. This is a great advantage of the optical potential formalism. In all of the applications made so far, $\phi_0(1\text{-}N)$ has been chosen to be the Hartree-Fock wavefunction of the target. Thus the P-space consists of the static-exchange configurations. Since it is impossible to use a complete expansion in Q-space, it becomes quite important to choose the correlating orbitals and configurations to reflect the physics of the low-energy scattering process. A primary consideration is, of course, an accurate represention of the polarization of the target by the incident electron and the subsequent back-reaction of the polarized target on the electron. An elegant way to accomplish this is to use polarized orbitals extracted from a coupled Hartree-Fock calculation. These functions accurately represent the dipole distortion of the molecular target in the presence of an electric field. By adding a further set of diffuse atomic functions to represent the "continuum" electron we can adequately span the space of the electrons. The Q-space is constructed by taking antisymmetrized products of the polarized and scattering orbitals in which an occupied and scattering function are singly excited. These configurations are all single and double excitations away from the static-exchange reference set. However they do not include any double excitations of the core electrons. Such excitations would correlate the core electrons, an effect which we wish to exclude from present calculations. Thus we try to place our efforts on the differential correlations induced by the incident electron rather than the full (N+1) electron problem. In most cases this has been quite satisfactory in bringing the calculations into good agreement with experiment (17,18). However a more general treatment is needed in which target and induced correlations are treated in a balanced fashion. Such a treatment is currently under investigation and will be reported upon in later publications.

Before closing this section it is worth mentioning a new development in the treatment of electronically inelastic collisions which is related to the above discussion. Recently (30) we have shown that it is possible and practical to place the off-diagonal channel-channel interactions in separable form. This leads rather naturally to a formalism in which it is only necessary to solve inhomogeneous, elastic scattering equations for each channel. The reduction in dimensionality from a coupled to a single-channel problem makes the equations much more tractable. The final solution to the coupled channel problem is determined by inverting a matrix whose dimension is the number of expansion functions used for the coupling matrix elements. Optical potentials may be incorporated in the coupling matrix with little additional effort. The method has been successfully applied to the 1s-2s-2p close-coupling equations in atomic hydrogen (30) and molecular applications are underway.

Now let us turn to the extension of the formalism to molecular photoionization.

Photoionization. The photoionization process,

$$AB + h\nu \rightarrow AB^{+} + e \tag{23}$$

can be characterized by the dipole matrix element between the initial bound state of AB and the electron-ion continuum wavefunction of the final state. Since the final state is no more than the scattering wavefunction for an electron on a molecular ion, it is quite easy to adapt the LAMOPT formalism to photoionization (29). To accomplish this two things are required: the replacement of the free-particle with the Coulomb Green's function and the calculation of the bound-free dipole matrix element. From the latter quantity, the angular distribution of the photoelectrons

$$\sigma_{\Gamma}(\Omega) = A_{00} + A_{20}\, P_{2}(\cos\,\theta) \quad , \tag{24}$$

and the total photoionization cross section σ_{Γ} can be calculated. Both quantities are simply related to

$$d_{\ell m} = \langle \Phi_{B}^{m_i} | rY_{1m''} | \Psi_{\ell m} \rangle_{m''} = m - m_i \tag{25}$$

where Φ^{m_i} ($\Psi_{\ell m}$) is the bound (continuum) orbital of the electron and m_i(m) is the azimuthal quantum number. In all of the applications of the LAM to photoionization we have calculated wavefunctions for both the bound and continuum electrons at the Hartree-Fock (HF) level. The bound orbitals are taken as solutions of the neutral Hartree-Fock equations. The continuum orbitals are calculated using the frozen-core approximation by which the HF orbitals of the neutral, target molecule are used to represent the ion core (FCHF). This leads to considerable simplification in the form of the dipole matrix elements reducing them to one-electron integrals. Since the bound and continuum electrons are not solutions of the same one-electron Hamiltonian it is necessary to introduce LaGrange undetermined multipliers to ensure orthogonality of the orbitals. This leads to a continuum improved virtual orbital (IVO) equation of the form

$$(h + \sum_i (2a_i J_i - b_i K_i) - E)|\Psi_{\ell m}\rangle = \sum_q \lambda_q |\Phi_q\rangle \quad (26a)$$

$$a_i, b_i = \text{coulomb, exchange orbital occupancies} \quad (26b)$$

$$\lambda_q = \text{LaGrange multiplier} \quad (26c)$$

for the continuum electron. This is solved using the LAM.

Now that we have described the formalism in some detail, let us look at our applications to resonant processes.

Applications

Electron + N_2 Scattering. The scattering of low-energy electrons from N_2 is one of the most thoroughly studied problems in molecular physics. The primary reason for this is that the cross section is dominated by a low-energy π_g (2.4 eV), shape resonance which has profound effects on the vibrational excitation spectrum. Under ordinary scattering situations the probability for nuclear excitation by low-energy electrons is quite small due to the large difference in mass of the two particles. However in a resonant process the electron can distort the charge distribution of the target sufficiently to cause great changes in the forces on the nuclei. It is these changes which cause the vibrational excitation not the direct impulsive force of the collision. The temporary capture of the incident electron into the low-lying, π_g anti-bonding orbital of N_2 was put forth as the explanation of the excitation mechanism (2). Since this orbital has a significant amount of valence character it can sufficiently change the potential seen by the nuclei during the collision. These ideas were expanded upon and refined by Herzenberg (32,33) and his collaborators using the complex eigenvalue techniques of Siegert. The calculations however remained semi-empirical and, although they gave good argreement with the experiments of Schultz (2)were dependent on a semi-quantitative adjustable parameter. With the development of the molecular R-matrix method it became possible to perform a first principles calculation of this process. However, in order to accomplish this it was necessary to generalize the R-matrix method to include nuclear motion. The generalization was accomplished by Schneider, LeDourneuf and Burke (34), using the Born-Oppenheimer approximation of the R-matrix levels of the (N+1) electron problem as the zeroth order approximation. The theory was successfully applied by Schneider, LeDourneuf and Lan (35) to the e+N_2 problem. The results of that calculation, which are shown in Fig. 1, demonstrated for the first time that an ab initio method could, within the context of the Born-Oppenheimer approximation, explain the resonant vibrational excitation process. Other calculations (6) followed which confirmed the results of the R-matrix study. Recently (31) we have performed a series of calculations using the LAM combined with the optical potential technique which unequivo-

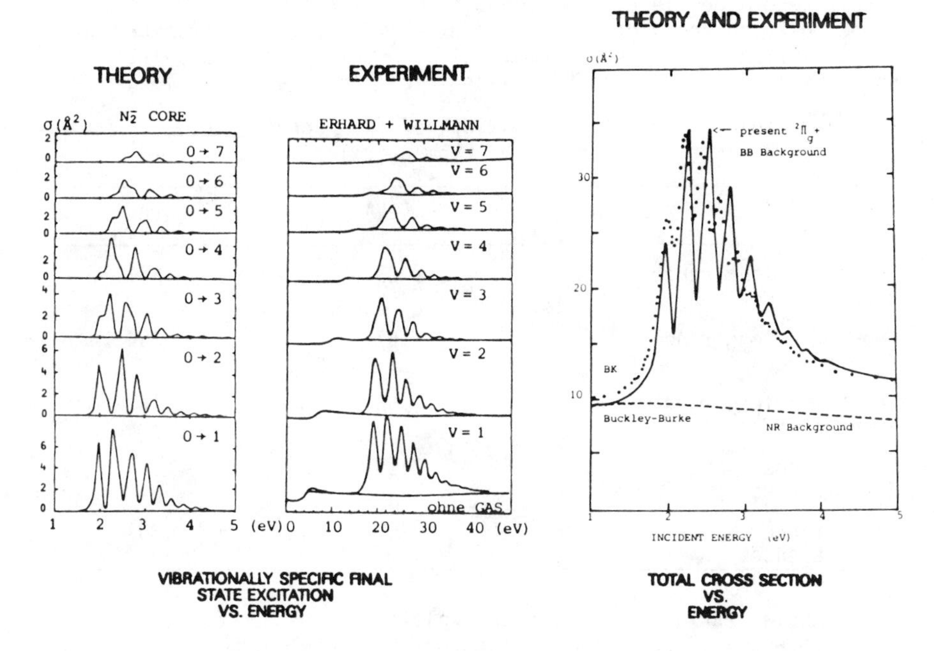

Figure 1. A comparison of experimental and theoretical vibrational excitation cross sections for N_2 scattering.

cally demonstrated that the π_g resonance is dominated by short-range distortion. These ideas were implicit in the R-matrix calculations (35) which were based on the negative ion SCF wavefunction of N_2. The N_2^- SCF wavefunction does not contain any excitation which destroys the Σ_g symmetry of the N_2 core. In fact to first order in perturbation theory the N_2^- SCF wavefunction can be obtained as a single-excitation CI using N_2 SCF orbitals. Hazi and coworkers (6) used this equivalence and the Feshbach formalism to obtain results in good agreement with the R-matrix calculations of Schneider, LeDourneuf and Lan. The recent LAM calculations given in Table I also show that one must be cautious in not overcorrelalating the negative ion wavefunction with respect to the neutral molecule. More extended double excitation CI calculations lowered the resonance position and width below the experimental value demonstrating these difficulties rather dramatically. Finally, the calculations show possible problems with semi-empirical theories using long-range

Table I. Position and Width of $^2\Pi$ Resonance in N_2 as a function of the type of calculation

Type of Calc.	E_R(eV)	Γ_R(eV)
Short and long range 3 references	2.03	.281
Short and long range 6 references	1.97	.264
Short and long range 19 references	1.66	.178
Short range 3 references	2.13	.314
Short range 19 references	2.07	.301

cutoff polarization potentials to explain the resonance (31). These potentials, which have adjustable parameters, may be tuned to reproduce the resonant features but it is dangerous to place too much emphasis on the forms of the interaction. The long-range polarization potential has only a minor effect on the resonance in N_2; it is mainly a short-range effect. The adjustment of the cutoff in the model potential mimics these short-range features in a crude but unfortunately unpredictable fashion. It may be unwise to rely on the predictions of these model potentials for other symmetries which may be dominated by quite different physical effects.

Electron + H_2 Scattering. The low-energy elastic scattering of electrons from H_2 shows a broad feature which is due to a p-wave shape resonance. In our calculations (17), no attempt was made to treat this in any special fashion. The calculation was the first one in which we included polarization and correlation using an optical potential and the intent was to explain the low-energy behavior of the cross section. The results, which are shown in

Fig. 2, are in excellent agreement with experiment (39,40) over a considerable energy range and clearly reproduce the broad shape resonance mentioned above. The optical potential was constructed from a set of one-electron functions incorporating the SCF distortions of the target in the presence of an electric field. These were coupled to the incident electron in the manner described earlier and results in an optical potential of about 400 spin eigenfunctions. The calculation required about 8 seconds of CRAY I time per energy.

Electron + H_2^+ Scattering. The first application of the LAMOPT to molecular ions was undertaken to try to resolve some differences in the results obtained for $e+H_2^+$ scattering by the Stieltjes calculations of Hazi (41) and those of the Japanese (42) using the Kohn variational method (18). The calculations were performed with a number of basis sets in order to understand any problems which might have affected either previous set of calculations. Our early results confirmed the position and width of the first resonance as given by the Stieltjes method but produced higher resonances in poor agreement with the Japanese and earlier close coupling calculations. Since the basis set used for these calculations was not designed to treat Rydberg like resonances, we modified it to include more diffuse orbitals of the proper symmetry and re-ran the calculations. The results, which are shown in Fig. 3 confirm the position and width of the lowest resonance and are in good agreement with the close-coupling calculations of Collins and Schneider (36) and the Kohn variational results for the second resonance. In addition a third resonance was found which is considerably lower than that of the Kohn calculation. Although we have not explored this Rydberg series of resonances any further, it is clear that the Kohn calculation for the third member of the series is much too high. In fact it lies above the ionization potential of the H_2^+ ion and must be an artifact of the poor basis used in the Kohn calculation. The quality of the results obtained by the optical potential approach for the $e+H_2^+$ scattering gives us much confidence in its application to more complicated problems. In addition the calculations demonstrate that we can deal with ionic as well as neutral systems and Feshbach as well as shape resonances with the formalism.

Photoionization of N_2. In this section we consider the following processes,

$$h\nu + N_2 \rightarrow N_2^+ \, (3\sigma_g^{-1}) \;+ e \begin{Bmatrix} k\sigma_u \\ k\pi_u \end{Bmatrix} \tag{27}$$

$$\rightarrow N_2^+ \, (\pi_{xu}^{-1}) \;+ e \begin{Bmatrix} k\sigma_g \\ k\pi_{xg} \\ k\delta_{xyg} \end{Bmatrix}$$

for the ground state of N_2 (29). The first process has a broad σ_u shape resonance, while the latter process leads to a spurious π_{xg}

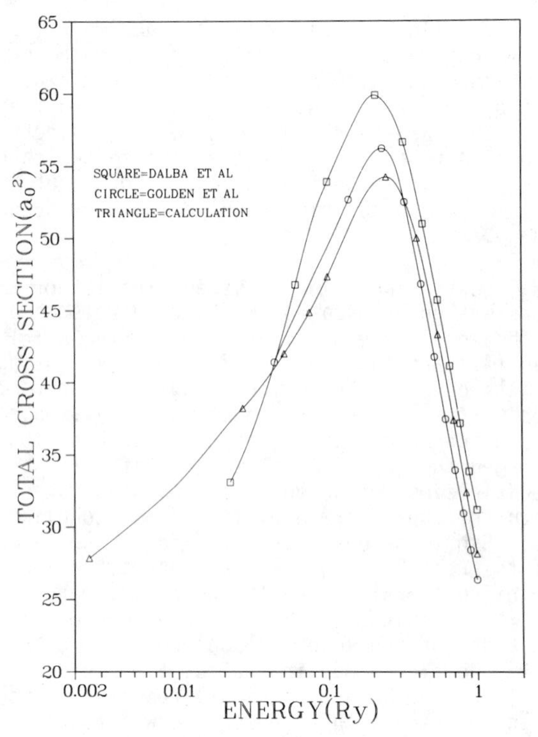

Figure 2. Comparison of theoretical and experimental total cross sections for e-H_2 scattering. Curves are as follows: (1) represents the effective optical potential; (2) represents the experiment by Golden _et al._, and (3) represents the experiment by Dalba _et al._

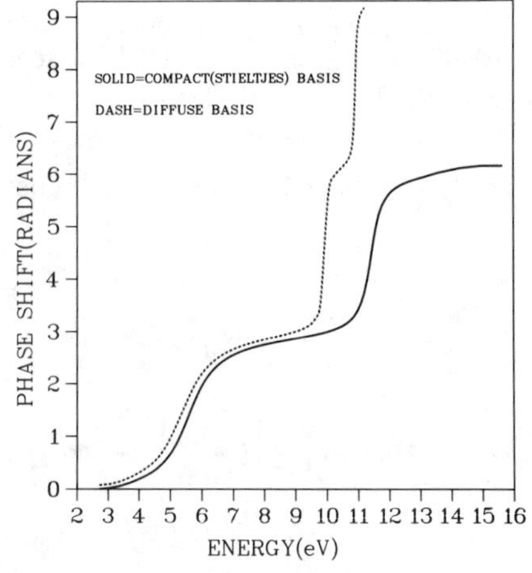

Figure 3. Eigenphase sums as a function of energy for the $^1\Sigma_g$ symmetry for e-H_2^+ collisions.

shape resonance at the frozen-core Hartree-Fock level (FCHF). Both calculations were performed with a number of basis sets to represent the exchange operator. In all cases very little sensitivity to the basis was observed in the scattering function. However sensitivities of the order of 10-20% have been observed in the cross section due to the inclusion or exclusion of diffuse orbitals in the construction of the occupied molecular orbitals. Evidently the more diffuse character of the integrand in the dipole matrix element is quite sensitive to small components in the bound orbitals. Similar conclusions have been observed by ONeil and Reinhardt in the photoionization of H_2 (37). The results of our calculations are shown in Figs. 4-6, where we compare with those of other approaches and experiment (43). The agreement between the LAM and Schwinger variational (SV) method is quite good. Reasonable agreement with the Stieltjes method is observed for ionization from the $3\sigma_g$ orbital. The ionization of the π_{xu} orbital of N_2 is one of the classic failures of the FCHF model. The HF potential improperly places a valence-like π_g orbital above the ionization continuum. Better calculations, such as those based on an random phase approximation (RPAE) model (5) or optical potential formulation can correct the difficulty and remove the spurious resonance. This has already been done with the RPAE and calculations using our optical potential approach will be undertaken in the near future when we can deal with sets of coupled, open channels.

Photoionization of NO. We consider photoionization of the 2π orbital of NO into $k\sigma$, $k\pi_x$ and $k\delta_{xy}$ continua (29). Our interest in this process stemmed from a desire to resolve the rather large differences between the Stieltjes (44) and SVM calculations (45). The results of our calculations and a comparison of theory and experiment (46,47) is shown in Figs. 7-8. We observe no structure in the individual partial photionization cross sections and must conclude that these features are an artifact of the imaging procedure or linear dependence in the basis set. Our calculations are in reasonable agreement with the SV method, showing a number of broad shape resonances whose position can vary slightly with the basis set. This is especially evident in the sharper σ resonance. As with N_2, the difference between the LAM and SVM results is primarily due to the inclusion of diffuse orbitals in the bound molecular orbitals. Very little sensitivity to basis was observed in the scattering solutions.

Photoionization of CO_2. We have considered the following processes for photoionization of ground state CO_2 (29):

$$h\nu + CO_2 \rightarrow CO_2^+ \begin{Bmatrix} 1\sigma_g^{-1} \\ 2\sigma_g^{-1} \\ 4\sigma_g^{-1} \end{Bmatrix} + e\begin{Bmatrix} k\sigma_u \\ \\ k\pi_{xu} \end{Bmatrix}$$

We are particularly interested in these processes since there are considerble differences between the results of the Stieltjes method on one hand and those of the SV and LA method. The LAM and SVM both

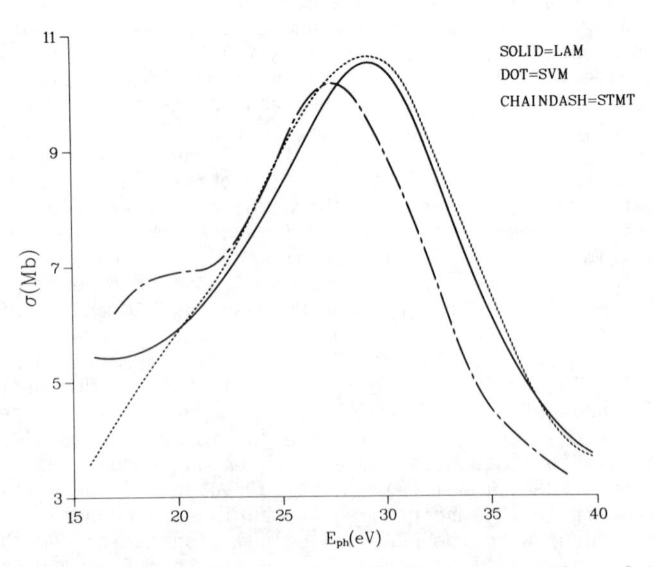

Figure 4. Total cross sections for the photoionization of N_2 leading to the $X^2\Sigma_g^+$ state of $N_2^+(3\sigma_g^{-1})$. Comparison of theoretical methods: solid line, LA; dashed line, SV; chain-dashed line, STMT.

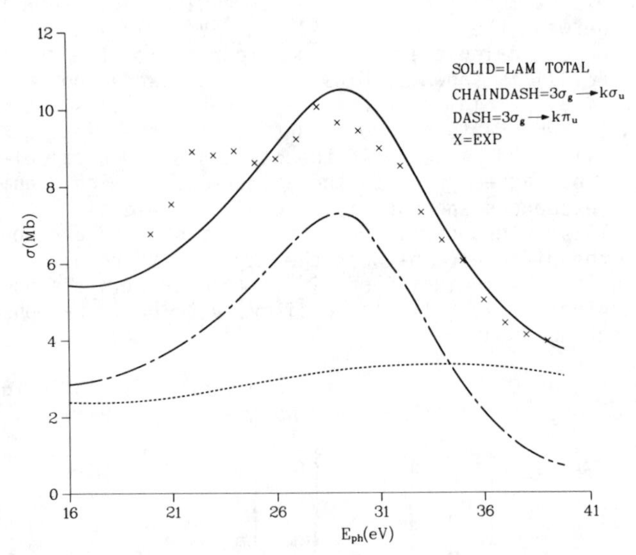

Figure 5. Partial and total cross sections for the photoionization of $N_2(N_2^+X^2\Sigma_g^+)$. Comparison of the LA method and experiment: solid line, total; chain-dashed line, $3\sigma_g \rightarrow k\sigma_u$; dashed line $3\sigma_g \rightarrow k\pi_u$; crosses. expt.

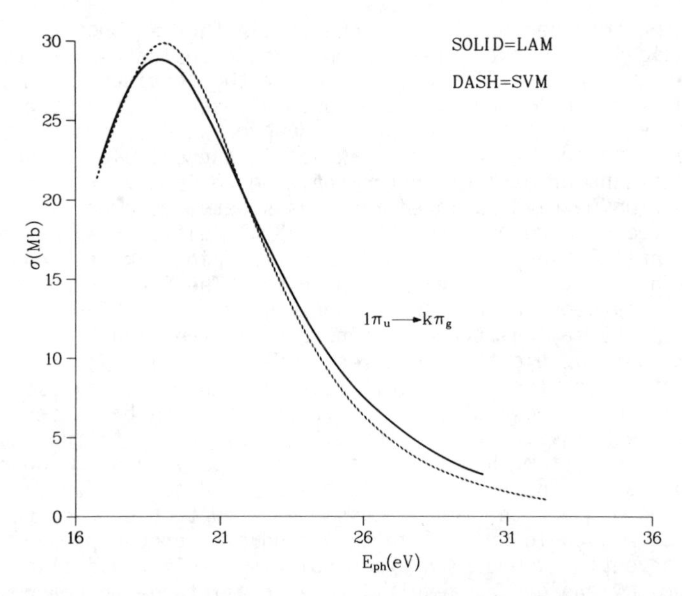

Figure 6. Partial cross sections for the photoionization of N_2 leading to the $A^2\Pi_u$ state of $N_2^+(1\pi_u^{-1})$. Comparison of theoretical methods for $1\pi_u \rightarrow k\pi_g$: solid line, LA; dashed line, SV.

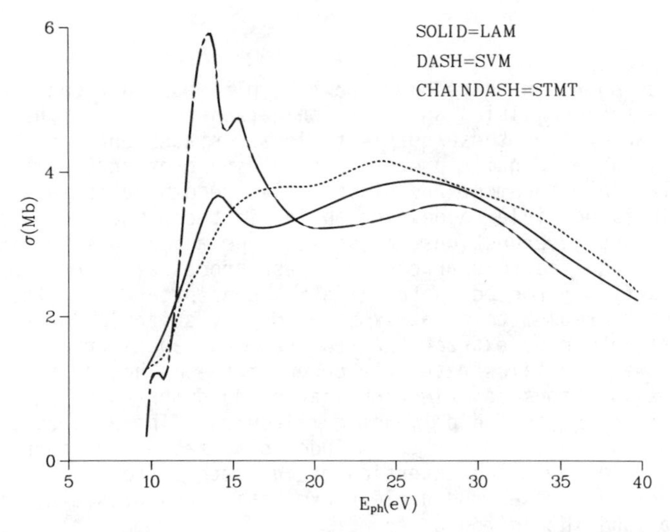

Figure 7. Total cross section for the photoionization of NO leading to the $X^1\Sigma^+$ state of $NO^+(2\pi^{-1})$. Comparison of theoretical methods: solid line, LA; dashed line, SV; chain-dashed line, STMT.

predict rather narrow shape resonances in the $1\sigma_g$ and $4\sigma_g$ ionizations while the Stieltjes approach gives rather broad, nonresonant shapes. The calculation of the continuum wavefunctions for these channels represents a most stringent test of the single-center expansion approach. However we have systematically increased the number of partial waves in the calculation until we are confident of a 5% or better convergence in the cross section. In addition we have explored many basis sets, more or less contracted or more or less diffuse. Small differences beween the SV and LA calculations can be noticed but nothing like the qualitative differences with the Stieltjes approach. The final results for the total cross section in these channels is given in Figs. 9-11. It is difficult to say why the agreement between the methods is so poor for $1\sigma_g$ and $4\sigma_g$ ejection processes. Earlier Stieltjes calculations on the $1\sigma_g$ photoionization by Daasch, Davidson, and Hazi (38) have shown a great sensitivity to both basis set and imaging technique. Perhaps the difficulty is due to the inability of the Stieltjes method to place enough eigenvalues in the resonant region. Since the resonances in the $1\sigma_g$ and $4\sigma_g$ ionization processes but not the $2\sigma_g$ process are at quite high electron energies, it could be difficult to produce a proper pseudospectrum with conventional Gaussian type orbitals. This would also explain why the low energy $2\sigma_g$ results agree in all three approaches. The resolution of these differences is not simple but we feel it is imperative to try to do so. At present the Stieltjes method is the only approach capable of dealing with complicated polyatomic species. Its reliability must be tested against other approaches where such tests are possible.

Conclusion

Resonance phenomena have been shown to play a significant role in many electron collision and photoionization problems. The long lived character of these quasi-stationary states enables them to influence other dynamic processes such as vibrational excitation, dissociative attachment and dissociative recombination. We have shown it is possible to develop *ab initio* techniques to calculate the resonant wavefunctions, cross sections and dipole matrix elements required to characterize these processes. Our approach, which is firmly rooted in the R-matrix concept, reduces the scattering problem to a matrix problem. By suitable inversion or diagonalization we extract the required resonance parameters. Finally we have illustrated the power of the method by calculating the cross sections for electron scattering or photoionization from a number of diatomic and polyatomic molecules. These calculations have been among the first to include polarization and correlation in an *ab initio* way. The extension of our methods to inelastic electronic processes and nuclear excitation and dissociation are underway and should appear soon.

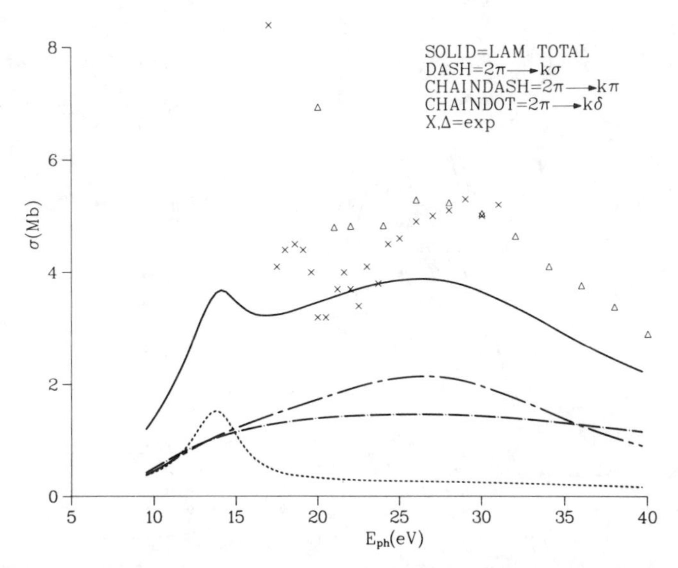

Figure 8. Partial and total cross sections for the photoionization of NO ($NO^+X^1\Sigma^+$). Comparison of the LA method and experiment: solid line, total; chain-dashed line, $2\pi \rightarrow k\pi$; dashed line $2\pi \rightarrow k\sigma$; dotted line, $2\pi \rightarrow k\delta$; crosses, expt.; triangles, expt.

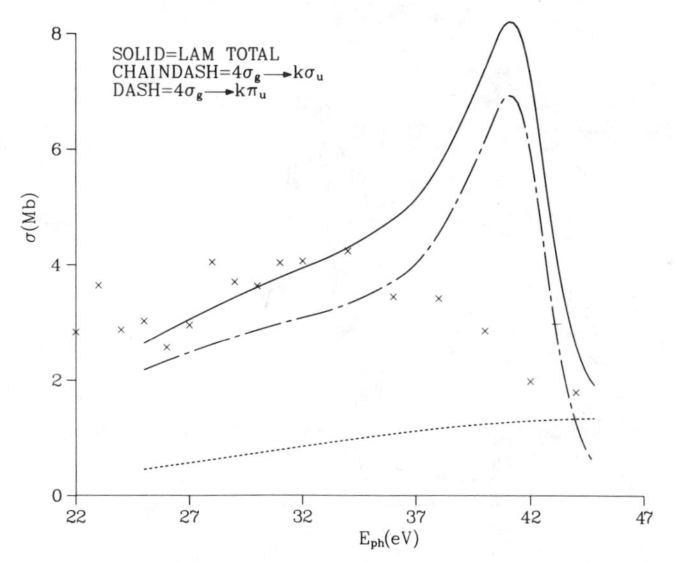

Figure 9. Partial and total cross sections for the photoionization of CO_2, ($CO_2{}^+C^2\Sigma_g{}^+$). Comparison of the LA method with experiment; solid line, total; chain-dashed line, $4\sigma_g \rightarrow k\sigma_u$; dashed line, $4\sigma_g \rightarrow k\pi_u$; crosses, expt.

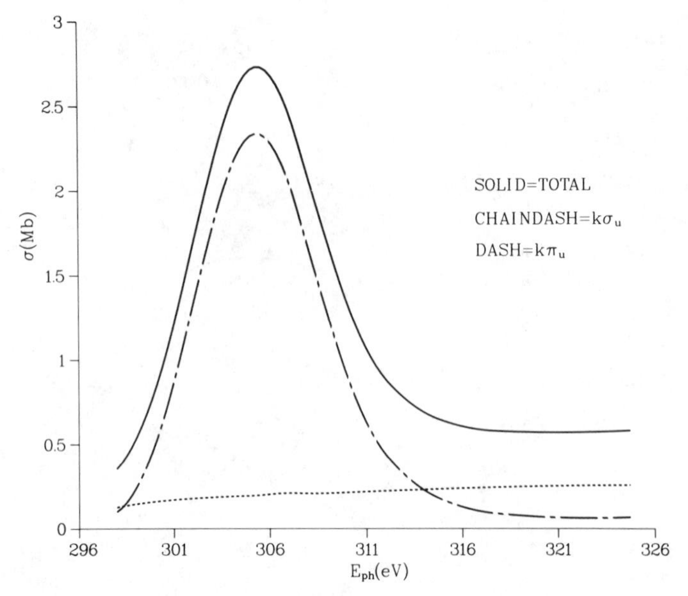

Figure 10. Partial and total cross sections for the photoionization of the $2\sigma_g$ orbital of CO_2 in the LA method.

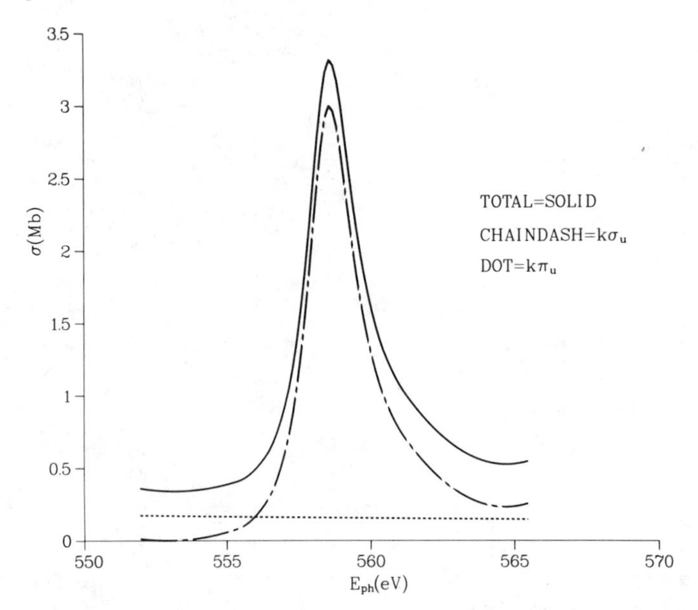

Figure 11. Partial and total cross sections for the photoionization of the $1\sigma_g$ orbital of CO_2 in the LA method.

Literature Cited

1. Burke P. G. In "Potential Scattering in Atomic Physics," Plenum Press, New York, 1977.
2. Schultz, G.J, Rev. Mod. Phys. 1973,45, 378, Ibid, 1973, 45, 423.
3. Taylor, H. S, Adv. Chem. Phys. 1970, 18, 91.
4. Dehmer J. L.; Dill, D. In "Electron-Molecule and Photon-Molecule Collisions"; Rescigno, T. N.; McKoy, B. V.; Schneider, B. I., Eds.; "Plenum Press, New York, 1979, p. 225.
5. Langhoff, P. W. In "Electron-Atom and Electron-Molecule Collisions"; Hinze, J. Ed.; Plenum Press, New York, 1983, p. 297.
6. Hazi, A. U. In "Electron-Atom and Electron-Molecule Collisions"; Hinze, J. Ed.; Plenum Press, New York, 1983, p. 103.
7. Schneider B. I.; LeDourneuf M.; Lan VoKy. Phys Rev. Letts. 1979, 43, 1926,.
8. Bardsley, J. N. In "Electron-Molecule and Photon-Molecule Collisions"; Rescigno, T. N.; McKoy, B. V.; Schneider, B. I., Eds.; Plenum Press, New York, 1979, p. 267.
9. Rescigno T. N. In "Electron-Atom and Electron-Molecule Collisions", Hinze, J. Ed.; Plenum Press, New York, 1983, p. 51.
10. Rescigno, T. N.; Orel A. E.; McCurdy, C. W. J. Chem. Phys. 1980, 73, 6347.
11. Burke, P. G.; Seaton, M. J. Methods Comput. Phys. 1971, 10, 1.
12. Nesbet R. L. In "Variational Methods in Electron-Atom Scattering Theory," Plenum Press, New York, 1980, p. 25.
13. Watson, D. K.; Lucchese, R. R.; McKoy, V.; Rescigno T. N. Phys. Rev. 1980, 20, 1474.
14. Schneider, B. I. In "Electron-Molecule and Photon-Molecule Collisions"; Rescigno, T. N.; McKoy, B. V.; Schneider, B. I., Eds.; Plenum Press, New York, 1979, p. 77.
15. Collins, L. A.; Schneider B. I. Phys. Rev. 1981, 24, 2387.
16. Schneider, B. I.; Collins, L. A. Phys. Rev. 1981, 24, 1264.
17. Schneider, B. I.; Collins, L. A. Phys. Rev. 1983, 24, 2847.
18. Schneider, B. I.; Collins, L. A. Phys. Rev. 1983, 28, 166.
19. Burke, P. G.; Noble, C. J.; Salvini, S. J. Phys. B. 1983, 4, L113.
20. Burke, P. G.; Robb, W. D. Adv. Atom. Molec. Phys. 1974, 11, 143.
21. Schneider B. I. Chem. Phys. Letts. 1975, 31, 237.
22. Bloch, C. Nucl. Phys. 1957, 4, 503.
23. Schneider B. I. Phys. Rev. 1975, A11, 1957.
24. Schneider B. I.; Morrison, M. A.; Ibid, 1977, A16, 1003.
25. Burke, P. G.; Mackey, I.; Shimamura, I. J. Phys. B. 1977, 10, 2497.
26. Light J. C.; Walker, R. B. J. Chem. Phys. 1976, 65, 4272.
27. Schneider B. I.; Walker, R. B. Ibid. 1971, 70, 2466.

28. An excellent set of computer codes for linear algebraic problems is the LINPACK routines, Society for Industrial and Applied Mathematics, Philadelphia, (1979).
29. Schneider B. I.; Collins, L. A. Phys. Rev. in press.
30. Collins, L. A.; Schneider, B. I. J. Phys. B in press.
31. Collins, L. A.; Schneider, B. I. submitted to Phys. Rev.
32. Birtwistle, B. T.; Herzenberg, A. J. Phys.B 1971, 4, 53.
33. Dube, L.; Herzenberg, A. Phys. Rev. 1979, A20, 194.
34. Schneider, B. I.; LeDourneuf, M.; Burke, P. G. J. Phys. B 1979, 12, L365.
35. LeDourneuf, M.; Lan VoKy.; Schneider B. I. In "Electron-Atom and Electron-Molecule Collisions"; Hinze, J., Eds., Plenum Press, New York,1983, p. 135.
36. Collins, L. A.; Schneider, B. I. Phys. Rev. 27, 1983, 101.
37. ONeil, S. V.; Reinhardt, W. P. J. Chem. Phys. 69, 1978, 2126.
38. Daasch W. R.; Davidson, E. U.; Hazi, A. U. J. Chem. Phys. 76, 1982, 6031.
39. Golden, D E.; Bandel, H. W.; Salerno, J. A. Phys Rev. 1966, 146, 40.
40. Dalba, G.; Fornasini, P.; Lazzizzera, I.; Ranieri, G.; Zecca, A. J. Phys B 1980, 13, 1481.
41. Hazi, A.; Derkits, C.; Bardsley, J. N. Phys. Rev. 1983 A27, 1751.
42. Takagi, H.; Nakamura, H. Phys Rev. 1983, A27, 691.
43. Plummer, E. W.; Gustafsson, T.; Gudat, W.; Eastman, D. E. Phys. Rev. 1977, A15, 2339.
44. Delaney, J. J.; Hillier, I. H.; Saunders, V. R. J. Phys. B 1982, 15, 1474.
45. Smith, M. E.; Lucchese, R. R.; McKoy, V. J. Chem. Phys. 1983, 79, 1360.
46. Southworth, S.; Truesdale, C. M.; Kobin, P. H.; Lind, D. W.; Brewer, W. D.; Shirley, D. A. J. Chem. Phys. 1982, 76, 143.
47. Brion, C. E.; Tan, K. H. In "Electron Spectroscopy and Related Phenomena," 1983, Vol. 23, 1.
48. Padial, N. T.; Csanak, G.; McKoy, V.; Langhoff, P. W. J. Chem. Phys. 1978, 69, 2992.

RECEIVED June 12, 1984

6

Dynamics of Molecular Photoionization Processes

D. L. LYNCH[1], V. McKOY[1], and R. R. LUCCHESE[2]

[1] Arthur Amos Noyes Laboratory of Chemical Physics, California Institute of Technology, Pasadena, CA 91125
[2] Department of Chemistry, Texas A & M University, College Station, TX 77843

Increasing amounts of data have shown that shape resonances play an important role in molecular photoionization. These resonances can lead to significant deviations of the vibrational branching ratios from Franck-Condon predictions and to vibrational state dependence of the photoelectron asymmetry parameters. They are one-electron in nature and their properties are determined primarily by the molecular core potential. These features suggest that their role in photoionization should be studied using realistic molecular potentials and photoelectron continuum states. We will first discuss the relevant aspects of the method we have developed for studying the electronic Hartree-Fock continuum states needed in molecular photoionization. We will then present the results of applications of this approach to resonant photoionization in several molecules including N_2, CO, CO_2, C_2H_2, and C_2N_2. Our emphasis will be on a comparison of these results both with experimental data and other theoretical predictions.

The main purpose of early studies of molecular photoelectron spectroscopy was to characterize a molecule in terms of a simple energy level scheme based on molecular orbitals (1). Although these early studies contributed significantly to our understanding of both molecular electronic structure and of photoelectron dynamics, they were carried out with traditional light sources and hence did not provide photoelectron spectra over a continuous range of photon energy. Measurements of these photoelectron spectra over a continuous and wide range of photon energies are obviously needed to characterize the dynamical aspects of the molecular photoionization process. Synchrotron radiation provides the intense tunable source of photons needed to study the continuous variation of molecular photoionization cross sections with photon energy. Such experiments utilizing synchrotron radiation, along with related theoretical developments, have led to remarkable progress in our understanding of the dynamics of molecular photoionization processes (2,3). Results to date have shown that shape, or single-particle, resonances play a very important role in

0097-6156/84/0263-0089$07.00/0

molecular photoionization. For example, in addition to the strong influence of shape resonances on the angular distributions of photoelectrons, they also lead to such important dynamical features in the photoelectron spectrum as non-Franck-Condon vibrational state distributions in molecular ions and a vibrational-state dependence in the photoelectron asymmetry parameters. It is hence not surprising that shape resonances are often the focal point of many current experimental and theoretical studies of molecular photoionization.

In this article we want to discuss several results of our recent studies of molecular photoionization. Our emphasis will quite naturally be on the role of shape resonances in these processes. These resonances are quasibound states which are formed by trapping of the photoelectron by the centrifugal barrier of the molecular force field. Moreover, the centrifugal barriers of these molecular force fields can involve very high angular momentum character. Such resonances are essentially one-electron in nature with charge densities primarily localized in the molecular core region and hence their properties are to a great extent determined by the molecular core potential. These characteristics of shape resonances along with the expected strong dependence of the wave function on energy and internuclear distance near the resonance position suggest that they should be studied using realistic molecular potentials and electronic continuum states. In our studies of molecular photoionization we will use the Hartree-Fock electronic continuum states of the molecular ion. Our approach for obtaining these molecular electronic continuum states has several important features which will be discussed below. With these continuum states we can examine the important characteristics of the resonances in some detail. For example, in addition to the usual enhancement of the cross section we can also obtain the photoelectron asymmetry parameters and eigenphase sums across the resonance region.

In the next section we will discuss the approach we have developed for obtaining the molecular Hartree-Fock continuum orbitals. We will discuss how our approach is based on the Schwinger variational method and how in its present form it can be viewed as a hybrid method that uses both the basis-set expansion techniques of quantum chemistry and the numerical single-center expansion techniques of atomic collision physics. We will then discuss the results of applications of this approach to study shape resonances in the photoionization of several molecules, e.g., N_2, CO, CO_2, C_2H_2, and C_2N_2. These results will also be compared with available experimental data and with the results of studies of these same systems by different methods and models.

Theoretical Developments

The rotationally unresolved, fixed-nuclei, photoionization cross section is given by

$$\sigma = \frac{4\pi^2\omega}{3c} \left| \langle \Psi_i | \vec{\mu} | \Psi_f \rangle \right|^2 \tag{1}$$

where $\vec{\mu}$ is the dipole moment operator and ω is the photon frequency. In Equation 1 Ψ_i represents the initial state of the molecule with N bound electrons and Ψ_f the final state with a photoelectron in the electronic continuum. Within the Born-Oppenheimer approximation Ψ_f can also be factored into electronic and nuclear components. The representation of Ψ_f requires a set of continuum orbitals describing the motion of the photoelectron in the nonspherical potential field of the molecular ion. In these studies we will use the Hartree-Fock continuum orbitals of the molecular ion which satisfy the one-electron Schrödinger equation (in atomic units)

$$(-\frac{1}{2}\nabla^2 + V_{N-1}(\underset{\sim}{r},R) - \frac{k^2}{2})\ \phi_{\underset{\sim}{k}}(\underset{\sim}{r},R) = 0 \tag{2}$$

where $\frac{k^2}{2}$ is the kinetic energy of the photoelectron, V_{N-1} is the Hartree-Fock (static-exchange) potential of the ion, and $\phi_{\underset{\sim}{k}}$ satisfies the appropriate boundary conditions. Most studies to date have used the frozen-core or sudden approximation and assumed that V_{N-1} is determined by the core orbitals of the neutral molecule. Here core refers to the molecular ion orbitals. Although the frozen-core approximation is generally adequate for photoionization out of valence or even inner-valence levels we will see that core relaxation or restructuring effects can be very important in resonant photoionization of K-shell electrons.

The nonspherical and nonlocal character of the molecular potential introduces several complications into the solution of the equation for $\phi_{\underset{\sim}{k}}$. These difficulties have stimulated the development of various approaches to the calculation of molecular photoionization cross sections. These approaches include the Stieltjes moment theory method which avoids the need for the direct solution of the collision equation by extracting photoionization cross sections from spectral moments of the oscillator strength distribution (4), the continuum multiple scattering model in which the scattering potential is approximated so that the scattering equations can be readily solved (5), and several methods for the direct solution of the Hartree-Fock equations for the continuum orbitals $\phi_{\underset{\sim}{k}}$ (6-11). We believe that results for several systems are beginning to show that there can be significant advantages in utilizing Hartree-Fock continuum orbitals explicitly in molecular photoionization studies.

Our approach to the solution of Equation 2 begins with the integral form of this equation. The integral form of Equation 2 is the Lippmann-Schwinger equation and is given by

$$\phi_{\underset{\sim}{k}} = \phi^{c}_{\underset{\sim}{k}} + G^{(-)}_{c} V \phi_{\underset{\sim}{k}} \tag{3}$$

where $\phi^{c}_{\underset{\sim}{k}}$ is the pure Coulomb scattering wave function, V is the molecular ion potential with the Coulomb component removed, and $G^{(-)}_{c}$ is the Coulomb Green's function with incoming-wave boundary conditions, i.e.,

$$V = V_{N-1} + \frac{1}{r} \tag{4a}$$

and

$$G^{(-)}_{c} = \frac{1}{2}(\nabla^2 + \frac{2}{r} + k^2 - i\varepsilon)^{-1} \tag{4b}$$

The continuum orbital $\phi_{\underset{\sim}{k}}(\underset{\sim}{r})$ can be expanded in terms of spherical harmonics of $\Omega_{\hat{k}}$, the direction of $\underset{\sim}{k}$, as

$$\phi_{\underset{\sim}{k}}(\underset{\sim}{r}) = \left(\frac{2}{\pi}\right)^{\frac{1}{2}} \sum_{\ell m} \frac{i^{\ell}}{k} \phi_{k\ell m}(\underset{\sim}{r})\, Y^{*}_{\ell m}(\Omega_{\hat{k}}) \qquad (5)$$

where $\phi_{k\ell m}(\underset{\sim}{r})$ are the partial wave scattering functions. Each $\phi_{k\ell m}(\underset{\sim}{r})$ satisfies its own Lippmann-Schwinger equation, i.e.,

$$\phi_{k\ell m} = \phi^{c(-)}_{k\ell m} + G^{(-)}_{c} V\phi_{k\ell m} \qquad (6)$$

where $\phi^{c(-)}_{k\ell m}$ is the Coulomb partial wave function, i.e.,

$$\phi^{c(-)}_{k\ell m} = e^{-i\sigma_{\ell}} F_{\ell}(\gamma;kr)\, Y_{\ell m}(\Omega_{\hat{r}}) \qquad (7)$$

In Equation 7, $F_{\ell}(\gamma;kr)$ is the regular Coulomb function with $\gamma = -1/k$ and σ_{ℓ} is the Coulomb phase shift (12).

We first obtain an approximate solution for $\phi_{k\ell m}$ of Equation 6 by assuming a separable approximation for the potential V of the form

$$V(\underset{\sim}{r},\underset{\sim}{r}') \simeq V^{S}(\underset{\sim}{r},\underset{\sim}{r}') = \sum_{i,j} \langle \underset{\sim}{r}|V|\alpha_{i}\rangle (V^{-1})_{ij} \langle \alpha_{j}|V|\underset{\sim}{r}'\rangle \qquad (8)$$

where the matrix $(V^{-1})_{ij}$ is the inverse of the matrix with elements $V_{ij} = \langle \alpha_{i}|V|\alpha_{j}\rangle$. With the approximate potential of the form of Equation 8, the solutions of Equation 6 are

$$\phi^{(0)}_{k\ell m}(\underset{\sim}{r}) = \phi^{c(-)}_{k\ell m} + \sum_{i,j} \langle \underset{\sim}{r}|G^{(-)}_{c}V|\alpha_{i}\rangle (D^{-1})_{ij} \langle \alpha_{j}|V|\phi^{c(-)}_{k\ell m}\rangle \qquad (9)$$

where the matrix $(D^{-1})_{ij}$ is the inverse of the matrix

$$D_{ij} = \langle \alpha_{i}|V - VG^{(-)}_{c}V|\alpha_{j}\rangle \qquad (10)$$

The use of a separable potential of the form of Equation 8 in Equation 6 to obtain solutions of the form of Equation 9 can be shown to be equivalent to using the functions $\alpha_{i}(\underset{\sim}{r})$ in the Schwinger variational principle for collisions (13). At this stage the functions $\alpha_{i}(\underset{\sim}{r})$ can be chosen to be entirely discrete basis functions such as Cartesian Gaussian (14) or spherical Gaussian (15) functions. We note that with discrete basis functions alone the approximate solution $\phi^{(0)}_{k\ell m}$ satisfies the scattering boundary condition. Such basis functions have been used successfully in electronic structure calculations for many years (14) and should be very effective in representing the multicenter nature of the scattering wave function and molecular potential in the near-molecular region. With suitably chosen discrete basis sets the continuum solutions given by Equation 9 can already provide quantitatively reliable photoionization cross sections (10,15-17) which, moreover, can be shown to be variationally stable at the Hartree-Fock level (16). The matrix elements and functions arising in the solution of Equation 9 must be evaluated by single-center expansion methods. We will return shortly to a discussion of several important features of the single-center expansion techniques which we use in the solution of Equation 9.

Although very useful, the $\phi^{(0)}_{k\ell m}$ of Equation 9 are solutions of the approximate potential V^S of Equation 8 and are not the actual continuum solutions of the Hartree-Fock potential V_{N-1}. In applications it can be important to use more accurate solutions than $\phi^{(0)}_{k\ell m}$ and to obtain the converged electronic continuum solutions of Equation 6. Thus, although such solutions are not always required, it is desirable to have a method for obtaining them. For this reason we have developed an iterative procedure for obtaining accurate solutions of Equation 6. The details of this procedure have been discussed elsewhere (18). Briefly the method begins by viewing the approximate continuum solutions of Equation 9 as new basis functions. These approximate numerical continuum solutions are now added to the initial set of discrete basis functions $\{\alpha_i\}$ to form an augmented basis. This augmented, but mixed, basis is then used in Equation 9 to generate an improved continuum solution, i.e., $\phi^{(1)}_{k\ell m}$. The iterative procedure is continued until the wave functions converge. When the wave functions do converge, it can be shown that they are solutions of the Lippmann-Schwinger equation for the potential V (18).

The solutions of Equation 6 can often be required to satisfy certain orthogonality constraints such as being orthogonal to several occupied orbitals of the molecular ion. The most straightforward approach to these orthogonality constraints is to include them by using a generalized Phillips-Kleinman pseudopotential in place of V in Equation 6. The details of this procedure have been discussed in Reference 10. An alternative approach to the inclusion of such orthogonality constraints is to introduce Lagrange undetermined multipliers into the system of equations. This procedure has been discussed elsewhere (11).

Finally, we have also developed an improved method for the study of photoionization cross sections which is based on the direct use of Schwinger-type variational expressions for the electric dipole matrix elements themselves (16). These more general variational expressions can also be iteratively improved. The analysis of these approaches revealed that the iterative Schwinger method which we have outlined above leads to variationally stable photoionization cross sections.

Solution of Equations

The matrix elements arising in the solution of Equation 9 for molecular systems have as yet no known analytic forms and hence must be evaluated by single-center expansion techniques. Once the single-center expansions of the functions in Equation 9, e.g., the basis functions $\alpha_i(\tilde{r})$, the Hartree-Fock potentials, and the Coulomb Green's functions, have been carried out, all the matrix elements can be obtained in terms of various radial integrals. Accurate results can be obtained using single-center expansions, as long as the dependence of the physical quantity of interest on the expansion parameters is carefully monitored. It is important to note that neither Equation 9 nor its iterated version requires the solution of coupled integro-differential equations.

For the expansion of the Hartree-Fock molecular orbitals we have used either Slater or Cartesian Gaussian functions. In addition to these basis functions we can also include spherical Gaussian functions in the initial scattering basis. A detailed discussion of the single-center expansion of Slater and Cartesian Gaussian functions has been given by Harris and Michels (19) and by Fliflet and McKoy (20), respectively. Spherical Gaussian functions, i.e.,

$$\chi(\underset{\sim}{r}) = N|\underset{\sim}{r} - \underset{\sim}{A}|^{\ell} \exp(-\alpha|\underset{\sim}{r} - \underset{\sim}{A}|^2)\, Y_{\ell m}(\Omega_{\underset{\sim}{r}-\underset{\sim}{A}}) \tag{11}$$

are particularly useful due to the ease with which functions with high ℓ can be computed. The single-center expansion of these functions can be obtained by noting that a spherical Gaussian function is the product of a simple s-type Cartesian Gaussian function and a solution to Laplace's equation, both of which have simple expansions about another center (20,21). With these expansions for the basis functions and the well-known expansion of $1/r_{12}$, we can easily obtain the single-center expansion of the Hartree-Fock potential of the molecular ion. Finally, the Coulomb Green's function can be readily expanded about a single center (22).

With these single-center expansions we use the extended Simpson's rule to evaluate the radial integrals arising in Equation 9. For those integrals involving a two-dimensional integrand with a discontinuity in the derivative at $r = r'$, e.g., due to the Green's function, we use a special integration formula for the regions containing the discontinuity (23). The integrals are evaluated on an interval $(0,r_{max})$. This interval $(0,r_{max})$ is subdivided into intervals in which different step sizes can be used. In general we use a step size of 0.01 a.u. in the region near the nuclei, and at large r we usually have at least 20 steps per half wave or $h < \pi/20\,k$ where h is the step size. Thus for k up to 1 a.u. we have h = 0.16 a.u. for the step size at large r. Between these two regions there are intervals with intermediate step sizes.

These single-center expansions must be truncated at some suitable maximum value. These maximum values, the step sizes on the grid, and r_{max} are the important parameters of the technique. These parameters can be enumerated as follows:

(1) ℓ_m, maximum ℓ in the expansion of scattering functions, variational basis functions, Green's function, and orbitals used in constructing orthogonality constraints,

(2) ℓ_s^x, maximum ℓ in the expansion of the scattering functions in the exchange terms,

(3) ℓ_i^x, maximum ℓ in the expansion of the occupied orbitals in the exchange terms,

(4) ℓ_i^d, maximum ℓ in the expansion of the occupied orbitals in the direct potential,

(5) λ_m^x, maximum ℓ in the expansion of $1/r_{12}$ in the exchange terms,

(6) λ_m^d maximum ℓ in the expansion of $1/r_{12}$ in the direct electronic potential,

(7) λ_m^n, maximum ℓ in the expansion of $1/r_{12}$ in the electron-nuclei electrostatic interaction,

(8) ℓ_p, maximum ℓ in the expansion of $\phi_{\underset{\sim}{k}}(\underset{\sim}{r})$ in Equation 5,

(9) r_m, maximum grid point,

(10) h_i, step sizes used in the grid.

To insure the accuracy of a calculation using these single-center expansions, the convergence of the photoionization cross section with respect to each of these parameters must be checked. The rate of convergence of a calculation will depend on the nature of the system being studied. One useful empirical result which has been observed is that the peak position in the photoionization cross section due to a shape resonance in the σ continuum approaches its asymptotic value as $(\ell_i^d)^{-3}$. This relationship has been verified for shape resonances in the photoionization of N_2 (10) and CO_2 (15).

Applications

We now discuss the results of applications of this procedure to study the effect of shape resonances on the photoionization cross sections of various molecules. Our examples will be chosen from the photoionization of both inner and valence orbitals of N_2, CO, CO_2, C_2H_2, and C_2N_2. In most of these studies the wave function of the final state is obtained in the frozen-core Hartree-Fock (FCHF) approximation. In this model the final-state wave function is described by a single electronic configuration in which the ionic core orbitals are constrained to be identical to the Hartree-Fock orbitals of the neutral molecule and the photoelectron wave function is a continuum Hartree-Fock orbital of the ion core potential. This frozen-core or "sudden" approximation should work best for photoionization of valence orbitals and, moreover, leads to some important computational simplifications. It can be, however, a very poor approximation for K-shell photoionization. In several of our studies we have considered the effects of correlation in the initial state by comparing the cross sections obtained using both Hartree-Fock (HF) and configuration-interaction (CI) initial-state wave functions. To estimate the effects of correlation in the final state we can compare the dipole length and velocity forms of the photoionization cross section.

As a first example we discuss the well-known shape resonance feature which occurs in the $3\sigma_g \rightarrow k\sigma_u$ channel leading to the $X^2\Sigma_g^+$ state of N_2^+. This shape resonance has been studied extensively by several methods (10,24,25). Our calculated cross sections using both HF and CI initial-state wave functions and the length and velocity forms of the oscillator strength are shown in Figure 1. The single-center expansion parameters used in these studies were $\ell_m = \ell_i^d = \lambda_m^d = 30$, $\ell_s^x = \lambda_m^x = 20$, $\ell_i^x = 16(1\sigma_g)$, $10(2\sigma_g)$, $10(3\sigma_g)$, $15(1\sigma_u)$, $9(2\sigma_u)$, $9(1\pi_u)$, $\lambda_m^n = 60$, $\ell_p = 7(k\sigma_u)$, $6(k\sigma_g, k\pi_u, k\pi_g)$, $r_m = 64$ a.u., and $h_i = 0.01 - 0.16$ a.u., with 800 grid points. The initial basis set contained a mixed basis of 18 Cartesian and spherical Gaussian functions in the σ_u symmetry while the π_u basis contained 8 such

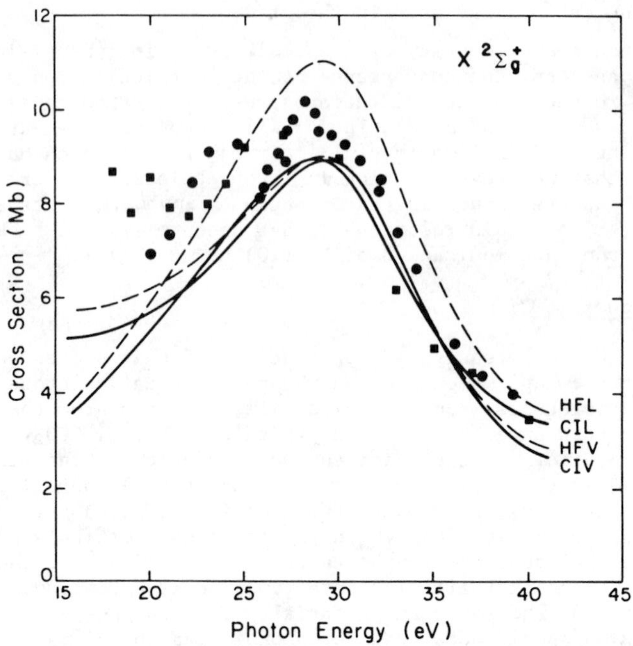

Figure 1. Photoionization cross section for production of the $X^2\Sigma_g^+$ state of N_2^+ with different initial state wave functions (Hartree-Fock or configuration interaction, i.e., HF or CI) and in the length and velocity forms (L or V); ●, experimental data of E. W. Plummer, T. Gustafsson, W. Gudat, and D. E. Eastman, Phys. Rev. A 15, 2339 (1977); ■, experimental data of A. Hamnett, W. Stoll, and C. E. Brion, J. Electron. Spectrosc. Relat. Phenom. 8, 367 (1976). Reproduced with permission from Reference 10, Copyright 1981, American Physical Society.

functions (10). For the Hartree-Fock orbitals we used a contracted Gaussian basis of the form (9s5p2d/4s3p2d) (10,26).

The results of Figure 1 show that the calculated cross sections are in good agreement with the available experimental data and that inclusion of initial-state correlation effects reduces the difference between the length and velocity forms of the cross section and, moreover, brings them into good agreement with experimental data. The differences between these FCHF results and the measured cross sections below 23 eV are attributed to autoionization, the effects of which are not included in the FCHF model. We have also studied the sensitivity of this resonant cross section to the parameters of the single-center expansions. An analysis of this dependence showed that the truncation of the partial wave expansions leads to a shift of less than 0.5 eV in the peak position of the resonance. These cross sections changed only slightly upon iterative improvement of the photoelectron continuum wave function (10). Figure 2 shows the resonant behavior of the σ_u eigenphase sum along with the corresponding cross sections. These eigenphase sums are useful in identifying resonant behavior in cases where the effect of the resonance on the vibrationally unresolved photoionization cross sections is weak (17). In Figure 3 we show our calculated photoelectron asymmetry parameters for the $X^2\Sigma_g^+$ state of N_2^+ along with the measured values (10). This asymmetry parameter essentially defines the angular distributions of photoelectrons ejected from freely rotating molecules. Both the length and velocity forms of this parameter are shown. The agreement with experiment is good, except for energies below 23 eV where autoionization is important.

In Figure 4 our calculated cross sections for the $3\sigma_g$ level are compared with the results of the Stieltjes moment theory (STMT) approach (25) and of the continuum multiple scattering model (CMSM) (24). Although there are some differences in the results of these methods, the essential resonant nature of this cross section is reproduced by all three approaches.

Figure 5 shows the calculated and experimental branching ratios for the $\nu' = 1$ to $\nu' = 0$ level for photoionization leading to the $X^2\Sigma_g^+$ level of N_2^+. This branching ratio, which has been scaled up by a factor of 100 for convenience in Figure 5, represents the population of ions in the $\nu' = 1$ level relative to that of the $\nu' = 0$ level. In the Franck-Condon approximation this branching ratio should be independent of photon energy. The non-Franck-Condon behavior of this branching ratio, as first predicted by Dehmer et al. (27), is due to the strong dependence of the photoelectron matrix element on internuclear distance in the region of a resonance. Our calculated branching ratios are in good agreement with the experimental data of West et al. (28). The discrepancy in these branching ratios below a photon energy of 23 eV can be attributed to autoionizing resonances. The large branching ratios predicted by the multiple scattering model is certainly an indication of the poor representation of the R dependence of the molecular ion potential in this model. It is important to note that substantial non-Franck-Condon behavior of this branching ratio occurs at photon energies well above the peak position of the resonance (27).

In Figure 6 the total photoionization cross section leading to the $X^2\Sigma_g^+$, $A^2\Pi_u$, and $B^2\Sigma_u^+$ states of N_2^+ is shown. The agreement

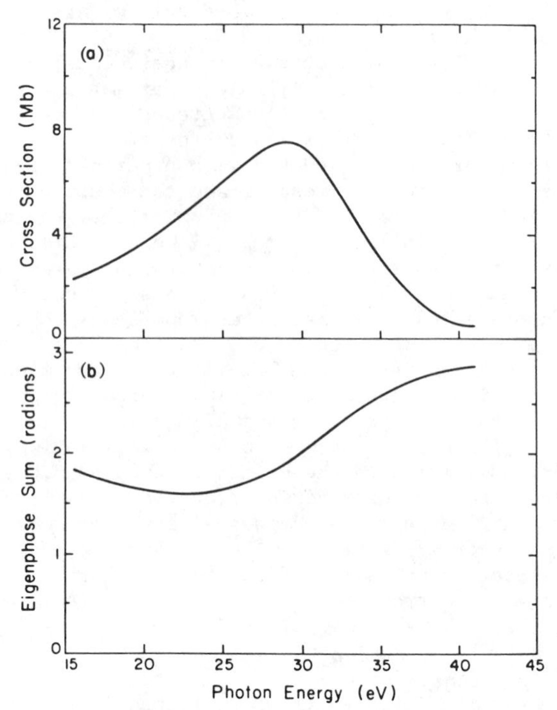

Figure 2. Photoionization cross section (a) and eigenphase sum, modulo π, (b) for the $3\sigma_g \rightarrow k\sigma_u$ channel of N_2.

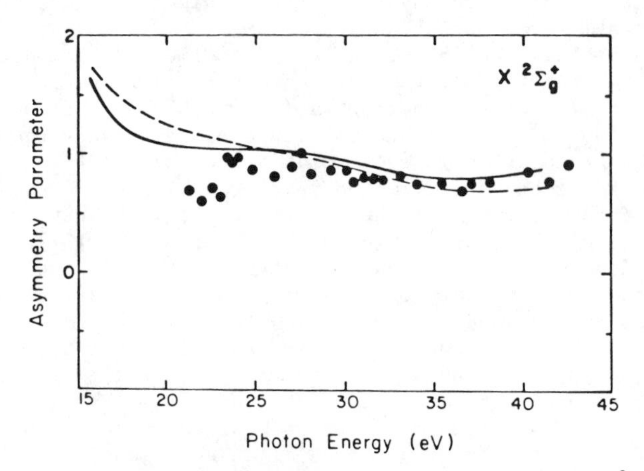

Figure 3. Photoelectron asymmetry parameters for the $X^2\Sigma_g^+$ state of N_2^+:—— and – – –, length and velocity forms respectively using a CI initial state; •, experimental data of G. V. Marr, J. M. Morton, R. M. Holmes, and D. G. McCoy, J. Phys. B 12, 43 (1979).

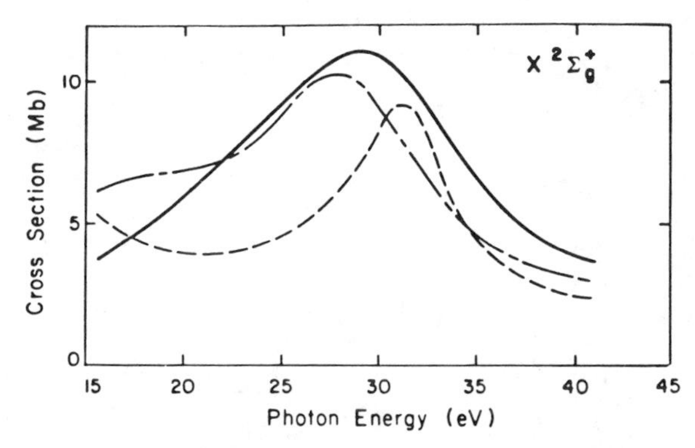

Figure 4. Comparison of photoionization cross sections for the $X^2\Sigma_g^+$ state of N_2^+:———,present FCHF results; ——— – ———, FCHF results of the STMT method (Reference 25); — — —, CMSM results of Reference 24.

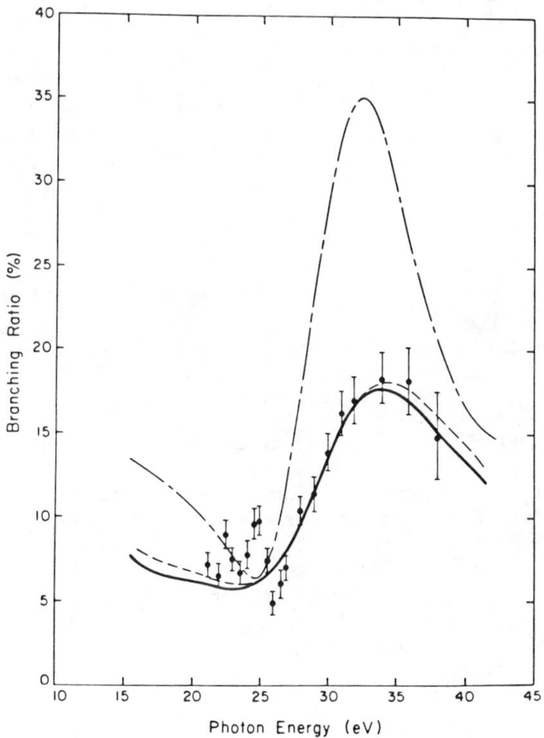

Figure 5. Vibrational branching ratios for the $\nu'= 1$ to $\nu'= 0$ levels of the $X^2\Sigma_g^+$ state of N_2^+ (see text):———, present FCHF results (dipole length);— — —, same but with dipole velocity form; ——— – ———, CMSM results of Reference 27; •, experimental data of Reference 28. (from R. Lucchese and V. McKoy, J. Phys. B. 14, L629 (1981)).Copyright 1981, Institute of Physics.

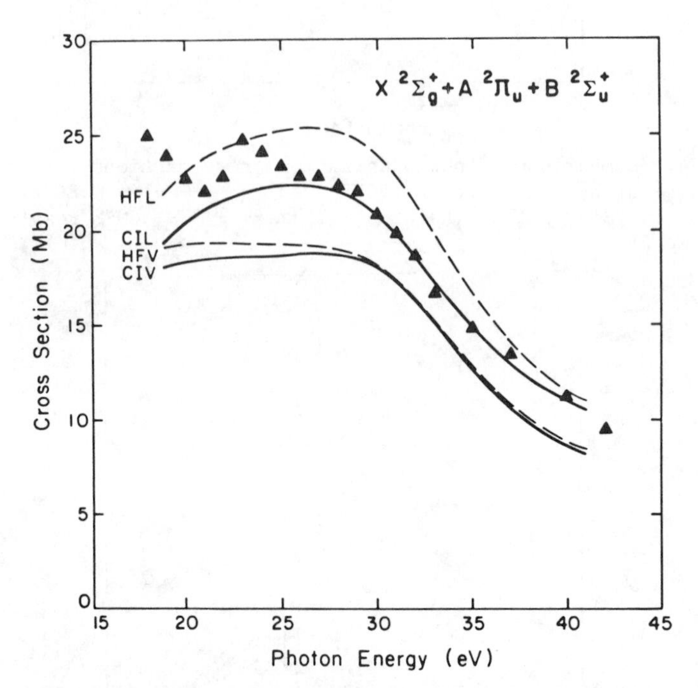

Figure 6. Total photoionization cross section for production of the $X^2\Sigma_g^+$, $A^2\Pi_u$, $B^2\Sigma_u^+$ states of N_2: present results (see caption for Figure 1); ▲, experimental data of Reference 29 corrected to include only the contributions from these three channels using the experimental branching ratios of A. Hamnett, W. Stoll, and C. E. Brion, J. Electron. Spectrosc. Relat. Phenom. 8, 367 (1976). Reproduced with permission from Reference 10, Copyright 1981, American Physical Society.

between the calculated and experimental cross sections (10,29) is good except for the features due to autoionization below 23 eV.

We have also studied the effect of this shape resonant σ_u continuum on photoionization out of the K-shell of N_2. The partial wave expansion parameters and initial scattering basis sets for these calculations are essentially the same as in the $3\sigma_g$ studies above. For the SCF orbitals we used the contracted Gaussian basis of the form (9s5p1d/4s3p1d) (26). Here, and elsewhere, we assume the experimental values for the ionization potentials, i.e., 409.9 eV for the N_2 K-shell. In Figure 7 we show our calculated frozen-core K-shell photoionization cross sections for N_2. These cross sections include contributions from the nearly degenerate $1\sigma_g$ and $1\sigma_u$ levels. The resonant nature of the cross section is very apparent. Figure 7 also shows the experimental K-shell photoionization cross sections obtained by (e,2e) measurements (30). These cross sections are in essential agreement with the recent synchrotron radiation data of Lindle (31). Our calculated cross sections are clearly in poor agreement with the experimental data. In particular, these frozen-core results lead to too narrow a resonance and concentrate too much of the cross section near threshold. The cross sections obtained by the STMT method within the same frozen-core model (32) clearly show a much broader feature than the present results and are apparently in better agreement with the available data. In principle these two methods should give the same cross sections. We have checked the convergence of our calculations and we believe that the partial wave expansions are adequate and should yield the converged cross sections. For example, our expansion parameters give a normalization of 0.9995 for the $1\sigma_g$ and $1\sigma_u$ orbitals. We believe that the STMT method has artificially broadened this resonance. This discrepancy between these resonant cross sections predicted by the two methods is analogous to what was seen in the $4\sigma_g \rightarrow k\sigma_u$ channel in photoionization leading to the $C^2\Sigma_g^+$ state of CO_2^+ where a prominent resonant feature in our results is essentially smoothed away by the STMT method (33).

To try to understand the origin of these differences between the calculated and measured cross sections we have studied the effects of core relaxation on these cross sections (34). Although the effects of restructuring of the ion core are not expected to be important for ionization of valence levels of atoms and molecules, it is well-known that they can be significant in inner-shell photoionization (35,36). For deep hole levels the relaxation of the core can lead to strong screening of the hole and a consequent decrease of the photoionization cross section near threshold. This effect can become even more important where there is resonant trapping of the photoelectron, as is the case here. The FCHF model completely neglects any rearrangement of the core. The other extreme is to assume that this core restructuring is complete during the passage of the photoelectron and that the appropriate potential field for the final state is that of the relaxed ion. In Figure 7 we show the K-shell photoionization cross sections obtained using a relaxed HF ionic core and the continuum orbitals of this core. This relaxation of the core has clearly decreased the cross section near threshold dramatically, broadened the resonance, and shifted it to higher energy. The resonance is now at too high an energy suggesting that the assumption of complete restructuring of the ionic core has led to excessive screening of the hole. In Figure 8 we show these relaxed core cross sections shifted

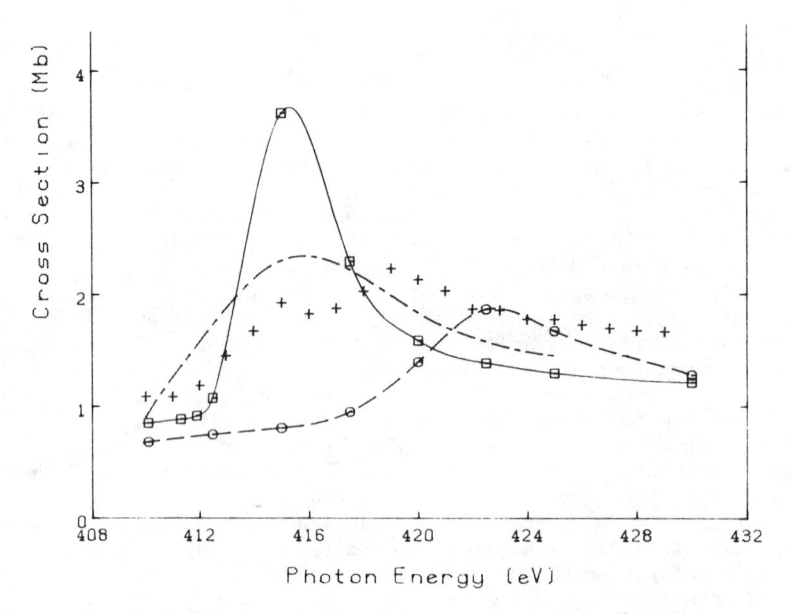

Figure 7. K-shell photoionization cross sections for N_2: ———, present FCHF results;— — —, present results with a relaxed ion core; —— – ——, FCHF results of the STMT method (Reference 32); +, experimental data of Reference 30.

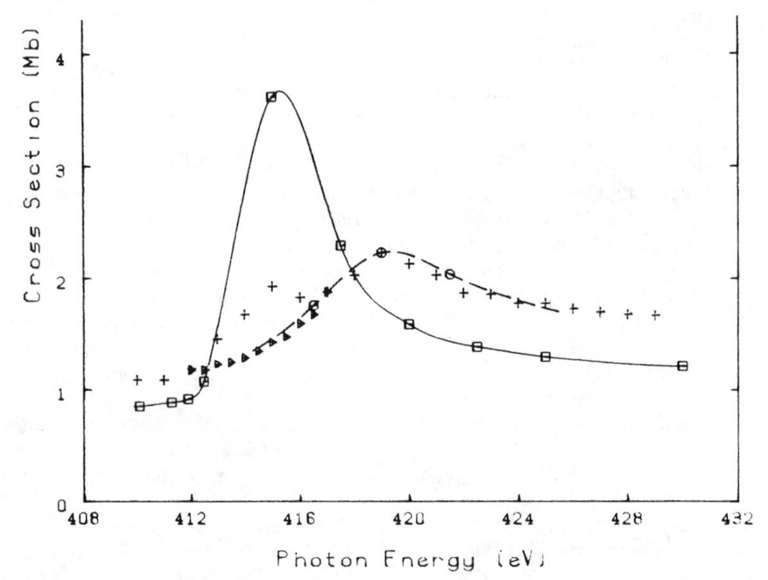

Figure 8. K-shell photoionization cross sections for N_2: ———, present FCHF results;— — —, present results with a relaxed ion core shifted so as to agree with the experimental data at 419.5eV; +, experimental data of Reference 30; △, estimated one-electron portion of the experimental cross sections (Reference 30).

so that their peak position coincides with that of the measured cross sections. In this figure we show the one-electron portion of the photoionization cross section as estimated from the experimental data by Kay et al. (30). The agreement between these cross sections is encouraging and obviously suggests that core relaxation is important in K-shell photoionization.

Figure 9 shows our calculated photoelectron asymmetry parameters using both the frozen and relaxed ionic cores, the results of the CMSM (37), and the experimental data of Lindle (31). Although the asymmetry parameters of the relaxed core model show a minimum at a higher energy than do the measured values, the overall shapes of these curves are quite similar around their minima. Figure 10 illustrates this by showing the asymmetry parameters of the relaxed core model shifted by the same amount as the cross sections were shifted in Figure 8.

Figure 11 shows the shape-resonant enhanced cross sections for photoionization of the 5σ level in CO (16). This resonance is analogous to the $3\sigma_g \rightarrow k\sigma_u$ feature in the photoionization of N_2 but the reduced symmetry of the continuum from σ_u to σ has led to a broadening of the resonance. In Figure 11 we compare these cross sections with those of the STMT (38) and CMSM (24) methods and with the experimental data. The various methods all show the expected resonance feature. There are, however, some differences of consequence between our results and those of the STMT method. The STMT cross sections suggest that the feature at 24 eV is due to a shape resonance whereas the present results suggest that it may be due to autoionization. A substantial part of the difference between these calculated cross sections arises from the use of Gaussian and Slater SCF bases in the STMT and our studies respectively (39).

We how look at the enhancement of the cross section for photoionization of the $4\sigma_g$ level of CO_2 leading to the $C^2\Sigma_g^+$ state. Figure 12 shows the fixed-nuclei and vibrationally averaged cross sections obtained by the present procedure and the CMSM (40) along with the fixed-nuclei results of the STMT method (41). The calculated cross sections are averaged over the symmetric stretching mode only. There are clearly significant differences between the cross sections given by the various methods. The CMSM leads to a very narrow resonance with a peak value which is dramatically reduced by vibrational averaging. On the other hand, the effect of vibrational averaging on the FCHF cross sections is much less. This resonance feature is essentially not evident in the STMT results which are again in better apparent agreement with the experimental data (42). It is puzzling that this resonant feature is not evident in the measured cross sections whereas it is very pronounced in the asymmetry parameters (43). This disagreement could arise from the neglect of vibrational averaging of the calculated cross sections over the other two vibrational modes of CO_2. There are obviously some important features of these cross sections which are not yet well understood.

As a final example we look at the role of shape resonances in the photoionization of C_2N_2 (44). The possibility of shape resonant trapping in this molecule can be quite interesting due to the C $\equiv$ N component and the C—C bond. Some line source studies (45) and synchrotron radiation measurements (46) of these cross sections have been carried out for photon energies below 24 eV. The CMSM has also been used to study both the photoionization cross sections and

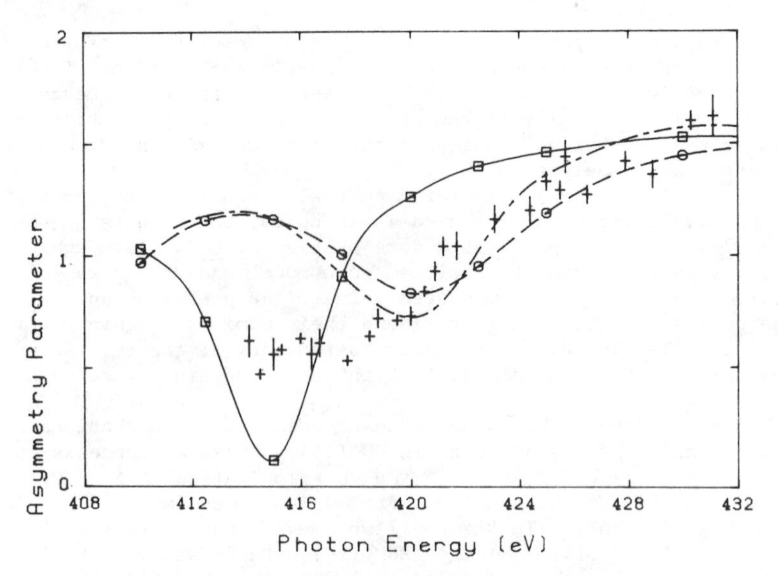

Figure 9. K-shell photoelectron asymmetry parameters in N_2: ———, present FCHF results; — — —, present results with a relaxed core; ___ _ ___, CMSM results of Reference 37; +, synchrotron radiation data of Reference 31.

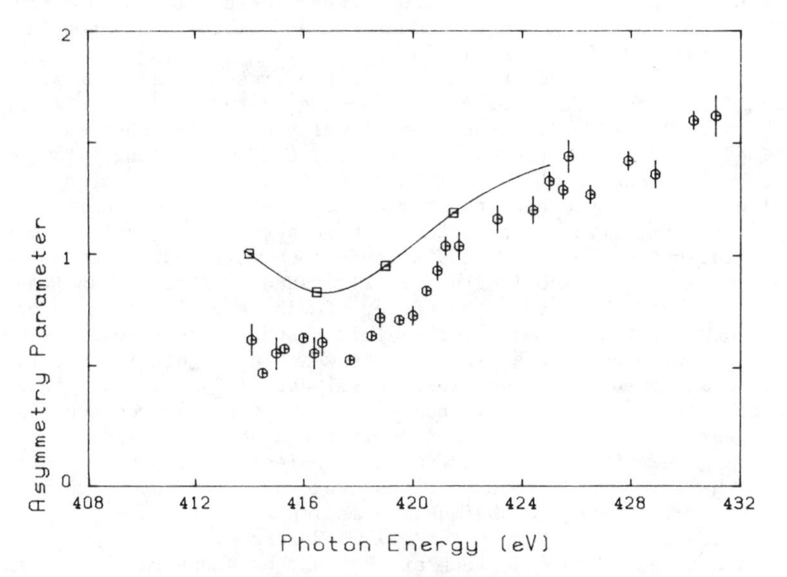

Figure 10. K-shell photoelectron asymmetry parameters in N_2: present results with a relaxed ion core shifted down in photon energy only by the same amount as the shift in the corresponding cross section in Figure 8; ●, synchrotron radiation data of Reference 31.

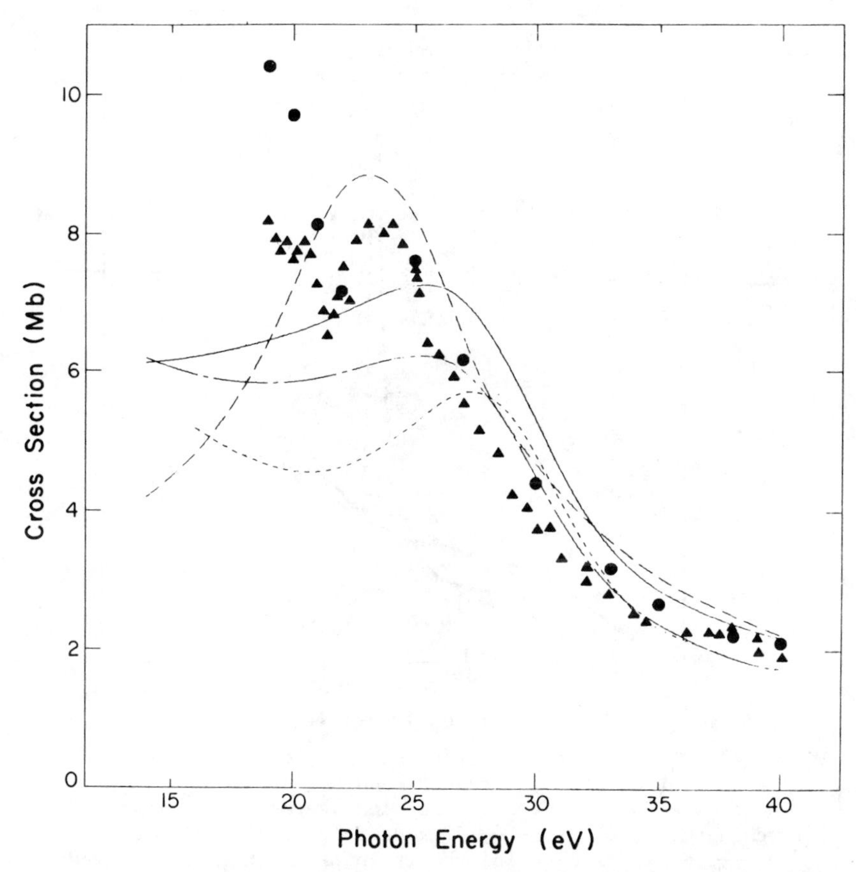

Figure 11. Photoionization cross section for production of the $X^2\Sigma^+(5\sigma^{-1})$ state of CO^+: ——— and —— – —— present results using the dipole length and velocity forms respectively; — — —, STMT results of Reference 38;- - - -, CMSM results of Reference 24. Reproduced with permission from Reference 16, Copyright 1983, American Physical Society.

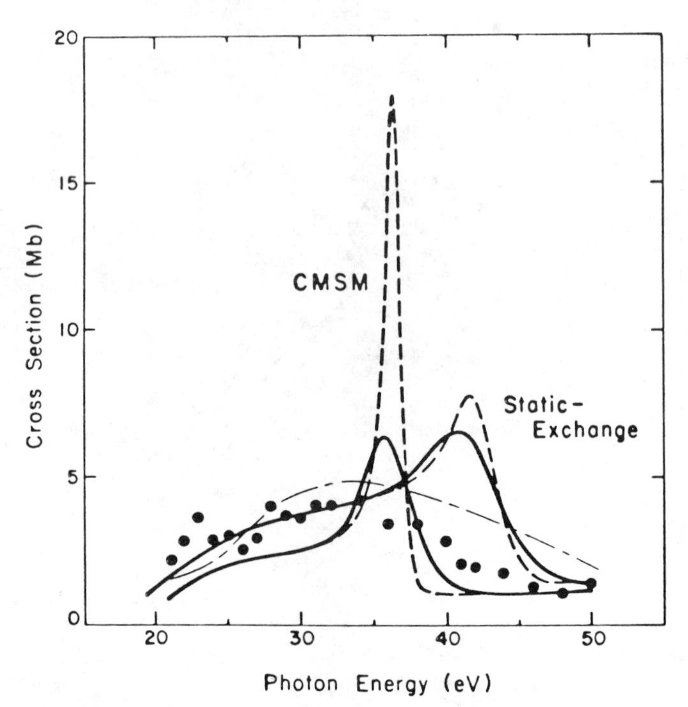

Figure 12. Photoionization cross section leading to the $C^2\Sigma_g^+(4\sigma_g^{-1})$ state of CO_2^+: present results and those of the CMSM of Reference 40: ———, cross section averaged over symmetric stretching mode; — — — —, fixed-nuclei cross sections; —— – ——, STMT fixed-nuclei results of Reference 41; •, experimental data of Reference 42. Reproduced with permission from Reference 15, Copyright 1982, American Physical Society.

photoelectron asymmetry parameters (45). These results indicate the presence of several shape resonances in this system (45). In Figure 13 we show our calculated cross sections for photoionization out of the $5\sigma_g$ level leading to the $A^2\Sigma_g^+$ state of $C_2N_2^+$ along with the individual contributions from the σ_u and π_u continua. These cross sections along with their eigenphase sums (44) show that there are low- and high-energy shape resonances in the σ_u continuum and an additional shape resonance in the π_u channel. There are no absolute measurements of the photoionization cross sections for the $A^2\Sigma_g^+$ state in this energy range. However, the line source studies of Kreile et al. (45) show substantial vibrational-state dependence of the photoelectron asymmetry parameters at 16.85 eV and 26.91 eV. This non-Franck-Condon behavior at 16.85 eV is certainly due to the low-energy σ_u resonance while we believe that the π_u resonance may be responsible for the behavior around 27 eV. The CMSM results (45) have previously identified these low- and high-energy σ_u shape resonances which we see in the present studies. The minima in the photoelectron asymmetry parameters around 18 eV and 32 eV in Figure 14 are associated with the low-energy σ_u and π_u shape resonances respectively. It is important to note that this double shape resonant behavior of the σ_u continuum of $C_2N_2^+$ has previously been seen in our studies of the photoionization of C_2H_2 (17). In fact, the photoionization cross sections for the $2\sigma_g$ level of C_2H_2 shown in Figure 15 indicate the presence of the double resonance structure (17). Future details of our studies of the photoionization cross sections of C_2N_2 will be discussed elsewhere (44).

Concluding Remarks

We have discussed some of our studies of the effects of shape resonances in the photoionization of linear molecules. We believe that a comparison of these results with those of other approaches to molecular photoionization and with experimental data show that there is substantial merit to explicitly using Hartree-Fock continuum states in studies of resonant molecular photoionization cross sections. Such studies provide very complete information on the photoionization process, e.g., the cross sections, the photoelectron asymmetry parameters, and eigenphases. Our approach does not involve the integration of coupled integro-differential equations and its extension to nonlinear triatomic molecules has almost been completed. With these electronic continuum orbitals, autoionizing resonances can also be studied. Such applications within the framework of the random-phase approximation (47) are underway.

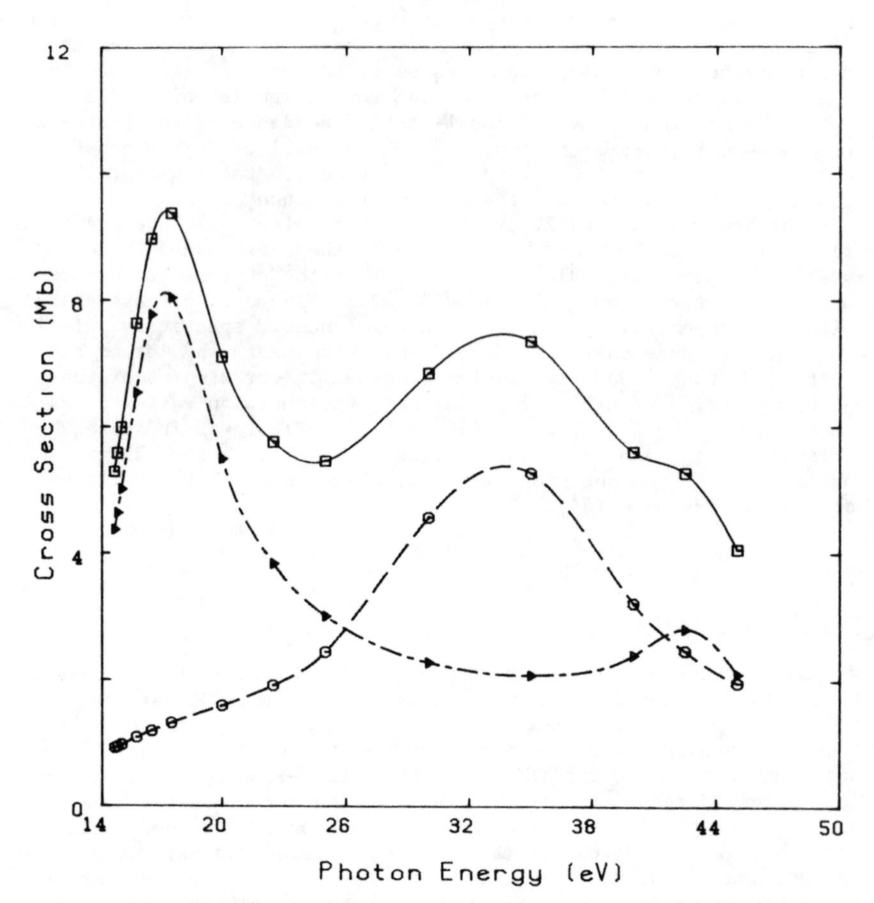

Figure 13. Photoionization cross sections for the $5\sigma_g$ level leading to the $A^2\Sigma_g^+$ state of $C_2N_2^+$: ——— – ——, $5\sigma_g \rightarrow k\sigma_u$ component; — — —, $5\sigma_g \rightarrow k\pi_u$ component; ———, $5\sigma_g$ cross sections (dipole length form).

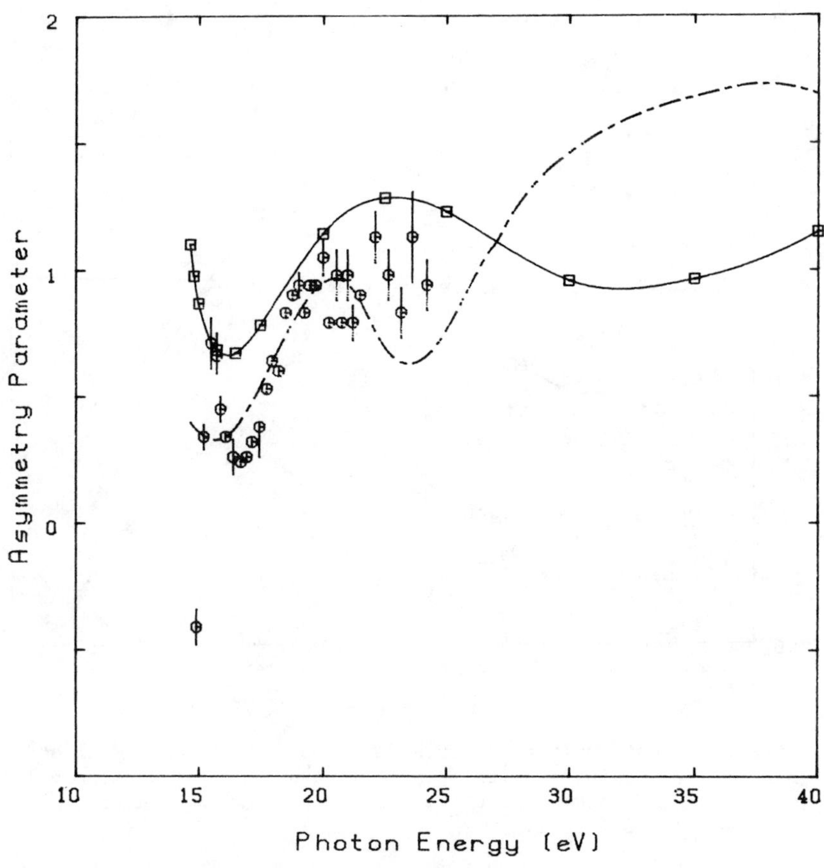

Figure 14: Photoelectron asymmetry parameters for the $5\sigma_g$ level of C_2N_2: ———, present results; — — —, CMSM results of Reference 45; •, experimental data ($\nu'' = 0 \rightarrow \nu' = 0$) of Reference 46.

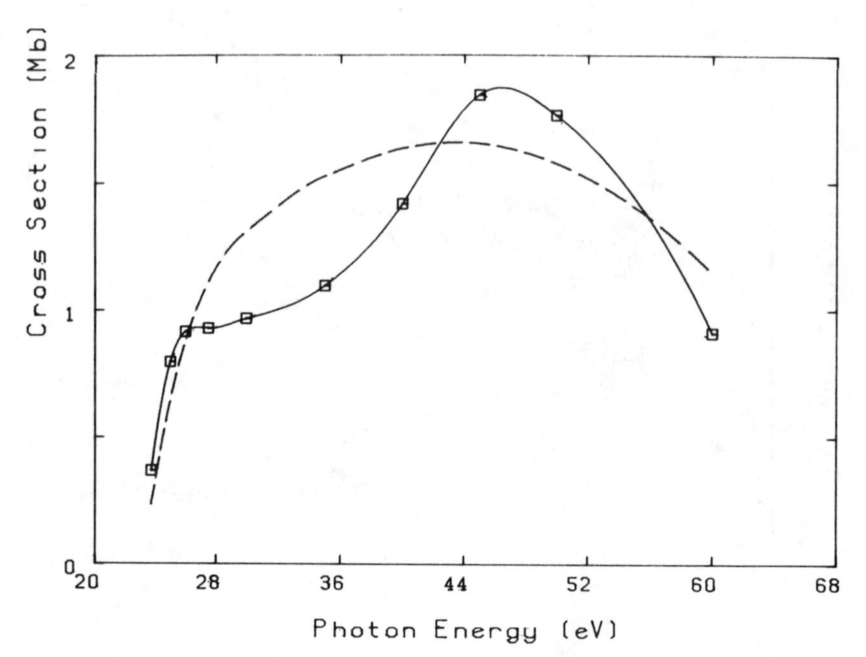

Figure 15. Photoionization cross sections for the $2\sigma_g$ level of C_2H_2: ———, present results (dipole length form) of Reference 17; —— – ——, STMT results of L. E. Machado et al. J. Electron. Spectrosc. 25, 1 (1982).

Acknowledgments

This material is based upon research supported by the National Science Foundation under Grant No. CHE-8218166. The authors acknowledge computing support from the National Center for Atmospheric Research (NCAR) which is sponsored by the National Science Foundation. This is Contribution No. 7025 from the Arthur Amos Noyes Laboratory of Chemical Physics, California Institute of Technology.

Literature Cited

1. See, for example, Turner, D. W.; Baker, A. D.; Brundle, C. R.; "Molecular Photoelectron Spectroscopy: A Handbook of He 584Å Spectra"; Wiley-Interscience, New York, 1970.
2. Dehmer, J. L.; Dill, D.; Parr, A. C. In "Photophysics and Photochemistry in the Vacuum Ultraviolet"; McGlynn, S.; Findley, G.; Huebner, R. , Eds.; Reidel Publishing Co., Holland, 1984, in press.
3. McKoy, V.; Carlson, T. A.; Lucchese, R. R. J. Phys. Chem., in press.
4. Langhoff, P. W. In "Electron-Molecule and Photon-Molecule Collisions"; Rescigno, T. N.; McKoy, B. V.; Schneider, B. Eds.; Plenum Press, New York, 1979, pp. 183-224.
5. Dill, D.; Dehmer, J. L. J. Chem. Phys. 1974, 61, 692.
6. Raseev, G.; Le Rouzo, H.; Lefebvre-Brion, H. J. Chem. Phys. 1980, 72, 5701.
7. Robb, W. D.; Collins, L.A. Phys. Rev. A 1980, 22, 2474.
8. Rescigno, T. N.; Orel, A. E. Phys. Rev. A 1982, 24, 1267.
9. Schneider, B. I.; Collins, L. A. Phys. Rev. A 1984, 29, 1695.
10. Lucchese, R. R.; Raseev, G.; McKoy, V. Phys. Rev.A 1982, 25, 2572.
11. Smith, M. E.; McKoy, V.; Lucchese, R. R. Phys. Rev. A 1984, 29, 1857.
12. Abramowitz, M. In "Handbook of Mathematical Functions"; Abramowitz, M.; Stegun, I. A. , Eds.; APPL. MATH. SERIES 55, National Bureau of Standards, Washington, D. C. 1964, p. 537.
13. See, for example, Miller, W. H. J. Chem. Phys. 1969, 50, 407.
14. See, for example, Dunning, T. H., Jr.; Hay, P. J. In "Methods of Electronic Structure Theory"; Schaefer III, H. F., Ed.; Plenum Press, New York, 1977, p. 1.
15. Lucchese, R. R.; McKoy, V. Phys. Rev. A 1982, 26, 1406;ibid.1992.
16. Lucchese, R. R.; McKoy, V. Phys. Rev. A 1983, 28, 1382.
17. Lynch, D.; Lee, M. T.; Lucchese, R. R.; McKoy, V. J. Chem. Phys. 1984, 80, 1907.
18. Lucchese, R. R.; Watson, D. K.; McKoy, V. Phys. Rev. A 1980, 22, 421.
19. Harris, F. E.; Michels, H. H. J. Chem. Phys. 1965, 43, S165.
20. Fliflet, A. W.: McKoy, V. Phys. Rev. A 1978, 18, 1048.
21. Morse, P. M; Feshbach, H. "Methods of Theoretical Physics, Part II"; McGraw-Hill, New York, 1953, p. 1271.
22. Newton, R. G. "Scattering Theory of Waves and Particles"; McGraw-Hill, New York, 1966, p. 431.
23. Lucchese, R. R.; Takatsuka, K.; McKoy, V. Phys. Repts., to be published.
24. Davenport, J. W. Phys. Rev. Lett. 1976, 36, 945; see also, Dehmer, J. L.; Dill, D.; pp. 225-265 of Reference 4.

25. Rescigno, T. N.; Bender, C. F.; McKoy, B. V.; Langhoff, P. W. J. Chem. Phys. 1978, 68, 970.
26. Dunning, T. H., Jr. J. Chem. Phys. 1971, 55, 3958.
27. Dehmer, J. L.; Dill, D.; Wallace, S. Phys. Rev. Lett. 1979, 43, 1005.
28. West, J. B.; Parr, A. C.; Cole, B. E.; Ederer, D. L.; Stockbauer, R.; Dehmer, J. L. J. Phys. B 1980, 13, L105.
29. Wight, G. R.; Van der Wiel, M. J.; Brion, C. E. J. Phys. B 1976, 9, 675.
30. Kay, R. B.; Van der Leeuw, Ph. E.; Van der Wiel, M. J. J. Phys. B 1977, 10, 2513.
31. Lindle, D. W. Ph. D. Thesis "Inner-Shell Photoemission from Atoms and Molecules Using Synchrotron Radiation"; University of California, Berkeley, 1983.
32. Rescigno, T. N.; Langhoff, P. W. Chem. Phys. Lett. 1977, 51, 65.
33. Lucchese, R. R.; McKoy, V. J. Phys. Chem. 1981, 85, 2166.
34. Lynch, D.; McKoy, V. Phys. Rev. A , to be published.
35. Kelly, H. P.; Carter, D. L.; Norum, B. E. Phys. Rev. A 1982, 25, 2052.
36. Amusia, M. Ya.; Ivanov, V. K.; Chernysheva, L. V. Phys. Lett. 1976, 59A, 191.
37. Dill, D.; Wallace, S.; Siegel, J.; Dehmer, J. L. Phys. Rev. Lett. 1979, 42, 411.
38. Padial, N.; Csanak, G.; McKoy, B. V.; Langhoff, P. W. J. Chem. Phys. 1978, 69, 2992.
39. Leal, E. P.; Lee, M. T.; Lucchese, R. R.; Lynch, D.; McKoy, V. J. Chem. Phys., to be published.
40. Swanson, R.; Dill, D.; Dehmer, J. J. Phys. B 1980, 13, L231.
41. Padial, N.; Csanak, G.; McKoy, B. V.; Langhoff, P. W. Phys. Rev. A 1981, 23, 218.
42. Brion, C. E.; Tan, K. H. Chem. Phys. 1978, 34, 141.
43. Carlson, T. A.; Krause, M. O.; Grimm, F. A.; Allen, J. D. Jr.; Mehaffy, D.; Keller, P. R.; Taylor, J. W. Phys. Rev. A 1981, 23, 3316.
44. Lynch, D.; Dixit, S.; McKoy, V. J. Chem. Phys., to be published.
45. Kreile, J.; Schweig, A.; Thiel, W. Chem. Phys. Lett. 1983, 100, 351.
46. Holland, D. M. P.; Parr, A. C.; Ederer, D. L.; West, J. B.; Dehmer, J. L. Int. J. Mass Spectr. and Ion Phys. 1983, 52, 195.
47. See, for example, Amusia, M. Ya; Cherepkov, N. A. Case Studies in Atomic Physics 1975, 5, 47.

RECEIVED June 11, 1984

7

Molecular Photoionization Resonances

A Theoretical Chemist's Perspective

P. W. LANGHOFF

Department of Chemistry, Indiana University, Bloomington, IN 47405

Progress is reported in theoretical studies of molecular partial-channel photoionization and electron-impact ionization cross sections, with particular reference to clarification of the origin and nature of resonances that appear in certain final-state molecular symmetries. These features are attributed to the presence of virtual valence orbitals, largely of σ* character, above ionization thresholds where they are merged into photoionization continua. A quantitative (Stieltjes) theory is described for computational studies of cross sections in both discrete and continuous spectral regions, and illustrative examples are reported. Three-dimensional graphical representations of Stieltjes orbitals in molecular point-group symmetry aid in identification and interpretation of resonance contributions as the σ* orbitals of Mulliken.

The intensities of molecular electronic transitions were first discussed comprehensively by Mulliken (1), who distinguished at the outset between Rydberg excitations, on the one hand, and intravalence excitations on the other (2). Although molecular Rydberg transitions are familiar from and closely related to atomic spectroscopy (3), molecular intravalence transitions have no direct atomic counterparts. Of particular interest in the present context are the $N \rightarrow V_{\sigma}(\sigma \rightarrow \sigma^*)$ and $N \rightarrow V_{\pi}(\pi \rightarrow \pi^*)$, or "normal to valence", transitions of Mulliken for conjugate pairs of bonding and antibonding orbitals (4). The transition dipole moments between these orbitals are proportional to the atomic separation between the bonded atoms, along which the polarization vector lies, producing very intense intramolecular "charge transfer" absorption spectra. Such transitions do not occur in atoms, since they would involve an electron jump between orbitals of the same shell, which is forbidden by Laporte's rule (3). In the present report, an account is given of such transitions, largely as they relate to molecular photoionization resonances.

0097-6156/84/0263-0113$07.50/0

The N→V transitions of Mulliken have proven somewhat elusive to experimental and theoretical identification and characterization. Difficulties associated with ultraviolet and vacuum ultraviolet (vuv) spectroscopy contribute to this circumstance (5). More generally, the propensity of $\sigma \rightarrow \sigma^*$ and $\pi \rightarrow \pi^*$ transitions to intermix, to mix strongly with Rydberg series, and to be positioned by such and other interactions at largely unexpected excitation energies, ranging over discrete, autoionizing, and even direct continuum intervals, has complicated their identification and interpretation. The situation is further complicated by the theorests inclination to treat discrete excitations and continuum photoionization processes as dissimilar phenomena requiring different computational approaches. In spite of these difficulties, considerable progress has been made recently in studies of the intensities of molecular electronic transitions, and in clarification of the role of Mulliken's orbitals in this connection.

Refinements in vuv spectroscopy (6), aided by the development of synchrotron radiation (7) and equivalent-photon electron-impact (8) tunable light sources, and closely related advances in photoelectron, fluorescence-yield, and electron-ion coincidence spectroscopy measurements of partial cross sections (9), have provided the complete spectral distributions of dipole intensities in many stable diatomic and polyatomic compounds. Of particular importance is the experimental separation of total absorption and ionization cross sections into underlying individual channel contributions over very broad ranges of incident photon energies.

Theoretical developments have provided the additional needed element for identification and clarification of Mulliken's intravalence transitions. It has proved possible employing so called Stieltjes methods to treat the discrete and continuum electronic states of molecules from the common perspective of a single theoretical framework, providing an effective merging of spectral and collision theories (10). In this way, wave functions are constructed as continuous functions of energy for both discrete and continuous intervals, providing a convenient basis for identification of intravalence transitions. In the present report, these methods are employed in identifying the intravalence transitions of Mulliken in selected diatomic and polyatomic molecules. In many cases, N→V transitions are seen to be positioned above ionization thresholds, corresponding to molecular photoionization resonances.

Brief descriptions are given in the following of needed aspects of cross sections, molecular orbitals, and of the more recently devised Stieltjes orbitals that have proved useful in spectral studies. Examples of the use of the Stieltjes formalism in identifying Mulliken valence orbitals in the cross sections of diatomic and polyatomic compounds are reported next. Also indicated are more general aspects of such intravalence transitions as they relate to electron-impact resonances in selected cases. The importance of dealing with both discrete and continuous spectral intervals on a common basis is emphasized throughout, particularly with reference to the clarification of the positionings of $\sigma \rightarrow \sigma^*$ and $\pi \rightarrow \pi^*$ excitations in molecular photoabsorption and ionization cross sections.

Experimental and Theoretical Considerations

Brief accounts are given in this section of basic aspects of photoabsorption and ionization (9), of the general natures of Rydberg and valence orbitals expected in diatomic and polyatomic molecules of interest (11,12), and of the Stieltjes orbitals that are conveniently employed in calculations and clarifications of the associated cross sections (10).

Absorption and Ionization. Cross sections for the photo excitation and ionization processes

$$M(i) + h\nu \rightarrow M(f) \quad (1a)$$

$$M(i) + h\nu \rightarrow M^{+}(f) + e^{-}(k) \quad (1b)$$

are directly proportional to the dipole transition strength or density (13)

$$f_{i\rightarrow f} = 2h\nu|\langle\phi_f|\hat{\mu}|\phi_i\rangle|^2 \quad (2a)$$

$$(df/d\varepsilon)_{i\rightarrow f} = 2h\nu|\langle k\phi_f|\hat{\mu}|\phi_i\rangle|^2 \quad (2b)$$

respectively. Here, M(i) is the target molecule in an initial vibronic state "i", M(f), $M^{+}(f)$ are corresponding final molecular and parent-ion states, respectively, produced by absorption of a photon $h\nu$, k is the outgoing electron kinetic energy, and $\hat{\mu}$ is the dipole moment operator. High-resolution spectral absorption studies are required to determine the states and intensities associated with the excitation process of Equation (1a), whereas, partial-channel studies involving tunable-source photoelectron spectral or fluorescence-yield measurements are needed to separate the generally degenerate ionization processes of Equation (1b). Corresponding conventional theoretical studies can provide the discrete and continuum states required in Equations (2). These are taken in the first approximation as one-electron excitation or ionization states, in which case ϕ_i refers to an occupied canonical Hartree-Fock orbital, and ϕ_f and $k\phi_f$ refer to one-electron discrete and continuum functions, respectively, determined in the field of the associated Koopmans hole state produced (14). More refined approximations to these wave functions must be employed in the presence of discrete-state configuration mixing or continuum-state channel coupling (15). In the present development Stieltjes methods described further below are used to study both discrete and continuum spectral regions on a common basis (10).

Rydberg and Valence Transitions. The conventional photoabsorption expressions of Equations (2) provide a general basis for studying the intensities of molecular electronic transitions provided suitable initial- and final-state wave functions are available. It is helpful in this connection, following Mulliken, to classify the unoccupied orbitals into which electrons may jump as Rydberg (R) or valence (V) types (4). Molecular Rydberg orbitals (R) are diffuse and are highly similar to their atomic counterparts, having principal quantum numbers larger than that of the outermost occupied valence shell of the relevant molecule. Their mean radii increase as n^2, their energies satisfy the familiar quantum defect formula, and their f numbers approach a Coulombic $1/n^3$ behavior with increasing n (3). Since interest attaches here on photoionization as well as absorption, it is useful to note that molecular Rydberg orbitals are closely related to their continuum Coulombic counterparts, associated with positive orbital energies (16), both sets joining together at the ionization threshold.

In contrast to Rydberg orbitals, unoccupied (virtual) valence orbitals (V) are relatively compact, being made up of valence-shell principal-quantum-number atomic orbitals. These arise from the splittings of valence-shell atomic orbitals that occur upon bond formation. As a consequence of their compact natures, excitations between occupied and unoccupied valence orbitals can have large transition moments. Since each molecule has only a small number of such orbitals, it is a simple matter to determine the V class of molecular orbitals for any compound from the atomic shells of the atoms comprising the molecule of interest, and from the total number of electrons that fill the molecular shells (4).

The unoccupied Rydberg and valence orbitals of stable compounds are generally regarded as relevant to the discrete excited states of molecules. However, they also give rise to autoionizing states when associated with the excitations of inner-valence orbitals, or with features below deeper (K,L,M,...) ionization edges in the x-ray portion of the spectrum that are associated with largely atomic inner shells. Furthermore, it must also be noted that virtual valence orbitals can be positioned *above* thresholds for valence-shell ionization, and, consequently, can contribute strongly to direct photoionization processes. This circumstance is perhaps not surprising when it is noted that valence orbitals have the approximate spatial dimensions of the molecular frame, and so their energies will depend in a sensitive manner upon bond lengths and the details of the molecular potential. Thus, in some cases V orbitals will appear well below Rydberg series, in some cases as intravalence interlopers, and in some cases above thresholds merged into photoionization continua. When it is further noted that in many cases two or more intravalence excitations are present in a single molecular symmetry, it is clear that the positioning of $N \rightarrow V$ contributions in molecular photoexcitation and ionization cross sections can be a sensitive and varied issue, making their spectroscopy very interesting and potentially complicated.

As an illustrative example of these circumstances, it is helpful to consider the valence-shell excitation spectrum of the HF molecule, in which case there is a single $4\sigma(\sigma^*)$ antibonding virtual valence orbital present. In Figure 1 is shown the relevant HF absorption cross section, obtained from recently reported electron energy-loss measurements (17). The state assignments shown in the figure are based on concomitant configuration-mixing calculations (17). These assignments indicate that the 4σ orbital contributes to the HF absorption spectrum through the largely single-configurational $1\pi \rightarrow 4\sigma$ excitation, and in the form of Rydberg-valence configurationally mixed states involving $3\sigma \rightarrow 4\sigma$ and $1\pi \rightarrow 3p\pi$ excitations. Further clarification of the natures of the states involved is provided by the associated potential energy curves depicted in Figure 2 (17). These show the $1\pi \rightarrow 4\sigma$ state as repulsive, in accordance with a bonding to antibonding transition, clarifying the width of the corresponding absorption band of Figure 1. The $3\sigma \rightarrow 4\sigma$ and $1\pi \rightarrow 3p\pi$ configuration mixing is seen from Figure 2 to correspond to an avoided crossing situation between the $(3\sigma \rightarrow 4\sigma)V_\sigma$ configurational state and the $1\pi \rightarrow 3p\pi$ Rydberg state. Since the V_σ configurational state, shown in the figure as a dashed curve, continues to rise in energy with decreasing nuclear separation, there is the suggestion that such $\sigma \rightarrow \sigma^*$ configurations can contribute to absorption processes above ionization thresholds, in which cases they correspond to continuum states. In order to pursue this circumstance, a method must be devised to treat discrete and continuum states on a common basis, to which attention is now addressed.

Stieltjes Orbitals. Conventional molecular orbitals, such as bound Hartree-Fock orbitals or the virtual excited-state orbitals obtained in the presence of static potential fields, have discrete negative energies and are unity normalized (11,12). By contrast, the positive energy scattering functions associated with ejected electrons have oscillatory spatial tails, and so are normalized differently, in the so-called Dirac delta-function sense, generally in the energy variable (16). The two sets, bound and continuum, are generally constructed using differing procedures, since different boundary conditions are involved. Little physical intuition of a type familiar to chemists has apparently been devised regarding scattering functions, such functions being largely the province of physicists. That is, the jargon and concepts of scattering theory, such as shape resonance, Feshbach resonance, channel coupling, and the like, are largely foreign to the language and ideas of bound-state studies. It is, in particular, not common practice to describe the continuum electronic states of molecules in terms of Rydberg and valence contributions, nor to clarify their spatial characteristics in terms of atomic compositions, largely as a consequence of the absence of theoretical procedures for constructing continuum states in such fashion.

Stieltjes orbitals provide a convenient basis for representation of both the discrete orbitals and continuum scattering functions of a molecule (10). They are spatially and spectrally localized eigenpackets that can be computed explicitly in Hilbert space for any bound-state or scattering energy. Consequently, they are particularly well suited for studies of molecular photoexcitation and ionization cross sections, and provide useful diagnostics of the underlying discrete or continuum Schrödinger states. They are written for a specified discrete or continuum energy ε in the explicit form (18)

$$\Phi_\varepsilon^{(n)} = \sum_{j=1}^{n} q_j(\varepsilon) v_j \quad , \tag{3}$$

where the v_j are an orthonormal set of (Lanczos) functions, and the $q_j(\varepsilon)$ are associated polynomials defined below. In the limit $n \rightarrow \infty$, the Stieltjes orbitals of Equation (3), when supplemented with appropriate normalization factors, converge to discrete or continuum Schrödinger states (19).

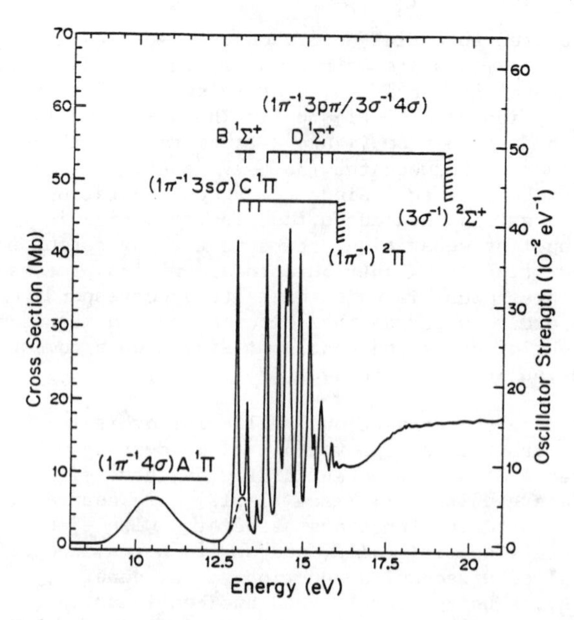

Figure 1. Valence-shell photoabosrption cross section in HF obtained from electron energy-loss measurements (17). Indicated assignments are based on configuration-mixing calculations, as discussed in the text.

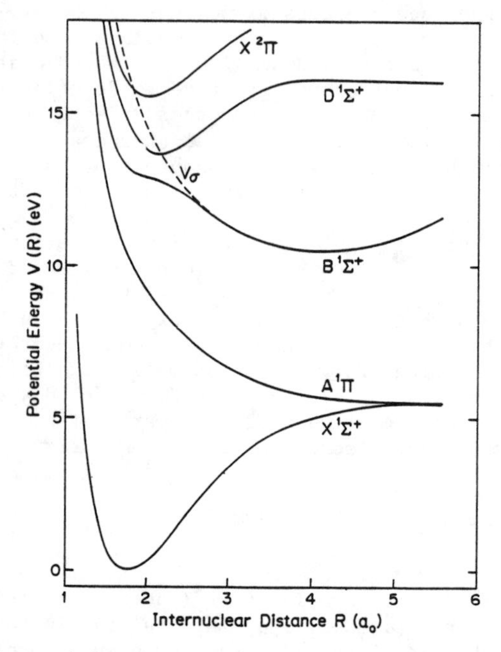

Figure 2. Potential energy curves for selected states in HF adopted from previously reported studies (17). The dashed curve provides an approximation to the $(3\sigma \rightarrow 4\sigma)V_{\sigma}$ configurational state energy.

The Lanczos basis is defined by the equations (10)

$$v_1 \equiv \Phi \quad , \quad \langle v_j | v_k \rangle = \delta_{jk} \tag{4a}$$

$$\beta_j v_{j+1} = (\hat{A}(\hat{H}) - \alpha_j) v_j - \beta_{j-1} v_{j-1} \tag{4b}$$

where Φ is a known test function, usually chosen to be $\hat{\mu}\phi_i$, and $\hat{A}(\hat{H})$ is an operator function of the Hamiltonian $\hat{H}$, usually chosen to be $(\hat{H}-\varepsilon_i)^{-1}$. Standard computational methods, involving basis sets, configurational state functions, molecular integrals and matric formation and algebra (11,12), are employed in obtaining the Lanczos functions v_j and matrix elements α_j, β_j of Equations (4) in body-frame point-group symmetry (19).

The polynomials $q_j(\varepsilon)$ are obtained from the equation

$$\beta_j q_{j+1}(\varepsilon) = (A(\varepsilon) - \alpha_j)\, q_j(\varepsilon) - \beta_{j-1} q_{j-1}(\varepsilon) \tag{5}$$

subject to the starting conditions $q_o=0$, $q_1=1$, where the recurrence coefficients α_j, β_j are the matrix elements of $\hat{A}(\hat{H})$ appearing in Equations (4). Since Equation (5) can be employed for any energy ε, it is clear that Stieltjes orbitals are obtained for the entire range of discrete and continuum excitations.

As a consequence of the orthonormality of the Lanczos basis, and similar properties of the polynomials $q_j(\varepsilon)$(10), convergence and other aspects of Stieltjes orbitals can be established (19). Of particular interest in the present context is the spectral composition of a Stieltjes orbital, which takes the form

$$\langle \Phi_\varepsilon | \Phi_{\varepsilon'}^{(n)} \rangle = \sum_{j=1}^{n} q_j(\varepsilon) q_j(\varepsilon') \tag{6}$$

where Φ_ε is a converged Schrödinger state. This quantity provides a measure of the spectral width of the Stieltjes orbital, provides lineshape functions for decomposition of the photo cross section (20) and gives the factors required for normalization of the Stieltjes orbitals in either the unity or Dirac delta-function sense (19).

The natures of the Lanczos functions and various properties of Stieltjes orbitals are best demonstrated by specific example. In Figure 3 are shown the first ten radial Lanczos functions for the 1s→kp spectrum of a hydrogen atom (18). These are obtained from solution of Equations (4) for j=1 to 10 employing the appropriate hydrogenic Hamiltonian in the operator $\hat{A}(\hat{H})$ and the 1s ground state orbital in the test function Φ. In this case, Equation (4b) can be solved for the v_j in terms of Laguerre functions with constant exponent, and the associated matrix elements α_j, β_j can be obtained analytically (18). It is seen from the figure that the Lanczos functions v_j extend to larger values of the radial coordinate with increasing value of j, suggesting that the spatial tails of discrete and continuum functions can be built up by the Stieltjes orbitals of Equation (3) with increasing order n.

In Figure 4 are shown Stieltjes orbitals for p waves in atomic hydrogen having energy ε=2 a.u. as a function of Stieltjes order n, in comparison with the corresponding exact regular Coulomb p wave (18). The former are constructed from Equation (3), incorporating the appropriate normalization factors, using the known Lanczos functions of Equations (4) and recursion to construct the polynomials of Equation (5). Since the Lanczos basis is regular at the origin, it is appropriate for convergence to regular p waves in atomic hydrogen. The value of ε=2 a.u. is chosen as a representative example, and is, in fact, an energy somewhat above the interval of strong ionization in hydrogen. Other energy functions are reported further below. Figure 4 illustrates the nature of convergence of Stieltjes orbitals for a fixed energy with increasing order, which is seen to occur first at small values of the radial coordinate. This is in accord with the spatial characteristics of the Lanczos functions of Figure 3, and suggests that low orders of Stieltjes orbitals can provide accurate representations of the local portions of Schrödinger states.

In Figure 5 are shown Stieltjes orbitals for p waves in atomic hydrogen for fixed order n=20 evaluated at a set of (Stieltjes) energies which are the characteristic eigenvalues of the $\hat{A}$ operator matrix in the corresponding 20-term Lanczos basis. These energies and associated Stieltjes orbitals define an orthonormal pseudospectrum that has optimal properties for representing the dipole cross section of atomic hydrogen (21). It is seen from Figure 5, which depicts only the 20th-order Stieltjes orbitals having positive energies up to $\sim$ 1 a.u., that convergence over the 10 a_o spatial interval shown has been adhieved. Since the dipole transition moments of Equations (1) extend only over the relatively small spatial interval ($\sim$ 10 a_o) in which the dipole test function Φ is nonzero (Figure 3), it is clear that the waves of Figure 5 can provide the necessary cross sectional information.

In Figure 6 are shown the lineshape functions of Equation (6), including appropriate normalization factors, corresponding to the Stieltjes orbitals of Figure 5. Also shown in the figure as a dashed curve is the dipole strength function or spectral density for atomic hydrogen. The 20th-order Stieltjes energies are seen to be distributed in accordance with the corresponding dipole strength, a higher density of energies appearing at threshold where the strength is largest. As a consequence, the Stieltjes orbitals in the continuum near threshold have smaller spectral widths than do those at higher energies, in which latter region the density of Stieltjes energies is lower. Since the characteristic Stieltjes energies are the generalized gaussian quadrature points of the corresponding dipole density, they are sure to appear in the spectrum where the cross section is large. Consequently, Stieltjes orbitals are expected to be particularly useful in diagnostic studies of resonance features in molecular photoionization cross sections, to which attention is now directed.

Molecular Photoionization Resonances

Representative examples of resonances in molecular partial-channel photoionization cross sections are described and discussed in this section. Comparisons are made between calculations, largely in static-exchange approximation (14,15), and measurements in selected cases, and Stieltjes orbitals are presented in three-dimensional

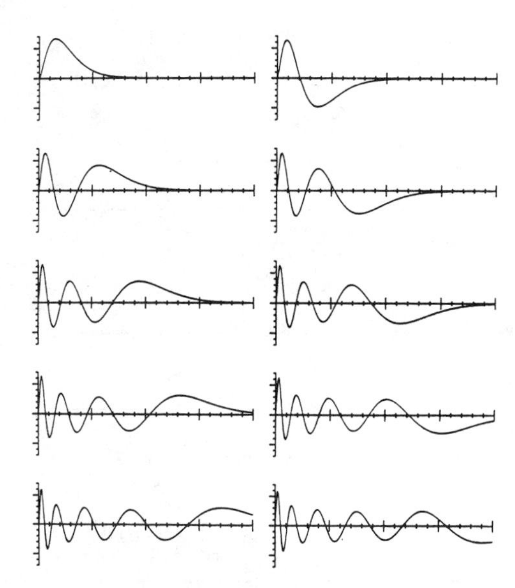

Figure 3. Radial Lanczos functions of Equations (4) for 1s→kp excitation and ionization of atomic hydrogen (18). The abscissa spans 24 a_o, the ordinate 0.07 a.u.

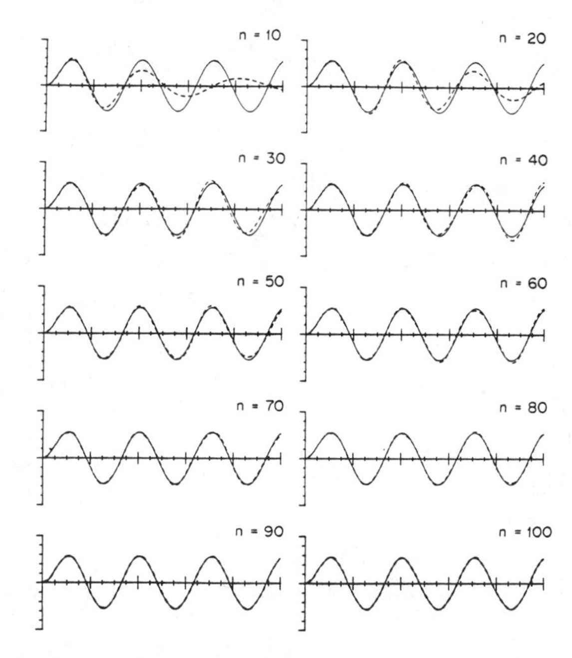

Figure 4. Stieltjes orbitals of Equation (3), including appropriate normalization factors, for ε=2 a.u. regular p waves in atomic hydrogen, in comparison with corresponding exact values (18). The abscissa spans 10 a_o, the ordinate 1 a.u.

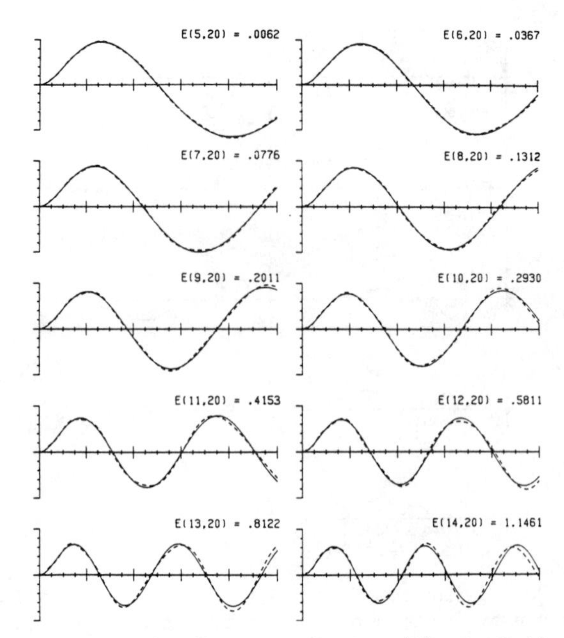

Figure 5. Stieltjes orbitals of Equation (3), including appropriate normalization factors, for regular p waves in atomic hydrogen evaluated at indicated Stieltjes energies, in comparison with corresponding exact values (18). Ordinate and abscissa as in Figure 4.

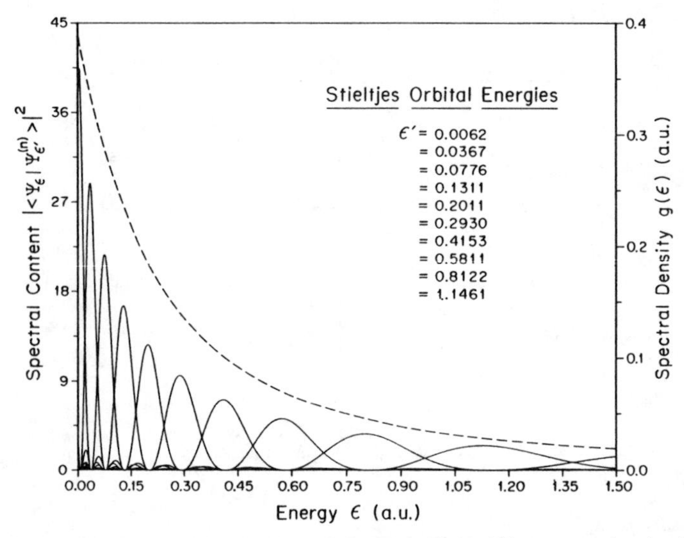

Figure 6. Lineshape functions of Equation (6) squared, including appropriate normalization factors, for the 20th-order Stieltjes orbitals of Figure 5. The dipole density for atomic hydrogen is shown as a dashed curve (18).

graphical forms as diagnostics of the σ^* natures of resonance wave functions.

<u>N_2, NO, and N_2O Molecules</u>. Photoexcitation and ionization processes in small nitrogen containing compounds have been subjects of considerable experimental and theoretical study (9). The $(3\sigma_g^{-1})X^2\Sigma_g^+$ channel photoionization cross section in N_2, in particular, has been the subject of considerable interest, due in part to the presence of a $\sigma\rightarrow\sigma^*$ shape resonance in the $3\sigma_g\rightarrow k\sigma_u$ polarization channel (22-25). Static-exchange calculations indicate that the $k\sigma_u$ functions in this compound are largely of Rydberg character in the threshold region, whereas $\sim$ 10-12 eV above threshold they acquire a distinct compact σ^* character (25). As a consequence, the corresponding photoionization cross section exhibits a pronounced maximum at these energies (22). The $\sigma\rightarrow\sigma^*$ transition in N_2 has gone unassigned for many years, since the σ^* orbital appears localized largely in the $k\sigma_u$ photoionization continuum, and contributes to the discrete spectral region only through V_σ-V_π mixing in the $b'^1\Sigma_u^+$ state. When coupling between the $3\sigma_g$ and $1\pi_u$ channels is neglected, the $1\pi_u\rightarrow k\pi_g$ cross section is found to include a spurious $\pi\rightarrow\pi^*$ contribution $\sim$ 3-4 eV above threshold (23). The effects of V_π-V_σ mixing lowers this peak into the discrete spectral region, where it correctly corresponds to the strong $^1\Sigma_g^+\rightarrow b'^1\Sigma_g^+$ resonance (24). It is of interest to note that the σ^* orbital in N_2 also contributes to the K-edge cross section (26), and to inner-valence channels associated with $2\sigma_g$ ionization (27). Since these aspects of photoionization of N_2 are so well documented in the literature (22-27), they are not reported explicitly here.

The open-shell multiplet structure of the NO^+ ion makes it a challenging system to study theoretically and experimentally (9). As in N_2, σ^* resonance features are expected in certain of the partial cross sections of NO. It is helpful to recall the spatial characteristics of occupied and virtual valence orbitals in NO as an aid in interpreting the partial cross sections and Stieltjes-orbital representations of the associated continuum functions in this case. In Figure 7 are shown the 1π, $2\pi(\pi^*)$, 4σ, 5σ, and $6\sigma(\sigma^*)$ canonical Hartree-Fock orbitals of NO constructed in a minimal cartesian gaussian basis set. These results provide satisfactory representations of the general spatial characteristics of the indicated orbitals, although the associated orbitals and energies, of course, do not correspond to Hartree-Fock limits. It should be noted that the unoccupied $6\sigma(\sigma^*)$ orbital appears at a positive energy, and on this basis alone can be expected to contribute to partial photoionization cross sections of $k\sigma$ final-state symmetry. Since the $2\pi(\pi^*)$ orbital is singly occupied in this case, it is bound and can be expected to contribute to the absorption cross section as a final state.

In Figure 8 are shown static-exchange calculations of the $(5\sigma^{-1})b^3\Pi$ cross section in NO, in comparison with corresponding measured values (28). A very pronounced resonance feature is evidently present in the $5\sigma\rightarrow k\sigma$ polarization component, with the maximum cross sectional value appearing at $\sim$ 8 eV above threshold, similar to the situation in N_2 (22-25). Although not reported here, a similar resonance feature is present in the measured and calculated $(5\sigma^{-1})A^1\Pi$ cross section, which can also be attributed to the $5\sigma\rightarrow k\sigma$ polarization component (28). In this case, however, the maximum of

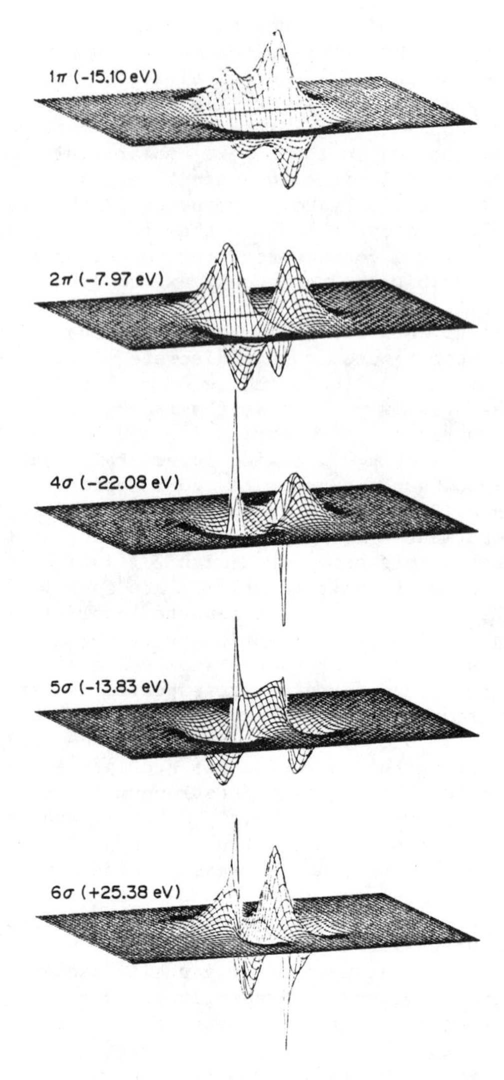

Figure 7. Canonical Hartree-Fock valence orbitals of NO constructed in a minimal cartesian gaussian basis set (28). Values shown refer to a 6 by 16 a_o plane, with N on the left, O on the right.

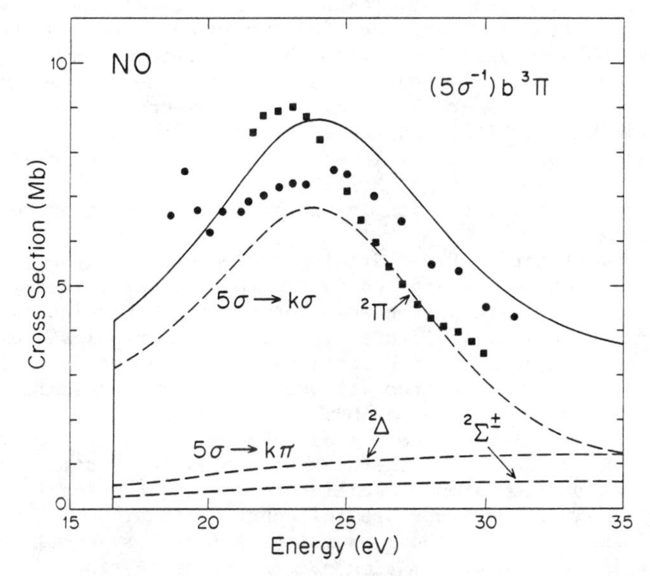

Figure 8. Partial-channel $(5\sigma^{-1})b\,^3\Pi$ photoionization cross section in NO obtained from static-exchange calculations and synchrotron-radiation measurements (28).

the cross section appears ∿ 4 eV above threshold, a multiplet-specific effect that can be attributed to the more attractive nature of the $(5\sigma^{-1})A^1\Pi$ potential relative to that for the $(5\sigma^{-1})b^3\Pi$ ionic state (28).

Three-dimensional graphical representations of 5th-order Stieltjes orbitals appropriate for $(5\sigma\rightarrow k\sigma)b^3\Pi$ ionization of NO are shown in Figure 9 as a diagnostic of the cross section of Figure 8. The orbitals shown refer to values in a plane of dimensions 6 by 16 a_o that includes the internuclear line, with the nitrogen nucleus at the left and the oxygen nucleus on the right. Stieltjes energies are shown to the left of the orbitals, each of which is unity normalized. Evidently, the 5th-order Stieltjes orbitals immediately above and below the ionization threshold are Rydberg in character, whereas the orbital at ∿ 25 eV associated with the resonance maximum is compact and of σ^* character. Indeed, comparison with the $6\sigma(\sigma^*)$ orbital of Figure 7 reveals a remarkable similarity between the resonance continuum Stieltjes orbital and the standard minimal basis set σ^* virtual valence orbital of NO. These results would seem to establish conclusively the presence of Mulliken's σ^* orbital in the continuum of NO, and to clarify the origin of the corresponding photoionization shape resonance.

A somewhat more refined Stieltjes diagnostic of the $(5\sigma\rightarrow k\sigma)b^3\Pi$ cross section of NO is reported in Figure 10. Here, 10th-order Stieltjes orbitals are presented at a series of energies that spans the resonance maximum of Figure 7, employing a graphical layout similar to that of Figure 9. A gradual variation in the spatial characteristic of the orbitals from diffuse Rydberg at threshold to compact valence near the resonance maximum is evident in the figure. Note that the contribution of the σ^* resonance to the Stieltjes orbitals is noticeably decreased at the highest energy, with still higher-energy waves taking on more oscillatory character which is not associated with virtual valence orbital contributions.

Studies of valence-shell photoionization cross sections in N_2O in static-exchange approximation aid in identification of the presence of the $8\sigma(\sigma^*)$, $9\sigma(\sigma^*)$, and $3\pi(\pi^*)$ virtual valence orbitals in the dipole spectrum of this compound (29). The σ^* orbitals are found positioned in the ionization continuum in a manner similar to that in CO_2 (30), with the $8\sigma(\sigma^*)$ orbital immediately above and the $9\sigma(\sigma^*)$ orbital ∿ 15 eV above threshold. Consequently, the $6\sigma\rightarrow k\sigma$ and $7\sigma\rightarrow k\sigma$ polarization components of the corresponding partial cross sections exhibit $8\sigma(\sigma^*)$ resonance maxima immediately above their ionization thresholds, in excellent accord with measured values (29). By contrast, the contribution of the $9\sigma(\sigma^*)$ orbital to these cross sections is found to be significantly smaller than that of the $8\sigma(\sigma^*)$ orbital on basis of both calculated and measured values (29). Construction of the generalized oscillator strength densities for $6\sigma\rightarrow k\sigma$ and $7\sigma\rightarrow k\sigma$ channels at various momentum transfer values helps to characterize the higher-lying $9\sigma(\sigma^*)$ orbital in these N_2O spectra. It is seen that the threshold $8\sigma(\sigma^*)$ resonance behaves like a dipole allowed state with increasing momentum transfer, its contribution to the cross sections decreasing rapidly, whereas the higher-lying $9\sigma(\sigma^*)$ resonance orbital behaves like a forbidden state in each channel, its contributions increasing with increasing momentum transfer. Finally, it is of interest to note that both static-exchange calculations and available experimental values suggest the presence of a resonance $3\pi(\pi^*)$

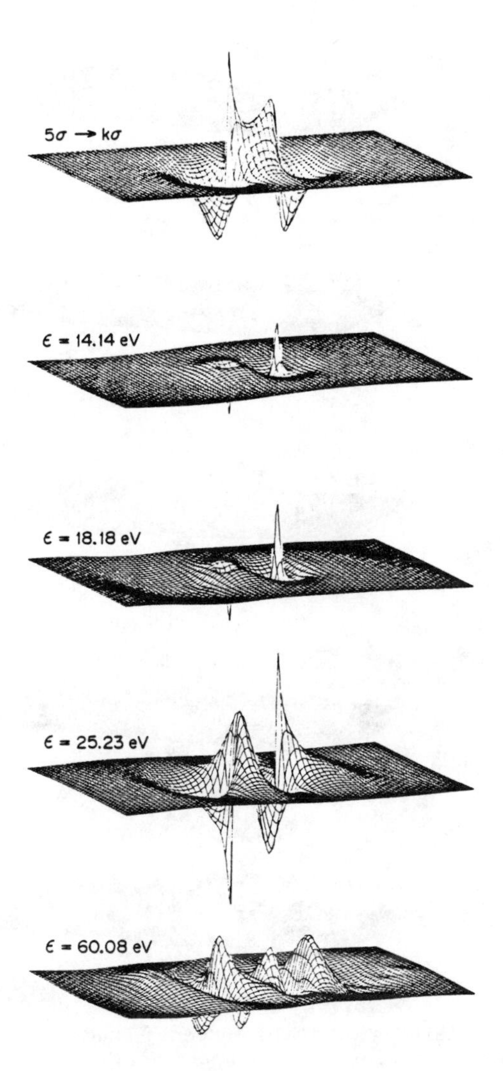

Figure 9. Occupied canonical Hartree-Fock and 5th-order Stieltjes orbitals for $(5\sigma \rightarrow k\sigma)b^3\Pi$ ionization of NO (28). Layout as in Figure 7.

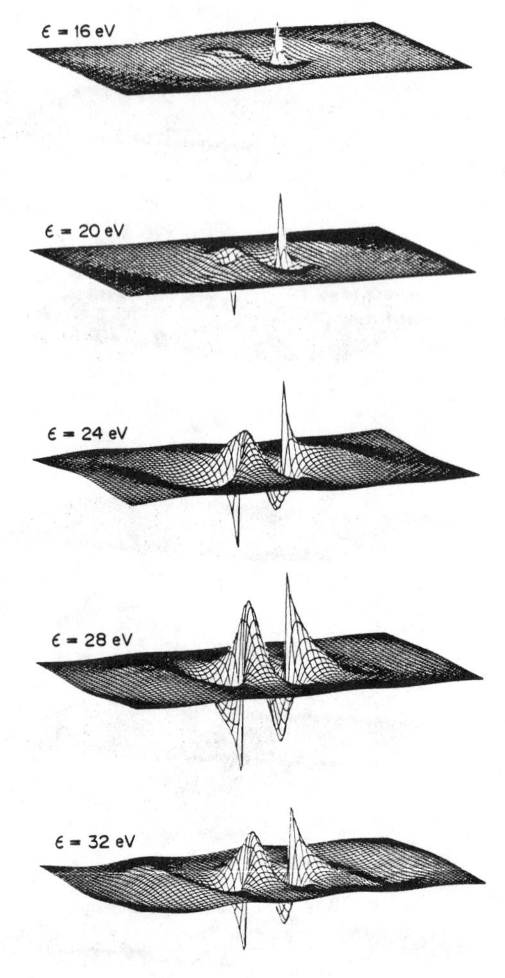

Figure 10. Stieltjes orbitals in tenth-order for $(5\sigma \rightarrow k\sigma)b\,^3\Pi$ ionization of NO in static-exchange approximation (28). Layout as in Figure 7.

contribution in the $2\pi \rightarrow k\pi$ polarization component of the $2\pi^{-1}$ cross section of N_2O (29). By contrast, the corresponding 1π channel shows no such photoionization resonance feature, the $3\pi(\pi^*)$ orbital appearing below threshold in this case. Since a π^* photoionization resonance has not been observed previously, further refined studies of the $2\pi^{-1}$ partial cross section in N_2O would be of interest.

CO, H_2CO, and CO_2 Molecules. Studies of photoionization of CO, H_2CO, and CO_2 molecules provide an opportunity to observe possible similarities of shape resonances in chromophoric C-O bonds (14). The $\sigma \rightarrow k\sigma$ polarization channels of CO photoionization cross sections are known to include strong features $\sim$ 10 eV above thresholds that are associated with shape resonances in scattering-theory parlance (31). Similar structures are found in $a_1 \rightarrow ka_1$ channels in H_2CO (32), and in $\sigma_g \rightarrow k\sigma_u$, $\sigma_u \rightarrow k\sigma_g$ channels in CO_2 (30). By contrast, $\pi \rightarrow k\pi$ channels in CO, and related channels in H_2CO and CO_2, exhibit broad structureless ionization cross sections, but are known to include strong $\pi \rightarrow \pi^*$ discrete resonances.

The $5\sigma \rightarrow k\sigma$ and $5a_1 \rightarrow ka_1$ resonance cross sections in static-exchange approximation shown in Figure 11 are found to be largely insensitive to the effects of configuration mixing (20). By contrast, $V_\pi - V_\sigma$ configuration mixing shifts the $\pi \rightarrow \pi^*$ resonance features in the $1\pi \rightarrow k\pi$ and $1b_1 \rightarrow kb_1$ cross sections into discrete spectral regions, resulting in nonresonant cross sections in these cases, as indicated in the figure (20). The situation in these two cases is largely similar to that in N_2, in which $V_\pi - V_\sigma$ configuration mixing is required to correctly position the π^* orbital in the spectrum. Helpful diagnostic of σ^* contributions to the $5\sigma \rightarrow k\sigma$ and $5a_1 \rightarrow ka_1$ cross sections of Figure 11 are obtained from single-channel static exchange calculations of Stieltjes orbitals.

In Figures 12 and 13 are shown 5th-order Stieltjes orbitals, and the relevant occupied orbitals, for the $5\sigma \rightarrow k\sigma$ and $5a_1 \rightarrow ka_1$ channels of Figure 11, presented as three-dimensional graphical representations in appropriate planes. The Stieltjes orbital energies obtained are seen to span the spectrum in each case in a sensible manner. Evidently, the two channels include highly similar σ^* orbitals $\sim$ 10 eV above threshold in each case. These are identified as $6\sigma(\sigma^*)$ and $6a_1(\sigma^*)$ virtual orbitals, respectively, by comparison with minimal basis-set calculations. This positioning of Mulliken's σ^* orbital is further confirmed in these cases by the absence of intravalence transitions in the calculated discrete excitation series. Similar features are also found in other ionization channels in these compounds in final-state $k\sigma$ and ka_1 symmetries (31,32). Moreover, a second $7a_1(\sigma^*)$ orbital and a valence $3b_2$ orbital in H_2CO, the former localized largely in the C-H bond, are found to contribute to certain partial channels in this compound (32).

In Figure 14 are shown the $5a_1 \rightarrow ka_1$ cross section of H_2CO in static-exchange approximation, and the lineshape functions of Equation 6 squared, including appropriate normalization factors, which provide spectral compositions of 10th-order Stieltjes orbitals. It is seen that the individual lineshapes include sidebands at higher and lower energies than their central maxima. These are separated by zero values at the energies of the maxima of the lineshapes associated with other Stieltjes orbitals. The sidebands are generally

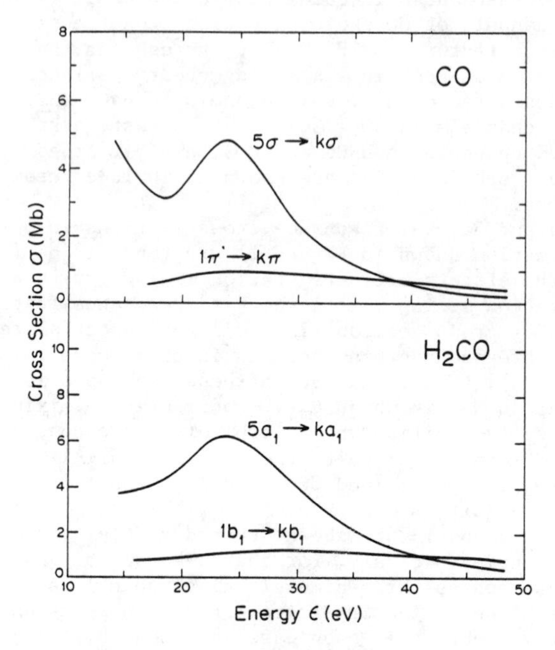

Figure 11. Polarization components of partial-channel photoionization cross sections of CO and H_2CO molecules as indicated, obtained from single- and coupled-channel static-exchange calculations (20).

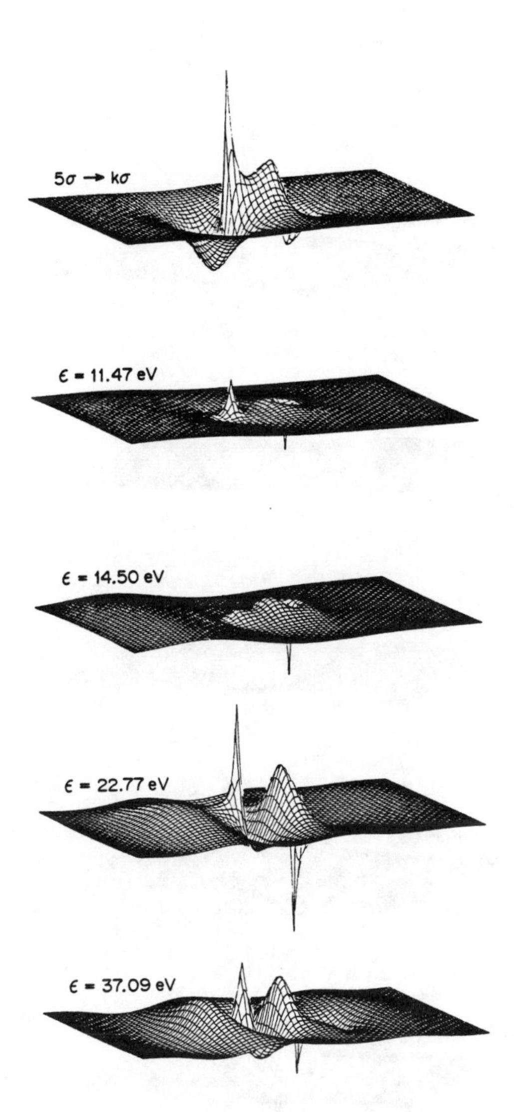

Figure 12. Occupied canonical Hartree-Fock and 5th-order Stieltjes orbitals in static-exchange approximation for $5\sigma \rightarrow k\sigma$ ionization of CO (20). Layout as in Figure 7, with the carbon nucleus on the left, the oxygen on the right.

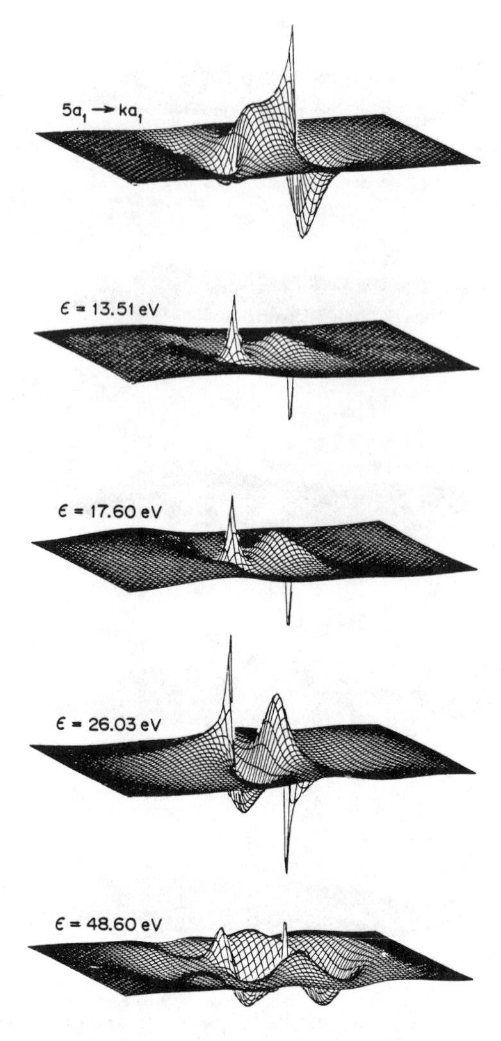

Figure 13. Occupied canonical Hartree-Fock and 5th-order Stieltjes orbitals in static-exchange approximation for $5a_1 \rightarrow ka_1$ ionization of H_2CO (20). Results shown refer to orbital values in a molecular plane of dimension 6 by 16 a_o, with the hydrogen nucleii on the left of the figure.

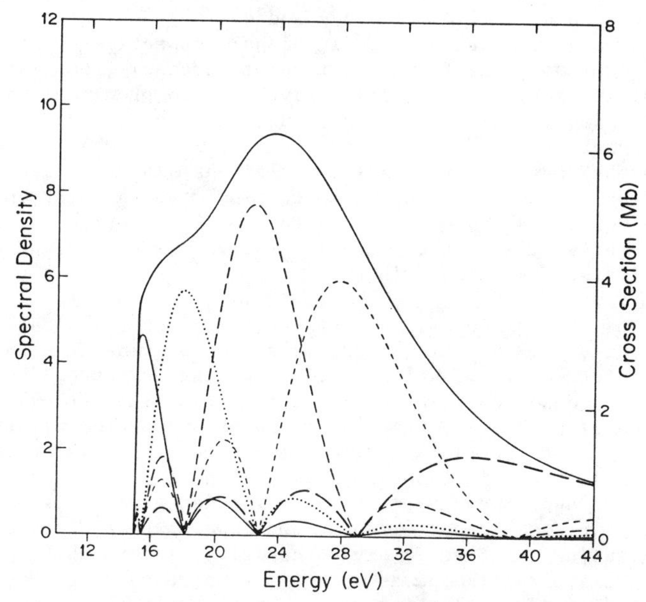

Figure 14. Spectral lineshapes of Equation (6) squared, include appropriate normalization factors, for 10th-order Stieltjes-orbital representation of the $5a_1 \rightarrow ka_1$ cross section of H_2CO (20).

required to allow for cancellations of the spatial tails of the scattering functions that comprise the square-integrable Stieltjes orbitals, and the zero points arise from the orthogonality of the Stieltjes orbitals evaluated at characteristic energies. Figure 14 illustrates the sense in which normalizable Stieltjes orbitals provide a representation of the indicated photoionization cross section of H_2CO, and shows the spectral compositions of the individual Stieltjes orbitals.

The contributions of σ^* and π^* orbitals to the cross sections of CO_2 are found to be largely similar to those in N_2O described above (30). In this case there are $5\sigma_g(\sigma^*)$ and $4\sigma_u(\sigma^*)$ orbitals present which are largely even and odd combinations, respectively, of the C-O bond σ^* orbitals, the former just above threshold, the latter higher lying in the spectrum. These give rise to prominent features in calculated and measured cross sections, although the degree of spectral concentration of the $4\sigma_u(\sigma^*)$ orbital is somewhat controversial at present, earlier Stieltjes calculations giving somewhat less localized results than more recent scattering wave calculations. As in CO and H_2CO, the $\pi \rightarrow \pi^*$ contribution to the CO_2 spectrum is localized below thresholds, although configuration mixing is required to obtain a satisfactory positioning.

H_2S, H_2CS, and CS_2 Molecules. Recent studies of photoionization cross sections of H_2S, H_2CS, and CS_2 molecules provide an opportunity for intercomparisons with results for the corresponding oxygen containing compounds. The substitution of sulfur for oxygen atoms has two generally important consequences on photoexcitation and ionization cross sections. First, the greater spatial extent of 3s and 3p atomic orbitals tends to make the photo cross sections more peaked at lower photon energies, the inner nodes giving rise to Cooper minima in some cases. Second, the lower orbital excitation energies in sulfur relative to oxygen can give rise to shake up states, or a failure of Koopmans approximation, even in the outer-valence region of ionization.

The $2b_1$, $5a_1$, and $2b_2$ orbital cross sections of H_2S shown in Figure 15 are seen to decrease rapidly with photon energy relative to the H_2O $1b_1$, $3a_1$ and $1b_2$ counterparts (33). By contrast, the $6a_1(\gamma^*)$ and $4a_1(\gamma^*)$ orbitals in these compounds are localized in discrete spectral regions as bound states.

The calculated cross sections of H_2CS and CS_2 further illustrate this trend, although in these cases final-state configuration mixing effects become more significant. Of particular interest in both compounds are the presence in the outer valence cross sections of strong features that arise from valence shell 3d orbitals, a behavior that is absent from the corresponding second-row compounds.

Final Remarks

The simple compounds discussed above illustrate some aspects of the contributions of σ^* and π^* orbitals to photo cross sections. Although σ^* orbitals can appear largely in continuous regions in certain cases, they can also be spectrally localized as bound states. The critical parameter is the bond length across which the $\sigma \rightarrow \sigma^*$ transition occurs, as is illustrated by the simple diatomic compounds

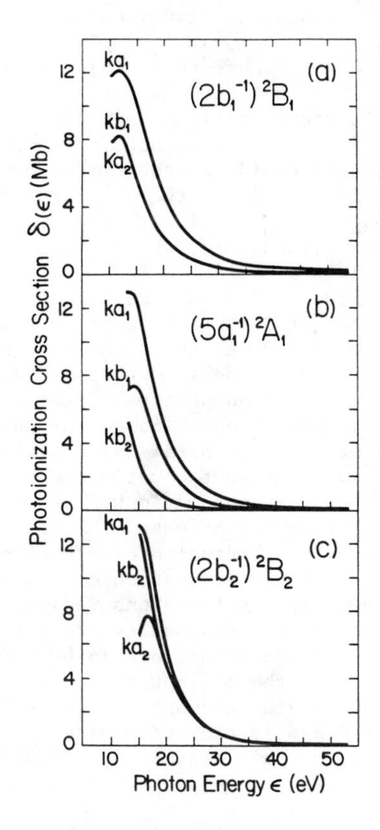

Figure 15. Partial-channel photoionization cross sections for $2b_1$, $5a_1$ and $2b_2$ orbitals of H_2S obtained in static-exchange approximation.

N_2, O_2, and F_2 (25,34,35,36). In the very short bond length triple-pair-bond N_2 molecule the $3\sigma_u(\sigma^*)$ orbital appears $\sim$ 10 eV above threshold (25), in the double-bond O_2 molecule the corresponding orbital is right at threshold (34), whereas in the single-bond F_2 molecule it is a bound orbital (35,36). This behavior is in accordance with the variation of energy levels in a square well, $E_n \sim (n/\ell)^2$, a model that is quantitatively as well as qualitatively relevant to these compounds.

Of course, further refinements of this orbital picture are required in most cases, as indicated in the foregoing discussions. There is strong V_σ-V_π configuration mixing in N_2, which plays a role in positioning both π^* and σ^* resonances. Because the σ^* orbital in O_2 falls so close to threshold, whether it is bound or not depends in a sensitive way on the details of the molecular potential. Recent studies show the $(3\sigma_u{}^{-1})b^4\Sigma_u^-$ multiplet ionization channel to include a σ^* contribution above threshold, whereas the corresponding $B^2\Sigma_u^-$ channel does not exhibit a resonance feature. Similarly, the O_2 K-edge cross section is sensitive to localized-hole and valence-shell-relaxation effects, resulting in a σ^* contribution just below threshold in this case. Moreover, although the σ^* orbital is bound in F_2, the V_σ configuration mixes strongly with a Rydberg series, spreading the $\sigma \rightarrow \sigma^*$ intensity over a broad spectral interval (36).

Although the interpretation of molecular photoionization shape resonances in terms of the σ^* orbitals of Mulliken described here is an elementary one, and requires the refinements indicated above, it has proved useful in various connections. Specifically, the number and final-state symmetry types of potential resonances in molecules are predicted simply by the virtual orbitals obtained from standard valence-basis considerations. The absence of strong intravalence transitions in discrete spectral regions indicates the possible appearance of photoionization resonances above threshold. Appropriate bond lengths play a central role in such discrete or continuum positionings, modified by the effects of configuration mixing. Stieltjes methods can provide a quantitative realization of these concepts, allowing the application of basis-function methods to the continuum and bound electronic states of molecules on a common basis.

Acknowledgments

It is a pleasure to acknowledge the cooperation of various coworkers, particularly M.R. Hermann, K. Greenwald, G.H.F. Diercksen, and S.R. Langhoff. Financial support provided by the Chemistry and International Programs Divisions of the National Science Foundation is gratefully acknowledged.

Literature Cited

1. Mulliken, R.S. J. Chem. Phys. 1939, 7, 14.
2. Mulliken, R.S. J. Chem. Phys. 1939, 7, 20.
3. Condon, E.U.; Odabasi, H. "Atomic Structure"; Cambridge University Press: Cambridge, 1980.
4. Mulliken, R.S. Accounts Chem. Res. 1976, 9, 7.
5. Samson, J.A.R. "Techniques of Vacuum Ultraviolet Spectroscopy"; Wiley: New York, 1967.

6. Samson, J.A.R. in "Handbuch der Physik"; Flugge, S.; Ed., Springer: Berlin, 1982; Vol. 31, pp. 187-237.
7. Koch, E.E.; Sonntag, B.F. in "Synchrotron Radiation"; Kunz, C., Ed.; Springer: Berlin, 1979.
8. Brion, C.E.; Hamnett, A. Adv. Chem. Phys. 1981, 42, 2.
9. Berkowitz, J. "Photoabsorption, Photoionization, and Photoelectron Spectroscopy"; Academic: New York, 1979.
10. Langhoff, P.W. in "Methods in Computational Molecular Physics"; Diercksen, G.H.F.; Wilson, S., Ed.; Reidel: Dordrecht, 1983, pp. 299-333.
11. Mulliken, R.S.; Ermler, W.C. "Diatomic Molecules"; Academic: New York, 1977.
12. Mulliken, R.S.; Ermler, W.C. "Polyatomic Molecules"; Academic: New York, 1981.
13. Fano, U.; Cooper, J.W. Rev. Mod. Phys. 1968, 40, 441.
14. Langhoff, P.W.; Padial, N.; Csanak, G.; Rescigno, T.N.; McKoy, B.V. Int. J. Quantum Chem. 1980, S14, 285.
15. Langhoff, P.W. in "Electron-Atom and Electron-Molecule Collisions"; Hinze, J., Ed.; Plenum: New York, 1983, pp. 297-314.
16. Bethe, H.A.; Salpeter, E.E. "Quantum Mechanics of One and Two-Electron Atoms"; Academic: New York, 1957.
17. Hitchcock, A.P.; Williams, G.R.J.; Brion, C.E.; Langhoff, P.W. Chem. Phys. 1984,
18. Hermann, M.R.; Langhoff, P.W. Phys. Rev. A 1983, 28, 1957.
19. Hermann, M.R.; Langhoff, P.W. J. Math. Phys. 1983, 24, 541.
20. Hermann, M.R.; Diercksen, G.H.F.; Fatyga, B.W.; Langhoff, P.W. Int. J. Quantum Chem. 1984,
21. Langhoff, P.W. Chem. Phys. Lett. 1973, 22, 60.
22. Rescigno, T.N.; Bender, C.F.; McKoy, B.V.; Langhoff, P.W. J. Chem. Phys. 1978, 68, 970.
23. Rescigno, T.N.; Gerwer, A.; McKoy, B.V.; Langhoff, P.W. Chem. Phys. Letters 1979, 66, 116.
24. Williams, G.R.J.; Langhoff, P.W. Chem. Phys. Letters 1981, 78, 21.
25. Hermann, M.R.; Langhoff, P.W. Chem. Phys. Letters 1981, 82, 242.
26. Rescigno, T.N.; Langhoff, P.W. Chem. Phys. Letters 1977, 51, 65.
27. Langhoff, P.W.; Langhoff, S.R.; Rescigno, T.N.; Schirmer, J.; Cederbaum, L.S.; Domcke, W.; von Niesses, W. Chem. Phys. 1981, 58, 71.
28. Hermann, M.R.; Langhoff, S.R.; Langhoff, P.W. Chem. Phys. Letters 1984,
29. Hermann, M.R. Ph.D. Thesis, Indiana University, Bloomington, 1984.
30. Padial, N.; Csanak, G.; McKoy, B.V.; Langhoff, P.W. Phys. Rev. 1981, A23, 218.
31. Padial, N.; Csanak, G.; McKoy, B.V.; Langhoff, P.W. J. Chem. Phys. 1978, 69, 2992.
32. Langhoff, P.W.; Orel, A.E.; Rescigno, T.N.; McKoy, B.V. J. Chem. Phys. 1978, 69, 4689.
33. Diercksen, G.H.F.; Kraemer, W.P.; Rescigno, T.N.; Bender, C.F.; McKoy, B.V.; Langhoff, S.R.; Langhoff, P.W. J. Chem. Phys. 1982, 76, 1043.

34. Langhoff, P.W.; Gerwer, A.; Asaso, C.; McKoy, B.V. Int. J. Quantum Chem. 1979, S13, 645.
35. Orel, A.E.; Rescigno, T.N.; McKoy, B.V.; Langhoff, P.W. J. Chem. Phys. 1980, 72, 1265.
36. Hitchcock, A.P.; Brion, C.E.; Williams, G.R.J.; Langhoff, P.W. Chem. Phys. 1982, 66, 435.

RECEIVED June 11, 1984

8

Shape Resonances in Molecular Fields

J. L. DEHMER

Argonne National Laboratory, Argonne, IL 60439

A shape resonance is a quasibound state in which a particle is temporarily trapped by a potential barrier (i.e., the "shape" of the potential), through which it may eventually tunnel and escape. This simple mechanism plays a prominent role in a variety of excitation processes in molecules, ranging from vibrational excitation by slow electrons to ionization of deep core levels by X-rays. Moreover, their localized nature makes shape resonances a unifying link between otherwise dissimilar circumstances. One example is the close connection between shape resonances in electron-molecule scattering and in molecular photoionization. Another is the frequent persistence of free-molecule shape resonant behavior upon adsorption on a surface or condensation into a molecular solid. The main focus of this article is a discussion of the basic properties of shape resonances in molecular fields, illustrated by the more transparent examples studied over the last ten years. Other aspects to be discussed are vibrational effects of shape resonances, connections between shape resonances in different physical settings, and examples of shape resonant behavior in more complex cases, which form current challenges in this field.

The last decade has witnessed remarkable progress in characterizing dynamical aspects of molecular photoionization (1-2) and electron-molecule scattering (2-3) processes. The general challenge is to gain physical insight into the processes occurring during the excitation, evolution, and decay of the excited molecular complex. Of particular interest in this context are the uniquely molecular aspects resulting from the anisotropy of the molecular field and from the interplay among rovibronic modes. Throughout this work, special attention is invariably drawn to resonant processes, in

0097-6156/84/0263-0139$07.50/0

which the excited system is temporarily trapped in a quasibound resonant state. Such processes tend to amplify the more subtle dynamics of excited molecular states and are often displayed prominently against non-resonant behavior in various physical observables.

One very vigorous stream of work has involved shape resonances in molecular systems. These resonances are quasibound states in which a particle is temporarily trapped by a potential barrier, through which it may eventually tunnel and escape. In molecular fields, such states can result from so-called "centrifugal barriers," which block the motion of otherwise free electrons in certain directions, trapping them in a region of space with molecular dimensions. Over the past ten years, this basic resonance mechanism has been found to play a prominent role in a variety of processes in molecular physics, and now takes its place alongside autoionizing and Feshbach resonances as central themes in the study of molecular photoionization and electron-molecule scattering processes. As discussed more fully in later sections, the expanding interest in shape resonant phenomena has arisen from a few key factors:

First, shape resonance effects are being identified in the spectra of a growing and diverse collection of molecules and now appear to be active somewhere in the observable properties of most small (nonhydride) molecules. Examples of processes which may exhibit shape resonant effects are X-ray and VUV absorption spectra, photoelectron branching ratios and photoelectron angular distributions (including vibrationally resolved), Auger electron angular distributions, elastic electron scattering, vibrational excitation by electron impact, and so on. Thus concepts and techniques developed in this connection can be used extensively in molecular physics.

Second, being quasibound inside a potential barrier on the perimeter of the molecule, such resonances are localized, have enhanced electron density in the molecular core, and are uncoupled from the external environment of the molecule. This localization often produces intense, easily studied spectral features, while suppressing non-resonant and/or Rydberg structure and, as discussed more fully below, has a marked influence on vibrational motion. In addition, localization causes much of the conceptual framework developed for shape resonances in free molecules to apply equally well to photoionization and electron scattering and to other states of matter such as adsorbed molecules, molecular crystals, and ionic solids.

Third, resonant trapping by a centrifugal barrier often imparts a well-defined orbital momentum character to the escaping electron. This can be directly observed, e.g. by angular distributions of scattered electrons or photoelectron angular distributions from oriented molecules, and shows that the centrifugal trapping mechanism has physical meaning and is not merely a theoretical construct. Recent case studies have revealed trapping of $\ell = 1$ to $\ell = 5$ components of continuum molecular wave functions. The purely molecular origin of the great majority of these cases is illustrated by the prototype system N_2 discussed in a later section.

Fourth, the predominantly one-electron nature of the phenomena lends itself to theoretical treatment by realistic, independent-electron methods (2,4-9), with the concomitant flexibility in terms of complexity of molecular systems, energy ranges, and alternative physical processes. This has been a major factor in the rapid exploration in this area. Continuing development of computational schemes also holds the promise of elevating the level of theoretical work on molecular ionization and scattering and, in so doing, to test and quantify many of the independent-electron results and to proceed to other circumstances such as weak channels, multiply-excited states, etc. where the simpler schemes become invalid.

A Dramatic Example Of Shape Resonant Behavior

Among the earliest and still possibly the most dramatic examples of shape resonance effects in molecules are the photoabsorption spectra of the sulfur K-shell (10-11) and L-shell (11-14) in SF_6. The sulfur L-shell absorption spectra of SF_6 and H_2S are shown in Figure 1 to illustrate the type of phenomena that originally drew attention to this area. In Figure 1 both spectra are plotted on a photon energy scale referenced to the sulfur L-shell ionization potential (IP) which is chemically shifted by a few eV in the two molecular environments, but lies near $h\nu \sim 175$eV. The ordinates represent relative photoabsorption cross sections and have been adjusted so that the integrated oscillator strength for the two systems is roughly equal in this spectral range, since absolute normalizations are not known. The H_2S spectrum is used here as a "normal" reference spectrum since hydrogen atoms normally do not contribute appreciably to shape resonance effects and, in this particular context, can be regarded as weak perturbations on the inner-shell spectra of the heavy atom. Indeed, the photoabsorption spectrum exhibits what appears to be a valence transition, followed by partially resolved Rydberg structure, which converges to a smooth continuum. The gradual rise at threshold is attributable to the delayed onset of the "$2p \rightarrow \varepsilon d$" continuum which, for second row atoms, will exhibit a delayed onset prior to the occupation of the 3d subshell. This is the qualitative behavior one might well expect for the absorption spectrum of this core level.

In sharp contrast to this, the photoabsorption spectrum of the same sulfur 2p subshell in SF_6 shows no vestige of the "normal" behavior just described. Instead three intense, broad peaks appear, one below the ionization threshold and two above, and the continuum absorption cross section is greatly reduced elsewhere. Moreover, no Rydberg structure is apparent, although an infinite number of Rydberg states must necessarily be associated with any molecular ion. Actually, Rydberg states were detected (15) superimposed on the weak bump below the IP using photograhic detection, but obviously these states are extremely weak in this spectrum. This radical reorganization of the oscillator strength distribution was interpreted (16) as a manifestation of potential barrier effects in SF_6, resulting in three shape-resonantly enhanced final state features of a_{1g}, t_{2g}, and e_g symmetry, in order of increasing energy. Another shape resonance feature of t_{1u} symmetry is prominent in the sulfur K-shell spectrum (10) and, in fact, is responsible for the weak feature just below the IP in Figure 1 owing to weak channel interaction. Hence four prominent features occur in the photoexcitation spectrum of SF_6 as a consequence of potential barriers caused by the molecular environment of the sulfur atom. Another significant observation (14) is that the SF_6 curve in Figure 1 represents both gaseous and solid SF_6, within experimental error bars. This is definitive evidence that the resonances are eigenfunctions of the potential well inside the barrier, and are effectively uncoupled from the molecule's external environment.

This beautiful empirical evidence had a strong stimulating effect in the study of shape resonances in molecular photoionization, just as early observations of the π_g shape resonance in elastic e-N_2 scattering did in the electron-molecule scattering field (3,17).

Basic Properties

The central concept in shape resonance phenomena is the single-channel, barrier-penetration model familiar from introductory quantum mechanics. In fact the name "shape resonance" means simply that the resonance behavior arises from the "shape", i.e. the barrier and associated inner and outer wells, of a local potential. The basic shape resonance mechanism is illustrated schematically (18) in Figure 2. There, an effective potential for an excited and/or unbound electron is shown to have an inner well at small distances, a potential barrier at intermediate distances, and an outer well (asymptotic form not shown) at large separations. In the context of molecular photoionization, this would be a one-dimensional abstraction of the effective potential for the photoelectron in the field of a molecular ion. Accordingly, the inner well would be formed by the partially screened nuclei in the molecular core and would therefore be highly anisotropic and would overlap much of the molecular charge distribution, i.e., the initial states of the photoionization process. The barrier, in all well documented cases, is a so-called centrifugal barrier. (Other forces such as repulsive exchange forces, high concentrations of negative charge, etc., may also contribute, but have not yet been documented to be pivotal in the molecular systems studied to date.) This centrifugal barrier derives from a competition between repulsive centrifugal forces and attractive electrostatic forces and usually (but not always) resides on the perimeter of the molecular charge distribution where the centrifugal forces can compete effectively with electrostatic forces. Similar barriers are known for d- and f-waves in atomic fields (19), however the ℓ (orbital angular momentum) character of resonances in molecular fields tends to be higher than those of constituent atoms owing to the larger spatial extent of the molecular charge distribution, e.g., see discussion in connection with N_2 photoionization below. The outer well lies outside the molecule where the Coulomb potential ($\sim -r^{-1}$) of the molecular ion again dominates the centrifugal terms ($\sim r^{-2}$) in the potential. We stress that this description has been radically simplified to convey the essential aspects of the underlying physics. In reality effective barriers to electron motion in molecular fields occur for particular ℓ components of particular

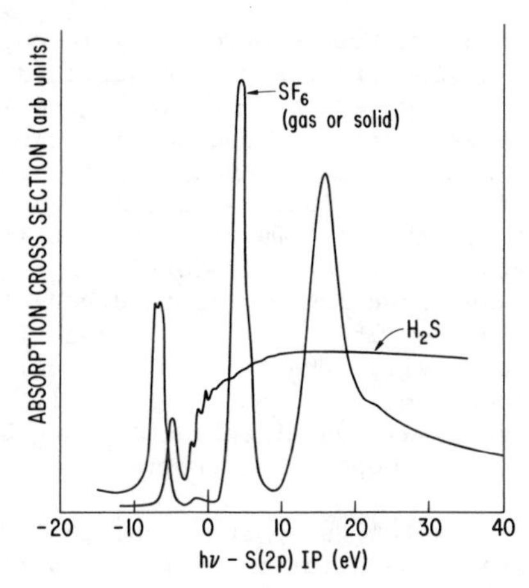

Figure 1. Experimental photoabsorption spectra of H_2S (taken from Ref. 13) and SF_6 (taken from Ref. 14) near the sulfur $L_{2,3}$ edge.

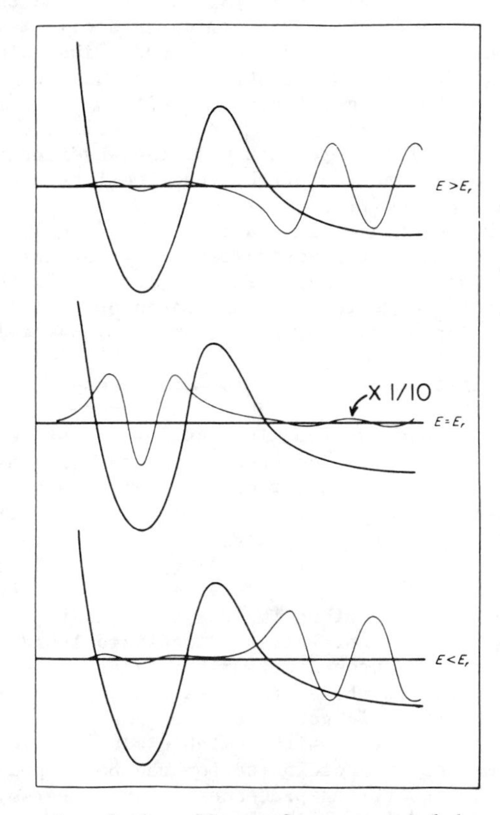

Figure 2. Schematic of the effect of a potential barrier on an unbound wave function in the vicinity of a quasibound state at E = E_r (adapted from Ref. 18). In the present context, the horizontal axis represents the distance of the excited electron from the center of the molecule. Reproduced with permission from Ref. 18. Copyright 1974, Academic Press.

ionization channels and restrict motion only in certain directions. Again, a specific example is decribed below.

Focusing now on the wave functions in Figure 2, we see the effect of the potential barrier on the wave mechanics of the photoelectron. For energies below the resonance energy $E < E_r$ (lower part of Figure 2), the inner well does not support a quasibound state, i.e., the wave function is not exponentially decaying as it enters the classically forbidden region of the barrier. Thus the wave function begins to diverge in the barrier region and emerges in the outer well with a much larger amplitude than that in the inner well. When properly normalized at large r, the amplitude in the molecular core is very small, so we say this wave function is essentially an eigenfunction of the outer well although small precursor loops extend inside the barrier into the molecular core.

At $E = E_r$ the inner well supports a quasibound state. The wavefunction exhibits exponential decay in the barrier region so that if the barrier extended to $r \to \infty$, a true bound state would lie very near this total energy. Therefore the antinode that was not supported in the inner well at $E < E_r$ has traversed the barrier to become part of a quasibound wave form which decays monotonically until it re-emerges in the outer well region, much diminished in amplitude. This "barrier penetration" by an antinode produces a rapid increase in the asymptotic phase shift by $\sim\pi$ radians and greatly enhances the amplitude in the inner well over a narrow band of energy near E_r. Therefore at $E = E_r$ the wavefunction is essentially an eigenfunction of the inner well although it decays through the barrier and re-emerges in the outer well. The energy halfwidth of the resonance is related to the lifetime of the quasibound state and to the energy derivative of the rise in the phase shift in well known ways. Finally for $E > E_r$ the wavefunction reverts to being an eigenfunction of the outer well as the behavior of the wavefunction at the outer edge of the inner well is no longer characteristic of a bound state.

Obviously this resonant behavior will cause significant physical effects: The enhancement of the inner-well amplitude at $E \sim E_r$ results in good overlap with the initial states which reside mainly in the inner well. Conversely, for energies below the top of the barrier but not within the resonance half width of E_r, the inner amplitude is diminished relative to a more typical barrier-free case. This accounts for the strong modulation of the oscillator strength distribution in Figure 1. Also, the rapid rise in the phase shift induces shape resonance effects in the photoelectron angular distribution. Another important aspect is that eigenfunctions of the inner well are localized inside the barrier and are substantially uncoupled from the external environment of the molecule. As mentioned above, this means that shape resonant phenomena often persist in going from the gas phase to the condensed phase, e.g., Figure 1, and, with suitable modification, shape resonances in molecular photoionization can be mapped (20) into electron-molecule scattering processes and vice versa. Finally, note that this discussion was focussed on total energies from the bottom of the outer well to the top of the barrier, and that no explicit mention was made of the asymptotic potential that determines the threshold for ionization. Thus valence or Rydberg states in this range can also exhibit shape resonant enhancement, even though they have true bound state behavior at large r, beyond the outer well.

We will now turn, for the remainder of this section, to the specific example of the well-known σ_u shape resonance in N_2 photoionization which was the first documented case (21) in a diatomic molecule and has since been used as a prototype in studies of various shape resonance effects as discussed below. To identify the major final-state features in N_2 photoionization at the independent-electron level, we show the original calculation (21-23) of the K-shell photoionization spectrum performed with the multiple-scattering model. This calculation agrees qualitatively with all major features in the experimental spectrum (24-25), except a narrow band of double excitation features, and with subsequent, more accurate calculations (26). The four partial cross sections in Figure 3 represent the four dipole allowed channels for K-shell (IP = 409.9 eV) photoionization. Here we have neglected the localization (27) of the K-shell hole since it doesn't greatly affect the integrated cross section and the separation into u and g symmetry both helps the present discussion and is rigorously applicable to the subsequent discussion of valence-shell excitation. (Note that the identification of shape resonant behavior is generally easier in inner-shell spectra since the problems of overlapping spectra, channel interaction, and zeros in the dipole matrix element are reduced relative to valence-shell spectra.)

The most striking spectral feature in Figure 3 is the first member of the π_g sequence, which dominates every other feature in the theoretical spectrum by a factor of ~30. (Note the first π_g peak has been reduced by a factor of 10 to fit in the frame.) The concentration of oscillator strength in this peak is a centrifugal barrier effect in the d-wave component of the π_g wavefunction. The final state in this transition is a highly localized state, about the size of the molecular core, and is the counterpart of the well-known (3,17) π_g shape resonance in e-N_2 scattering at 2.4 eV. For the latter case, Krauss and Mies (28) demonstrated that the effective potential for the π_g elastic channel in e-N_2 scattering exhibits a potential barrier due to the centrifugal replusion acting on the dominant ℓ=2 lead term in the partial-wave expansion of the π_g wavefunction. In the case of N_2 photoionization, there is one less electron in the molecular field to screen the nuclear charge so that this resonance feature is shifted (20) to lower energy and appears in the discrete. It is in this sense that we refer to such features as "discrete" shape resonances. The remainder of the π_g partial cross section consists of a Rydberg series and a flat continuum. The π_u and σ_g channels both exhibit Rydberg series, the initial members of which correlate well with partially resolved transitions in the experimental spectrum below the K-shell IP.

The σ_u partial cross section, on the other hand, was found to exhibit behavior rather unexpected for the K shell of a first-row diatomic. Its Rydberg series was extremely weak, and an intense, broad peak appeared at ~1 Ry above the IP in the low-energy continuum. This effect is caused by a centrifugal barrier acting on

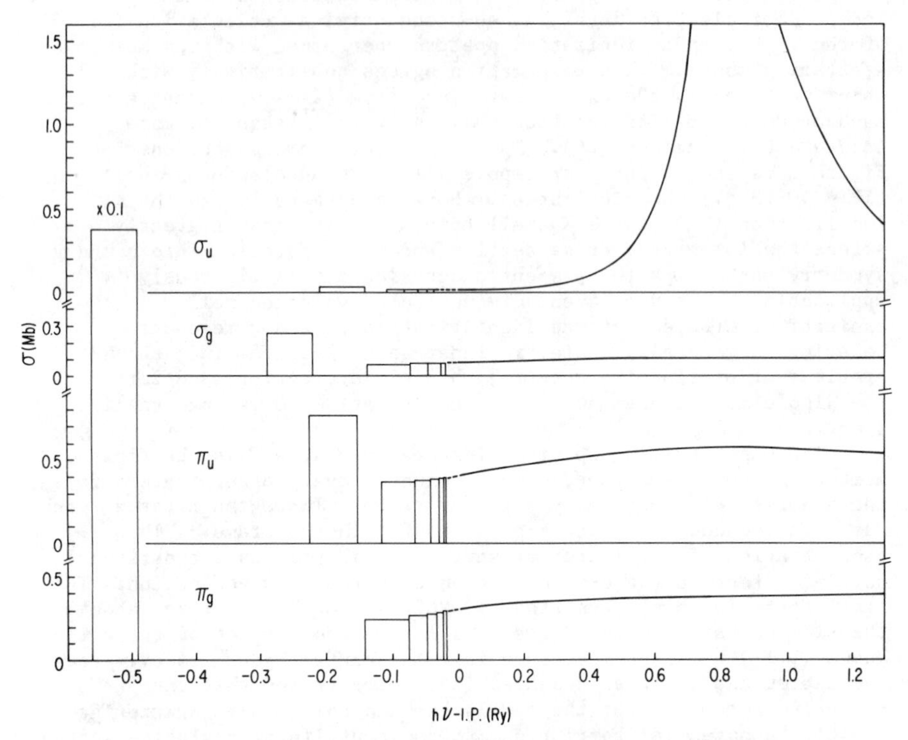

Figure 3. Calculated partial photoionization cross sections for the four dipole-allowed channels in K-shell photoionization of N_2 (Mb = 10^{-18} cm^2). Note that the energy scale is referenced to the K-shell IP (409.9 eV) and is expanded twofold in the discrete part of the spectrum.

the $\ell = 3$ component of the σ_u wavefunction. The essence of the phenomena can be described in mechanistic terms as follows. The electric dipole interaction, localized within the atomic K shell, produces a photoelectron with angular momentum $\ell = 1$. As this p-wave electron escapes to infinity, the anisotropic molecular field can scatter it into the entire range of angular momentum states contributing to the allowed σ- and π-ionization channels ($\Delta\lambda = 0, \pm 1$). In addition, the spatial extent of the molecular field, consisting of two atoms separated by 1.1 Å, enables the $\ell = 3$ component of the σ_u continuum wavefunction to overcome its centrifugal barrier and penetrate into the molecular core at a kinetic energy of ~1 Ry. This penetration is rapid, with a phase shift rise of ~π occurring over a range of ~0.3 Ry (21). These two circumstances combine to produce a dramatic enhancement of the photocurrent at ~1 Ry kinetic energy, with predominantly f-wave character.

The specifically molecular character of this phenomenon is emphasized by comparison with K-shell photoionization in atomic nitrogen and in the united-atom case, silicon. In contrast to N_2, there is no mechanism for the essential p-f coupling, and neither atomic field is strong enough to support resonant penetration of high-ℓ partial waves through their centrifugal barriers. (With substitution of "d" for "f," this argument applies equally well to the d-type resonance in the discrete part of the spectrum.) Note that the π_u channel also has an $\ell = 3$ component but does not resonate. This underscores the directionality and symmetry dependence of the trapping mechanism.

To place the σ_u resonance in a broader perspective and show its connection with high-energy behavior, we show, in Figure 4, an extension of the calculation in Figure 3 to much higher energy. Again the four dipole allowed channels in $D_{\infty h}$ symmetry are shown. The dashed line is two times the atomic nitrogen K-shell cross section. Note that the modulation about the atomic cross section, caused by the potential barrier extends to ~100 eV above threshold before the molecular and atomic curves seem to coalesce.

At higher energies, a weaker modulation appears in each partial cross section. This weak modulation is a diffraction pattern, resulting from scattering of the photoelectron by the neighboring atom in the molecule, or, more precisely, by the molecular field. Structure of this type was first studied over 50 years ago by Kronig (29) in the context of metal lattices. It currently goes by the acronym EXAFS (extended X-ray absorption fine structure) and is used extensively (30-31) for local structure determination in molecules, solids, and surfaces. The net oscillation is very weak in N_2, since the light atom is a weak scatterer. More pronounced effects are seen, e.g., in K-shell spectra (32) of Br_2 and $GeCl_4$. Our reason for showing the weak EXAFS structure in N_2 is to show that the low-energy resonant modulation (called "near-edge" structure in the context of EXAFS) and high-energy EXAFS evolve continuously into one another and emerge naturally from a single molecular framework, although the latter is usually treated from an atomic point of view.

Figure 5 shows a hypothetical experiment (22) which clearly demonstrates the ℓ character of the σ_u resonance. In this experiment, we first fix the nitrogen molecule in space and orient

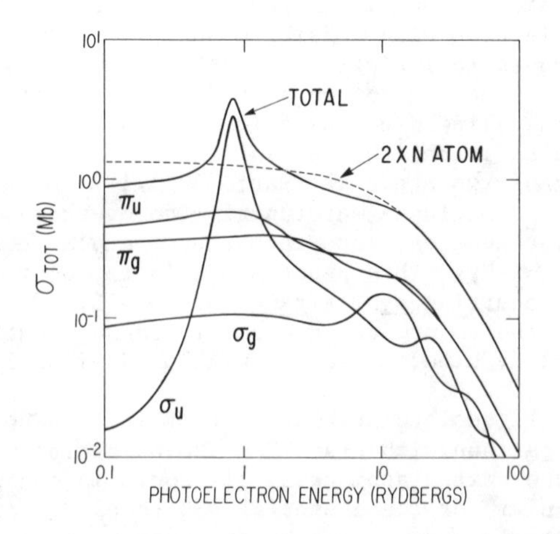

Figure 4. Calculated partial photoionization cross sections for the K-shell of N_2 over a broad energy range. The dashed line represents twice the K-shell photoionization cross section for atomic nitrogen, calculated using a Hartree-Slater potential.

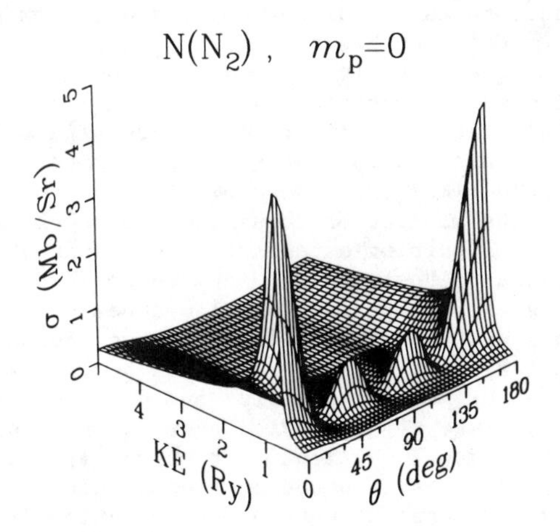

Figure 5. Calculated fixed-molecule photoelectron angular distribution for kinetic energies 0-5 Ry above the K-shell IP of N_2. The polarization of the ionizing radiation is oriented along the molecular axis in order to excite the σ continua and the photoelectron ejection angle, θ, is measured relative to the molecular axis.

the polarization direction of a photon beam, tuned near the nitrogen K edge, along the molecular axis. This orientation will cause photoexcitation into σ final states, including the resonant σ_u ionization channel. (Again hole localization is neglected for purposes of illustration.) Figure 5 shows the angular distribution of photocurrent as a function of both the excess energy above the K-shell IP and the angle of ejection, θ, relative to the molecular axis. Very apparent in Figure 5 is the enhanced photocurrent at the resonance position, KE ~ 1 Ry. Moreover the angular distribution exhibits three nodes, with most of the photocurrent exiting the molecule along the molecular axis and none at right angles to it. This is an f-wave ($\ell = 3$) pattern and indicates clearly that the resonant enhancement is caused by an $\ell = 3$ centrifugal barrier in the σ_u continuum of N_2. Thus the centrifugal barrier has observable physical meaning and is not merely a theoretical construct. Note that the correspondence between the dominant asymptotic partial wave and the trapping mechanism is not always valid, especially when the trapping is on an internal or off-center atomic site where the trapped partial wave can be scattered by the anisotropic molecular field into alternative asymptotic partial waves, e.g., BF_3 (33) and SF_6. Finally, note that the hypothetical experiment discussed above has been approximately realized by photionizing molecules adsorbed on surfaces. The shape-resonant features tend to survive adsorption and, owing to their observable ℓ-character, can even provide evidence (34-35) as to the orientation of the molecule on the surface.

The final topic in the discussion of basic properties of shape resonances involves eigenchannel contour maps (36), or "pictures" of unbound electrons. This is the continuum counterpart of contour maps of bound-state electronic wavefunctions which have proven so valuable as tools of quantum chemical visualization and analysis. Indeed, the present example helps achieve a physical picture of the σ_u shape resonance, and the general technique promises to be a useful tool for analyzing resonant trapping mechanisms and other observable properties in the future. The key to this visualization is the construction of those particlar combinations of continuum orbital momenta that diagonalize the interaction of the unbound electron with the anisotropic molecular field. These combinations, known as eigenchannels, are normalized with continuum boundary conditions and are the analogues of the eigenstates in the discrete spectrum, i.e., the bound states.

The f-dominated eigenchannel of σ_u symmetry in N_2 is an excellent case for which eigenchannel contour maps may be used to visualize shape resonant continuum states. In Figure 6, this eigenchannel is plotted at two energies, one below the resonance energy and the other near the center of the resonance. In Figure 6, the surfaces, whose contours have at large distance the three nodal planes characteristic of f orbitals, show clearly the resonant nature of the f-dominated σ_u eigenchannel. Below and above (not shown) the resonance energy, the ratio of the probability amplitude inside and outside the molecular core is typical of non-resonant behavior. But at the resonance energy there is an enormous enhancement of the wavefunction in the molecular interior; the wavefunction now resembles a molecular bound-state probability

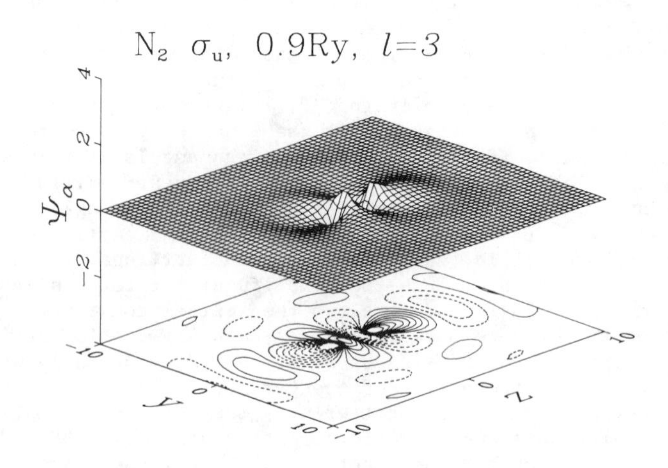

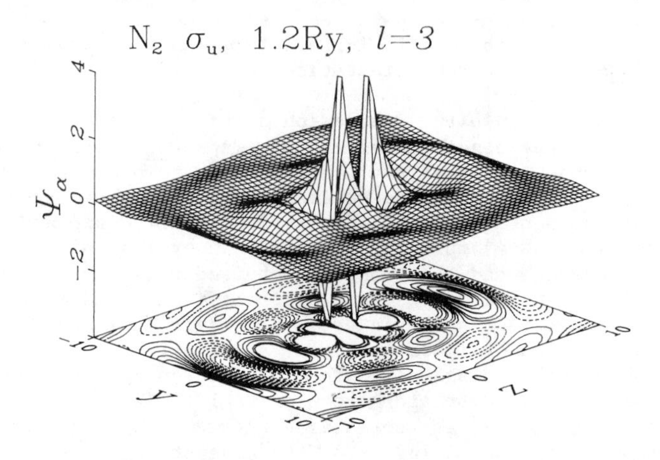

Figure 6. F-wave-dominated eigenchannel wavefunctions for nonresonant (top) and resonant (bottom) electron kinetic energies in the σ_u continuum of N_2. The molecule is in the yz plane, along the z axis, centered at y = z = 0. Contours mark steps of 0.03 from 0.02 to 0.29; positive = solid, negative = dashed. The lack of contour lines for 1.2 Ry near the nuclei is because of the 0.29 cutoff.

amplitude distribution. It is this enhancement, in the region occupied by the bound states, that leads to the very large increase in oscillator strength indicative of the resonance, and to the other manifestations discussed earlier and in the next section.

These eigenchannel plots are discussed more fully elsewhere (36); however, before leaving the subject, several points should be noted. First, the N_2 example that we have chosen is somewhat special in that there is a near one-to-one correspondence between the eigenchannels and single values of orbital angular momentum. Orbital angular momentum is, however, not a good quantum number in molecules and more generally we should not always expect such clear nodal patterns. More typically, earlier work (5,33,37-39) indicates that several angular momenta often contribute to the continuum eigenchannels (although a barrier in only one ℓ component will be primarily responsible for the temporary trapping that causes the enhancement in that and coupled components), and this means that the resulting eigenchannel plots will be correspondingly richer. Second, the dominant ℓ we have discussed pertains to the region outside the molecular charge distributions. The orbital momentum composition of these wavefunctions is more complicated in the molecular interior, as seen, e.g., in Figure 6. Nonetheless, continuity and a dominant ℓ may, as in the case of N_2, cause the emergence of a distinct ℓ pattern, even into the core region. Third, while these ideas were developed (36) in the context of molecular photoionization, the continuum eigenchannel concept carries over without any fundamental change to electron-molecule scattering. Finally, while we have used one-electron wavefunctions here, obtained with the multiple-scattering model, we emphasize that the eigenchannel concept is a general one and we look forward to its use in the analysis of more sophisticated, many-electron molecular continuum wavefunctions.

Shape-Resonance-Induced Non-Franck-Condon Effects

Molecular photoionization at wavelengths unaffected by autoionization, predissociation, or ionic thresholds has been generally believed to produce Franck-Condon (FC) vibrational intensity distributions within the final ionic state and v-independent photoelectron angular distributions. We now discuss a recent prediction (40-41) that shape resonances represent an important class of exceptions to this picture. These ideas are illustrated with a calculation of the $3\sigma_g \rightarrow \varepsilon\sigma_u, \varepsilon\pi_u$ photoionization channel of N_2, which accesses the same σ_u shape resonance discussed above at approximately $h\nu \sim 30$ eV, or ~14 eV above the $3\sigma_g$ IP. The process we are considering involves photoexcitation of N_2 X $^1\Sigma_g^+$ in its vibrational ground state with photon energies from the first IP to beyond the region of the shape resonance at $h\nu \sim 30$ eV. This process ejects photoelectrons leaving behind N_2^+ ions in energetically accessible states. As we are interested in the ionization of the $3\sigma_g$ electron, which produces the X $^2\Sigma_g^+$ ground state of N_2^+, we are concerned with the photoelectron band in the range 15.5 eV $\leqslant$ IP $\leqslant$ 16.5 eV. The physical effects we seek involve the relative intensities and angular distributions of the v = 0-2 vibrational peaks in the X $^2\Sigma_g^+$ electronic band, and, more specifically, the departures of these observables from behavior predicted by the FC separation.

The breakdown of the FC principle arises from the quasibound nature of the shape resonance, which, as we discussed earlier, is localized in a spatial region of molecular dimensions by a centrifugal barrier. This barrier and, hence, the energy and lifetime (width) of the resonance are sensitive functions of internuclear separation R and vary significantly over a range of R corresponding to the ground-state vibrational motion. This is illustrated in the upper portion of Figure 7 where the dashed curves represent separate, fixed-R calculations of the partial cross section for N_2 $3\sigma_g$ photoionization over the range $1.824a_0 \leq R \leq 2.324a_0$, which spans the N_2 ground-state vibrational wavefunction.

Of central importance in Figure 7 is the clear demonstration that resonance position, strength, and width are sensitive functions of R. In particular, for larger separations, the inner well of the effective potential acting on the $\ell = 3$ component of the σ_u wavefunction is more attractive and the shape resonance shifts to lower kinetic energy, becoming narrower and stronger. Conversely, for lower values of R, the resonance is pushed to higher kinetic energy and is weakened. This indicates that nuclear motion exercises great leverage on the spectral behavior of shape resonances, since small variations in R can significantly shift the delicate balance between attractive (mainly Coulomb) and repulsive (mainly centrifugal) forces which combine to form the barrier. In the present case, variations in R, corresponding to the ground-state vibration in N_2, produce significant shifts of the resonant behavior over a spectral range several times the fullwidth at half maximum of the resonance calculated at $R = R_e$. By contrast, nonresonant channels are relatively insensitive to such variation in R, as was shown by results (42) on the $1\pi_u$ and $2\sigma_u$ photoionization channels in N_2.

Thus, in the vicinity of a shape resonance, the electronic transition moment varies rapidly with R. This parametric coupling was estimated in the adiabatic-nuclei approximation by computing the net transition moment for a particular vibrational channel as an average of the R-dependent dipole amplitude, weighted by the product of the initial- and final-state vibrational wavefunctions at each R (40,43),

$$D_{v_f v_i} = \int dR \chi_{v_f}^{\dagger}(R) D^{-}(R) \chi_{v_i}(R).$$

The vibrational wavefunctions were approximated by harmonic-oscillator functions and the superscript "minus" denotes that incoming-wave boundary conditions have been applied and that the transition moment is complex. Note that even when the final vibrational levels v_f of the ion are unresolved (summed over), vibrational motion within the initial state $v_i = 0$ can cause the above equation to yield results significantly different from the $R = R_e$ result, because the R dependence of the shape resonance is highly asymmetric. This gross effect of R averaging can be seen in the upper half of Figure 7 by comparing the solid line (R-averaged result, summed over v_f) and the middle dashed line ($R = R_e$). Hence, even for the calculation of gross properties of the whole, unresolved electron band, it is necesary to take into account vibrational-motion effects in channels exhibiting shape resonances. As we stated earlier, this is generally not a critical issue in nonresonant channels.

Effects of nuclear motion on individual vibrational levels are shown in the bottom half of Figure 7. Looking at the partial cross sections in Figure 7, we see that the resonance position varies over a few volts depending on the final vibrational state, and that higher levels are relatively more enhanced at their resonance position than is $v_f = 0$. This sensitivity to v_f arises because transitions to alternative final vibrational states preferentially sample different regions of R. In particular, $v_f = 1,2$ sample succesively smaller R, governed by the maximum overlap with the ground vibrational state, causing the resonance in those vibrational channels to peak at higher energy than that for $v_f = 0$. The impact of these effects on branching ratios is clearly seen in Figure 2 of (40), where the ratio of the higher v_f intensities to that of $v_f = 0$ is plotted in the resonance region. There we see that the ratios are slightly above the FC factors (9.3%, $v_f = 1$; 0.6%, $v_f = 2$) at zero kinetic energy, go through a minimum just below the resonance energy in $v_f = 0$, then increase to a maximum as individual $v_f > 0$ vibrational intensities peak, and finally approach the FC factors again at high kinetic energy. Note the maximum enhancement over the FC factors is progressively more pronounced for higher v_f, i.e., 340% and 1300% for $v_f = 1,2$, respectively.

Equally dramatic are the effects on $\beta(v_f)$ discussed in (40). Especially at and below the resonance position, the β's vary greatly for different final vibrational levels. Carlson first observed (44-45) that, at 584 Å, the $v_f = 1$ level in the $3\sigma_g$ channel of N_2 had a much larger β than the $v_f = 0$ level even though there was no apparent autoionizing state at that wavelength. This is in semiquantitative agreement with the theoretical calculation (40) which gives $\beta(v_f = 0) \sim 1.0$ and $\beta(v_f = 1) \sim 1.5$. Although the agreement is not exact, we feel this demonstrates that the "anomalous" v_f dependence of β in N_2 stems mainly from the σ_u shape resonance which acts over a range of the spectrum many time its own ~5 eV width. The underlying cause of this effect is the shape-resonance-enhanced R dependence of the dipole amplitude, just as for the vibrational partial cross sections. In the case of $\beta(v_f)$, however, both the R dependence of the phase and of the magnitude of the complex dipole amplitude play a crucial role, whereas the partial cross sections depend only on the magnitude.

The theoretical predictions discussed above were soon tested in two separate experiments. In Figure 8 the branching ratio for production of the v = 0 and 1 vibrational levels of N_2^+ X $^2\Sigma_g^+$ is shown. The dash-dot line is the original prediction (40). The solid dots are the recent measurements (46) in the vicinity of the shape resonance at $h\nu \sim 30$ eV. The conclusion drawn from this comparison is that the observed variation of the vibrational branching ratio relative to the FC factor over a broad spectral range qualitatively confirms the prediction; however, subsequent calculations (7,48) with fewer approximations have achieved better agreement based on the same mechanism for breakdown of the FC separation. The dashed and solid curves are results based on a

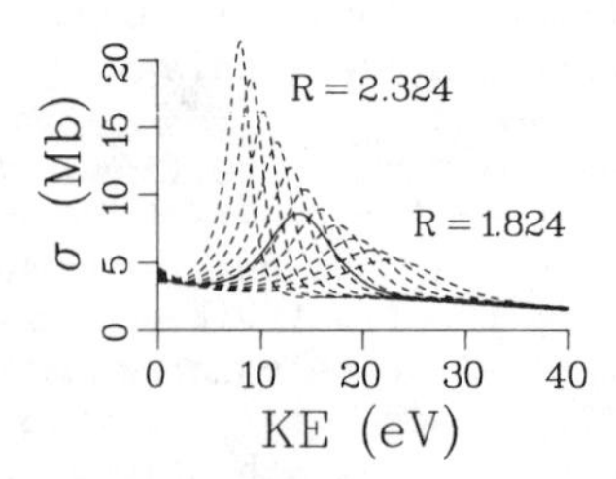

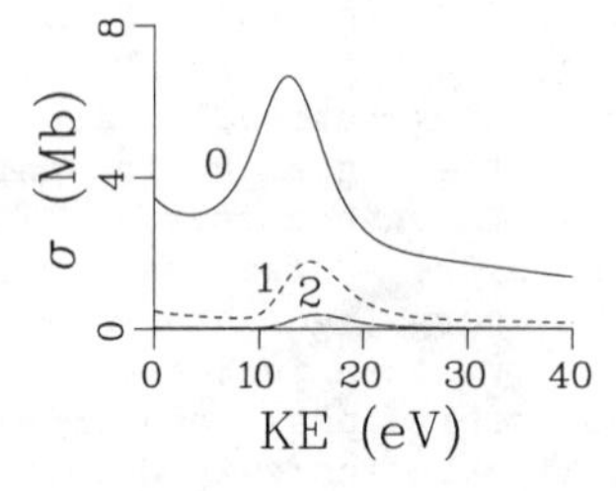

Figure 7. Cross sections for photoionization of the $3\sigma_g$ (v_i = 0) level of N_2. Top: fixed-R (dashed curves) and R-averaged, vibrationally unresolved (solid curve) results. Bottom: results for resolved final-state vibrational levels, v_f = 0-2.

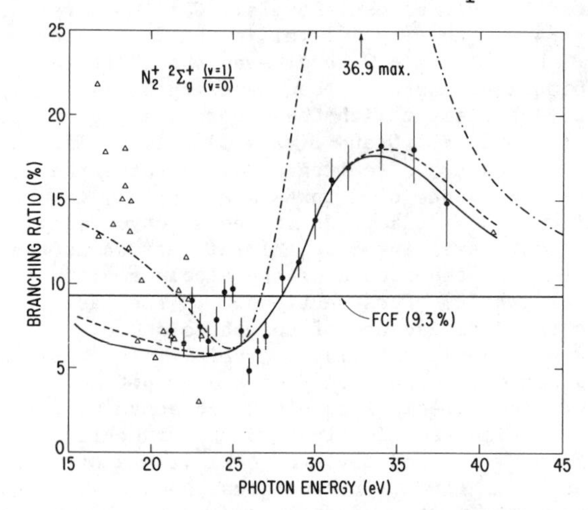

Figure 8. Branching ratios for production of the v = 0,1 levels of N_2^+ X $^2\Sigma_g^+$ by photoionization of N_2: ●, Ref. 46; Δ, Ref. 47; -•-•-, multiple scattering model prediction from Ref. 40; ———, frozen-core Hartree-Fock dipole length approximation from Ref. 48; - - -, frozen-core Hartree-Fock dipole velocity approximation from Ref. 48.

Schwinger variational treatment (48) of the photoelectron wavefunction. The two curves represent a length and velocity representation of the transition matrix element, both of which are in excellent agreement with the data. This is an outstanding example of interaction between experiment and theory, proceeding as it did from a novel prediction, through experimental testing, and final quantitative theoretical agreement in a short time. Also shown in Figure 8 are data in the 15.5 eV $\leqslant h\nu \leqslant$ 22 eV region which are earlier data (47) obtained using laboratory line sources. The apparently chaotic behavior arises from unresolved autoionization structure.

The angular distribution asymmetry parameters, β, for the v = 0,1 levels of N_2^+ X $^2\Sigma_g^+$ over roughly the same energy region are reported in (49). In the region above $h\nu \sim 25$ eV, this data also shows qualitative agreement with the predicted (40) v-dependence of β caused by the σ_u shape resonance. In this case the agreement is somewhat improved in later calculations (48), mainly for v = 1, however the change is less dramatic than for the branching ratios.

Finally, note that we have illustrated the vibrational effects of shape resonances in the context of molecular photoionization; however, a rather analogous mechanism makes shape resonances extremely efficient in inducing vibrational excitation in electronically-elastic electron-molecule scattering (3,17,41).

Connections Between Shape Resonances In Electron-Molecule Scattering And In Molecular Photoionization And Related Connections

At first glance, there is little connection between shape resonances in electron-molecule scattering (e + M) and those in molecular photoionization ($h\nu$ + M). The two phenomena involve different numbers of electrons and the collision velocities are such that all electrons are incorporated into the collision complex. Hence, we are comparing a neutral molecule and a molecular negative-ion system. However, although the long-range part of the scattering potential is drastically different in the two cases, the strong short-range potential is not drastically different since it is dominated by the interactions among the nuclei and those electrons common to both systems. Thus, shape resonances which are localized in the molecular core substantially maintain their identity from one system to another, but are shifted in energy owing to the difference caused by the addition of an electron to the molecular system. This unifying property of shape resonances thus links together the two largest bodies of data for the molecular electronic continuum -- $h\nu$ + M and e + M -- and although these resonances shift in energy in going from one class to another and manifest themselves in somewhat different ways, this link permits us to transfer information between the two. This can serve to help interpret new data and even to make predictions of new features to look for experimentally. Actually, this picture (20) was surmised empirically from evidence contained in survey calculations on e + M and $h\nu$ + M systems and, in retrospect, from data. These observations can be summarized as follows: By and large, the systems $h\nu$ + M and e + M display the same manifold of shape resonances, only those in the e + M system are shifted $\sim$ 10 eV to higher electron energy. Usually, there is one shape resonance per symmetry for a subset of the symmetries available. The shift depends on the symmetry of the state, indicating, as one would expect, that the additional electron is not uniformly distributed. Finally, there is substantial proof that the ℓ-character is preserved in this process, although interaction among alternative components in a scattering eigenchannel can vary and thus alter the ℓ mixing present.

There are several good examples available to illustrate this point, e.g., N_2, CO, CO_2, BF_3, SF_6, etc. In general, one can start from either the neutral or the negative ion system, but, in either case, there is a preferred way to do so: In the $h\nu + M$ case, it is better to examine the inner-shell photoabsorption and photoionization spectra. Shape resonances almost invariably emerge most clearly in this context. Additional effects, discussed briefly at the end of this section, frequently make the role of shape resonances in valence-shell spectra more complicated to interpret. In the e + M case, a very sensitive indicator of shape resonance behavior is the vibrational excitation channel. Vibrational excitation is enhanced by shape resonances (3,17), and is typically very weak for non-resonant scattering. Hence, a shape resonance, particularly at intermediate energy (10-40 eV) (41,50), may be barely visible in the vibrationally and electronically elastic scattering cross section, and yet be displayed prominently in the vibrationally inelastic, electronically elastic cross section.

Two examples will help illustrate these points. In e - SF_6 scattering, the vibrationally elastic scattering cross section has been calculated theoretically (51) and shown to have four shape resonances of a_{1g}, t_{1u}, t_{2g}, and e_g symmetry at approximately 2, 7, 13, and 27 eV, respectively. The absolute total cross section measured by Kennerly et al. (52) shows qualitative agreement, except that no clear sign of the e_g resonance is present. (This resonance might be more evident in the vibrational excitation spectrum, which is not available.) Hence, using the guidelines given above, one would expect shape resonance features in the $h\nu + M$ case at -8, -3, 3, and 17 eV (on the kinetic energy scale) to a very crude, first approximation. Indeed, the K- and L-shell photoabsorption spectra of SF_6 show such intense features, as discussed in an earlier section.

Using N_2, we reverse the direction of the mapping, and start with $h\nu + N_2$, which was discussed extensively in earlier sections. Here a "discrete" shape resonance of π_g symmetry and a shape resonance of σ_u symmetry are apparent in the K-shell spectrum (24), e.g., Figure 3. These occur at ~ -9 and 10 eV on the kinetic energy scale (relative to the ionization threshold). Hence, one would look for the same set of resonances in e-N_2 scattering at ~ 1 and ~ 20 eV electron scattering energies. The well-known π_g shape resonance (17) is very apparent in the vibrationally elastic cross section; however, there is only a very broad bump at ~ 20 eV (53). As noted above, the vibrationally inelastic cross section is much more sensitive to shape resonances, and, indeed, the σ_u shape resonance in e-N_2 scattering has been established theoretically and experimentally by looking in this channel (41, 54-58). Several other excellent examples exist, but we will conclude by pointing out that the connections between e-CO_2 and $h\nu$ - CO_2 resonances have been recently discussed (39) in detail, including a study of the eigenphase sums in the vicinity of the σ_u shape resonance in the two systems.

Finally, we note similar connections and additional complications upon mapping from inner-shell to valence-shell hν + M spectra. On going from deep inner-shell spectra to valence-shell spectra, shape resonances in hν + M also shift approximately 1-4 eV toward higher kinetic energy, due to differences in screening between localized and delocalized holes as well as other factors. As mentioned above, several complications arise in valence-shell spectra which can tend to obscure the presence of a shape resonance compared to their more straightforward role in inner-shell spectra. These include greater energy dependence of the dipole matrix element, interactions with autoionizing levels, strong continuum-continuum coupling between more nearly degenerate ionization channels, strong particle-hole interactions, etc. So, for the most transparent view of the manifold of shape resonance features in hν + M, one should always begin with inner-shell data.

Progress And Prospects

In summary, we have used prototype studies on N_2 to convey the progress made in the study of shape resonances in molecular fields, particularly in molecular photoionization. This included the identification of shape resonant features in photoionization spectra of molecules and the accrual of substantial physical insight into their properties, many of which are peculiar to molecular fields. One recent example has been the prediction and experimental confirmation of the role of shape resonances in producing non-Franck-Condon effects in vibrational branching ratios and photoelectron angular distribution.

What this discussion has failed so far to convey is the already extensive body of work that has developed around these basic themes. Even in an early interpretation (16) of shape resonance effects in X-ray spectra, it was obvious that the phenomena would be widespread as over ten molecules, or local molecular moieties, were observed (13,16) to have shape resonant behavior. At the present, it is not difficult to identify over two dozen examples of molecules exhibiting the effects discussed above in one or more final state symmetries [references for the following examples are cited in (1), and the inner-shell cases are listed according to molecule in the bibliography given by Hitchcock (59)]. These include simple diatomics (N_2, O_2, CO, NO), polyatomics with subgroups related to the first row diatomics (HCN, C_2N_2, CH_3CN), triatomics (CO_2, CS_2, OCS, N_2O, SO_2) and more highly-coordinated molecules and local molecular environments (SF_6, $SO_4^=$, SF_5CF_3, SF_2O_2, SF_2O, BF_3, SiF_4, $SiCl_4$, $SiF_6^=$, SiO_2, NF_3, CF_4). There is no doubt that many cases have been overlooked in this list and that many examples will be identified in the future as the exploration of molecular photoionization dynamics continues, particularly with the increasing utilization of synchrotron radiation sources.

Several examples from this body of literature serve both to emphasize some of the interesting complications that can arise and to caution against assuming the manifestations of shape resonances will always conform to the independent-electron concepts used above to explain the fundamentals of the subject in connection with N_2 photoionization: (i) For example, in the isoelectronic molecule CO, much of what was said for N_2 might be expected to apply with suitable modifications to account for the loss of inversion symmetry. However for photoionization of the outermost 5σ orbital (the counterpart of the $3\sigma_g$ in N_2) the vibrational branching ratios (60) and β's (61) for the weaker v≥2 channels deviate qualitatively from the calculations (62). This has been postulated (62) to arise from channel interaction with weak, doubly-excited autoionizing states in the region of the shape resonance. In any case, some departure from the simplified picture is drastically altering the otherwise anticipated behavior of these weak vibrational channels. (ii) In the case of O_2 photoionization, an analogous σ_u shape resonance is expected (63-65), but its identification in the VUV photoionizaton spectrum has been complicated by the existence of extensive autoionization structure in the region of interest. Recent work (66) using variable-wavelength photoelectron measurements and a multichannel quantum defect analysis of the principal autoionizing Rydberg series has sorted out this puzzle, with the result that the σ_u shape resonance was established to be approximately where expected, but was not at all clearly identifiable without the extensive analysis used in this case. (iii) In the case of CO_2 a σ_u shape resonance of completely different origin was expected (37,67) for photoionization of the $4\sigma_g$ orbital, leading to the $\tilde{C}\ ^2\Sigma_g^+$ state of CO_2^+. This resonance, however, was not apparent in partial cross section measurements on this channel (68). Nevertheless predictions (38,67) of a shape-resonant feature in the corresponding β was confirmed (69) and work in several laboratories (37,38,67,69-74) has now converged to qualitative agreement for this observable. In addition, very recent measurements have shown that this resonance is observable in the partial cross sections, but is shifted to lower energy (75). Future experimental work on vibrational branching ratios (38) and v-dependent β's (38) would greatly aid in the further study of this case. (iv) The most dramatic display of shape resonance phenomena is in SF_6 which exhibits four prominent resonantly enhanced features (a_{1g}, t_{1u}, t_{2g}, and e_g symmetries) in its inner-shell spectra (10,12,16). However the role of the t_{2g} and e_g shape resonances in valence-shell spectra is poorly established. The experimental and theoretical evidence is too involved to summarize here, but we do note that strong evidence (76), associated with the behavior of the t_{2g} resonance, exists for strong channel interaction among valence-shell photoionization channels, and the failure to clearly observe (77-78) the e_g resonance also suggests substantial departures from the elementary ideas described above. (v) Finally, returning to the case of N_2, we note that the photoionization of the $2\sigma_g$ orbital should also access the σ_u shape resonance. However for this inner-valence orbital, extensive vibronic coupling leads to a breakdown (79) of the single-particle model leading to the observation (80) of tens of "satellite" vibronic states in the photoelectron spectrum instead of a single peak due to ionization of the $2\sigma_g$ orbital. Nevertheless if the sum of this complicated structure is summed and plotted versus photon energy, the resonant enhancement re-emerges.

These five cases are excellent examples of the additional challenges that can arise in the study of shape resonance phenomena. They should not diminish the simplicity and power of the fundamental shape resonance dynamics but, rather, should show how the fundamental framework showcases more complicated (and interesting) photoionization dynamics which, in turn, require a more sophisticated framework for full understanding.

Another form of progress is measured by the applicability of ideas to other observables, or, more broadly, to other subfields: (i) We have already touched upon the close connection (20) between shape resonance phenomena in molecular photoionization and electron-molecule scattering. (ii) Shape resonances in adsorbed molecules are now used rather extensively (34-35) as a probe of the geometry and electronic properties of adsorption sites. (iii) As discussed in connection with the inner-shell spectra of SF_6, free-molecule concepts concerning localized states carry over to the condensed phase. In such cases a local "molecular" point of view can often provide more direct physical insight into photoexcitation dynamics of solids than a band-structure approach. (iv) Also noted above, shape resonances are often low energy precursors to EXAFS structure occurring from ~100 eV to thousands of eV above inner-shell edges. In fact, such resonant features are very sensitive to local structure and may be very useful for local structure determination. (v) An intimate connection also exists with antibonding valence states in quantum chemistry language (81). This was dramatically demonstrated over ten years ago, when Gianturco et al. (82) interpreted the shape resonances in SF_6 using unoccupied virtual orbitals in an LCAO-MO calculation. This conection is a natural one since shape resonances are localized within the molecular charge distribution and therefore can be realistically described by a limited basis set suitable for describing the valence MO's. However, the scattering approach used in the shape-resonance picture is necessary for analysis of various dynamical aspects of the phenomena. (vi) Finally, shape resonances have been used as characteristic features in the analysis of such diverse subjects as electron optics (5) of molecular fields and hole localization (27) in inner-shell ionization, and as the cause for molecular alignment in photoionization, leading to anisotropy in the angular distribution (83) of Auger electrons from the decay of K-shell holes.

Looking to the future, there are several enticing prospects for significant progress. For instance, the present set of known and characterized resonance features is only the tip of the iceberg. The joint theoretical and experimental efforts to extend our knowledge of these useful states to other molecules and to other detection channels should be very fruitful and present new issues to be resolved. Another theme, so natural in research, is the study of those cases in which our prevalent ideas break down. Several examples were cited above, but this is expected to be a major growth area owing to the expansion both in detailed experimental studies and in the variety of computational methods capable of treating inter- and intra-channel coupling and non-Born-Oppenheimer effects. In this context, the study of weak channels, such as weak vibrational channels, should be most useful in highlighting the departures from the independent-particle, abiabatic framework used above. Another obvious example is the potential for more active investigations of shape resonantly localized excitations in other contexts, such as adsorbed molecules, solids, and biological molecules. In these and unforseen ways, the expansion, refinement, and unification of these recent developments in the study of shape resonances in molecular fields will provide a stimulating theme in molecular physics in the coming years.

Acknowledgments

The perspective and many of the ideas presented here resulted from a very rewarding, ten-year collaboration with Dan Dill (Boston U.), whom I gratefully acknowledge, and from contributions by several other co-workers, over the years, who are very evident in the list of references. This work was supported by the U.S. Department of Energy and the Office of Naval Research.

Literature Cited

1. Dehmer, J. L.; Dill, D.; Parr, A. C. In "Photophysics and Photochemistry in the Vacuum Ultraviolet"; McGlynn, S.; Findley, G.; Huebner, R., Eds.; D. Reidel: Dordrecht, Holland, 1984; in press, and references therein.
2. See related articles, this volume.
3. Lane, N. F. Rev. Mod. Phys. 1980, 52, 29 and references therein.
4. Dill, D.; Dehmer, J. L. J. Chem. Phys. 1974, 61, 692.
5. Dehmer, J. L.; Dill, D. In "Electron-Molecule and Photon-Molecule Collisions"; Rescigno, T.; McKoy, V.; Schneider, B., Eds.; Plenum: New York, 1979; p. 225.
6. Langhoff, P. W. In "Electron-Molecule and Photon-Molecule Collisions"; Rescigno, T. N.; McKoy, V.; Schneider, B., Eds.; Plenum: New York, 1979, p. 183.
7. Raseev, G.; Le Rouzo, H.; Lefebvre-Brion, H. J. Chem. Phys. 1980, 72, 5701.
8. Lucchese, R. R.; McKoy, V. Phys. Rev. A 1981, 24, 770.
9. Lucchese, R. R.; Raseev, G.; McKoy, V. Phys. Rev. A 1982, 25, 2572.
10. LaVilla, R. E.; Deslattes, R. D. J. Chem. Phys. 1966, 44, 4399.
11. LaVilla, R. E. J. Chem. Phys. 1972, 57, 899.
12. Zimkina T. M.; Fomichev, V. A. Dokl. Akad. Nauk SSSR 1966, 169, 1304; Sov. Phys. Dokl. 1966, 11, 726.
13. Zimkina T. M.; Vinogradov, A. C. J. Phys. (Paris) Colloq. 1971, 32, 3.
14. Blechschmidt, D.; Haensel, R.; Koch, E. E.; Nielsen, U.; Sagawa, T. Chem. Phys. Lett. 1972, 14, 33.
15. Nakamura, M.; Morioka, Y.; Hayaishi, T.; Ishiguro, E.; Sasanuma, M. "Third International Conference on Vacuum Ultraviolet Radiation Physics"; Physical Society of Japan: Tokyo, 1971; paper 1pA1-6.
16. Dehmer, J. L. J. Chem. Phys. 1972, 56, 4496.

17. Shulz, G. J. Rev. Mod. Phys. 1973, 45, 422.
18. Child, M. S. "Molecular Collision Theory"; Academic: New York, 1974; p. 51.
19. Fano U.; Cooper, J. W. Rev. Mod. Phys. 1968, 40, 441.
20. Dehmer J. L.; Dill, D. In "Symposium on Electron-Molecule Collisions"; Shimamura, I.; Matsuzawa, M., Eds.; University of Tokyo Press: Tokyo, 1979, p. 95.
21. Dehmer, J. L.; Dill, D. Phys. Rev. Lett. 1975, 35, 213.
22. Dill, D.; Siegel, J.; Dehmer, J. L. J. Chem. Phys. 1976, 65, 3158.
23. Dehmer, J. L.; Dill, D. J. Chem. Phys. 1976, 65, 5327.
24. Kay, R. B.; van der Leeuw, Ph.E.; van der Wiel, M. J. J. Phys. B 1977, 10, 2513.
25. Hitchcock, A. P.; Brion, C. E. J. Electron Spectrosc. 1980, 18, 1.
26. Rescigno, T. N.; Langhoff, P. W. Chem. Phys. Lett. 1977, 51, 65.
27. Dill, D.; Wallace, S.; Siegel, J.; Dehmer, J. L. Phys. Rev. Lett. 1978, 41, 1230; 1979, 42, 411.
28. Krauss, M.; Mies, F. H. Phys. Rev. A 1970, 1, 1592.
29. Kronig, R. de L. Z. Physik 1931, 70, 317; 1932, 75, 191.
30. "Synchrotron Radiation: Techniques and Applications"; Kunz, C., Ed.; Springer-Verlag: Berlin, 1979.
31. "Synchrotron Radiation Research"; Winick, H.; Doniach, S., Eds.; Plenum: New York, 1980.
32. Kincaid, B.; Eisenberger, P. Phys. Rev. Lett. 1975, 34, 1361.
33. Swanson, J. R.; Dill, D.; Dehmer, J. L. J. Chem. Phys. 1981, 75, 619.
34. Gustafsson, T.; Plummer, E. W.; Liebsch, A. In "Photoemission and the Electronic Properties of Surfaces"; Feuerbacher, B.; Fitton, B.; Willis, R. F., Eds.; J. Wiley: New York, 1978.
35. Gustafsson, T. Surface Science 1980, 94, 593.
36. Loomba, D.; Wallace, S.; Dill, D.; Dehmer, J. L. J. Chem. Phys. 1981, 75, 4546.
37. Swanson, J. R.; Dill, D.; Dehmer, J. L. J. Phys. B 1980, 13, L231.
38. Swanson, J. R.; Dill, D.; Dehmer, J. L. J. Phys. B 1981, 14, L207.
39. Dittman, P. M.; Dill, D.; Dehmer, J. L. Chem. Phys. 1983, 78, 405.
40. Dehmer, J. L.; Dill, D.; Wallace, S. Phys. Rev. Lett. 1979, 43, 1005.
41. Dehmer, J. L.; Dill, D. In "Electronic and Atomic Collisions"; Oda N.; Takayanagi, K., Eds.; North-Holland: Amsterdam, 1980; p. 195.
42. Wallace, S. Ph.D. Thesis, Boston University, Boston, 1980.
43. Chase, D. M. Phys. Rev. 1956, 104, 838.
44. Carlson, T. A. Chem. Phys. Lett. 1971, 9, 23.
45. Carlson, T. A.; Jonas, A. E. J. Chem. Phys. 1971, 55, 4913.
46. West, J. B.; Parr, A. C.; Cole, B. E.; Ederer, D. L.; Stockbauer, R.; Dehmer, J. L. J. Phys. B 1980, 13, L105.
47. Gardner, J. L.; Samson, J. A. R. J. Electron Spectrosc. 1978, 13, 7.
48. Lucchese, R. R.; McKoy, V. J. Phys. B 1981, 14, L629.

49. Carlson, T. A.; Krause, M. O.; Mehaffy, D.; Taylor, J. W.; Grimm, F. A.; Allen, J. D. J. Chem. Phys. 1980, 73, 6056.
50. Dill, D.; Welch, J.; Dehmer, J. L.; Siegel, J. Phys. Rev. Lett. 1979, 43, 1236.
51. Dehmer, J. L.; Siegel, J.; Dill, D. J. Chem. Phys. 1978, 69, 5205.
52. Kennerly, R. E.; Bonham, R. A.; McMillan, J. J. Chem. Phys. 1979, 70, 2039.
53. Kennerly, R. E. Phys. Rev. A 1980, 21, 1876.
54. Dehmer, J. L.; Siegel, J.; Welch, J.; Dill, D. Phys. Rev. A 1979, 21, 101.
55. Pavlovic, Z.; Boness, M. J. W.; Herzenberg, A.; Shulz, G. J. Phys. Rev. A 1972, 6, 676.
56. Truhlar, D. G.; Trajmar, S.; Williams, W. J. Chem. Phys. 1972, 57, 3250.
57. Chutjian, A.; Truhlar, D. G.; Williams, W.; Trajmar, S. Phys. Rev. Lett. 1972, 29, 1580.
58. Rumble, J. R.; Truhlar, D. G.; Morrison, M. A. J. Phys. B 1981, 14, L301.
59. Hitchcock, A. P. J. Electron Spectrosc. 1982, 25, 245.
60. Stockbauer, R.; Cole, B. E.; Ederer, D. L.; West, J. B.; Parr, A. C.; Dehmer, J. L. Phys. Rev. Lett. 1979, 43, 757.
61. Cole, B. E.; Ederer, D. L.; Stockbauer, R.; Codling, K.; Parr, A. C.; West, J. B.; Poliakoff, E. D.; Dehmer, J. L. J. Chem. Phys. 1980, 72, 6308.
62. Stephens, J. A.; Dill, D.; Dehmer, J. L. J. Phys. B 1981, 14, 3911.
63. Gerwer, A.; Asaro, C.; McKoy, V.; Langhoff, P. W. J. Chem. Phys. 1980, 72, 713.
64. Raseev, G.; Lefebvre-Brion, H.; Le Rouzo, H.; Roche, A. L. J. Chem. Phys. 1981, 74, 6686.
65. Dittman, P. M.; Dill, D.; Dehmer, J. L. J. Chem. Phys. 1982, 76, 5703.
66. Morin, P.; Nenner, I.; Adam, M. Y.; Hubin-Franskin, M. J.; Delwiche, J.; Lefebvre-Brion, H.; Giusti-Suzor, A. Chem. Phys. Lett. 1982, 92, 609.
67. Grimm, F.; Carlson, T. A.; Dress, W. B.; Agron, P.; Thomson, J. O.; Davenport, J. W. J. Chem. Phys. 1980, 72, 3041.
68. Gustafsson, T.; Plummer, E. W.; Eastman, D. E.; Gudat, W. Phys. Rev. A 1978, 17, 175.
69. Carlson, T. A.; Krause, M. O.; Grimm, F. A.; Allen, J. D.; Mehaffy, D.; Keller, P. R.; Taylor, J. W. Phys. Rev. A 1981, 23, 3316.
70. Langhoff, P. W.; Rescigno, T. N.; Padial, N.; Csanak, G.; McKoy, V. J. Chim. Phys. 1980, 77, 589.
71. Grimm, F. A.; Allen, J. D.; Carlson, T. A.; Krause, M. O.; Mehaffy, D.; Keller, P. R.; Taylor, J. W. J. Chem. Phys. 1981, 75, 92.
72. Lucchese, R. R.; McKoy, V. J. Phys. Chem. 1981, 85, 2166.
73. Padial, N.; Csanak, G.; McKoy, V.; Langhoff, P. W. Phys. Rev. A 1981, 23, 218.
74. Lucchese, R. R.; McKoy, V. Phys. Rev. A 1982, 26, 1406.
75. Roy, P.; Nenner, I.; Adam, M. Y.; Delwiche, J.; Hubin-Franskin, H. J.; Lablanquie, P.; Roy, D. Chem. Phys. Lett. to be published.

76. Dehmer, J. L.; Parr, A. C.; Wallace, S.; Dill, D. Phys. Rev. A 1982, 26, 3283.
77. Gustafsson, T. Phys. Rev. A 1978, 18, 1481.
78. Levinson, H. J.; Gustafsson, T.; Soven, P. Phys. Rev. A 1979, 19, 1089.
79. Cederbaum, L. S.; Domcke, W.; Schirmer, J.; von Niessen, W. Phys. Scr. 1980, 21, 481.
80. Krummacher, S.; Schmidt, V.; Wuilleumier, F. J. Phys. B 1980, 13, 3993.
81. See, e.g., Langhoff, P., this volume
82. Gianturco, F. A.; Guidotti, C.; Lamanna, U. J. Chem. Phys. 1972, 57, 840.
83. Dill, D.; Swanson, J. R.; Wallace, S.; Dehmer, J. L. Phys. Rev. Lett. 1980, 45, 1393.

RECEIVED June 11, 1984

9

Temporary Negative Ion States in Hydrocarbons and Their Derivatives

K. D. JORDAN[1] and P. D. BURROW[2]

[1]Department of Chemistry, University of Pittsburgh, Pittsburgh, PA 15260
[2]Department of Physics and Astronomy, University of Nebraska, Lincoln, NB 68588

Electron scattering experiments, in particular electron transmission spectroscopy (ETS), have provided a wealth of data on the temporary negative ion states of polyatomic molecules. Following brief introductions to the transmission technique and to the characteristics of resonances, we examine the spectra of several "simple" unsaturated hydrocarbons and discuss shape resonances arising from the temporary occupation of π^* orbitals and, in the case of ethylene, the Feshbach resonances associated with Rydberg orbitals. The importance of long-range interactions in anion states is illustrated with data in several non-conjugated dienes. The classification of higher lying resonances in benzene is discussed with regard to their "shape" and "core-excited" characteristics. Finally we examine resonances associated with σ^* orbitals and the decay of such states in the dissociative attachment channel.

Evidence for resonances in the cross sections for electron scattering from polyatomic molecules, including hydrocarbons, can be found in the literature as far back as the late 1920's (1,2). The authors of these papers, however, were unaware that the pronounced low energy peaks in the cross sections of molecules such as ethylene and acetylene were due to temporary negative ion formation. Haas (3), in 1957, was apparently the first to observe that strong vibrational excitation accompanied such a peak, and to invoke an unstable negative ion complex as the means through which the excitation takes place.

Since the renaissance of electron scattering studies beginning in the late 1950's, temporary anion states have been observed in all atoms and virtually all small molecules which have been examined. For the most part, such studies were carried out by physicists who focused their attention on atoms, diatomics and a few selected triatomic molecules. Much of this work concentrated on achieving a detailed understanding of the decay modes of temporary anions into

0097–6156/84/0263–0165$06.00/0

the available channels, in particular the angular scattering distribution, excitation of vibrational levels and electronic states, and fragmentation by dissociative attachment (4).

Beginning in 1965, a number of groups initiated studies of temporary anion formation in hydrocarbons (5-8), with extensive contributions from the groups of Compton (9) and Christophorou (10) at Oak Ridge. Except for the work of Hasted and coworkers (7), these studies relied primarily on the trapped-electron (11) or SF_6 scavenger methods (12) for the detection of temporary negative ion states. Both of these methods are sensitive to the production of slow electrons, typically those with energies less than 100 meV, which are produced just above the threshold for an inelastic process. For the observation of temporary anions, these data present some problems of interpretation. To yield a signal, resonances lying below the electronically excited states of a molecule must decay strongly into highly excited vibrational levels of the ground state of the neutral molecule which are nearly coincident with the resonance. The technique is therefore less sensitive to resonances of short lifetimes since these decay with smaller probability into the required high vibrational levels. Higher lying resonances, furthermore, may be confused with or overlap peaks arising from the excitation of electronic states of the neutral molecule. Finally, little information about the profiles of the resonances is derivable with these methods, and the energy resolution is not generally sufficient to provide information on the vibrational structure of the resonances.

Despite reservations concerning the use of these methods to locate and characterize resonances in complex molecules, they provided ample evidence for temporary negative ions in a great variety of hydrocarbons. It must be said, though, that in comparison to the widespread adoption of photoelectron spectroscopy (PES) by chemists during this same period of time, the work on resonances appears to have made rather little impact on the chemical community as a whole, and surprisingly little work was initiated by chemists in this area. In our view, this was in part a consequence of the methods utilized, which did not permit a "global" picture of the temporary negative ion states of molecules. This in turn made it difficult to establish the connection between anion energies and the electronic structure, as described by molecular orbitals. Indeed, the complementary relationship between the cation energies determined by photoelectron spectroscopy and anion energies deduced from electron scattering measurements was not fully appreciated.

Electron Transmission Spectroscopy (ETS)

In 1971 a new variation on the transmission method was introduced by Sanche and Schulz (13). The technique incorporated the trochoidal monochromator of Stamatovic and Schulz (14) and a modulation scheme to obtain the derivative of the electron current transmitted through a gas cell. This combination provides a relatively simple and very sensitive means of locating resonances as they appear in the total scattering cross section. In particular the energy resolution (20-50 meV), which is substantially better than that found in most trapped electron and SF_6 scavenger studies, is sufficient to observe the vibrational structure possessed by anions long lived enough to undergo appreciable nuclear distortion. Furthermore the apparatus is simpler than electrostatically focused instruments and much less sensitive to the difficulties encountered in studies of reactive compounds.

Briefly, the trochoidal monochromator, which incorporates an axial magnetic field and a crossed electric field for energy dispersion, produces a beam of current, typically 10^{-8} to 10^{-9} A, which is directed into a gas cell. The gas density in the cell is adjusted so that the current arriving at the electron beam collector is reduced to approximately e^{-1} of its initial value at energies where the scattering is large. The rejection of the scattered electrons takes place primarily at a retarding electrode following the gas cell, where a potential barrier reflects electrons whose axial velocities have been reduced below a selected value. In the absence of the potential barrier at the cell exit, some scattered electrons will be rejected by other means such as multiple scattering and collisions with the aperture edges. At low energies an additional rejection mechanism becomes significant: Electrons which are elastically scattered into a cone around 180° can re-enter the monochromator where they disperse and are lost (15). The transmitted beam thus reflects the loss due to differential elastic scattering rather than the total cross section. As we illustrate later, the visibility of vibrational structure in a resonance is often enhanced in the "backscattering" mode. Finally, in the innovation introduced by Sanche and Schulz (13), a small ac voltage is applied to a cylinder within the gas cell thus modulating the energy of the electrons, and the derivative with respect to energy of the unscattered or "transmitted" beam is detected.

Following studies in hydrocarbons by Sanche and Schulz (16) and Nenner and Schulz (17), we began applying ETS to chemical problems (18) in 1975, being particularly motivated to correlate the anion states of hydrocarbons with their electronic structure. Electron transmission studies are now being carried out by a number of other groups in the U.S. and in Europe. A bibliography of published work is available from the authors.

In this paper we first review the characteristics of resonances in molecules, and then discuss recent experimental results in polyatomic systems, with examples chosen to illustrate the types of information that can be provided by ETS. We discuss some of the problems uncovered, and the role that other electron scattering methods and theory could play in solving these problems.

Resonance Characteristics

Temporary anion states may be broadly classified either as shape resonances or core-excited resonances (4). The former are well described by a configuration in which the impacting electron attaches to an atom or molecule in one of the originally unoccupied orbitals. In the latter, electron capture is accompanied by electronic excitation, giving rise to a temporary anion with a two-particle-one-hole (2p-1h) configuration. One can further distinguish core-excited resonances into those in which the resonance lies energetically below its parent state and those in which it lies above. The former are referred to as Feshbach resonances and the latter as core-excited shape resonances since they can also be viewed as resulting from electron attachment to an excited state of the neutral molecule. Although all three types of resonances appear in ET spectra, shape resonances are dominant at low energies in unsaturated molecules, and we will devote most of our discussion to these.

Next, we review briefly the connection between the lifetimes of temporary anions and orbital symmetry. In the case of an atom, the polarization and centrifugal terms combine to give an electron-atom interaction of the form:

$$V(r) = -\frac{\alpha}{2r^4} + \frac{\ell(\ell+1)}{2r^2}$$

at sufficiently large distances from the atom. Here α denotes the polarizability and ℓ the angular momentum associated with the orbital into which the electron is captured. When $\ell \neq 0$, the resonance has a finite lifetime due to tunneling of the electron through the angular momentum barrier. ETS measurements in the group IIa and IIb atoms provide examples of pure p-wave and d-wave shape resonances (19). The usual view is that in the absence of an angular momentum barrier, there is no temporary electron capture or time delay of the scattered electron.

The extension of this picture to shape resonances in molecules requires that the single value of ℓ be replaced by an infinite sum over angular momentum components. However, if the molecular symmetry is sufficiently high, resonances may be well characterized by only one or two ℓ values. For example, the $^2\Pi_g$ shape resonance of N_2 is well described by a d-wave angular scattering distribution (4). CO, on the other hand, has important ℓ=1 and ℓ=2 components in its angular distribution, the former arising from the different size of the oxygen and carbon 2p orbitals. One important consequence of the lower symmetry of CO is that its $^2\Pi$ shape resonance has a shorter lifetime than that of N_2, as reflected, for example, by the much weaker vibrational structure in the ET spectrum (20). Molecules such as benzene and ethylene have sufficiently high symmetry that their shape resonances should be dominated by a single angular momentum component. Indeed, from the limited angular scattering studies that have been carried out (21,22), this appears to be the case. On the other hand, many polyatomic molecules with no overall symmetry have been found by ETS to have shape resonances (20). This implies that the wavefunctions of the "extra" electrons in these systems have a sizeable mixture of components with ℓ>0, and it indicates that the unsaturated portion of the molecule possesses local symmetry.

For most of the polyatomic molecules we have studied, the shape resonances have lifetimes in the range of 10^{-13} to 10^{-15} seconds, although resonances at energies close to 0 eV may have much longer lifetimes (4). The longer-lived shape resonances typically display well defined structure due to nuclear motion, whereas those in which the electron detaches in a time short compared to that required for appreciable motion of the nuclei are broad and featureless, with widths of the order of 1 eV or more. Theoretical calculations (23) have shown that structure due to nuclear motion may still be weakly visible in cases in which the electron detaches in a time as short as one tenth of the period of the particular vibration.

Core-excited resonances, which are the counterpart to shake-up processes in PES, usually occur at higher energies and are generally less prominent than shape resonances. The two types of core-excited resonances have very different lifetimes; the core-excited shape resonances have short lifetimes because they decay rapidly into their "parent" states, and do not typically display vibrational structure. On the other hand, Feshbach resonances usually have long lifetimes and sharp vibrational structure, since their decay must be accompanied by electron rearrangement (4). In general, 2p-1h core-excited resonances in which the particle orbitals derive from valence orbitals lie above one or more of their parent states, while those derived from Rydberg orbitals lie below (4). This characteristic ordering follows from the relative amounts of Coulomb repulsion in the two cases.

Ethylene, Butadiene, and Hexatriene

Electron transmission data in the sequence of molecules -- ethylene, butadiene, and hexatriene -- illustrate many of the features of temporary anion states in unsaturated compounds. Our most recent measurements (24) in these compounds are shown in Figure 1, which displays the derivative of transmitted current as a function of electron energy. Ethylene, the prototypical alkene, has been studied frequently using transmission methods (7,16,18). The shape resonance near 1.7 eV was assigned to electron capture into the $b_{2g}(\pi^*)$ orbital by Bardsley and Mandl (25). In contrast to earlier studies, our work (18) showed weak undulations due to nuclear motion which we attributed to excitation of the C-C stretch mode. The weak structure is consistent with an anion lifetime shorter than a period of this vibration. Walker et al. (22) have studied the excitation of the vibrational levels of the electronic ground state proceeding through this resonance and found that the C-C stretch is the dominant mode which is excited. Theoretical calculations (26) have shown that the equilibrium structure of the anion is highly nonplanar. It thus appears that electron detachment occurs sufficiently rapidly that the molecule has little opportunity to undergo out-of-plane distortion.

The ET spectrum of trans-butadiene shows two well-defined resonances which we attributed (18) to occupation of the two empty π^* orbitals. The lower resonance lies below that of ethylene and exhibits sharper structure. Figure 2 shows these data on an expanded energy scale for both "high rejection" conditions in which the signal derives from the total scattering cross section, and "low rejection" which reflects the differential elastic scattering near 180° (15). The symmetric C-C vibrations of the anion are the most pronounced, but there is evidence for low frequency out-of-plane modes as well. The upper resonance lies above that of ethylene and is featureless.

These results provide a nice confirmation of expectations based on π MO theory and show also the interplay between symmetry, resonance energy, and anion lifetime. Although the first orbital of butadiene has a leading partial wave of $\ell=1$, and that of ethylene has $\ell=2$, the substantially lower energy of the former results in a slower rate of tunneling. The increased lifetime makes possible the

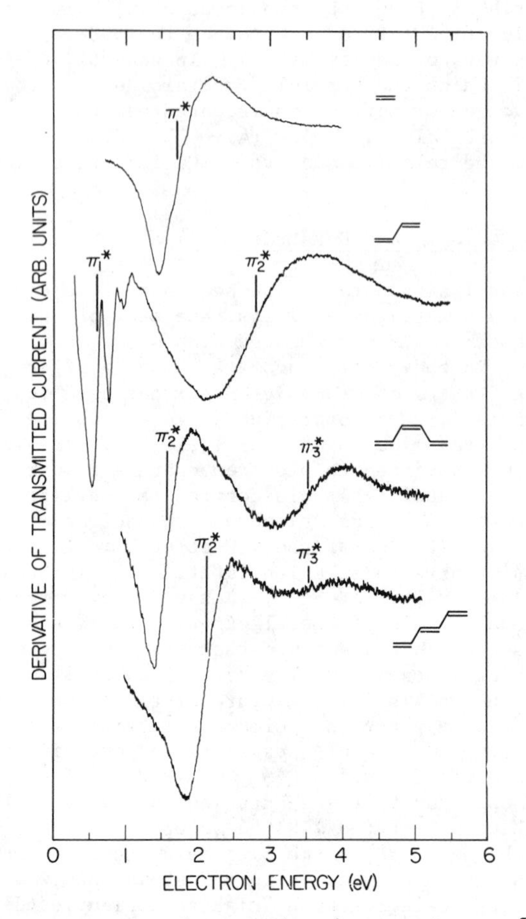

Figure 1. The derivative of transmitted current as a function of electron energy in ethylene, butadiene, cis- and trans-hexatriene (from Ref. 24).

appearance of the low-frequency modes in the spectrum. The ethylene anion, with its shorter lifetime, simply does not live long enough to permit its low-frequency out-of-plane modes to appear in the ET spectrum.

In the lower portion of Figure 1, the ET spectra (27) of the cis and trans isomers of hexatriene, separated chromatographically, are displayed. Two pronounced shape resonances are seen in the spectra of each isomer. Hexatriene might be expected to have three shape resonances, one associated with each of its empty π^* orbitals. The ground state anion, however, is stable, and the two features observed in the ET spectra therefore correspond to the excited states. The energy of the first resonance of the cis isomer is 0.55 eV lower than that of the trans isomer. This was unexpected since the π IP's of the cis and trans isomers agree to better than 0.1 eV (28). We have attributed the "extra" stability of this anion of cis-hexatriene to a strong C_2-C_5 bonding interaction in the π_2^* orbital. This interaction is much more important in the first excited anion state than in the ground or second excited anion states because of the larger charge densities at the C_2 and C_5 atoms. Since the filled π_2 orbital also has large densities at these sites, we concluded that long-range through-space interactions are more important in the anion state due to the more extended nature of the anion wavefunction. This interpretation was supported by calculations. The importance of such interactions has been noted for several other anions (29-33).

At higher energies, other types of resonances are found in these molecules. In ethylene, narrow Feshbach resonances above 6.6 eV were first observed by Sanche and Schulz (16). In Figure 3 we show the spectrum in this region measured at somewhat higher resolution. The first Rydberg state of the neutral molecule is characterized by the $C_2H_4^+$ ground state core plus a $3s(a_{1g})$ electron. Sanche and Schulz (16) suggested that the first Feshbach resonance is formed by adding a second 3s electron to this Rydberg state.

The spectrum serves to illuminate several of the characteristics of core-excited resonances. The "doublet" structure, repeated at intervals of approximately 170 meV, is characteristic of the ν_2 C-C symmetric stretch. The spacing between the first and second features in each pair is 60 meV. We attribute these to two quanta of the CH_2 torsional mode, i.e. $2\nu_4$. These characteristic energies in the anion are quite close to those of the Rydberg "parent" state, as expected. The existence of the low frequency modes is a clear indication of the long lifetime of the Feshbach resonance relative to that of the $^2B_{2g}$ shape resonance discussed previously.

At energies above that of the singlet Rydberg state at 7.11 eV, indicated by S in Figure 3, the vibrational levels of the $\pi^{-1}(3s)^2$ resonance may decay into the $\pi^{-1}3s$ parent state thus shortening the lifetime and broadening the structure. At the time these data were taken, however, it was not clear why the third doublet, which lies below the singlet Rydberg state, was also broadened. This puzzle was resolved by data of Wilden and Comer (34) who located the companion triplet Rydberg state at 6.98 eV (with $2\nu_4$ at 7.03 eV), indicated by T on Figure 3, into which the resonances may also decay.

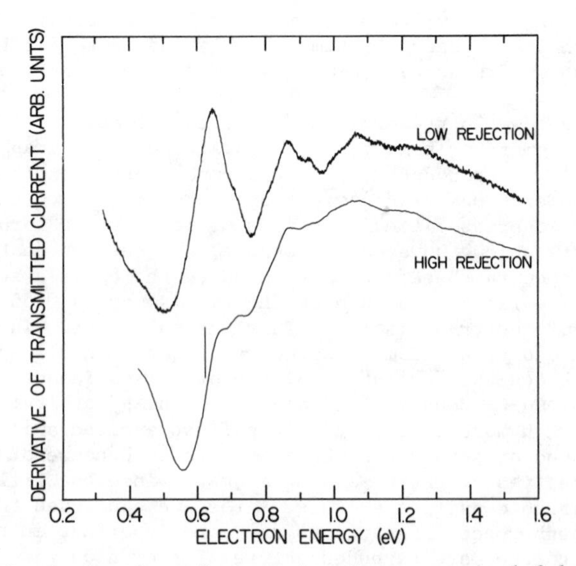

Figure 2. The derivative of transmitted current in 1,3-butadiene. The curve marked "low rejection" is obtained by rejecting only those elastically scattered electrons which return to the monochromator. The "high rejection" data is derived by rejection of all scattered electrons (from Ref. 24).

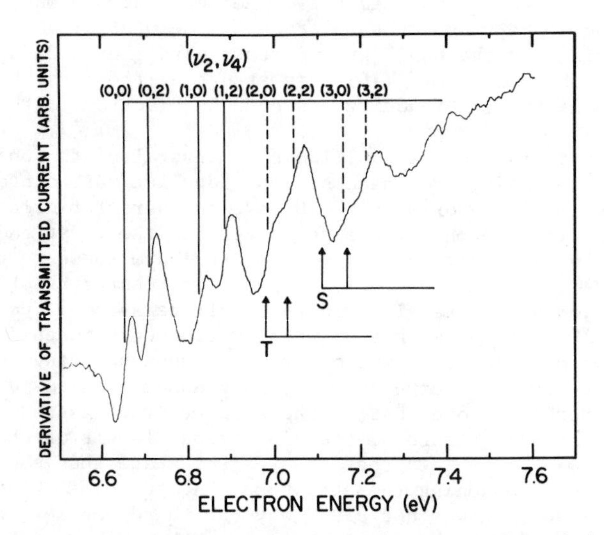

Figure 3. The derivative of transmitted current in ethylene showing the resonances associated with the lowest singlet (S) and triplet (T) Rydberg states.

We have not detected Feshbach resonances in butadiene or hexatriene, and in general, it appears that such resonances are much weaker in larger hydrocarbons than in di- and triatomic molecules (16). Most hydrocarbons, including butadiene and hexatriene, however, display broad resonances at high energy which could be due either to shape resonances with attachment into σ^* orbitals or to core-excited resonances in which electron capture into a π^* orbital is accompanied by $\pi \rightarrow \pi^*$ excitation. Examples of such resonances will be discussed later. A definitive assignment of these broad states awaits a study of their decay channels by electron energy loss methods.

Nonconjugated Dienes

The problem of long range interactions between widely separated subunits in molecules is of importance in many areas of chemistry and biology. Although many studies have appeared characterizing these interactions in excited states and cationic species, very little has been done in gas-phase anions. For these reasons we have undertaken a study of such interactions in nonconjugated dienes and diones. Several of the dienes we have examined (35) are shown below:

(1) 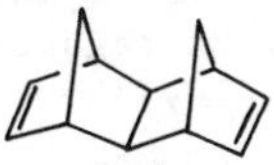 (2) 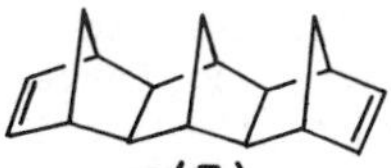 (3)  (4) 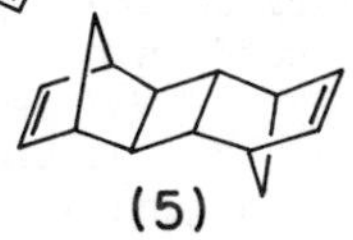 (5)

The excellent agreement between the measured splittings of the anion states and the values obtained from Koopmans' theorem (i.e., derived from the energies of the appropriate virtual orbitals of SCF calculations on the neutral molecules) supports the use of an orbital picture for interpreting these experiments. In compounds (1) and (4), the two double bonds are sufficiently close that much of the splitting between the π^* orbitals arises from direct, or through-space, interactions. On the other hand, the π^* orbitals of (2) and (5) are split by 0.8 eV and 0.6 eV, respectively. These splittings are over an order of magnitude greater than those which could arise from through-space interactions in these compounds. Rather, they arise from the through-bond interaction due to the $\sigma - \pi^*$ and $\sigma^* - \pi^*$ mixing made possible by the nonplanarity of the molecules.

To understand better the nature of these interactions, we consider the $\pi^+ = (\pi_a + \pi_b)$ and $\pi^- = (\pi_a - \pi_b)$ delocalized orbitals formed from the π_a and π_b localized orbitals. In the absence of through-bond interaction, the π^+ and π^- orbitals would be nearly degenerate. The π^+ orbital is symmetric with respect to the plane bisecting the molecule "parallel" to the double bonds, while the π^- orbital is antisymmetric. Consequently the π^+ orbital can mix only with symmetric σ orbitals and the π^- orbital with antisymmetric σ orbitals. Since the symmetric and antisymmetric orbitals are at appreciably different energies, this mixing causes a splitting between the π^+ and π^- orbitals. A detailed analysis of the results for (2) shows that the π^+ orbital is stabilized because it mixes more strongly with σ^* orbitals than with σ orbitals, while the π^- orbital is destabilized due to a greater mixing with σ orbitals than with σ^* orbitals.

Theoretical work suggests that through-bond splittings of a few tenths of an eV are possible even when the double bonds are separated by 9 - 10 Å as in (3). Unfortunately, we have not been able to confirm this by ETS since the widths of the resonances are much greater than the expected splittings. It is possible that measurements of the excitation functions of the vibrational levels or the angular scattering distribution could provide information on the splittings in these cases.

Shape and Core-excited Resonances in Benzene

The benzene molecule and its derivatives have been extensively studied using a variety of spectroscopic techniques. Not only does benzene serve as a prototype for aromatic systems, but the degeneracy of its highest occupied and lowest unoccupied molecular orbitals makes it an important system for the study of substituent and Jahn-Teller effects.

The shape resonances of benzene and some of its derivatives have been explored by a number of investigators using ETS (16,17,20, 36). With the exception of questions regarding pseudo-Jahn-Teller problems (37), for example in the alkyl-benzenes, substituent effects on the doubly-degenerate $^2E_{2u}$ ground state anion near 1.1 eV can be generally understood in terms of inductive and mesomeric effects.

In the remainder of this section we consider the higher lying resonances in benzene. In general, the simple classification of resonances involving valence orbitals as either "shape" or "core-excited" appears to break down as the anion energy increases (38). For example, the parentage of the 1^2B_{2g} "shape" resonance at 4.8 eV is not clear-cut, since it is known to decay into electronically excited states of benzene as well as the ground state (39,40). It has been suggested previously that this anion must be described by an admixture of both shape and core-excited configurations (17).

In Figure 4 we show the transmission spectrum of benzene at higher energies. The lowest feature is the second "shape" resonance ($^2B_{2g}$). Above this lie several smaller features which must correspond primarily to core-excited states. Detailed studies of the decay channels of these resonances have not yet been carried out, although some information is available. The vertical arrows in Figure 4 indicate the energies at which maxima occur in the excitation functions for ν_1 of the X^1A_{1g} ground electronic state of benzene, as observed by Azria and Schulz (50), and for the lowest vibrational level of the first triplet excited state ($^3B_{1u}$) as measured recently by Allan (41). The decay of the resonances into a number of other excited states has also been observed (41). The shading is a very crude indication of the half-width of the peaks occurring in these cross sections. Note that the peaks in the excitation functions will generally correlate with the midpoints of the structures in the transmission spectrum.

This comparison shows that a given high-lying resonance may decay into either the ground state or an electronically excited

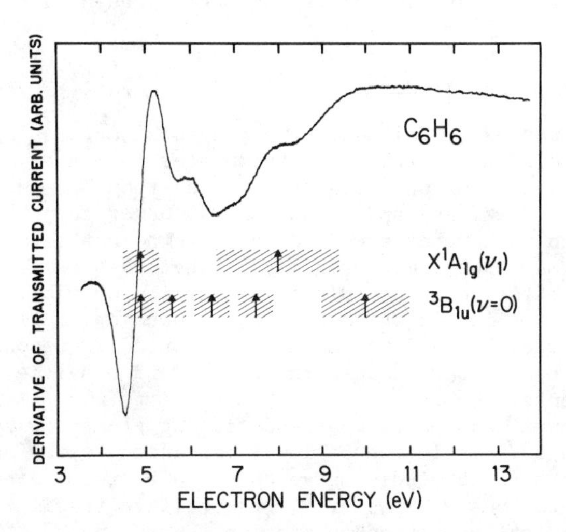

Figure 4. The derivative of transmitted current in benzene. The largest feature is the $^2B_{2g}$ resonance at 4.8 eV. The vertical arrows locate the maxima for decay into the X $^1A_{1g}$ ground state and the $^3B_{1u}$ excited state of benzene. The shaded regions show roughly the half-widths of these peaks (from Ref. 24).

state, or both. In more complex molecules such as naphthalene or styrene, the $\pi \rightarrow \pi^*$ transitions occur at still lower energies, and the number of core-excited resonances is even greater. Thus, accounting for the structure in the total cross sections is a considerable challenge for theorists since a number of configurations are required for an adequate description of such resonances. A discussion of these points and conjectures concerning their relationship to the lifetime of the anions have been presented elsewhere (38).

Resonances Associated with σ^* Orbitals

The shape resonances described in the previous sections have all been associated with π^* orbitals. Resonances formed by attachment into the σ^* orbitals of unsubstituted hydrocarbons typically occur at high energy ($E > 5$ eV) and are usually broader than π^* resonances. In compounds in which they are hard to discern in the total cross section, they may appear more readily in the cross sections for vibrational excitation, since direct excitation of vibration is weak, at least for levels which are not allowed optically, and hence the interfering background is small. Such broad σ^* resonances, for example, have been observed in methane (42) and ethylene (22) using such measurements.

We have examined the role of substituent groups on hydrocarbons to learn which systems introduce low-lying anion states associated with σ^* orbitals. ETS studies have shown that groups containing first row heteroatoms such as F, N, or O fail to stabilize the σ^* orbitals sufficiently to produce prominent resonances at low energy in the ET spectra. The situation is quite different for heterocompounds substituted with second, third, and fourth row atoms; all of these compounds are found to have well defined σ^* resonances below 4 eV.

In Figure 5, the ET spectra of CH_3Cl, CH_2Cl_2, $CHCl_3$, and CCl_4 are reproduced (43). Each of these molecules shows one or two resonances which we attribute to C-Cl σ^* orbitals. Support for this assignment is provided by the number of resonances which appear and the trends in their relative energies as the number of Cl atoms increases, as well as by theoretical calculations. CH_3Cl has a single broad resonance, centered at 3.5 eV. As expected, CH_2Cl_2 has two resonances corresponding to the bonding and antibonding combinations of the two localized C-Cl σ^* orbitals. Both $CHCl_3$ and CCl_4 should each have only two low-lying anion states, doubly degenerate in the former and triply degenerate in the latter. The ET spectrum of $CHCl_3$ shows two resonances as expected, but that of CCl_4 displays only one low-lying resonance. In this case the increasing stability brought about by Cl substitution causes the $^2A_1$ anion to be bound, and thus only the $^2T_2$ anion is seen in the spectrum. There is no sign of vibrational structure in any of the resonances. In the absence of other data, it is not clear whether the widths derive primarily from the short lifetimes or from the repulsiveness of the anion curves in the Franck-Condon region.

The anion states appearing in the ET spectra may also give rise to stable anion fragments via the dissociative attachment process. Because of the large electron affinity of the chlorine atom, the σ^*

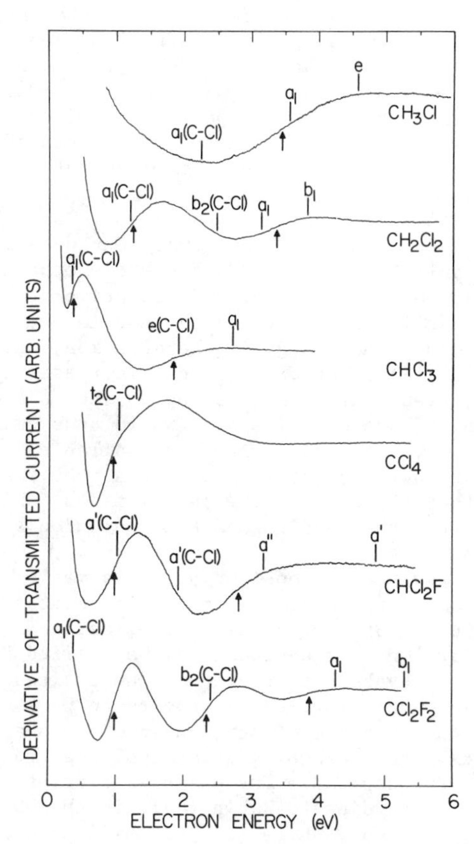

Figure 5. The four upper curves show the derivative of transmitted current in the chloro-substituted methanes. The vertical arrows indicate the experimental midpoints of the resonances. The vertical lines locate the theoretical anion energies and show the orbital symmetries. The theoretical energies are normalized to the experimental data only at the $^2A_1$ resonance in $CHCl_3$. Reproduced with permission from Ref. 43. Copyright 1982, American Institute of Physics.

resonances in the chloromethanes described above may decay readily in this channel. In the simplest picture, the yield of stable anions is viewed as arising from a cross section for attachment modulated by an escape probability which will be influenced strongly by the lifetime of the resonance and the rate of motion on the anion surface. The escape probability may modify considerably both the shape and position of a resonance as viewed by ETS. As the resonance lifetime shortens, the yield of stable anion fragments is reduced, and the maximum in the yield is shifted to lower energies. One should not therefore expect the energies of the peaks determined in the ETS and DA measurements to agree. Indeed, in the chloromethanes the first peak for Cl^- production occurs at much lower energy than that for electron capture into the resonance.

In substituted unsaturated compounds both σ^* and π^* resonances may appear at low energy in the ET spectra. Our data in the chloroethylenes (44) reveal that the σ^* resonances lie typically 1 - 2 eV above the ground state π^* resonance. Dissociative attachment in the unsaturated compounds is more interesting than in the chloromethanes described above because the energy of the $^2\Sigma$ anion, which correlates with the $R(^2\Sigma) + Cl^-(^1S)$ asymptote, rapidly decreases as the C-Cl bond is stretched, resulting in a crossing of the $^2\Sigma$ and $^2\Pi$ surfaces. Out-of-plane motion of the chlorine atom would thus cause the surfaces to undergo an avoided crossing. The available dissociative attachment data for the chloroethylenes show a peak in the Cl^- production close to that for forming the $^2\Pi$ anion (45) which we believe can be understood from the coupling of the two anion surfaces in this energy region.

Over the past few years, considerable information (29,36,46,48) has been acquired on σ^* resonances in molecules containing other hetero atoms, including Si, Ge, P, As, Sb, S, Se, and Br. One problem in which the data on σ^* resonances may be applied concerns the degree of valence vs. Rydberg character of the excited states of neutral molecules. In particular, there appears to be a direct correspondence between the presence of pronounced σ^* resonances in the ET spectrum and the existence of well-defined valence transitions into the σ^* orbitals of the neutral. For example, singlet valence transitions are not seen, according to Robin (49), in the absorption spectra of HF, H_2O, NH_3, or CH_4, while such transitions are quite intense in HCl, H_2S, PH_3, and SiH_4. The bonds involving group II and heavier hetero atoms are longer and weaker than those of the corresponding first row compounds. As a result, the associated σ orbitals are less strongly bound, and the σ^* orbitals are less strongly antibonding. As a consequence the σ^* orbitals lie lower in energy and hence are more valence-like for the second row compounds.

Conclusions

In this review, we have shown that electron transmission spectroscopy provides a useful and relatively simple way to survey the temporary negative ion states of complex molecules. The utility of the technique has been further enhanced by a number of recent developments which we briefly mention here.

i) The mechanisms by which the scattered electrons are rejected has been clarified (15), and a better understanding of certain artifacts

which may appear in the spectra has been achieved. As shown in Figure 2, it has been found that the visibility of vibrational structure in some low-lying resonances may be improved by rejecting only those electrons elastically scattered back into the monochromator.
ii) McMillan and Moore (50) have carried out a study of the monochromator and used the results to design an improved instrument.
iii) Preliminary studies of ETS using a free jet, rather than a static gas cell, have been performed (51). The instrumental resolution, particularly in the wings of the electron energy distribution, was improved enough to observe the doublet structure due to spin-orbit splitting of the $^2\Pi_g$ resonances in O_2.
iv) The feasibility of detecting the temporary anion states of free radicals by ETS was demonstrated at an early stage by measurements in atomic hydrogen (52) and atomic oxygen (53). This work has not yet been followed by studies in molecular radicals, although ETS has been applied to vibrationally excited species by Michejda et al. (54), who detected vibrationally excited levels of N_2 by observing the "hot bands" in the ET spectrum of N_2 which had passed through a microwave discharge.

In spite of these developments, it is important to realize that ETS provides no information on the decay channels of temporary anions. Further progress toward a detailed understanding of resonances in complex molecules will require careful studies of the vibrational levels and electronically excited states of the neutral molecule which are formed upon electron detachment. These measurements provide essential data related to the electronic configuration of the anion and the distortion it undergoes during its lifetime. The angular distributions of the scattered electrons also yield information on the anion configuration, although they may be useful in practice only if the molecular symmetry is high.

The apparatus required to study decay processes is much more complex and expensive than that used in ETS. Recently, however, Allan (41) has modified the "standard" transmission spectrometer to include a two-stage trochoidal energy analyzer which permits energy loss and excitation function measurements to be carried out on those electrons which scatter in the forward direction. A similar apparatus has been constructed in our group at Pittsburgh. This experimental arrangement provides a relatively simple and inexpensive means for measurements of energy loss and excitation functions in chemically interesting systems.

Relatively little data are available on dissociative attachment processes in polyatomic molecules. Such processes are intrinsically complicated in large molecules because they may occur through direct dissociation along a repulsive anion surface or through predissociation via the coupling of two electronic states. Measurements of the absolute yield of the charged fragments and the internal energy distributions of the fragments as a function of electron impact energy will contribute to our understanding of the shapes of the anion potential surfaces and the lifetimes of the resonances.

In our concluding remarks we would like to turn to the role of theory in the study of temporary anions in polyatomic molecules.

At the simplest level, theoretical work has proven useful in assigning the anion states and in determining the relative importance of various "chemical interactions" in polyatomic molecules. In our own work, we have utilized Koopmans' theorem together with the virtual orbitals from SCF calculations on the closed-shell neutral molecules. Strictly speaking, such a procedure is not legitimate since the variational principle is not applicable to temporary anion states. Both the anion energies and the energies of the virtual orbitals depend strongly on the basis set being employed, and for very flexible basis sets the wavefunctions would "collapse" back to those of the neutral molecule plus a free electron. Nonetheless, we have found that provided "uniformly good" basis sets of moderate size are employed, such as the 6-31G* basis sets of Pople and coworkers (55), the relative orbital energies along a series of related compounds generally correlate quite well with the anion energies measured experimentally. As useful as such calculations have been, they cannot provide any insight into the anion lifetimes or the profiles of the resonances to be expected in the scattering cross sections.

Full scattering calculations have not yet been performed for most of the polyatomic molecules considered here, and L^2 techniques such as the coordinate rotation method which give good agreement with experiment for the energies and widths of the shape resonances of Mg (56,57) and N_2, have not yet been successfully applied to a molecule such as ethylene. The method which has proven most successful to date is the multiple-scattering $X\alpha$ model. The application to resonances was originally pioneered by Dill and Dehmer (58), and it has been applied in a slightly different form by Bloor and coworkers (59,60) to a variety of polyatomic molecules. Recently, Tossell and Davenport (61) have also used this method to characterize shape resonances in inorganic and organometallic species. However, as currently formulated, the $X\alpha$ method is not able to provide information on core-excited resonances. Hence, it could not be utilized to unravel the multitude of anion states appearing in a molecule such as naphthalene. Nor can it provide theoretical estimates of the vibrational structure in resonances or the population of vibrational levels of the neutral molecule following the decay of the resonance.

Acknowledgments

The research described here has been carried out with the support of the National Science Foundation. We gratefully acknowledge the contributions of our coworkers -- J. A. Michejda, N. S. Chiu, A. Modelli, A. R. Johnston, A. E. Howard, L. Ng, V. Balaji and T. M. Stephen.

Literature Cited

1. Stuart, H. A. "Molekülstruktur"; Springer Verlag: Berlin 1934, p. 52.
2. Massey, H. S. W. "Electronic and Ionic Impact Phenomena"; Oxford University Press: London, 1969; Vol. 2.
3. Haas, R. Z. für Phys. 1957, 148, 177.
4. Schulz, G. J. Rev. Mod. Phys. 1973, 45, 378, 423.
5. Bowman, G. R.; Miller, W. D. J. Chem. Phys. 1965, 42, 681.
6. Compton, R. N.; Christophorou, L. G.; Huebner, R. H. Phys. Letters 1966, 23, 656.

7. Boness, M. J. W.; Larkin, I. W.; Hasted, J. B.; Moore, L. Chem. Phys. Letters 1967, 1, 292.
8. Brongersma, H. H. Thesis, Leiden, 1968.
9. Compton, R. N.; Huebner, R. H. "Advances in Radiation Chemistry"; Wiley-Interscience: New York, 1970; p. 281.
10. Christophorou, L. G. "Atomic and Molecular Radiation Physics"; Wiley-Interscience: London, New York, 1971.
11. Schulz, G. J. Phys. Rev. 1958, 112, 150.
12. Curran, R. K. J. Chem. Phys. 1963, 38, 780.
13. Sanche, L.; Schulz, G. J. Phys. Rev. A 1972, 5, 1672.
14. Stamatovic, A.; Schulz, G. J. Rev. Sci. Instr. 1970, 41, 423.
15. Johnston, A. R.; Burrow, P. D. J. Elect. Spectrosc. and Relat. Phenom. 1982, 25, 119.
16. Sanche, L.; Schulz, G. J. J. Chem. Phys. 1973, 58, 479.
17. Nenner, I.; Schulz, G. J. J. Chem. Phys. 1975, 62, 1747.
18. Burrow, P. D.; Jordan K. D. Chem. Phys. Letters 1975, 36, 594.
19. Burrow, P. D.; Michejda, J. A.; Comer, J. J. Phys. B: Atom. Molec. Phys. 1976, 9, 3225.
20. Jordan, K. D.; Burrow, P. D. Acc. Chem. Research 1978, 11, 341.
21. Wong, S. F.; Schulz, G. J. Phys. Rev. Letters 1975, 35, 1429.
22. Walker, I. C.; Stamatovic, A.; Wong, S. F. J. Chem. Phys. 1978, 69, 5532.
23. Birtwistle, D. T.; Herzenberg, A. J. Phys. B: Atom. Molec. Phys. 1971, 4, 53.
24. Burrow, P. D.; Michejda, J. A.; Jordan, K. D., in preparation.
25. Bardsley, N.; Mandl, F. Rept. Progr. Phys. 1968, 31, 471.
26. Paddon-Row, M. N.; Rondan, N. G.; Houk, K. N.; Jordan, K. D. J. Am. Chem. Soc. 1982, 104, 1143.
27. Burrow, P. D.; Jordan, K. D. J. Am. Chem. Soc. 1982, 104, 5247.
28. Beez, M.; Bieri, G.; Bock, H.; Heilbronner, E. Helv. Chim. Acta 1973, 56, 1028.
29. Burrow, P. D.; Jordan, K. D. Chem. Phys. Letters 1975, 36, 594.
30. Chiu, N. S.; Burrow, P. D.; Jordan, K. D. Chem. Phys. Letters 1979, 68, 121.
31. Burrow, P. D.; Ashe, A. J., III; Bellville, D. J.; Jordan, K. D. J. Am. Chem. Soc. 1982, 104, 425.
32. Staley, S. W.; Giordan, J. C.; Moore, J. H. J. Am. Chem. Soc. 1981, 103, 3638.
33. Staley, S. W.; Bjorke, M. D.; Giordan, J. C., McMillan, M. R.; Moore, J. H. J. Am. Chem. Soc. 1981, 103, 7057.
34. Wilden, D. G.; Comer, J. J. Phys. B: Atom. Molec. Phys. 1980, 13, 1009.
35. Balaji, V.; Jordan, K. D.; Burrow, P. D.; Padden-Row, M. N.; Patney, H. K. J. Am. Chem. Soc. 1982, 104, 6849.
36. Giordan, J. C.; Moore, J. H. J. Am. Chem. Soc. 1983, 105, 6541.
37. Jordan, K. D.; Michejda, J. A.; Burrow, P. D. J. Am. Chem. Soc. 1976, 98, 1295.
38. Jordan, K. D.; Burrow, P. D. Chem. Phys. 1980, 45, 171.
39. Smyth, K. C.; Schiavone, J. A.; Freund, R. S. J. Chem. Phys. 1974, 61, 1782, 1789.
40. Azria, R.; Schulz, G. J. J. Chem. Phys. 1975, 62, 573.
41. Allan, M. Helvetica Chim. Acta 1982, 65, 2008.
42. Rohr, K. J. Phys. B: Atom. Molec. Phys. 1980, 13, 4897.

43. Burrow, P. D.; Modelli, A.; Chiu, N. S.; Jordan, K. D. J. Chem. Phys. 1982, 77, 2699.
44. Burrow, P. D.; Modelli, A.; Chiu, N. S.; Jordan, K. D. Chem. Phys. Letters 1981, 82, 270.
45. Christophorou, L. G. In "Photon, Electron, and Ion Probes of Polymer Structure and Properties"; Dwight, D. W.; Fabish, T. J.; Thomas, H. R., Eds.; ACS SYMPOSIUM SERIES No. 162, American Chemical Society: Washington, D.C., 1981; pp. 11-34.
46. Burrow, P. D.; Jordan, K. D., unpublished data.
47. Giordan, J. C. J. Am. Chem. Soc. 1983, 105, 6544.
48. Modelli, A.; Jones, D.; Distefano, G. Chem. Phys. Letters 1982, 86, 434.
49. Robin, M. B. "Higher Excited States of Polyatomic Molecules"; Academic: New York, 1974; Vol. I.
50. McMillan, M. R.; Moore, J. H. Rev. Sci. Instr. 1980, 51, 944.
51. Stephen, T. M.; Burrow, P. D., unpublished data.
52. Sanche, L.; Burrow, P. D. Phys. Rev. Lett. 1972, 29, 1639.
53. Spence, D.; Chupka, W. A. Phys. Rev. A 1974, 10, 71.
54. Michejda, J. A.; Dubé, L. J.; Burrow, P. D. J. Appl. Phys. 1981, 52, 3121.
55. Hehre, W. J.; Ditchfield, R.; Pople, J. A. J. Chem. Phys. 1972, 56, 2257.
56. McCurdy, C. W.; Lauderdale, J. G.; Mowrey, R. C. J. Chem. Phys. 1981, 75, 1835.
57. Donnelly, R. A.; Simons, J. J. Chem. Phys. 1980, 73, 2858.
58. Dill, D.; Dehmer, J. L. J. Chem. Phys. 1974, 61, 692.
59. Bloor, J. E.; Sherrod, R. E.; Grimm, F. A. Chem. Phys. Letters 1981, 78, 351.
60. Bloor, J. E.; Sherrod, R. E. Chem. Phys. Letters 1982, 88, 389.
61. Tossell, J. A.; Davenport, J. W. J. Chem. Phys. 1984, 80, 813.

RECEIVED June 11, 1984

10

Negative Ion States of Three- and Four-Membered Ring Hydrocarbons

Studied by Electron Transmission Spectroscopy

ALLISON E. HOWARD and STUART W. STALEY

Department of Chemistry, University of Nebraska, Lincoln, NB 68588-0304

Electron transmission spectra have been obtained for the C_3-C_6 cis-cycloalkenes (**3-6**) and cycloalkanes (**7-10**). The somewhat low value for the first negative ion state of **3** (1.73 eV) can be understood on the basis of compensating effects due to the short C=C bond and the "reversed" polarity of **3** relative to **4-6**. The lowest * resonances observed in **3**, **4**, **7**, and **8** have been assigned on the basis of ab initio 6-31G molecular orbital calculations and by consideration of the components of the angular momentum associated with each resonance.

Three- and four-membered ring compounds have been of great interest to chemists for over a half century. This is due primarily to the fact that the high degree of ring strain in these compounds, particularly in cyclopropane and cyclopropene and their derivatives, confers upon them certain properties usually associated with "unsaturated" or olefinic compounds. Such properties originate from the character of the "frontier orbitals", i.e., the higher occupied and lower unoccupied molecular orbitals. Although the occupied orbitals have been extensively investigated by photoelectron spectroscopy, essentially no research has been reported on the unoccupied orbitals of small-ring compounds.

This paper summarizes the first investigations of three- and four-membered ring compounds by the technique of electron transmission spectroscopy (ETS). We will briefly discuss two general areas associated with the negative ion states of small ring hydrocarbons: a) π^* states in cyclopropene and related molecules, and b) negative ion states associated with the system in cycloalkenes and cycloalkanes. Of special interest is an analysis which we believe allows us to assign orbital symmetries to the σ^* negative ion states of cyclopropane, cyclopropene, cyclobutane and cyclobutene.

0097-6156/84/0263-0183$06.00/0

Experimental Method

Electron transmission spectroscopy (ETS) is an invaluable technique for obtaining gas-phase electron affinities (1,2). In this experiment, a monoenergetic beam of electrons (FWHM = 20-30 meV) is allowed to interact with molecules in a static cell. The electrons then pass through a retarding region and those electrons which possess an axial velocity sufficient to overcome the retarding potential are collected. The spectra discussed in this paper were obtained under high rejection conditions (3); that is, they approximate the total electron scattering cross-sections for the molecules of interest.

A sharp decrease in the transmitted current is seen at energies which correspond to the energies of the negative ion states of these molecules. In the context of Koopmans' theorem (4), these resonances are associated with the negative of the SCF molecular orbital energies of the neutral molecule. In order to accentuate the variations in the transmitted current, a small (20-50 meV) AC voltage (1) is applied to the collision chamber. Thus, the derivative of the transmitted current with respect to energy is obtained.

The $^2P_{3/2}$ resonance in argon (5) was used to calibrate the spectra. We have found this to be a useful calibration gas due to the narrow width, symmetrical profile, and relatively low energy of the resonance.

The relative errors are estimated to be ± 0.8% of the attachment energy. The actual errors associated with these energies are probably much larger, on the order of ± 0.05 eV below 2 eV and ± 0.10 eV above 2 eV.

The π^* ($^2A_2$) States of the Temporary Negative Ions of Cyclopropene and Cyclobutene.

The first negative ion states for the *cis*-cycloalkenes (**3-6**), as well as for the ethylene (**1**) and *cis*-2-butene (**2**), are associated with

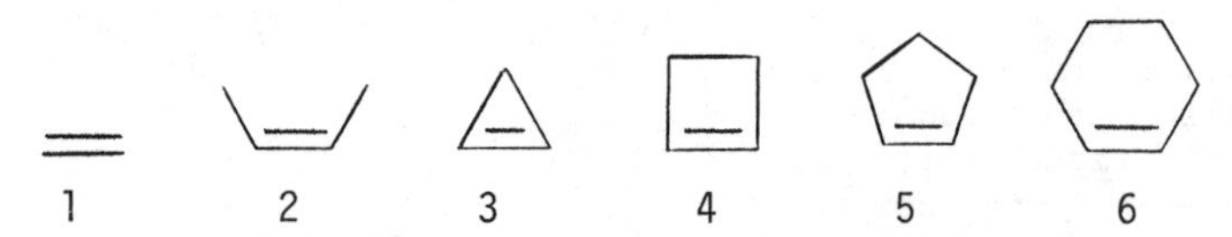

temporary capture of an electron into the π^* orbital. The electron transmission spectra of **2-6** are presented in Figure 1 and the values for the resonances are given in Table I. Note that the π^* orbital is destablized on substitution of **1** with two methyl groups, as in **2**, or with two or more methylene groups in a ring, as in **4-6**. This is in accord with the commonly (2,7,12) (but not universally (13)) observed destabilization of π^* orbitals by alkyl substituents.

The effect of hyperconjugation on spectroscopic parameters associated with the π and π^* states of **1-6** is illustrated by the data in Table II. In particular, the attention of the reader is directed to the relative values of the Coulomb integral (J) which is equal to $\langle\pi^2(1)|e^2/r_{12}|\pi^{*2}(2)\rangle$, the mutual repulsion of two electrons (1 and 2) in two orbitals (π and π^*). This integral can be calculated from

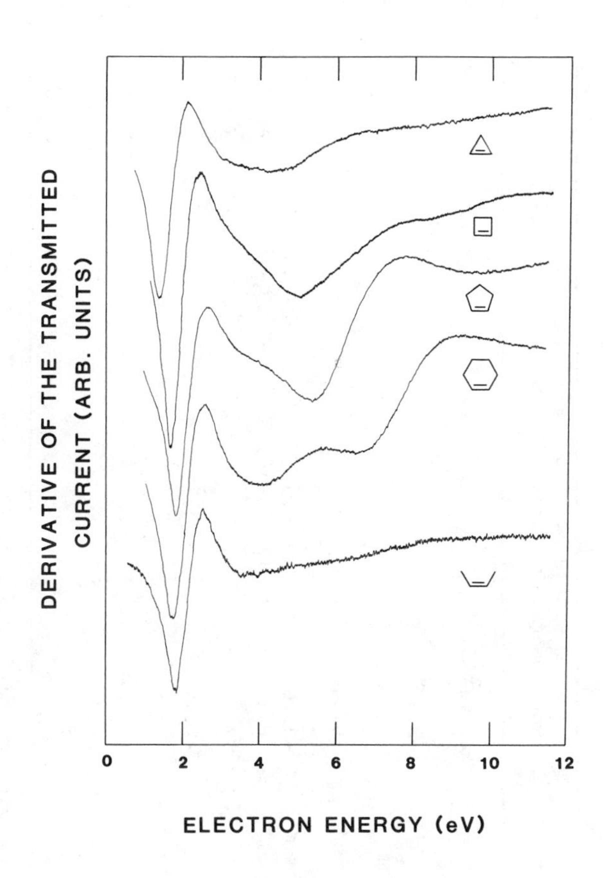

Figure 1. Electron transmission spectra of **2-6**.

Table I. Observed and calculated electron affinities of small and medium ring cycloalkenes and related compounds

Compound	Orbital	Electron Affinity (eV) Experimental	Calcd. (STO-3G)[a]	Calcd (6-31G)[a]	6-31G Orbital Symmetry (ℓ)
ethylene (1)[b]	π*	-1.74[c]	- 8.64	-4.72	
cis-2-butene(2)	π*	-2.16[d]			
cyclopropene (3)[e]	π*	-1.73	- 8.81	-4.71	a_2(3,2)
	σ*	-5.50	-13.70	-7.53	b_2(2,1)
cyclobutene (4)[f]	π*	-2.00	- 8.63	-4.94	a_2(3,2)
	σ*	-6.27	-16.02,-16.15	-7.89,-8.11	b_1(2,1)
					b_2(2,1)
cyclopentene (5)[g]	π*	-2.14	- 8.71	-5.07	
	σ*	-6.32			
cyclohexene (6)[h]	π*	-2.13[i]	- 8.81	-5.10	
	σ*	-4.84			
	σ*	-7.70			

[a]Koopmans' Theorem approximation. [b]Structure: Ref. (6). [c]-1.78 eV; Ref. (2,7). [d]-2.22 eV, Ref. (2,7). [e]Structure: Ref. (8). [f]Structure: Ref. (9). [g]Structure: Ref. (10). [h]Structure: Ref. (11). [i]-2.07 eV; Ref. (2).

Table II. Spectroscopic parameters[a] associated with the π and π* orbitals of **1-6**.

Compound	IP[b]	EA[c]	T_1[d]	S_1[e]	J[f]	2K[g]
ethylene (**1**)	10.51	-1.74	4.25	7.65	8.00	3.40
<u>cis</u>-2-butene (**2**)	9.29	-2.16	4.20	7.12	7.25	2.92
cyclopropene (**3**)	9.86	-1.73	4.16	7.19	7.43	3.03
cyclobutene (**4**)	9.43	-2.00	4.23	7.03	7.20	2.80
cyclopentene (**5**)	9.18	-2.14	4.15	7.00	7.17	2.85
cyclohexene (**6**)	9.12	-2.13	4.24	6.81	7.01	2.57

[a] In eV; see text. [b] Ref. (14). [c] This work. [d] Ref. (15). [e] Ref. (14b).
[f] $J = IP - EA - T_1$. [g] $2K = S_1 - T_1$.

the relationship $J = IP - EA - T_1$, where IP is the π ionization potential, EA is the π* electron affinity, and T_1 is the energy of the vertical π → π* singlet triplet transition. Note that the value of J decreases in the order **1** > **2**, **4**, **5** > **6**, i.e., as the region in space over which π and π* are delocalized through hyperconjugation increases.

Interestingly, the π* orbital in **3** has a node bisecting the C=C bond and therefore cannot interact with the CH_2 group through hyperconjugation, whereas the corresponding π orbital can so interact. This is in accord with the value of J for **3**, which is about 0.2 eV greater than those for **2**, **4**, and **5**. The value of 2K, where K is the "exchange" integral ($\langle\pi(1)\pi^*(1)|e^2/r_{12}|\pi(2)\pi^*(2)\rangle$), which corresponds to the electrostatic repulsion between equal overlap charge densities due to electrons (1) and (2), decreases in the same order as does that of J. Note that the values of J and 2K in Table II probably have uncertainties of ± 0.1 eV or more associated with them.

Since the π* orbital in **3** does not interact with the CH_2 orbitals, it might at first seem not very surprising that the EA of **3** is about equal to that of **1**. However, this observation assumes considerable interest when it is recognized that the C=C bond in **3** (1.296 Å) (8) is 0.04 Å shorter than that in **1** (1.339 Å) (6). In fact, the π* orbital of ethylene is predicted by <u>ab initio</u> calculations (6-31G basis set) to be destabilized by 0.29 eV on shortening the C=C bond from 1.339 to 1.296 Å. In addition, the <u>relative</u> order of the π* orbital energies for **1**, **3**, and **4** (Table I) is not reproduced at the STO-3G basis set level, but is at the 6-31G level. This can be traced to the effect of the hyperconjugative donation of electron density from the π orbital to the pseudo-π* CH_2 orbital which leads to a "reversed" polarity in **3** (μ = 0.45 D with the positive end toward the double bond) (16) compared with **4-6**. The resultant decreased screening of the π* orbital by the bonding electrons leads to a lower electron affinity than would be expected on the basis of the

C=C bond length alone. This represents the first time that a "π-inductive" or screening effect has been demonstrated as a factor in determining electron affinities.

σ* Resonances in Three- and Four-Membered Rings

In this section, we shall refer to resonances which are associated with virtual orbitals which are antibonding at σ bonds as σ* resonances, even though these orbitals may have overall π* symmetry. The spectra of **7-10** are given in Figure 2 and the energies of the observed σ* resonances are listed in Tables I and III along with their

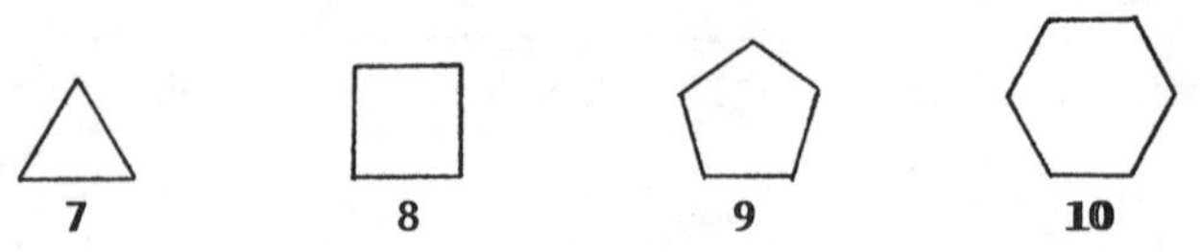

symmetry designations and the angular momentum quantum number (ℓ) associated with the "leading waves" (lowest ℓ values) (11) in the resonances. The resonances were assigned to specific orbitals on the basis of a) the correlation (r = 0.98) between resonance energies and *ab initio* 6-31G orbital energies, as shown in Figure 3, and b) consideration of the ℓ values associated with each orbital. Orbitals with an ℓ = 0 contribution were excluded because the corresponding resonances are expected (but not necessarily required) to have too short of a lifetime for observation.

Drawings of the 6-31G orbitals for cyclopropane (**7**) are given in Figure 4. The first resonance is not associated with the lowest (a_1') σ* orbital calculated at 6.64 eV which has ℓ values of 2 and 0, but with the a_2'' orbital at 6.97 eV (ℓ = 3,1). The second resonance in **7** appears to be associated with the e" orbitals (ℓ = 3,2) calculated to be at 10.35 eV (on the basis of the correlation in Figure 3), rather than with the e' orbitals (ℓ = 2,1) at 8.21 eV or the a_2' orbital (ℓ = 5,3) at 8.62 eV. Whether this assignment is correct and, if so, for what reasons, awaits further investigation.

If our interpretation of the σ* data in Tables I and III is correct, this suggests a direction for future research. Thus, if one wants to gain insight into the conjugating ability of three-membered rings, a study of the first σ* resonance in substituted cyclopropanes may prove unrewarding, but a study of the corresponding resonance in cyclopropenes substituted at C_3 may be very worthwhile.

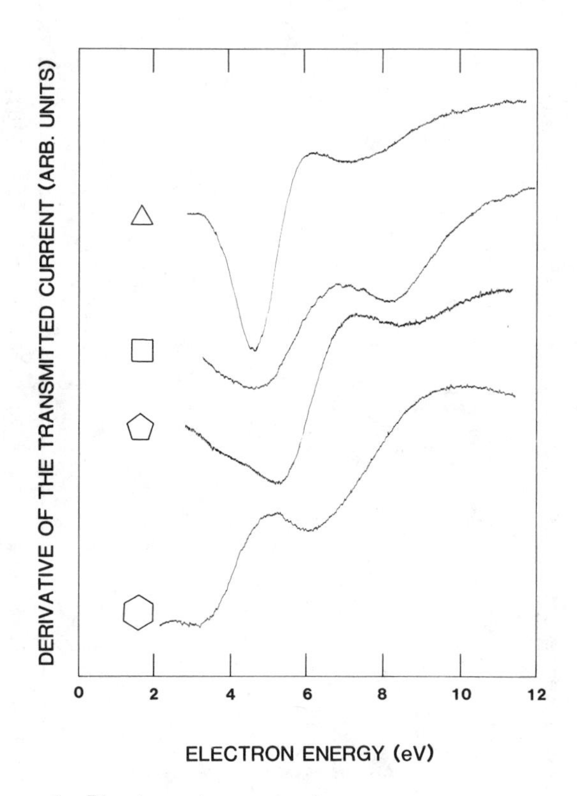

Figure 2. Electron transmission spectra of **7-10**.

Table III. Observed and calculated electron affinities of small and medium ring cycloalkanes.

Compound	Electron Affinity (eV) Experimental	Calcd. (STO-3G)[a]	Calcd (6-31G)[a]	6-31G Orbital Symmetry (ℓ)
cyclopropane (**7**)[b]	-5.29	-13.34	-6.97	$a_2''(3,1)$
	-8.78	-15.57	-8.21,-8.62	$e'(2,1),a_2'(5,3)$
cyclobutane (**8**)[c]	-5.80	-14.68	-6.44,-7.64	$b_2(2,1),e(2,1)$
	-9.34	-17.43,-18.34	-9.44,-9.47,-9.68	e,b_2,e(all 2,1)
cyclopentane (**9**)	-6.14			
	-9.69			
cyclohexane (**10**)	-4.11			
	-7.75			

[a] Koopmans' Theorem approximation. [b] Structure: Ref. (17). [c] Structure: Ref. (18).

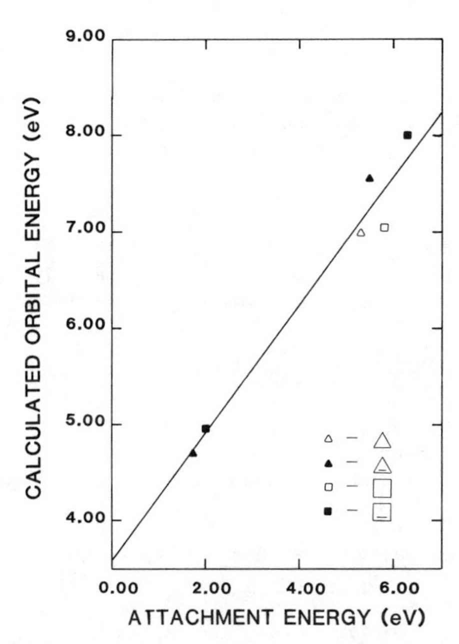

Figure 3. Correlation of <u>ab initio</u> 6-31G orbital energies with experimental attachment energies (AE = - vertical EA) for **3**, **4**, **7** and **8**.

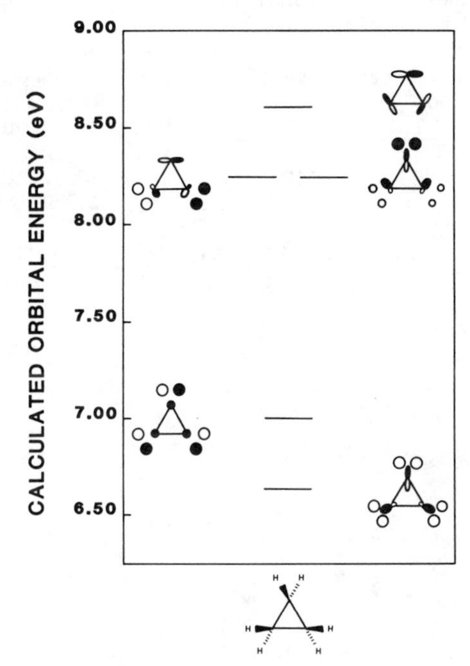

Figure 4. <u>Ab initio</u> 6-31G antibonding orbitals for cyclopropane (**3**).

Acknowledgments

This research was supported by the National Science Foundation (Grant CHE 81-10428). We especially appreciate the long-standing cooperation of Prof. Paul Burrow during the course of this work.

Literature Cited

1. Sanche, L.; Schulz, G.I. Phys. Rev. A 1972, 5, 1672.
2. Jordan, K.D.; Burrow, P.D. Acc. Chem. Res. 1978, 11, 341.
3. Johnston, A.R.; Burrow, P.D. J. Electron Spect. Relat. Phenom. 1982, 25, 119.
4. Koopmans, T. Physica 1934, 1, 104.
5. Read, F.H. J. Phys. B 1968, 1, 893.
6. Duncan, J.L.; Wright, I.J.; van Lerberghe, D. J. Mol. Spectrosc. 1972, 42, 463.
7. Jordan, K.D.; Burrow, P.D. J. Am. Chem. Soc. 1980, 102, 6882.
8. Stigliani, W.M.; Laurie, V.W.; Li, J.C. J. Chem. Phys. 1975, 62, 1890.
9. Bak, B.; Led, J.J.; Nygaard, L.; Rastrup-Andersen, J.; Sørensen, G.O. J. Mol. Struct. 1969, 3, 369.
10. Rathjens, G.W. J. Chem. Phys. 1962, 36, 2401.
11. Chiang, J.F.; Bauer, S.H. J. Am. Chem. Soc. 1969, 91, 1898.
12. Staley, S.W.; Giordan, J.C.; Moore, J.H. J. Am. Chem. Soc. 1981, 103, 3638.
13. Jordan, K.D.; Michejda, J.A.; Burrow, P.D. J. Am. Chem. Soc. 1976, 98, 1295.
14. (a) Heilbronner, E.; Bischof, P. Helv. Chim. Acta 1970, 53, 1677; (b) M.B. Robin, "Higher Excited States of Polyatomic Molecules", Vol. 2, Academic Press: New York, 1975; p. 24.
15. (a) Sauers, I.; Grezzo, L.A.; Staley, S.W.; Moore, J.H., Jr. J. Am. Chem. Soc. 1976, 98, 4218; (b) Sato, Y.; Satake, K.; Inouye, H. Bull. Chem. Soc. Jpn. 1982, 55, 1290.
16. (a) Kasai, P.H.; Meyers, R.J.; Eggers, D.F.; Wiberg, K.B. J. Chem. Phys. 1959, 30, 512; (b) Wiberg, K.B.; Ellison, G.B.; Wendoloski, J.J.; Pratt, W.E.; Harmony, M.D. J. Am. Chem. Soc. 1978, 100, 7837.
17. Butcher, R.J.; Jones, W.J. J. Mol. Spectrosc. 1973, 47, 64.
18. Cremer, D. J. Am. Chem. Soc. 1977, 99, 1307.

RECEIVED June 11, 1984

11

Anion Resonance States of Organometallic Molecules

JUDITH C. GIORDAN[1], JOHN H. MOORE, and JOHN A. TOSSELL

Department of Chemistry, University of Maryland, College Park, MD 20742

The technique of electron transmission spectroscopy has only recently been used to measure the energies of anion states of organometallics, following a long and very significant history of application to the study of di- and triatomics and unsaturated hydrocarbons. Two classes of investigation have been carried out: the study of transition metal compounds such as $Cr(CO)_6$, $Fe(CO)_5$, ferrocene, and other metallocenes; and the study of organic species such as benzene on which organometallic and other main group organo ligands have been substituted. Theoretical calculations, the most successful of which employed the multiple scattering Xα formalism, have been used to describe the scattering process and assign anion states in the transition metal compounds as well as model compounds such as SiH_4. Most recently mass spectrometric techniques have been used to identify fragments arising from dissociative anion states.

Electron scattering experiments, once considered valuable only to study atoms and small molecules, are now being used to aid in understanding the structure and dynamics of large organic and even organometallic molecules. Over the last two decades the ionization potentials of a great variety of organic, main group organo, and organometallic molecules have been determined by photoelectron spectroscopy (PES). These energies have been shown to correlate with the energies of occupied molecular orbitals. Since essentially all descriptions of molecular spectra and reactions between molecules can be explained using a molecular orbital picture, these data have proven quite useful. To complete this picture, however, the

[1]Current address: Polaroid Corporation, Waltham, MA 02254.

0097-6156/84/0263-0193$06.00/0

chemist must also have available a measure of the energies of unoccupied orbitals. Molecular electron affinities can provide this data. Over the past decade, the technique of electron transmission spectroscopy (ETS), which allows for the measurement of these electron affinities, has been developed and effectively applied in the study of a great number of organic systems. As we will illustrate this technique is now finding application to transition metal and main-group organic species. This is especially exciting since we now have a tool to directly probe the energies of those unoccupied orbitals which are invoked in ligand field splitting, through-space interaction, and d-orbital bonding arguments. Concommitant with these experiments, molecular orbital methods are being developed to aid in interpreting the results. In this summary focus will be given to examples of organometallic and main-group-organic species which illustrate the thrust of our work in this area. Overall, we wish to use ETS in conjunction with other spectral, computational, thermochemical, and photochemical data, some of which are already available, to study metal-ligand interactions and their implications for ligand substitution and catalytic reactions.

In the following we present the results of studies of the anion states of the first row metallocenes, a series of transition metal hexacarbonyls and a variety of main-group organo ligands substituted on benzene. Work on the transition metal hexacarbonyls and metallocenes provides benchmark data to which that for other complexes may be compared. The work on the substituted benzenes anticipates the study of the electron transmission spectra of more complex organometallics such as $Cr(CO)_5L$ (where L is a two-electron-donor ligand). In this case the π-system of benzene serves as a probe of the empty acceptor orbitals of L. For example, by studying para-bis-(dimethylphosphino)benzene (and trimethylphospine, as well) we gain knowledge of the energies of the unoccupied orbitals associated with the $-PMe_2$ group and the nature of the interaction of the $-PMe_2$ group orbitals with the benzene π* orbitals.

Experiment and Theory

Electron transmission spectroscopy (1) is the experiment conjugate to PES. Whereas PES measures the energy needed to remove an electron from an occupied orbital, ETS measures the energy of a negative ion state arising from electron capture into an unoccupied orbital. The ETS experiment involves measurement of the transparency of a gas to an electron beam as a function of energy. The transparency depends in an inverse fashion upon the electron scattering cross section. Temporary negative ion formation occurs with large cross section over only a narrow energy range. Since the negative ion promptly decays by giving up the trapped electron, the formation and decay process appears as a sharp fluctuation in the electron-scattering cross section. The process, as well as the corresponding feature in the transmission vs. electron energy spectrum, is referred to as a "resonance".

The electron spectrometer consists of an electron source followed by an electron monochromator, a gas cell, and an electron collector (2). In practice the first derivative of the transmitted current as a function of energy is recorded since the derivative is sensitive to the abrupt change in transmitted current associated with a resonance (3). The energy of a resonance is known as an "attachment energy" (AE) and, with respect to the derivative spectrum, is defined as the point vertically midway between the minimum and maximum which characterize the resonance. For the present purposes an attachment energy may be identified with the negative of the corresponding electron affinity (EA).

The chief limitation of ETS is that it gives only the energy associated with unstable negative ions. That is, only negative electron affinities can be obtained with ETS.

In the derivative electron transmission spectrum there is always a sharp feature near 0 eV which corresponds to the abrupt turning on of the electron current at threshold (see Figs. 1 and 4). The existence of an anion state which is marginally stable or marginally unstable will frequently be manifest in a significant broadening of the threshold feature.

To obtain information on dissociative attachment processes originating in unstable anions produced in the electron transmission spectrometer, a time-of-flight spectrometer has recently been appended to our apparatus. The resolution of this device is sufficient to determine both the mass and kinetic energy of negative ion fragments.

The complexity of the organometallic molecules we have studied prevents us from using qualitative, symmetry-based, MO models for assignment of the observed resonances. Rather, we have employed the Multiple Scattering Xα (MS-Xα) method (4) which has proven valuable for assignment of the PES and UV spectra of molecules containing transition metal atoms (5). In early studies on the d^6 transition metal hexacarbonyls (6) and the first transition series metallocenes (7), we employed the bound-state version of the MS-Xα method, stabilizing the anion by enclosure in a spherical charge shell. In later work, the continuum MS-Xα method (8) has been applied to $Cr(CO)_6$ (9) and to a series of tetrahedral C and Si compounds (10).

Metallocenes

Derivative electron transmission spectra for the 3d metallocenes (7) are presented in Fig. 1. Very similar results have been obtained by Modelli, et al. (11). Occupations of the predominantly M3d highest occupied orbitals and ground electronic states are given in Table I.

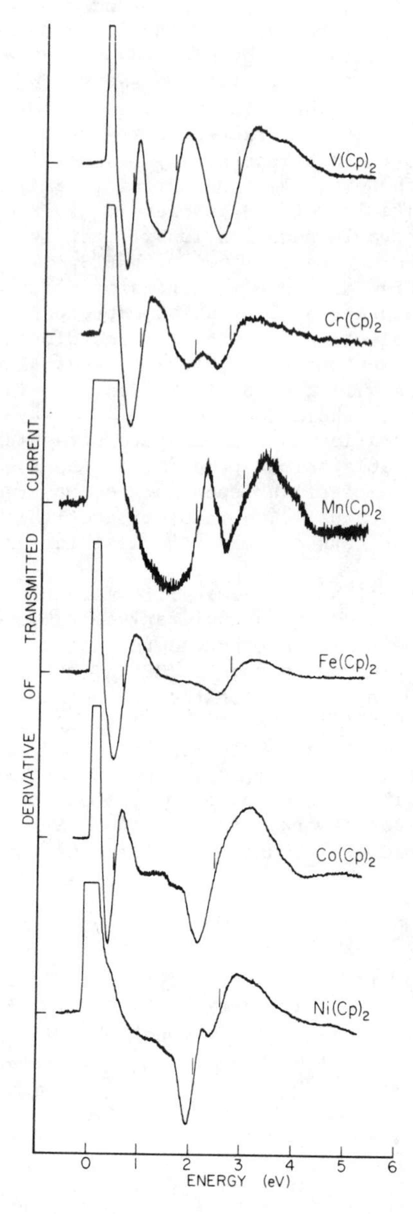

Figure 1.Derivative electron transmission spectra of the 3d metallocenes, $M(Cp)_2$ (M=V, Cr, Mn, Fe, Co, Ni) (from ref. 7). Each spectrum was calibrated against that of N_2.

Table I. Orbital Occupations and Ground Electronic States of Metallocenes

$V(Cp)_2$	$(3e_{2g})^2(5a_{1g})$	$^4A_{2g}$
$Cr(Cp)_2$	$(3e_{2g})^3(5a_{1g})$	$^3E_{2g}$
$Mn(Cp)_2$	$(3e_{2g})^4(5a_{1g})$	$^2A_{1g}$
	or $(3e_{2g})^2(5a_{1g})(4e_{1g})^2$	$^6A_{1g}$
$Fe(Cp)_2$	$(3e_{2g})^4(5a_{1g})^2$	$^1A_{1g}$
$Co(Cp)_2$	$(3e_{2g})^4(5a_{1g})^2(4e_{1g})$	$^2E_{1g}$
$No(Cp)_2$	$(3e_{2g})^4(5a_{1g})^2(4e_{1g})^2$	$^3A_{2g}$

A correlation diagram showing the observed AE's is presented in Fig. 2. The results have been interpreted using stabilized bound state MS-Xα calculations. A qualitative MO scheme for the $M(Cp)_2$ compounds is presented in Fig. 3 (where $Cp=C_5H_5$, the cyclopentadienyl group). For the Fe, Co, and Ni compounds the calculated attachment energies for the $4e_{1g}$ M3d-orbital and for the $4e_{2g}$ and $3e_{2u}$ Cp π^*-orbitals are given in Table II. For the $4e_{1g}$ orbital, calculation

Table II. Calculated Attachment Energies (eV) for Occupation of $4e_{1g}$, $4e_{2g}$, and $3e_{2u}$ Orbitals in $Fe(Cp)_2$, $Co(Cp)_2$ (Spin-Polarized Case), and $Ni(Cp)_2$

	$Fe(Cp)_2$	$Co(Cp)_2$	$Ni(Cp)_2$
$4e_{1g}$	1.15	0.93[a],-0.47[b]	-0.13
$4e_{2g}$	1.19	1.16	1.14
$3e_{2u}$	3.26	3.49	3.68

[a]average AE for singlet states ($^1A_{1g}$ + $^1E_{2g}$).
[b]AE for $^3A_{2g}$ state.

and experiment agree to within 0.5eV and the observed stabilization in going from Fe to Ni is reproduced. The Cp π^* type orbitals of e_{2u} symmetry also have a calculated AE close to the experimental value, but the $4e_{2g}$ are considerably lower in energy with substantial Rydberg character. Later MS-Xα calculations (11) using a slightly larger partial wave basis set obtained an additional e_{2g} state with AE close to that for the $3e_{2u}$, in agreement with experiment. In the latter study, spin-polarized MS-Xα calculations were also performed on $V(Cp)_2$ and $Cr(Cp)_2$ yielding the calculated AE of Table III (8).

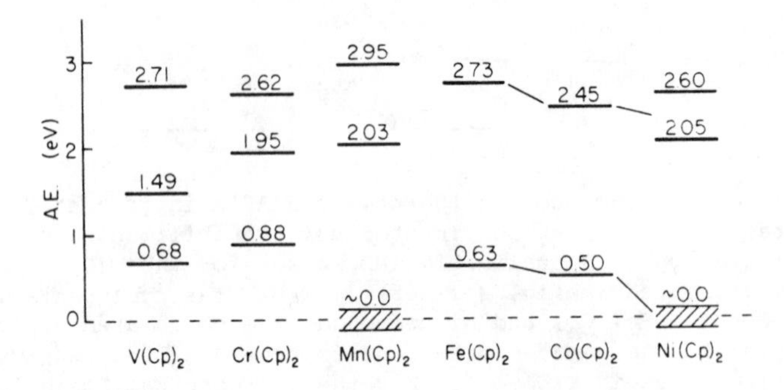

Figure 2. Correlation diagram giving attachment energies of the 3d metallocenes (from ref. 7).

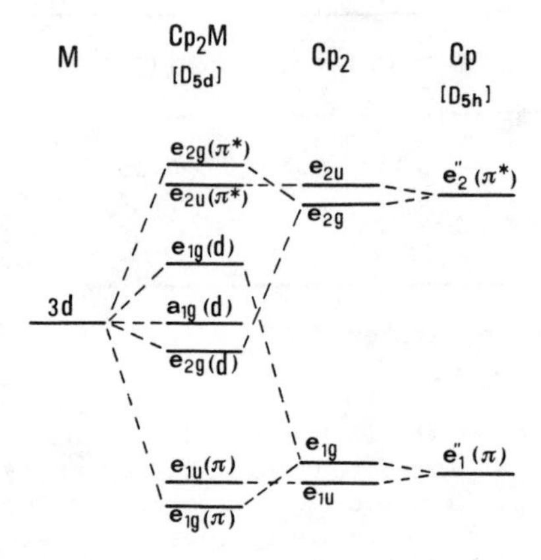

Figure 3. Qualitative diagram of the frontier orbitals of $M(Cp)_2$ and its fragments (from ref. 11).

Table III. Calculated Attachment Energies (eV) for Occupation of Empty Orbitals in $V(Cp)_2$, $Cr(Cp)_2$, and $Fe(Cp)_2$

	$V(Cp)_2$[a]	$Cr(Cp)_2$[b]	$Fe(Cp)_2$
$5a_{1g}$	0.14	-0.58	
$3e_{2g}$	0.33	-0.19	
$4e_{1g}$	1.35,2.89	1.14,2.20	1.26
$4e_{2g}$	2.88,3.04	2.94,3.05	2.98
$3e_{2u}$	2.66,2.71	2.80,2.85	2.78

[a] quintet and triplet states, respectively.
[b] quartet and doublet states, respectively.

It is interesting that in $V(Cp)_2$ there are triplet state anions arising from $5a_{1g}$ and $3e_{2g}$ orbitals lying just above threshold and both quintet and triplet E_{1g} anion states occurring in the 1-3 eV range. Unfortunately, a complete spectral interpretation has not yet been performed for $Mn(Cp)_2$, for which the ground electronic state is not definitively known (12). Nonetheless, the following general characteristics of the M3d orbitals emerge from the $M(Cp)_2$ data: (1) states of different spin multiplicity may differ in energy by more than 1 eV, (2) spin averaged AE's for the $4e_{1g}$ orbital becomes less positive across the series, (3) this drop is smaller from V to Fe since the decreasing M-to-ring distance (13) tends to destabilize the $4e_{1g}$ and larger from Fe to Co and Ni since the increasing M-to-ring distance augments the general M3d stabilization effect. It is also apparent that the transition state approach (14) used in refs. 7 and 11 facilitates the comparison of ETS results with other spectral quantities. For example, in $Ni(Cp)_2$ the MS-Xα ground state orbital eigenvalue for the $4e_{1g}$ orbital is -2.8 eV, decreasing to -5.4 eV in the photoionization transition state (15) (0.5 electrons removed from $4e_{1g}$). In the electron attachment transition state (0.5 electrons added to $4e_{1g}$) the eigenvalue becomes -0.1 eV. The calculated ionization potential (IP) and electron affinity (EA) are thus 5.4 and 0.1 eV, respectively. The calculated IP compares well with the experimental value of 6.4 eV while the broadening of the ETS threshold feature is consistent with an anion state just below threshold. It therefore appears that the bound-state MS-Xα method yields accurate energetics for both occupied and unoccupied M3d orbitals of the metallocenes.

Transition Metal Hexacarbonyls

This series affords the chemist an opportunity to see the potential of electron scattering experiments such as ETS for studying metal-ligand bonding. Using the energies of the ligand-based carbon monoxide orbitals and primarily metal-based orbitals of the d^6 hexacarbonyls as reference points, ligands (L) on the metal can be varied leaving 1 to 5 carbon monoxides in place. The effect of this perturbation on the energy of the CO-based 2π orbital as measured by ETS can then be assessed in terms of increased or decreased bonding between the metal and the carbonyl as a function of the new ligand(s). Ultimately, what is being probed is the ablity of the ligand to hinder CO substitution on the molecule, i.e., ligand exchange or catalytic behavior. To assess the efficacy of this concept, resonances in the electron transmission spectra of these molecules must be assigned to negative ion states or unfilled molecular orbitals in each compound.

Derivative electron transmission spectra are presented in Fig. 4 for the transition metal hexacarbonyls $M(CO)_6$, (M=Cr, Mo, W). In addition, Fig. 5 shows as a function of electron impact energy the total negative ion current from dissociative attachment to $Cr(CO)_6$. MS-Xα orbital energies calculated for the ground states of $Mo(CO)_6$ and CO are shown in Fig. 6. The broadness of the threshold peak in the spectrum is indicative of an attachment process occurring near threshold. This is quite evident in the ion current spectrum of Fig. 5. Stabilization electron attachment transition state calculations indicated that a number of $M(CO)_6^-$ anion states [e.g., those produced by adding an electron to the lowest empty t_{1u} ($9t_{1u}$) orbital of $Cr(CO)_6$] were actually bound. The ETS peaks therefore corresponded to occupation of higher-energy empty orbitals. For example, the feature below 1 eV in the ETS and ion current spectrum of $Cr(CO)_6$ was attributed to electron capture into the $4t_{2g}$ (second lowest empty t_{2g}) and the $6e_g$ (lowest empty e_g) with calculated AE's of 0.9 and 0.2 eV, respectively, while peaks A and B were attributed to occupation of the $3t_{2u}$ and $7e_g$ orbitals and C to occupation of the $5t_{2g}$ orbital. Subsequent restricted Hartree-Fock (RHF) calculations (16) indicated that the lowest energy $Cr(CO)_6^-$ anion obtained by occupation of the $9t_{1u}$ orbital lies 1.5 eV above threshold. However, the RHF calculations gave no explanation for the threshold ETS or for the maximum around 0.4eV in the ion current spectrum.

Of course the ETS experiment can only be formulated correctly as a scattering experiment. Any bound-state approach, whether MS-Xα or RHF, must be treated with caution. We have therefore performed continuum MS-Xα calculations of elastic electron scattering cross sections for $Cr(CO)_6$ using the methods of ref. 8. Elastic scattering cross sections for several different choices of scattering potential were calculated and a decomposition of the cross section by the symmetry of the continuum electron for a representative potential is given in Fig. 7. The choice of scattering potential for such a calculation is difficult (17). We have initially used potentials obtained from self-consistent electron attachment transition states subtracting out the stabilizing shell potential. The calculated cross sections appear to be in reasonable agreement with the ion current results and their interpretation is similar to that from our bound state calculations: e.g., the maximum in ion current at about 0.4eV is associated with the $4t_{2g}$ orbital and that around 1.6eV with the $3t_{2u}$ orbital. Nonetheless, it is apparent that several channels, both resonant and nonresonant, contribute to the total cross section and the calculated variation of total cross section in the 1-3 eV region is insufficient to quantitatively explain the ETS. This discrepancy may be associated with the small partial wave basis (sp only) used for C and O. It is well known that

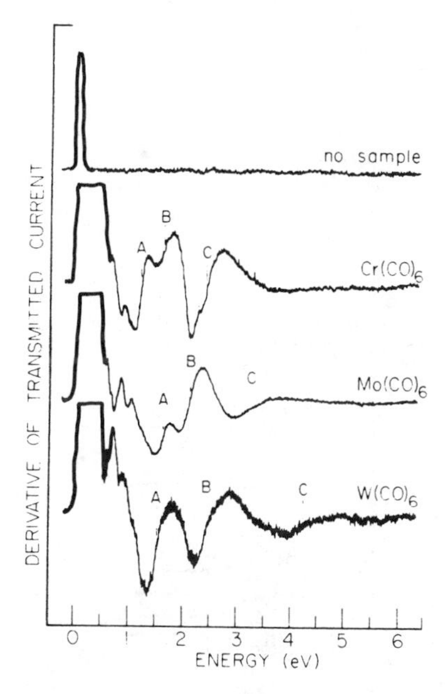

Figure 4. Derivative electron transmission spectra of $Cr(CO)_6$, $Mo(CO)_6$ and $W(CO)_6$ (from ref. 6).

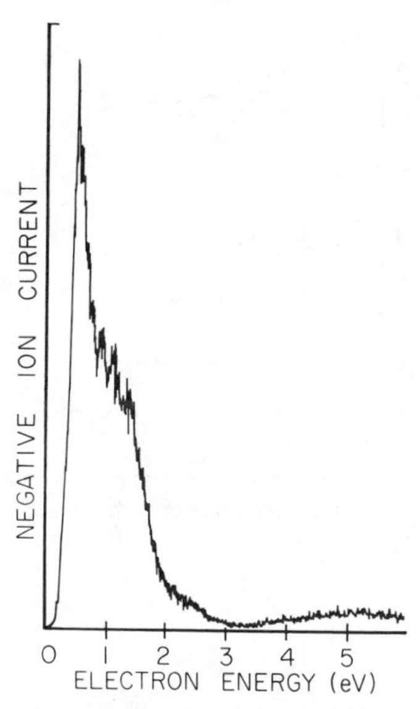

Figure 5. Ion current vs. electron impact energy on $Cr(CO)_6$. The ions are essentially all $Cr(CO)_5^-$ (from ref. 9).

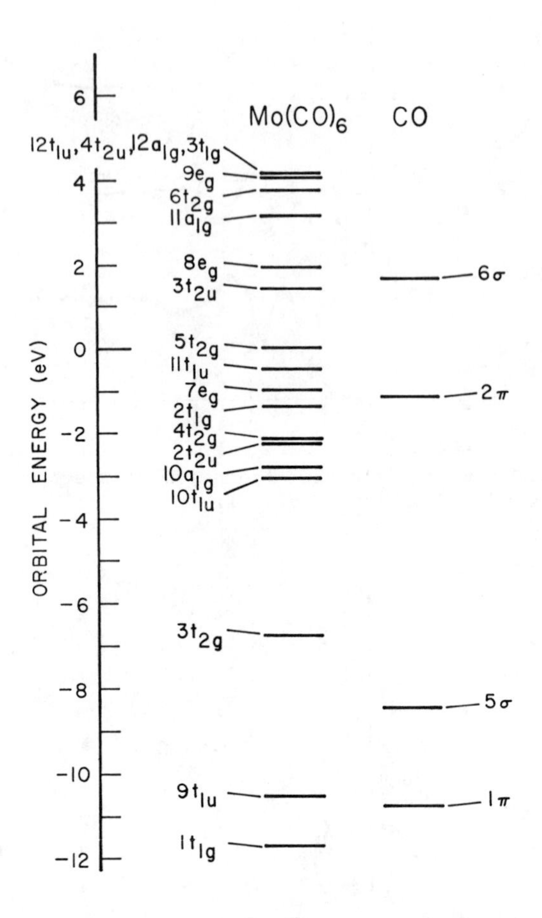

Figure 6. Calculated orbital energies in the ground states of neutral $Mo(CO)_6$ and CO. The highest occupied orbitals are the $3t_{2g}$ and 5π, respectively (from ref. 6).

electron scattering resonances in molecules such as CO often involve higher partial waves (d or f). The present continuum MS-Xα results for $Cr(CO)_6$ must therefore be considered preliminary.

The case of $Cr(CO)_6$ illustrates the extreme difficulty of obtaining a convincing ETS assignment using calculation alone. The MS-Xα method is of course limited by its use of a muffin-tin potential. For unoccupied orbitals a significant fraction of the electron density occurs in the interatomic region, the least accurately treated part of the molecule. In general, diffuse orbitals tend to be overstabilized by the constant interatomic potential. This problem can be reduced to some extent by the use of overlapping atomic spheres (as we have done) but the influence of degree of overlap on energetics is not negligible (18). Schemes which employ more accurate MS-Xα type potentials (19) give orbital energies somewhat different from conventional values. The RHF scheme also encounters severe difficulty in the description of anion resonances due to basis set limitations and correlation effects, which seem to be larger in Hartree-Fock calculations than in local exchange calculations. In general, one would expect MS-Xα to give molecular anions which are too stable and RHF to give them as not stable enough, so that the correct value would lie somewhere in between. We may use experimental PES (20) and UV energies (21) and calculated changes in orbital eigenvalue to estimate the EA of the $Cr(CO)_6$ $9t_{1u}$ orbital as shown in Fig. 8. The IP of the $2t_{2g}$ orbital corresponds to its eigenvalue in a transition state in which its occupation number is 5.5 (and all other orbitals have their ground-state occupations). Addition of 0.5 electrons to the $9t_{1u}$ orbital raised the calculated $2t_{2g}$ eigenvalue by 2.2 eV. The $2t_{2g} \rightarrow 9t_{1u}$ excitation energy corresponds to the $9t_{1u} - 2t_{2g}$ orbital energy difference in a transition state with occupations $2t_{2g}^{5.5}\ 9t_{1u}^{0.5}$. Addition of 0.5 electrons to the $2t_{2g}$ orbital raises the calculated $9t_{1u}$ orbital eigenvalue by 1.9 eV. This analysis indicates that the $9t_{1u}$ orbital lies very close to threshold and that the MS-Xα value indeed lies below experiment and the RHF value above. It is clear that a definitive interpretation of the ETS of $Cr(CO)_6$ must await more rigorous continuum MS-Xα and Hartree-Fock CI calculations but that such calculations alone, without a comprehensive consideration of other spectral data, cannot be definitive.

Main-Group Compounds

In order to better understand bonding in main-group organo compounds and to obtain data on main-group species which can act as ligands in transition metal complexes, study of the negative ion states of various saturated and unsaturated Group IV, V, VI, and VII hydrocarbons was undertaken. One example of this work is the series of para-disubstituted benzenes:

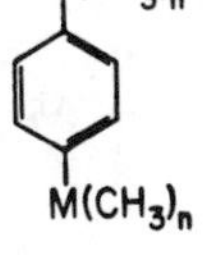

$M(CH_3)_n$

$M(CH_3)_n$

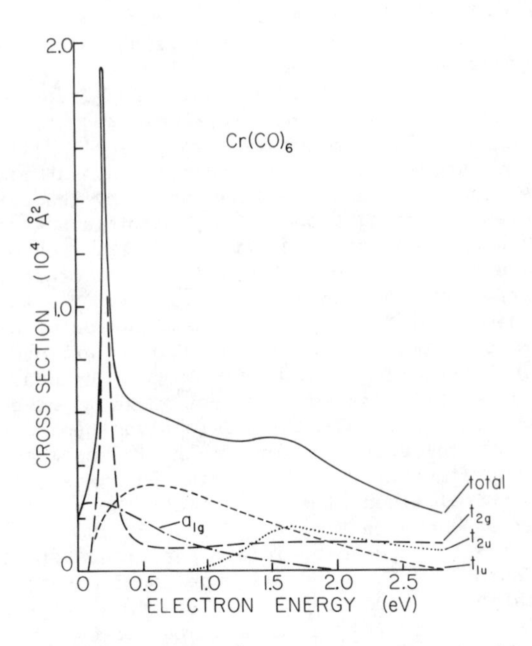

Figure 7. Decomposition of continuum MS-Xα cross section by symmetry of continuum electron for the scattering potential given by occupation of the $3t_{2u}$ orbital (e_g and t_{1g} channels give $\sigma < 30Å^2$ and are not shown), (from ref. 9).

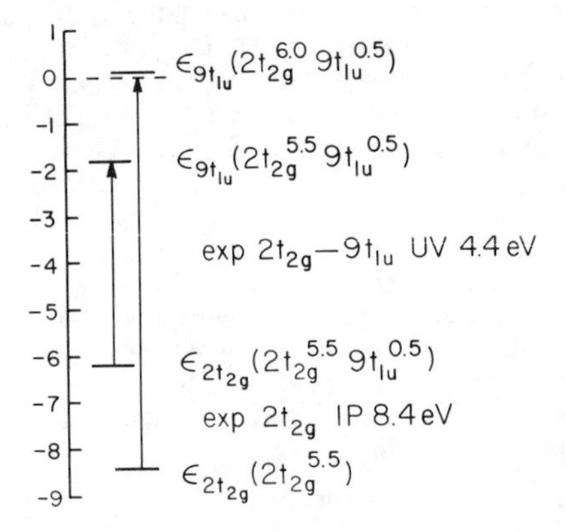

Figure 8. Estimation of EA of $Cr(CO)_6$ $9t_{1u}$ orbital using experimental $2t_{2g}$ IP and $2t_{2g} \rightarrow 9t_{1u}$ excitation energy and calculated orbital energy changes with occupation number from MS-Xα calculations.

and their parent model compounds $M(CH_3)_{n+1}$ and MH_{n+1}, where n=3 for M=C,Si,Ge,Sn; n=2 for M=N,P,As; n=1 for M=O,S; and n=0 for M=F,Cl,Br. Here we test the efficacy of employing anion state energies determined from ETS measurements to aid in explaining (1) the effect of lone pair participation in bonding, (2) the extent of d-orbital participation of row 2, 3, and 4 elements in bonding, (3) bonding of 1st row <u>versus</u> 2nd and 3rd row elements, and (4) the ability of these ligands to participate in metal-to-ligand back-bonding. In the parent compounds, resonances associated with ligand σ^* orbitals are evident. In the substituted benzenes the effect of such σ^* orbitals are manifest as perturbations of the benzene π^* orbital energies while in metal carbonyls the effect can be seen in the splitting of the t_{2g} orbital.

As an example of our observations on the ligand model compounds, we show in Fig. 9 a correlation diagram giving the energies of the σ^* resonances in the series SiH_4 - HCl and $SiMe_4$ - MeCl. An immediate observation is that the empty σ^*-orbitals are destabilized by substitution of Me for H. Thus, $P(Me)_3$ should be a weaker acceptor than PH_3 in compounds such as $Cr(CO)_5PH_3$, as is indeed inferred on the basis of PES (<u>22a</u>) and IR evidence (<u>22b</u>). Analysis of the properties of substituted carbonyls such as $Cr(CO)_5L$, where L is a two electron donor ligand such as PH_3, shows a correlation between ligand AE and Cr3d-L π-bonding effects. Representative data are given in Table IV. When L is a strong π acceptor (e.g., PMe_3)

Table IV. Correlation of Properties of $Cr(CO)_5L$ with Attachment Energies of L

L	AE(eV) of L	"t_{2g}" splitting (eV)[c]	"π acceptance" of L (mdyn/Å)[d]
CO	2.0[a]	0	0.74
NMe_3	4.8[b]	0.31	0.0
PH_3	4.8[b]	0.13	-----
PMe_3	3.1	0.14	0.48
SMe_2	3.3	0.20	0.15

[a] ref 1
[b] present result
[c] ref. 23
[d] ref. 24

effective symmetry at the Cr atom is nearly octahedral and the occupied metal 3d orbital (of t_{2g} symmetry in octahedral symmetry) is almost unsplit. When L is a poor π-acceptor (e.g., NMe_3) the splitting of this orbital is substantial (<u>23</u>). IR and Raman spectra may also be used to generate scales of σ-donor and π-acceptor strength for ligand, L (<u>24</u>). CO is found to have the strongest π-acceptor character and N bases the weakest. Again, a general correlation is observed between the AE of the ligand and its π-acceptor character. Thus, measurements of AE's for free ligands may help to understand other bonding properties.

In order to address the question of d-orbital participation in the bonding schemes of molecules containing Si, P, and the like, MS-Xα calculations have been performed. As an example of the interpretation of this data we will focus on the series (C,Si)X_4 (10) where X=H,F,Cl. This work was initiated to provide an interpretation of the ETS of SiH_4 (25), which exhibits a resonance at 2.1 eV whose shape suggests a cross section which rises quickly around 2 eV and then levels out. The t_2 symmetry σ^*-orbital responsible for the change in cross section shown in Fig. 10 has a significant amount of Si p and d character, being much more localized than the corresponding orbital in CH_4. This orbital also yields the dominant peak in the SiL x-ray absorption spectrum (XAS) of SiH_4 (26) and a correspondence of SiL XAS, Si 2p IP, and ETS resonance energy may be drawn using the transition state approach as shown in Fig. 11, which is similar in character to Fig. 8. Both the bound state MS-Xα approach and the calculated scattering cross-section yield values (2.1 and 2.4 eV respectively) virtually identical with experiment. For $Si(Me)_4$ the AE of this t_2 orbital increases to 3.9 eV and in para-bis(trimethylsilyl)benzene the π^* (e_{2u}) resonance in benzene at 1.09 eV is split, with one component of the π^*-orbital stabilized to 0.54 eV by hyperconjugative interaction of the empty $-Si(Me)_3$ π^* with the ring π^*-orbital with nonzero electron density at the 1 carbon (27). Thus, understanding of the σ^* resonance in SiH_4 helps to clarify our interpretation of features in the more complicated spectra of $-Si(Me)_3$ substituted benzenes. Although calculations have not been performed for $Si(Me)_4$ itself, comparison of the Si2p IP's (28) and SiL X-ray absorption spectra (29) of SiH_4 and $Si(Me)_4$ indicates that the t_2 orbital lies about 2 eV higher (closer to threshold) in $Si(Me)_4$, consistent with the ETS results.

Assessment of the lone-pair interaction in these systems can be made by noting the perturbation to the benzene π^*-orbital upon substitution with a Group V moiety (Fig. 12). In going from the N to the P compound there is a stabilization of the σ^*-orbital which yields a stabilizing hyperconjugative interaction with the appropriate ring π^*-orbital, lowering its AE. In the N compound this orbital is destabilized by interaction with the N lone-pair and its energy is raised above that of the noninteracting π^*-component. A similar effect is seen in the As compounds. The change in energy seen is probably a result of both reduced π^*-lone pair interactions and increased σ^*-π^* interactions in the P and As compounds.

In summary, we find ETS to be a valuable tool for the elucidation of the energetics and other properties of low-energy unoccupied orbitals in organometallic compounds and their fragments. Such studies complement PES studies of occupied orbitals and UV, X-ray absorption, and electron energy loss studies of transitions to low-energy unoccupied orbitals yielding a complete picture of valence orbital interactions in the compound.

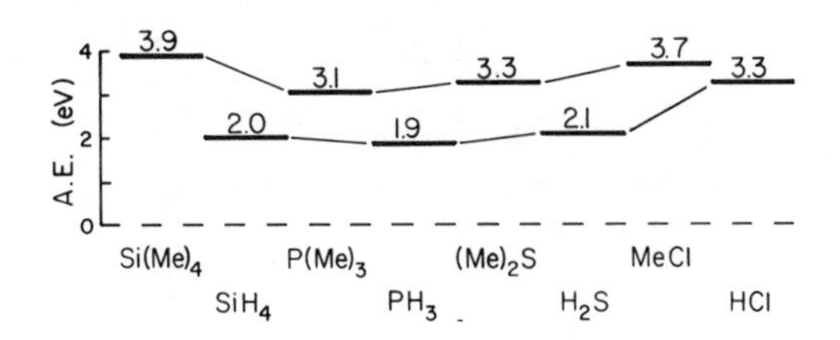

Figure 9. Attachment energies from ETS for $Si(H,Me)_4^-(H,Me)Cl$.

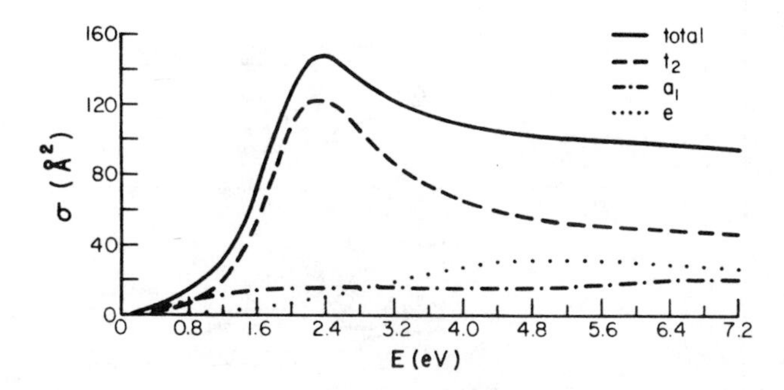

Figure 10. Continuum MS-Xα elastic electron scattering cross sections for SiH_4, total and by symmetry of continuum electron (from ref. 10).

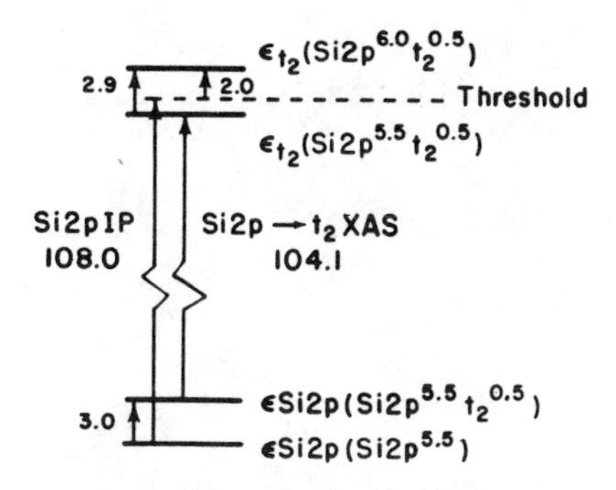

Figure 11. Relationship between core orbital ionizations potential, X-ray absorption energies and ETS resonance energies for SiH_4 using transiton state model (from ref. 10).

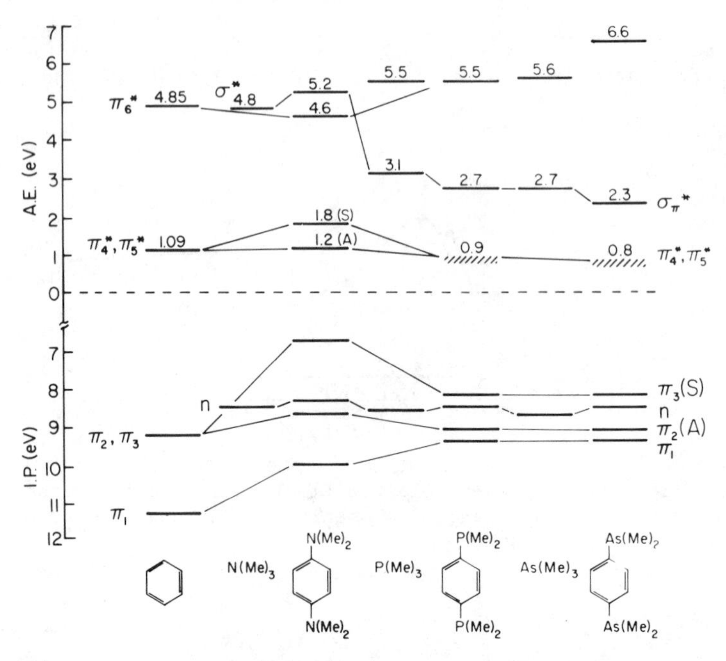

Figure 12. Attachment energies (AE) and ionization potentials (IP) of group V substituted benzenes and substituent parent compounds.

Acknowledgments

We thank NSF (Grant No. CHE-81-21125) and the Computer Science Center, University of Maryland for support of this work, the ACS for permission to reproduce Fig. 1, 2, and 4-7, the AIP for permission to reproduce Fig. 10 and 11, and Elsevier Science Publishers for permission to reproduce Fig. 3.

Literature cited

1. Jordan, K.D.; Burrow, P.E. Acc. Chem. Res., 1978, 11, 344: Schulz, G.J., Rev. Mod. Phys., 1973, 45, 379; Schulz, G.J., Rev. Mod. Phys., 1973, 45, 423.
2. Stamatovic, A.; Schulz, G.J., Rev. Sci. Inst., 1970, 41, 423; McMillan, M.R.; Moore, J.H., ibid., 1980, 51, 944.
3. Sanche, L.; Schulz, G.J., Phys. Rev., 1972, 45, 1672.
4. Johnson, K.H., Adv. Quantum Chem., 1973, 7, 143.
5. Johnson, K.H., Annu. Rev. Phys. Chem., 1975, 26, 39: Case, D.A., Annu. Rev. Phys. Chem., 1982, 33, 151.
6. Giordan, J.C.; Moore, J.H.; Tossell, J.A., J. Am. Chem. Soc., 1981, 103, 6632.
7. Giordan, J.C.; Moore, J.H., Tossell, J.A.; Weber, J., J. Am. Chem. Soc., 1983, 105, 3431.
8. Davenport, J.W.; Ho, W., Schrieffer, J.R., Phys. Rev., 1978, B17, 3115; Dill, D.; Dehmer, J.L. J. Chem. Phys., 1974, 61, 692.
9. Tossell, J.A.; Moore, J.H.; Olthoff, J.K., J. Am. Chem. Soc., 1984, 106, 823.
10. Tossell, J.A.; Davenport, J.W., J. Chem. Phys., 1984, 80, 813.
11. Modelli, A.; Foffani, A.; Guerra, M.; Jones, D.; Distefano, G., Chem. Phys. Lett., 1983, 99, 58.
12. Ammeter, J.H.; Bucher, R.; Oswald, N., J. Am. Chem. Soc., 1974, 96, 7833.
13. Haaland, A., Acc. Chem. Res., 1979, 2, 415.
14. Slater, J.C., "Quantum Theory of Molecules and Solids"; McGraw-Hill; New York, 1974, vol. 4.
15. Goursot, A.; Penigault, E.; Weber, J., Nouv. J.Chim., 1979, 3, 675.
16. Vanquickenborne, L.G.; Verhulst, J., J. Am. Chem. Soc., 1983, 105, 1769.
17. Rumble, J.R., Jr.; Truhlar, D.G., J. Chem. Phys., 1980, 72, 3206.
18. Aizman, A.; Case, D.A., Inorg. Chem., 1981, 20, 528.
19. Watari, K.; Leite, J.R.; DeSiqueira, M.L.,J. Phys. Chem. Solids, 1982, 43, 1053.
20. Higginson, B.R.; Lloyd, D.R.; Burroughs, P.; Gibson, D.M., Orchard, A.F., J. Chem. Soc. Far. Trans., II, 1973, 69, 1659.
21. Beach, N.A.; Gray, H.B., J. Am. Chem. Soc., 1968, 90, 5713.
22. (a) Yarborough, L.W.; Hall, M.B., Inorg. Chem., 1978, 17, 2269; (b) Huheey, J.E. "Inorganic Chemistry", 3rd ed., Harper and Row, New York, 1983, p. 432-438.
23. Cowley, A.H., Prog. Inorg. Chem., 1979, 26, 45.
24. Graham, W.A.G., Inorg. Chem., 1968, 7, 315.
25. Giordan, J.C., J. Am. Chem. Soc., 1983, 105, 6544.
26. Friedrich, H.; Sonntag, B.; Rabe, P.; Butscher, W. and Schwarz, W.H.E., Chem. Phys. Lett., 1979, 64, 360.
27. (a) Modelli, A.; Jones, D.; Distefano, G., Chem. Phys. Lett. 1982, 86, 434. (b) Giordan, J.C.; Moore, J.H.; J. Am. Chem. Soc., 1983, 105, 6541.
28. Perry, W.B.; Jolly, W.L., Inorg. Chem., 1974, 13, 1211.
29. Schwarz, W.H.E., Chem. Phys., 1975, 9, 157; Fomichev, V.V.; Zinkima, T.M., Vinogrodov, A.S.; Edvokimov, A.M., J. Struct. Chem., 1970, 11, 626.

RECEIVED July 10, 1984

12

Vibrational-Librational Excitation and Shape Resonances

In Electron Scattering from Condensed N_2, CO, O_2, and NO

L. SANCHE and M. MICHAUD

MRC Group in Radiation Sciences, Faculty of Medicine, University of Sherbrooke, Sherbrooke, Quebec, Canada J1H 5N4

Low-energy (0.2-30 eV) electron scattering from multilayer films of N_2, CO, O_2, and NO, condensed on a metal substrate near 20 K, has been investigated using hemispherical electron spectrometers and one-dimensional multiple scattering theory. Comparisons of experimental electron energy loss spectra with those generated theoretically indicate that the former are composed of peaks which result from single and multiple vibrational losses convoluted with multiple librational phonons having a mean energy of about 8 meV. Strong and broad peaks in the energy loss functions of the v=1 to 3 vibrational levels of ground state N_2, CO, and O_2 are interpreted to arise from the formation of transient anions. In N_2 and CO, all of the gas-phase shape resonances are observed, whereas in O_2, only the $^2\Pi_u$ and $^4\Sigma_u^-$ states seem to exist. The $^2\Pi_g$ state of N_2^- exhibits vibrational structure and a relaxation shift of 0.7 eV in the solid. Nitric oxide dimerizes upon condensation and we observe two broad resonances at 11.6 and 14.2 eV in the ν_1 and ν_5 decay modes of the dimer. They appear to be associated with the splitting of a single NO^- state via "through space" orbital interaction.

Most of our knowledge on low energy (0.2-30 eV) electron interactions with matter is derived from experiments involving electrons colliding with gaseous targets composed of uncorrelated atoms and molecules. Only a few experiments to date have directly probed with low energy electrons more complex states of matter namely, clusters (1,2) and thin-film molecular solids (3-9). These experiments have been particularly successful in describing modifications in the characteristics of transient anion states introduced by a change of phase. In clusters, these transitory states have been studied via dissociative electron attachment reactions (1,2) whereas, in thin-film molecular solids, electron energy loss spectra and excitation functions have been measured (3-9). We elabo-

0097-6156/84/0263-0211$06.00/0

rate here on recent knowledge derived from electron scattering in disordered films formed by condensing N_2, CO, O_2, and NO gas on a cold metal substrate. Particular emphasis is placed on the occurrence of shape resonances (i.e., transient anions consisting of a ground state molecule plus an electron captured into a usually unfilled orbital) in these films and on multiple scattering effects which must be understood for a proper interpretation of the spectra.

The apparatus and conditions of operation are briefly described in the next section. In the following sections we outline the major ingredients of a phenomenological multiple scattering theory into which elastic and inelastic scattering probabilities and energy losses can be fed in order to generate electron energy loss spectra. As an example, electron energy loss spectra in N_2 films are compared with the theoretical spectra. From such comparisons, it is shown that multiple scattering contributions can be identified as well as the major loss mechanisms (e.g., vibrational excitation, librational phonons,..., etc.). Attempts can also be made to estimate the magnitude of total scattering cross sections but this requires measurement of energy-loss amplitudes as a function of film thickness in addition to mathematical analysis. Further examples of vibrational excitation are shown in CO and O_2 films. Shape resonances are identified as strong maxima in the excitation functions of condensed N_2, CO, O_2 and NO. Anion states in these solids seem to have symmetries similar to their gas-phase counterparts with the exception of those in condensed NO. Nitric oxide molecules dimerize upon condensation, so that some of the structures in the excitation functions of condensed NO arise from the formation of $(NO)_2$ states. We conclude in the last section with comments on the relaxation energies of condensed-phase transient anions and an outlook on excitation functions of more complex molecules such as CO_2.

Experiment

The energy loss spectra and the energy-dependence of vibrational energy losses reported herein were measured by high resolution electron-energy-loss spectroscopy. The spectrometer has recently been described in detail (7). The essential features of the experiment are the following: a monoenergetic electron beam emerging from an hemispherical monochromator is incident on a molecular film; within a well-defined solid angle, a small portion of the backscattered electrons are energy analysed at 45° from the plane of the substrate by an hemispherical analyser. The novel feature of this spectrometer (7) resides in the entrance optics of the analyser and exit optics of the monochromator which consist of three cylinder and three aperture lenses operated in a "double zoom" configuration. Programming these lenses allows us to keep nearly constant transmission in the analyser and constant current density on the target over wide energy limits (1-30 eV). The angle of the incident beam can be varied between 14 and 70° from the film normal. The film is formed by condensing the gas on a clean metal substrate maintained near 20 K by an electrically-isolated mechanical contact with the cold head of a closed-cycle refrigerated cryostat. This latter is mounted on a welded bellows equipped with precision adjusting screws. The arrangement allows xyz positioning of the sample. The cold head temperature is measured by an hydrogen bulb thermometer. The apparatus is housed in a bakeable ion-cryopumped UHV system (10) capable of sustaining working pressures in the 10^{-11} torr range.

The film thicknesses were estimated within 40% accuracy from the number of molecules which deposited on the substrate. This number could be calculated from geometrical considerations and gas kinetic theory, as previously described (11). Experiments were performed with films about 50 Å thick. Above approximately 20 Å the spectra were found to be independent of thickness. The beam resolution was 10-20 meV with corresponding currents of $5x10^{-10}$ to $5x10^{-9}$ A. The incident electron energy was calibrated with respect to the vacuum level by measuring the onset of electron transmission through the film and the onset of total reflection from the film surface. The accuracy was ±0.15 eV for the primary electron energy and ±2 meV for energy-loss measurements. Angular distribution studies of the elastic current indicated that N_2, CO and O_2 films were preferentially oriented with respect to the plane of the substrate. $(NO)_2$ films were highly disordered.

Theory

What is needed to interpret electron energy loss spectra of multilayer films is an expression which can provide, for incident electrons at a fixed energy, the energy distribution of electrons scattered out of the film in terms of elastic and inelastic mean free paths (MFP) or scattering probabilities. With such an expression one hopes to obtain an estimate of these MFP and a suitable description of the scattering mechanisms responsible for the spectral features.

An analytical solution of the general problem of electron scattering by multiple targets held between two infinite interfaces is not in general available. Nevertheless, this type of multiple scattering problem has been approached by several investigators mainly in connection with electron transfer in mobility, photoelectron and Auger electron experiments. In order to simplify the analysis, most investigators have formulated approximate solutions valid for the specific problem they were considering. The energy loss spectrum or energy distribution of inelastically scattered particles in a solid has been calculated by Landau (12) for the case of small-angle elastic and inelastic scattering where inelastic scattering cross sections are assumed to be energy independent. The Landau theory (12) is therefore well suited to describe high-energy electron scattering by solids. A more complete formulation of this problem is to be found in the recent work of Toogard and Sigmund (13) on Auger and photoelectron emission from solids. These authors included the possibility of large-angle elastic scattering but inelastically scattered particles were still considered to move in the direction of the primary beam. Since these theories are not applicable to our measurements where only backscattered electrons are detected, we recently developed a unidimensional multiple scattering theory applicable to large-angle inelastic scattering (7). The main ingredients are outlined below.

We consider an electron beam of current intensity I_o incident on a vacuum-solid boundary, at an angle θ_i from its normal. The backscattered current distribution per unit solid angle emerging at an angle θ_d from the normal is described by $I(E_i,E,\theta_i,\theta_d)$. This intensity depends on the primary electron energy E_i, the energy loss E, and the angles θ_i and θ_d. In the energy loss mode, the measured current is proportional to I for a fixed solid angle defined by the aperture of the analyser entrance optics. This intensity distribution can be developed into a sum of partial distributions related to a finite number of collisions. We let the backscattered current distribution per unit solid angle arising from all single collision sites $\vec{r}_1$ be represented by a function I_1. I_2 represents the contribution of double-collision combinations from all possible sites $\vec{r}_1$ and $\vec{r}_2$ and I_n the contribution arising from multiple combinations from n sites. The backscattered current distribution per unit solid angle $I(E_i,E,\theta_i,\theta_d)$ may then be written as the sum of all I_n currents;

$$\text{i.e.,}\quad I = I_1+I_2+I_3+\ldots = \Sigma_{n=1}^{\infty} I_n \tag{1}$$

Each current I_n may be expressed as a product of probabilities for scattering at sites $\vec{r}_{j-1}$ and traveling without scattering between different sites. For randomly oriented molecules or crystallites the probability of an electron to travel with no intervening collision from a point $\vec{r}_{j-1}$ to a site $\vec{r}_j$ can be simply given by Lambert's law

$$P(\vec{r}_j - \vec{r}_{j-1}, E_{j-1}) = \exp\,[-\alpha(E_{j-1})|\vec{r}_j - \vec{r}_{j-1}|] \tag{2}$$

where $\alpha(E_{j-1})$ is the sum of the total probability per unit length (i.e., the inverse MFP).

For example, the probability per unit solid angle for an electron incident on the film at an angle θ_i to lose energy $E=E_i-E_1$ in a single collision at $\vec{r}_1$, and escape the film at an angle θ_d will be given by the product of the probabilities for the electron to travel a distance ℓ_1 from the film-vacuum interface to $\vec{r}_1$ (i.e., $\exp[-\alpha(E_i)\ell_1]$), to lose energy E and be scattered at an angle $\theta=180^o-\theta_i-\theta_d$, in the element $d\ell_1$ at $\vec{r}_1$ (i.e., $Q(E_i,E,\theta)d\ell_1$ where Q is a collision probability per unit length per unit solid angle), and to travel a distance ℓ_2 from $\vec{r}_1$ to the film-vacuum interface (i.e., $\exp[-\alpha(E_i-E)\ell_2]$). Thus we have,

$$dI_1 = I_o\ \exp[-\alpha(E_i)\ell_1]Q(E_i,E,\theta)d\ell_1\ \exp[-\alpha(E_i-E)\ell_2]$$

If we let $\ell_1 = h/\cos\theta_i$ and $\ell_2 = h/\cos\theta_d$, where h is the shortest distance between $\vec{r}_1$ and the solid film-vacuum interface we have for the last expression integrated on h over the film thickness L,

$$I_1 = \frac{I_o Q(E_i,E,\theta)\cos\theta_d}{\alpha(E_i)\cos\theta_d+\alpha(E_i-E)\cos\theta_i}\left\{1-\exp-(\alpha(E_i)L/\cos\theta_i+\alpha(E_i-E)L/\cos\theta_d)\right\} \tag{3}$$

At large thicknesses (i.e., for large values of L) the exponential term vanishes and the scattered current I_1 becomes independent of thickness. For small thicknesses the exponential term in equation 3 may be expressed by the first two terms of its Taylor expansion. This gives

$$I_1 = (I_o/\cos\theta_i)Q(E_i,E,\theta)L \tag{4}$$

Expressions for the other current distributions per unit solid angle I_2, $I_3 \ldots I_n$ cannot easily be obtained unless other simplifying assumptions are made, namely that, (1) the film is a semi-infinite solid (i.e., $L \to \infty$); (2) all electrons are assumed to move at an average angle with respect to the film normal; (3) the scattering probability per unit length for any given energy loss is constant (i.e., we assume an average electron energy dependence in the range where the transition is energetically possible; outside this range the inelastic cross section is null); (4) the elastic scattering probability is constant (i.e., unchanged after an inelastic event). The first assumption is justified experimentally for films having thicknesses greater than about 20 Å. The second one also appears justified if the number of collisions, including quasielastic ones, is high enough or if the film constituents are sufficiently disordered so as to produce a nearly isotropic distribution approximated by an average direction. This condition reduces the problem to one dimension. The third and fourth assumption are not expected to be too stringent as long as the energy of inelastically scattered electrons lies within a range where the MFP do not change too abruptly. With these assumptions equation 1 may be written as a sum of multiple integrals

$$I(E) = I_o \int_o^\infty P(x_1-x_o)\ Q(E)\ P(x_o-x_1)\ \varepsilon dx_1 +$$
$$I_o \int_o^\infty\int_o^\infty\int_o^\infty P(x_1-x_o)Q(T)P(x_2-x_1)Q(E-T)P(x_o-x_1)\varepsilon^2 dx_1 dx_2 dT + \ldots \tag{5}$$

where in each terms, I_o is the incident current intensity and x_j (j=0,1,2) is a one dimensional variable replacing the three dimensional variable $\vec{r}_j$. Q(E) and Q(E-T) are the scattering probabilities per unit lenght for collisions to be in the forward or backward direction with a loss of energy equal to E and E-T, respectively. ε is a phenomenological parameter averaging the angular dependence. The one dimensional version of equation 2, becomes

$$P(x_j-x_{j-1}) = \exp[-\alpha\varepsilon\,|x_j-x_{j-1}|]$$

where the inverse mean free path $\alpha = \int_o^\infty 2Q(E)dE$.

The solution of I(E) is purely a question of reducing equation 5 to a simpler expression via a series of mathematical transformations including Fourier transforms. With these transformations, the integrations on position in equation 5 may be elimitated to yield (7),

$$I(E) = (I_o/2)(Q(E)/\alpha) + (I_o/2)\int_o^\infty(Q(T)/\alpha)(Q(E-T)/\alpha)dT +$$
$$(5I_o/8)\int_o^\infty\int_o^\infty(Q(T)/\alpha)(Q(T')/\alpha)Q(E-T')dT'dT + \ldots \tag{6}$$

This series provides the intensity of electrons having loss energy E via processes whose total scattering probability per unit length is 2Q. It can easily be evaluated for a large number of terms. In the present calculation, we limited the series to one hundred terms (i.e., we considered that a single electron could collide up to 100 times at different sites within the film).

Electron energy loss spectra

Electron energy loss spectra for electrons of various energies (E_i) incident at an angle of 14° from the normal of N_2, CO and O_2 films are shown in Figs. 1A, 2 and 3, respectively. The spectra in the insert of Fig. 2 were recorded in the specular direction (i.e., at 45° incidence). The vertical gains in each curve or portion of a curve are referenced to the elastic peak, with the exception of those in Fig. 1A where the first energy loss peak is the reference. The horizontal bars in the right of Fig. 1 represent the zero intensity base line. All electron energy loss spectra exhibit a series of peaks which can be ascribed to vibrational excitation of ground state N_2, CO and O_2. In O_2 films, additional peaks are due to intra-molecular vibrational excitation of the electronically excited states $a^1\Delta_g$ and $b^1\Sigma_g^+$. It can readily be seen that at certain impact energies (E_i=3 eV, 2.5 eV and 9 eV for N_2, CO and O_2, respectively) the intensities of vibrational energy losses are greatly increased (i.e., up to two orders of magnitude). Such strong enhancements in the vibrational intensities are interpreted to result from the formation of transient anions at particular energies. A discussion on their presence in molecular solids is deferred to the next section.

Despite the apparent similarity between the spectra in Fig. 1A, 2, and 3 and those obtained with the same molecules in the gas-phase (14) two major differences exist: in the solid, the amplitudes of energy-loss peaks decrease much less and their widths enlarge with increasing vibrational quantum number. These differences are most pronounced in N_2 films and least noticeable in condensed O_2. As seen in Figs. 1A, 2, and 3, they are also strongly dependent on the primary beam energy, E_i. These changes can be explained by multiple scattering effects using qualitative arguments or calculated spectra generated by equation 6 of the last section. The scattered intensity in single-collision gas-phase experiments would result only from the first term in equation 6. The other terms in equation 6 represent multiple interactions which, in a solid target, necessarily increase the scattered intensity. Furthermore, the magnitude of this additional contribution depends on the product of terms representing the ratio Q/α of the scattering probabilities per unit length for a particular loss to the inverse total mean free path. When this ratio is large the multiple scattering terms in equation 6 contribute significantly to the probed intensity I(E). Since peaks resulting from multiple vibrational losses exhibit no anharmonicity they occur at slightly different energies than the single-loss peaks of the corresponding overtone frequencies. Because of the many possible combinations of multiple and single vibrational and phonon losses, the overall effect leads to an appreciable increase of the width of peaks

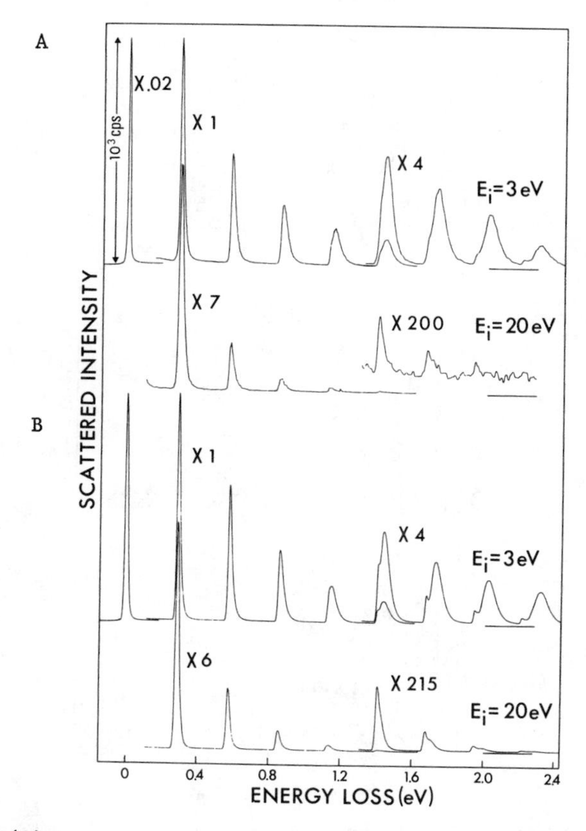

Figure 1. (A) Energy-loss spectra for 3 and 20 eV electrons incident on a 50 Å N_2 films at 14° from the normal. (B) Mathematical simulation of the spectra in (A).

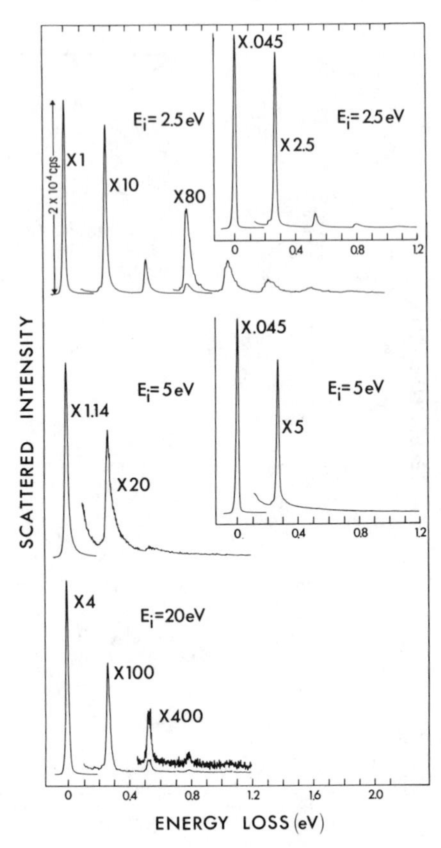

Figure 2. Energy loss spectra for 2.5, 5, and 20 eV electrons impinging on a 50 Å CO film. The curves shown in the inserts were recorded in the specular direction. The others were recorded with a beam incident at 14° from the surface normal.

attributed to intramolecular vibrations. These effects are most apparent in the E_i=3 and 2.5 eV curves in Figs. 1A and 2, respectively, where the vibrational peaks broaden as a function of vibrational quantum number. At these energies the ratio Q/α is large ($\simeq$ .45) since only vibrational and phonon transitions are energetically possible. However, when the incident energy is increased to 20 eV many electronic transitions become energetically possible and contribute to increase α; whereas, the vibrational excitation probability which contributes significantly to Q is sensibly smaller than at E_i=3 and 2.5 eV. We therefore observe a sharpening of the vibrational peaks which exhibit an anharmonicity characteristic of ground state N_2 and CO. Similarly, in O_2 (Fig. 3), peaks in all of the spectra are relatively sharp because the ratio Q/α always remains small principally due to the occurence of low-lying electronic states. The largest broadening of vibrational peaks is observed in the 4-7 eV and 4-6 eV incident-energy region in N_2 and CO, respectively, (e.g., the middle curve in Fig. 2) where energy losses are dominated by phonon transitions. At these energies, the vibrational excitation cross section is small and no electronic excitation is possible, but Q/α remains high because Q is now the largest part of α which is composed principally of the total phonon and elastic scattering cross sections. For the case of electron scattering from a CO film in the specular direction at E_i=5 eV shown in the lower insert of Fig. 2, the v=1 vibrational loss peak remains sharp. In this direction, vibrational excitation occurs mostly via a long-range interaction with the permanent dipole of the CO molecule (15) and short-range multiple scattering events are less prominent.

In Fig. 1B, we show curves calculated with equation 6 which reproduce quite faithfully the amplitudes, widths, positions, and line shapes the experimental peaks in the upper curves. In these calculations additionally to the vibrational losses, the quasi-elastic losses (i.e. librational plus translational phonons) of solid α-N_2 (16,17) which are dominated at 4 and 8 meV by librational excitation were considered. Contributions from acoustical phonons, inhomogeneous broadening and combination losses were neglected. The energies and relative amplitudes of seven phonon losses were taken from optical measurements (18,19) to initiate our calculations. Good fits were obtained for a particular class of distributions of amplitudes corresponding to a single 8 meV effective libron normalised to the sum of amplitudes. The distribution used to generate the curves in Fig. 1B is composed of one libron at 4 meV, one at 8 meV, and five between 10-14 meV with relative amplitudes in the ratios 4:2:1, respectively. The total scattering probability of all phonons was taken equal to 1.5 of the sum of the vibrational scattering probabilities. The relative vibrational probabilities were chosen close to the experimental values and the sum of the total fixed at 18% of α. The total elastic scattering probability was taken to be 55% of α. The sharp peaks on the low-energy side of the last four multiple-scattering vibrational peaks are produced in the calculation by single vibrational losses. These peaks also appear in the experimental spectrum where they are broader and not so well resolved. This discrepancy could only be resolved either by introducing in the theory a reflection coefficient at the film-vacuum interface or by splitting the probability of vibrational losses into vibrational-librational combinations. The calculated curve for 20 eV incident energy was generated by reducing the total scattering probability for v=1 from 9 to 2% of α while keeping the relative vibrational amplitudes close to the experimental values and assuming that 25% of the collisions produce electronic excitation. The total librational and elastic scattering probabilities were both fixed at 36% of α. Comparing this curve with the experimental spectrum we conclude that, at E_i=20 eV, the vibrational peaks are composed of a single vibrational loss with some tailing due to multiple vibrational collisions; these peaks are also further broaden by multiple librational losses. Similar calculations can be made to explain the general features of the spectra in Figs. 2 and 3.

The variation in the integrated amplitude of each energy loss peak exhibits approximately the same general trend as a function of thickness: a linear rise, at small coverages, followed by a rapid change in slope and a saturation of current intensities at larger thicknesses. The variation in the integrated amplitude of the intensity of the v=1 level of ground state N_2 as a function of the average number of monolayers in the film is shown in Fig. 4. Although multiple phonon losses contribute to the peak width, the behavior of this curve is well approximated by equation 3 derived for single inelastic losses. The behavior at small and large thicknesses, is given by equation 4 and by suppressing the exponential term in equation 3, respectively. These two behaviors are delineated by a region in Fig. 6 where the slope changes rapidly. We can take advantage of this change of slope to estimate the sum of the total vibrational cross section. If we consider the v=1 loss integrated over a given energy range, only collisions that either take electrons out of the energy range of integration or out of the film contribute to change the magnitude of the v=1 intensity with thickness. Thus, since quasi elastic multiple scattering is not resolved and therefore integrated under the v=1 peak in our experiment, only inelastic vibrational excitation and backscattering into vacuum is contributing to the sum of the effective total collision probability per unit length which produces the curve in Fig. 6. With these arguments and considering that equation 3 holds in the case of the v=1 loss, we make the approximation $\alpha(E_i) \simeq \alpha(E_i - E) \simeq nQ_T$ where n is the density of molecule in the solid and Q_T the total vibrational cross section. Then, equating the low and high thickness behavior of equation 3 we have,

$$Q_T \simeq (1/nL)(1/\cos\theta_i + 1/\cos\theta_d)^{-1}.$$

A value for nL is obtained by multiplying the number of monolayer (1.75) deduced graphically from Fig. 6 at the intersect of the extrapolation of the low and high thickness behavior of equation 3 by the number of molecules in a monolayer. We consider a monolayer from the closed-pack (111) face of the face-centered-cubic crystal structure of α-N_2 (20,21). This gives 7.2×10^{14} molecules/cm^2. With $\theta_i = 14^{\circ}$ and $\theta_d = 45^{\circ}$, the sum of the total vibrational excitation cross section in the solid at 2 eV is estimated to be 3.3×10^{-16} cm^2. Despite the crudeness of the approximation, this value is

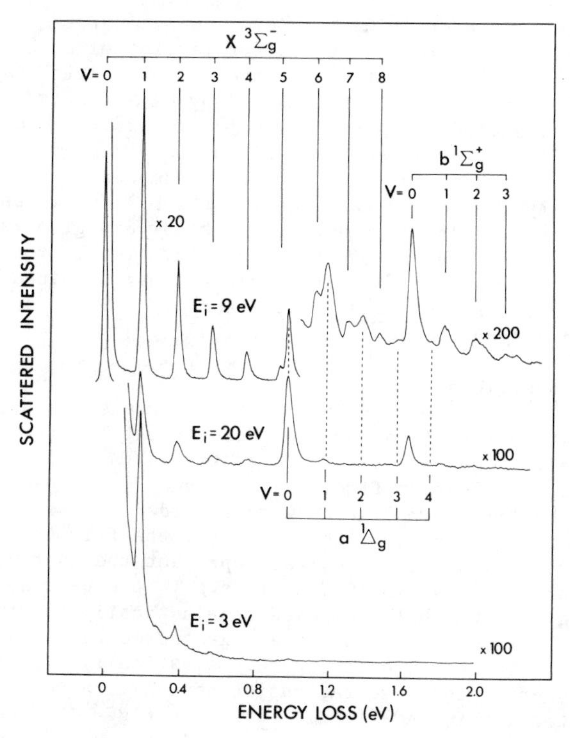

Figure 3. Energy-loss spectra for electrons of 9, 20 and 3 eV primary energies incident on a multilayer disordered O_2 film.

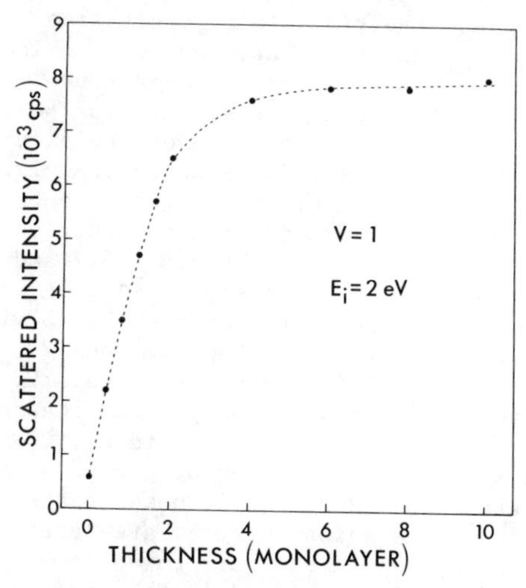

Figure 4. Intensity of electron energy losses to the fundamental frequency of N_2 as a function of film thickness.

close to the value of 7×10^{-16} cm^2 for the sum of the total vibrational excitation cross sections for levels v=1 to 8 in the gas (11) at E_i=2.5 eV calculated from the sum of total cross sections for all vibrational states (22) and by calibrating v=1 on the rotationally summed value of 2.0×10^{-16} cm^2 (23). We expect such a similarity in the cross sections since the width of the $^2\Pi_g$ resonance is practically unchanged in the condensed phase (4). From the previously determined ratio of vibrational to librational excitation probabilities at resonance, we now can give a value of 5×10^{-16} cm^2 for the sum of the total librational cross sections. Interestingly, this value is the same as that obtained from close-coupling calculations (24) for the sum of all rotational cross sections in gaseous N_2.

Excitation functions:

The excitation functions for the three lowest vibrational energy losses (v=1,2, and 3) of ground state N_2, O_2, and NO within the $(NO)_2$ dimer are shown in Figs. 5, 6, and 7, respectively. The curves in the first two figures were recorded in specular direction and those in Fig. 7 at 14° incidence from the film normal. In the O_2 and $(NO)_2$ films the v=0 curves represent the energy dependence of the "vibrationally elastic" peak (e.g., the peak at 0 eV energy loss in Fig. 3) which is composed of elastically and inelastically scattered electrons. The latter are generated by losses to phonons that cannot be resolved from purely elastically scattered electrons. Low-frequency vibrations of the dimer also contribute to the v=0 intensity in $(NO)_2$ films (6). In Fig. 6, the excitation functions of the electronic state $^1\Delta_g$ in the solid and of v=1 in the gas phase are also included. All vibrational excitation functions in Figs. 5 to 7 exhibit broad "humps" which indicate that at particular energies the vibrational excitation cross section of ground state N_2, O_2 and $(NO)_2$ are enhanced. These maxima in the excitation functions are interpreted as due to the formation transient anions having symmetries similar to their gas-phase counterpart. The most striking example of the occurrence of shape resonances in molecular solids is shown in Fig. 5. The four "humps" in the upper curve of this figure can be correlated (5) with similar maxima in the v=1 excitation function of gaseous N_2 at 2.3, 8, 13.7 and 22 eV (25). The 2.3, 13.7 and 22 eV structures are ascribed (26-28) to the anion states $^2\Pi_g$, $^2\Delta_g$ and $^2\Sigma_g^+$. Similarly in the excitation function of the v=1 to 3 vibrational levels of condensed CO, two maxima at about 2.5 and 20 eV can be correlated (7) with the $^2\Pi$ and $^2\Sigma$ states of gaseous CO (25-29). The relationship between the results in the two phases of O_2 is also apparent from Fig. 6. Our observations in N_2, CO and O_2 are summarized in Table I where the energies of gas-phase anions are listed and compared with the energies of those in the solid having a similar symmetry. The anion energies are taken in the v=1 decay channel. The differences in the gas- to solid-phase anion energies or relaxation shifts (Δ) are listed in the last column of Table I. We note that the energy of the resonance in O_2 films (Fig. 6) increases with increasing vibrational quantum number of the energy loss. This effect has been interpreted (3) to arise from an admix-

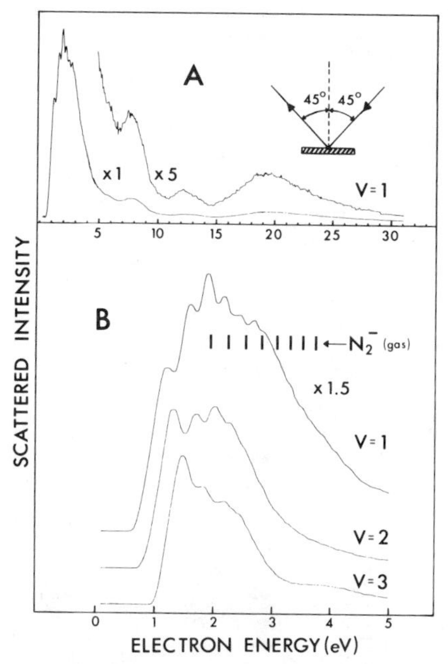

Figure 5. (A) Energy dependence of vibrational excitation intensity of the first vibrational level of ground-state N_2 in a multilayer molecular nitrogen film. (B) Excitation function for the first three vibrational levels of N_2 in the 0-5 eV region.

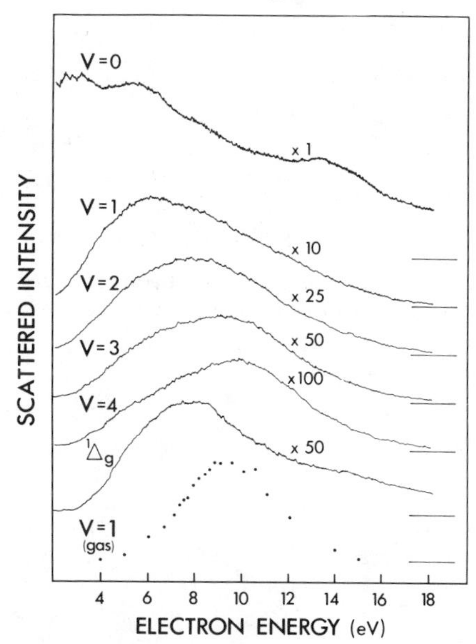

Figure 6. Energy dependence of the reflectivity (top curve) and of vibrational and electronic excitation processes in a multilayer O_2 film.

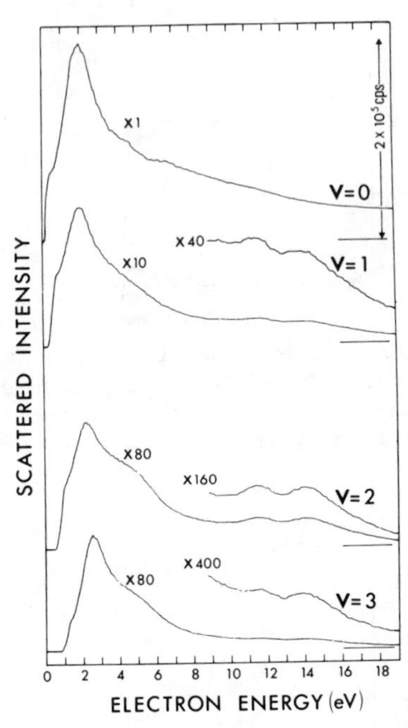

Figure 7. Electron excitation function of quasi-elastic losses (v=0) and vibrational energy losses to the v=1, 2, and 3 $\nu_1+\nu_5$ modes of $(NO)_2$.

ture of the $^2\Pi_u$ and $^4\Sigma_u^-$ anion states of O_2^- whose relative contributions would vary according to decay channel. The electron excitation functions recorded between 0-19 eV in $(NO)_2$ (Fig. 7) reveal the presence of two broad "humps" at 11.6 ± 0.2 and 14.2 ± 0.2 eV and a strong maximum near 2 eV. The latter is present in the "quasi elastic" channel (i.e., v=0) and in the v=1, 2, and 3 vibrational excitation functions of the in-phase (ν_1) and out-off phase

TABLE I. ENERGIES AND SYMMETRIES OF GAS- AND SOLID-PHASE TRANSIENT ANIONS

ANION		ENERGY[a]		
		Gas[b]	Solid[c]	Δ
N_2^- ($^2\Pi_g$)		1.93[d]	1.23[d]	0.7
	N_2^- (?)	8.0	6.2	1.8
N_2^-($^2\Delta_g$)		13.7	12.6	1.1
	N_2^-($^2\Sigma_u^+$)	22.0	19.7	2.3
CO^-($^2\Pi$)		1.5[d]	0.8[d]	0.7
	CO^-($^2\Sigma^+$)	21.0	19.0	2.0
O_2^-($^2\Pi_g$)		-0.5	---	---
	O_2^-($^2\Pi_u$+ $^4\Sigma_u^-$)	9.3	6.0	3.3

a: Energy of maxima in v=1 excitation function unless otherwise stated;
b: Energies from ref. 22;
c: Present work;
d: Energy of first vibrational peak of anion;
Δ: Relaxation shift.

(ν_5) vibrations of NO molecules within the dimer (6). The broad "humps" are exclusively related to vibrational excitation of the ν_1 and ν_5 modes. In matrix experiments with films of Ar containing 5% NO, only one broad resonance around 12 eV is observed (6) in the v=2 decay channel. At this concentration dimer formation is much reduced and scattering is believed occur predominently from isolated NO molecules. We therefore interpreted (6) the 11.6 and 14.2 eV resonances as derived from a splitting, via "through space" interaction, of unfilled degenerate π_u2p NO orbitals. No clear-cut evidence of the presence $(NO)_2$ anions derived from the lowest unfilled π_g2p orbital of the NO molcule was found (6) from the strong maximum near 2 eV.

Further evidence of the formation of transient anions in the condensed phase is provided by resolving the vibrational structure of the $^2\Pi$ state of N_2^- in the v=1, 2, and 3 decay channels in the solid. From this result, shown in Fig. 5b, we can estimate (4) the average vibrational energy (0.29 eV) and the lifetime (≃ 3×10^{-15} sec.) of the resonance. These values, are close to those found in the gas phase (0.27 eV and 3.5×10^{-15} sec (30), respectively). Furthermore, the shift in the energy of the oscillatory structure with decay channel in the three excitation functions of Fig. 1(b) can be explained by finite lifetime oscillator models (30-33) such as the "boomerang" model (30-31)). Such an energy dependence on decay channel indicates that the lifetime of the resonance is comparable to the vibrational period. These findings suggest that the short-range part of the e^{-}-N_2 potential well is not significantly modified in the solid and consequently, that the anion retains essentially the $^2\Pi_g$ symmetry.

Conclusions

In this work, we have tried to describe some aspects of inelastic electron scattering from molecular solid films including vibrational and librational excitation, multiple scattering effects and transient anion formation. Since the scattering cross sections involved in the primary energy range of interest (1-30 eV) are high ($\simeq 10^{-16}$-10^{-15} cm^2), the target films ($\simeq$ 50Å) must be considered as "thick" by the electron beam. It is therefore necessary to resort to multiple scattering theory in order to properly interpret energy-loss spectra. Despite the limitations imposed on the measurements by multiple collisions, it appears possible to identify in such spectra contributions from individual major energy-loss mechanisms.

By measuring over wide energy limits the excitation functions of intramolecular vibrational losses in N_2, CO, O_2, and NO films it was possible to study the influence of phase on the energies of transient negative ions of these molecules. From our most recent results (7) in more complex condensed molecules, it appears that the present data represent the simplest cases where the anion symmetry is not appreciably perturbed by the surrounding matrix. In other words, the resonant electron truly localizes on a single molecular site during the anion lifetime and does not contribute to the formation of a delocalized state or a wide band within the solid. This is not so obvious in more complex systems. For example in condensed CO_2, we find numerous and intense structures in the excitation functions, but their amplitudes and energies change considerably according to the measured intramolecular vibrational loss. It seems for CO_2 that the symmetry rules of the transitions involved, and possibly the anion symmetry, are different from those in the gas phase (7).

As shown in Table I, the relaxation energy of each anion observed can be deduced by comparison with the gas-phase energy. From the values listed, there appears to exist no particular rule or law by which one could predict relaxation energies. This result may be due to the complexity of the energy shift caused by the phase change which may be divided into two components: changes in the intramolecular relaxation energy of the negative ion and an energy change coming from extramolecular forces. This latter energy change not only results from the screening of the negative charge by the polarization of the surrounding molecules but also from the influence of the surrounding matrix on the anion molecular orbital configuration. The latter shift is similar to a chemical shift or "cage" effect in photoemission. If the interaction of surrounding molecules with the negative ion is strong (e.g., if the wavefunction of the extra electron strongly overlaps with neighbors) we may expect the anion to reduce its molecular symmetry or even lose its identity. A breakdown in the molecular symmetry of a transient anion will automatically modify the partial wave content of the spherical harmonic expansion of the resonant electron wavefunction. The lower symmetry is most likely to result in a wave packet containing more spherical harmonics of lower angular momentum. Thus, the extra electron can tunnel more easily through its centrifugal barrier and the energy and the lifetime of the anion are reduced. If the configuration interaction is small, the energy shift should result mainly from screening, but its magnitude may depend on the lifetime of the transient state, since the polarization of the medium is related to the time of residence of the electron at a particular molecular site. For short-lived anions ($<10^{-15}$ sec) the localization time may be too short to allow complete electronic polarization of the surrounding medium. In this case, the intramolecular relaxation related to reorganization of the electronic orbital of the negatively charged molecule may also be incomplete. Thus when the lifetime changes from gas to solid, the total relaxation energy could also be modified via the intramolecular relaxation energy term. We also note that when the MFP is small, electrons interact with the first few layers of the film. Consequently, the contribution of the polarization energy to the measured relaxation shift is smaller than in the bulk due to the reduction of polarizable nearest neighbors in the vicinity of the surface.

Literature Cited

1. C.E. Klots and R.N. Compton, J. Chem. Phys. 1978, 69, 1644.
2. C.E. Klots and R.N. Compton, J. Chem. Phys., 1978, 69, 1636.
3. L. Sanche and M. Michaud, Phys. Rev. Lett., 1981, 47, 1008.
4. L. Sanche and M. Michaud, Phys. Rev., 1983, 27, 3856.
5. L. Sanche and M. Michaud, Chem. Phys. Lett., 1981, 84, 497.
6. L. Sanche and M. Michaud, J. Chem. Phys., in press.
7. L. Sanche and M. Michaud, **to be published.**
8. J.E. Demuth, D. Schmeisser, and Ph. Avouris, Phys. Rev. Lett., 1981, 47, 1166.
9. D. Schmeisser, J.E. Demuth, and Ph. Avouris, Phys. Rev. B, 1982, 26, 4857.
10. M. Michaud and L. Sanche, J. Vac. Sci. Technol., 1980, 17, 274.
11. L. Sanche, J. Chem. Phys., 71, 4860.
12. L. Landau, J. Phys. (Moscow), 1944, 8, 201.
13. S. Toogard and P. Sigmund, Phys. Rev., 1982, 25, 4452.
14. J.W. Boness and G.J. Schulz, Phys. Rev. A, 1973, 8, 4452.
15. J.C. Bertolini and J. Rousseau, Ann. Phys. (Fr), 1980, 5, 115.
16. M. Brigth, A. Ron, and O. Schnepp, J. Chem. Phys., 1969, 51, 1318.
17. J.E. Cahill and G.E. Leroi, J. Chem. Phys., 1969, 51, 1324.
18. M. Brith, A. Ron, and O. Schnepp, J. Chem. Phys., 1969, 51, 1318.
19. J.E. Cahill and G.E. Leroi, J. Chem. Phys., 1969, 51, 1324.

20. L.H. Bolz, M.E. Boyd, F.A. Mauer and H.S. Perser, Acta Cryst., 1959, 12, 249.
21. L. Vegard, Z. Physik, 1930, 61, 185.
22. G.J. Schulz, Phys. Rev., 1964, 135, A988.
23. K. Jung, Th. Antoni, R. Muller, K-H. Kochem. and H. Ehrhart, J. Phys. B, 1982, 15, 3535.
24. D.G. Truhlar, M.A. Brandt, A. Chutjian, S.K. Srivastava and S. Trajmar, J. Chem. Phys., 1976, 65, 2962.
25. G.J. Schulz, in "Principles of Laser Plasmas", Bekefi G., Ed.; Wiley: New York, 1976, p.33.
26. D.Dill and J.L. Dehmer, Phys. Rev. A, 1977, 16, 1423.
27. J.R. Rumble Jr., D.G. Truhlar, and M.A. Morrison, J. Chem. Phys., 1983, 79, 1846.
28. K. Onda and D.G. Truhlar, J. Chem. Phys., 1980, 72, 5249.
29. A. Chutjian, D.G. Truhlar, W. Williams, and S. Trajmar, Phys. Rev. Lett., 1972, 29, 1580.
30. D.T. Bertwistle and A. Herzenberg, J. Phys., 1971, 4, 53.
31. L. Dubé and A. Herzenberg, Phys. Rev. A, 1979, 20, 194.
32. J.W. Gadzuk, J. Chem. Phys. , 1983, 79, 3982.
33. W. Domcke and L.S. Cederbaum, Phys. Rev. A, 1977, 16, 1465.

RECEIVED June 11, 1984

VAN DER WAALS COMPLEXES

13

Vibrational Predissociation of Small van der Waals Molecules

ROBERT J. LE ROY

Guelph-Waterloo Centre for Graduate Work in Chemistry, University of Waterloo, Waterloo, Ontario N2L 3G1, Canada

The nature of vibrationally and rotationally predissociating states of atom-diatom Van der Waals molecules and the fundamental considerations governing their predissociation are discussed. Particular attention is focussed on the influence of the potential energy surface and the information about it which might be extracted from accurate measurements of predissociation lifetimes. Most of the results discussed pertain to the molecular hydrogen-inert gas systems, and details of previously unpublished three-dimensional potential energy surfaces for diatomic hydrogen with krypton and xenon are presented.

A discussion of Van der Waals molecules is a natural component of a treatise on resonance phenomena, for a variety of reasons. The most obvious of these is simply the fact that transitions involving both compound-state and shape resonance levels figure prominently in the spectra of these species. Indeed, in many cases the metastable nature of the final states of such transitions is a key feature of the manner in which they are observed (1-3). Moreover, observations of structure due to resonances in scattering cross sections can provide detailed information regarding intermolecular potential energy functions (4).

Van der Waals molecules are themselves of great current interest because of the information which may be extracted from them about both intermolecular forces and intramolecular and intermolecular dynamics. With regard to the former, studies of the discrete spectra of Van der Waals molecules have already provided multidimensional potential energy surfaces of unparalleled sophistication and accuracy for two classes of atom-diatom complexes, the molecular hydrogen-inert gas (5-10) and hydrogen halide-inert gas (11-14) systems. At the same time, vibrational predissociation of Van der Waals molecules is proving to be an extremely revealing means of studying intramolecular energy transfer and relaxation processes (1-3,15,16). Such studies are of particular interest because of the

0097-6156/84/0263-0231$09.00/0

large differences (up to two orders of magnitude!) in the characteristic frequencies associated with the different modes of motion. It seems reasonable to hope that such large differences would make a description of the dynamics based on a separation of variables be fairly realistic. Such a separable picture would help provide an intuitive physical understanding of the magnitude and trends of properties such as the overall predissociation rate and the associated branching ratios, and of their dependence on the nature of the system.

Vibrational predissociation is of course intimately related to vibrationally and rotationally inelastic collisions. Both processes involve coupling between the same initial and final state channels, but they differ in that the inelastic processes must occur at energies above the appropriate internal motion excitation threshold and are observed in collision experiments, while predissociation (usually) occurs at energies below this threshold and is observed spectroscopically. While the present paper focusses most attention on the phenomenon of predissociation, the nature of the information contained in these two types of experiments will be compared.

The present chapter mainly discusses the simplest class of atom-diatom Van der Waals molecules, the molecular hydrogen-inert gas complexes. While experimental information on the vibrational predissociation of these species is as yet relatively limited, our knowledge of the potential energy surfaces which govern their dynamics (9,10) is unequalled for any other systems. Moreover, the small reduced mass and large monomer level spacings make accurate calculations of their properties and propensities relatively inexpensive to perform. For these reasons, these species have come to be treated as prototype systems in theoretical studies of vibrational predissociation (17-25).

In the following, we first begin with a qualitative description of the origin and nature of resonance behaviour and vibrational predissociation in Van der Waals molecules. This is followed by a brief outline of how accurate vibrational predissociation calculations for simple atom-diatom complexes can be performed. Of course, performance of such calculations presumes an accurate knowledge of the potential energy surface of the species in question. Our presentation of a summary of selected results regarding vibrational predissociation of atom-diatom systems is therefore preceeded by a description of the characteristics of some of these surfaces, which includes details regarding previously-unreported potential energy functions for molecular hydrogen interacting with krypton and xenon.

Resonances and Predissociation of Van der Waals Molecules

For the case of an atom-diatom Van der Waals molecule, the two possible types of resonance phenomena are schematically illustrated in Figure 1. There, $V_1(R)$ represents the potential energy between the atom and the ground state diatom, while $V_2(R)$ is that between an atom and a diatom in excited vibrational-rotational state (v',j'). The function labelled χ_b is the radial wavefunction for a truly bound state, while χ_q is that for a quasibound or shape resonance level which may predissociate by tunneling past the effective barrier of the supporting potential. Although χ_q is actually a

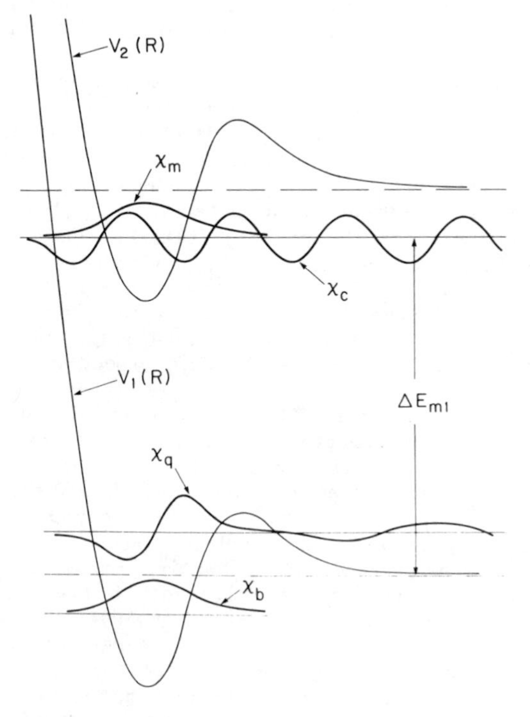

Figure 1.
Schematic representation of atom-diatom predissociation; $V_2(R)$ is the potential for an excited state $(v',j')=(0,0)$ diatom interacting with an atom, and $V_2(R)$ that for a ground state diatom $(v,j)=(0,0)$.

continuum wavefunction, the potential barrier effectively modulates its amplitude in the classically accessible region over the potential well; near resonance this "internal amplitude" has a local maximum whose width (as a function of energy) correlates with the resonance lifetime or collisional time delay maximum (26). It is this modulation of the wave function amplitude at small distances which causes spectroscopic transitions involving these continuum levels to yield a discrete spectrum.

Transitions involving such orbiting or shape resonance states were observed in the discrete infrared spectra of the hydrogen-inert gas Van der Waals complexes (27-29), and were included in the analyses which determined the potential energy surfaces of these systems (5-10). Moreover, observations of structure due to shape resonances in low energy total scattering cross sections have proved to be a very sensitive way of determining their vibrationally averaged effective spherical potentials (4). Indeed, the quality of our knowledge of this aspect of these potentials is dramatically illustrated by the almost exact agreement between the curves for H_2-Kr and H_2-Xe obtained from scattering (4) and spectroscopy (9-10) (see Figures 25 and 27 of Ref. (4)). In view of the fact that the potentials being compared have different functional forms and were based on different types of experimental data, this agreement is a remarkable demonstration of how accurately this aspect of these potential surfaces is known.

One interesting feature of these two ways of observing orbiting resonances is their complementarity, in that a resonance observed by one technique is not readily accessible to the other. This was clearly demonstrated for the molecular hydrogen-inert gas systems, where the quasibound levels observed spectroscopically are those with relatively small tunneling predissociation widths ($\Gamma<1$ cm^{-1}) which lie well below the potential barrier maxima (6,27-29), while those observed in scattering are relatively broad and lie at energies very near to or above these peaks (6,30). This is a natural consequence of the fact that the best energy resolution readily achievable in a scattering experiment is of the order of a few percent, while very broad lines are difficult to observe as part of a discrete spectrum. As a result, this complementarity should also apply to possible observations of the Feshbach or compound-state resonances discussed below.

The second type of metastable level behaviour, that of most interest here, is also illustrated in Figure 1. There, χ_m represents the radial wave function for an atom bound to an internally excited diatom by potential curve $V_2(R)$, while χ_c is an isoenergetic continuum state wave function associated with the interaction $V_1(R)$ between the atom and the ground state diatom. If the atom-diatom potential surface did not depend on the relative orientation or internal bond length of the diatom, these two states would be completely independent. Vibrational predissociation therefore corresponds to the mixing of these two types of zeroth-order behaviour by the internal coordinate dependence of the intermolecular potential, giving rise to decay of the metastable states represented by χ_m.

Of course, the exact wave functions in this region are pure continuum functions, consisting of a linear combination of a function (similar to χ_c) associated with the open channel $V_1(R)$, with a bound-type function (like χ_m) associated with the closed channel $V_2(R)$. As the energy is varied across the width of one of these resonance states, the amplitude of the closed-channel portion of this wavefunction passes through a local maximum with a Breit-Wigner lineshape, in exact analogy with the behaviour of the "internal amplitude" of the wave functions associated with shape resonances.

In the present discussion, vibrational predissociation is taken to include both cases in which the monomer vibrational quantum number changes (v'>0 in Figure 1), as well as the phenomenon of predissociation by internal rotation, corresponding to v'=0 and j'>0 in Figure 1. Although these two types of processes usually have very different rates for a given species (17-19,22,25), their theoretical descriptions are essentially identical, differing only in which features of the potential energy surface contribute most to the coupling between the open and closed channels. Moreover, since the internal rotation of the diatom monomer is in effect merely a very loose bending vibration of the complex, it is quite appropriate for these processes to share the same name.

Calculating Resonance Energies and Widths

Accurate Calculations. Accurate calculations of the energies and widths of vibrationally predissociating levels of atom-diatom complexes must start with the full Hamiltonian

$$\mathbf{H}(\underline{R},\underline{r}) = -(\hbar^2/2\mu)R^{-1}(\partial^2/\partial R^2)R + \ell^2/2\mu R^2 + V(R,r,\theta) + \mathbf{H}_d(\underline{r}) \quad (1)$$

where $\underline{R}$ is the axis of the complex, $\underline{r}$ that of the diatom and θ the angle between them, $\mu=m_a m_d/(m_a+m_d)$ is the effective reduced mass associated with the interaction of an atom of mass m_a with a diatom of mass m_d, ℓ^2 is the total angular momentum operator associated with the rotation of $\underline{R}$, $V(R,r,\theta)$ is the atom-diatom interaction potential, and $\mathbf{H}_d(\underline{r})$ is the vibration-rotation Hamiltonian for the isolated diatom.

The exact eigenfunctions of this Hamiltonian may be expanded as

$$\Psi_\alpha^{JM}(\underline{R},\underline{r}) = r^{-1}R^{-1} \sum_{v''}\sum_{a''} \phi_{v''j''}(r)\ \Phi^J_{a''}(\hat{\underline{R}},\hat{\underline{r}})\ \chi^{J\alpha}_{v''a''}(R) \quad (2)$$

where J and M are the quantum numbers for the total angular momentum and its space-fixed projection, and α is an index identifying a particular eigenstate of the system. In the present context it is convenient to associate α with the set of "zero-coupling-limit" quantum numbers $\{v,j,n,\ell,J\}$, where n is associated with the stretching vibration of the Van der Waals bond R. The functions $\{\phi_{vj}(r)\}$ are the radial eigenfunctions for the vibration of the free diatom, and $\{\chi_{va}(R)\}$ are a complete set of angular basis functions with quantum numbers collectively identified as "a".

In the Arthurs-Dalgarno (31) space-fixed representation, $a=\{j,\ell\}$ and $\Phi^J_a(\underline{R},\underline{r})$ is the total angular momentum function, defined as an appropriate linear combination of products of spherical harmonic functions for the rotation of $\underline{R}$ and $\underline{r}$ (10). Alternately, in the bodyfixed coordinate system: $a=\{J,\Omega\}$, the operator ℓ^2 is replaced by $(\mathbf{J}-\mathbf{j})^2$, and the angular basis functions are parity adapted linear combinations of products of spherical harmonic functions for the rotation of r, times the usual symmetric top eigenfunctions for the rotation of $\underline{R}$. In either case, substituting Equation 2 into Equation 1 and taking inner products with the internal motion basis functions $\{\phi_{v'j'}(r)\ \Phi_{a'}(\hat{\underline{R}},\hat{\underline{r}})\}$ yields the usual set of coupled differential equations for an atom-diatom system,

$$\{-(\hbar^2/2\mu)d^2/dR^2 + U(v',a';v',a';J|R) - E_\alpha^J\}\ \chi_{v'j'}^{J\alpha}(R) \qquad (3)$$

$$= -\sum_{v''}{}'\sum_{a''}{}'\ U(v',a';v'',a'';J|R)\ \chi_{v''j''}^{J\alpha}(R)$$

where E_α^J is the total energy and $U(v',a';v'',a'';J|R)$ are the matrix elements of the internal motion basis functions with the operator U:

$$\mathbf{U}(\underline{R},\underline{r}) = \mathbf{H}_d(r) + \ell^2/2\mu R^2 + V(R,r,\theta) \qquad (4)$$

In either angular basis, the matrix elements of $\mathbf{H}_d(\underline{r})$ are diagonal and equal to the vibration-rotation energies of the free diatom; in the space-fixed representation the matrix elements of $\ell^2/2\mu R^2$ are also diagonal in v', j' and ℓ' and equal to the simple centrifugal term $\ell'(\ell'+1)\hbar^2/2\mu R^2$.

The formal scattering theory for describing compound-state resonances such as the vibrationally predissociating states of interest here, is well established (see, e.g., (32-33) and references therein). For an isolated narrow resonance associated with closed channel m, the S-matrix element between (open) channels j and j' is given by (33)

$$S_{jj'}(E) = [S_d(E)]_{jj'} - ig_{mj}g_{mj'}/(E-E_m+i\Gamma_m/2) \qquad (5)$$

where E_m is the resonance energy and Γ_m its total width, g_{mj} is a complex number whose squared modulus is the partial width $\Gamma_{mj}=|g_{mj}|^2$ which determines the flux of dissociation products into channel j (and whose sum over j equals Γ_m), and $[S_d(E)]_{jj'}$ is a slowly varying contribution due to the direct scattering (i.e., scattering which does not involve channel m). A given predissociating level may then be characterized by performing close-coupled scattering calculations for a range of energies surrounding the resonance, and determining the resonance parameters from simultaneous least-squares fits of the resulting S-matrix elements to Equation 5 (33). The quality of the resulting fit provides a measure of the validity of the isolated narrow resonance approximation for this state. Of course, if there is only one open channel, the S-matrix consists of the single term $S=\exp[2i\eta(E)]$, where the open channel phase shift $\eta(E)$ is a sum of a Breit-Wigner line shape function plus a slowly-varying background phase (32).

While most of the "exact" results discussed herein were obtained in the above manner, a number of other equally valid procedures may also be used (34-38). One of these, the complex coordinate rotation method is described in detail in the following chapter.

Together with the method outlined above, the Siegert quantization scheme discussed by Atabek and Lefebvre (36), and the secular equation method (19,39), it describes the predissociation purely as a scattering phenomenon which is completely characterized by the properties of the (continuum) wavefunction at the energy in question.

In practise, however, predissociating states will usually be observed as structure in a photodissociation spectrum, in which context the observed structure also depends on the initial state wavefunction and the transition dipole moment. A close-coupling procedure which takes account of the latter considerations was introduced some time ago by Shapiro (37), and has recently been applied to the description of vibrational predissociation of Van der Waals molecules (22,23,38). However, although it yields a detailed spectrum, its use also presumes knowledge of the transition dipole function, information which is even less readily available than that regarding the potential energy surface required to define the wavefunctions. In any case, for the molecular hydrogen-inert gas systems discussed in most detail here, its predictions regarding resonance energies and widths (22,23) are essentially identical to those obtained (assuming the same coupled channel basis set) using the S-matrix fits described above (19,24,25). Thus, while Shapiro's artificial channel method (37,38) may in principal be a more appropriate procedure for describing vibrationally predissociating levels, differences between its predictions and those of pure scattering theory treatments such as that described above, may be difficult to observe.

Approximation Methods. Most approximate treatments of predissociation are based on a partitioning of the total Hamiltonian $\mathbf{H}$ into two parts. The first of these, $\mathbf{H}_0$, provides a zeroth-order description of the dynamics of the system, and its discrete eigenvalues are estimates of the energies of all bound and metastable states. The remainder of the Hamiltonian, $\mathbf{H}' = \mathbf{H} - \mathbf{H}_0$, couples the discrete eigenfunctions associated with metastable levels of $\mathbf{H}_0$ to isoenergetic continuum wavefunctions associated with the open channels, leading to predissociation. Within the isolated narrow resonance approximation, the partial width (full width at half maximum) associated with predissociation of metastable state m into open channel j is given by the golden-rule expression

$$\Gamma_{mj} = 2\pi \left| \int \Psi_m \, \mathbf{H}' \, \Psi_c d\tau \right|^2 \tag{6}$$

where Ψ_m is the unit normalized bound-state eigenfunction of $\mathbf{H}_0$ associated with the metastable state, and Ψ_c the isoenergetic continuum eigenfunction of $\mathbf{H}_0$ associated with open channel j.

The simplest approximate methods are those in which $\mathbf{H}_0$ includes only the diagonal matrix elements of $\mathbf{U}$, so that $V(R,r,\theta)$ is replaced by a set of vibrationally averaged radial channel potentials $V(v,a;\, v,a;J|R)$ depending only on R. This effectively neglects the coupling terms on the right hand side of Equation 3 and collapses the wavefunction expansion of Equation 2 to a single product term. If this is the only approximation made, the result is the "distortion" approximation of Levine et al. (40,41) in which the operator $\mathbf{H}'$ comprises the off-diagonal matrix elements of $\mathbf{U}$ appearing on the right hand side of Equation 3. As might be expected, the quality of the results obtained in this way are quite sensitive to the choice of the angular basis functions, $\{\Phi_a^J(\hat{R},\hat{r})\}$ (19); moreover, the identity of the most appropriate basis function set may well change from case to case.

The two most obvious possible choices of angular basis functions for use in this approach are the space-fixed and body-fixed representations mentioned above. Beswick and Requena (17) also suggested use of an "intermediate" approximation in which the wave function remains a simple product, but its angular part is a (fixed) linear combination of body-fixed (or space-fixed) basis functions defined by the requirement that they diagonalize the matrix representation of **U** at some particular distance R*. However, the utility of this type of approach depends critically on a propitious choice of the distance R* and on the assumption that the character of the resulting locally-optimized angular wavefunction does not vary drastically across the physically accessible range of R. In practise these conditions will rarely be fully satisfied.

Cruder versions of the distortion approximation are obtained upon applying simplifying approximations to the potential function retained in $\mathbf{H}_0$ (15,41-46). From the viewpoint of computational ease and accuracy they have, respectively, little and nothing to offer. However, if the approximations made allow the matrix elements of Equation 6 to be evaluated analytically, the value of the information obtained regarding trends and functional dependence on the parameters of the system may more than compensate for the loss of absolute accuracy.

This is the case for the harmonic-internal-vibration plus Morse interaction model of Beswick and Jortner (15,43,44) and Ewing (45,46). According to it, the partial width of Equation 6 is proportional to $\exp\{-\sqrt{2\mu\Delta E_{mj}}/b\hbar\}$, where ΔE_{mj} is the energy gap between the metastable level energy E_m and the asymptote of the open channel effective potential (see Figure 1), and b is the range parameter of both the Morse potential and the simple exponential radial part of the coupling function in **H'**. Ewing showed (46) that, all else being equal, the Franck-Condon overlap considerations incorporated into this model could give rise to predissociation lifetimes for systems of interest differing by as much as 20 orders of magnitude! In reality, of course, all else is never equal, and (as will be seen below) even small changes in the supporting potentials and/or coupling functions can cause significant changes in metastable level widths. However, this approach does provide a fairly transparent explanation of qualitative features of the predissociation process, which is of considerable value.

A number of more sophisticated approximate methods are described and tested in Refs. (19), (25) and (33). As with the procedures discussed above, they focus their attention on finding approximate solutions to Equation 3, which is the basis of any full quantum mechanical description of the problem. While this type of approach is appropriate for cases in which the strength of the interchannel coupling is not significantly larger than the monomer level spacings, it is not likely to be so useful for "strong-coupling" systems such as $He-I_2$ or the ethylene dimer, where this not the case. Although consideration of the latter type of system is beyond the scope of the present article, it is encouraging to note that new types of approximation methods are now being developed (47,48) which appear to offer considerable hope of providing reliable means for exploiting the wealth of data now available for such systems (2,16).

Atom-Diatom Potential Energy Surfaces

General Considerations. Since the intermolecular potential determines the details of all aspects of the molecular dynamics, we preceed our discussion of the results of various predissociation calculations with an examination of some of the potential energy surfaces used. Such surfaces depend on the atom-diatom distance R, the diatom bond length r, and the angle between these two axes θ. In general, they may be expanded in terms of Legendre functions of the angular coordinate and powers of a diatom stretching coordinate ξ:

$$V(R,r,\theta) = V(R,\xi,\theta) = \sum_{\lambda}\sum_{k} \xi^k\, P_\lambda(\cos\theta)\, V_{\lambda k}(R) \tag{7}$$

The stretching variable may be chosen in a variety of ways (49), but in the present paper it is assumed to have the form $\xi=(r-r_0)/r_0$ where r_0 is some fixed reference length.

Other ways of representing the θ- and ξ-dependence of such potentials are certainly possible (13,14,50), and for strongly anisotropic systems some of those other methods are, at the very least, more economical with regard to the number of parameters required to specify a realistic surface. However, the sum of products in Equation 7 retains some significant advantages. The first of these is the fact that when performing calculations using close-coupling (exact) or related approximate methods, it is most convenient if the potential is expressed as a linear expansion in a set of known angular functions, such as the Legendre functions of Equation 7. A potential whose θ-dependence is defined in another manner would usually be expanded in this form before being used in practical calculations. Such expansions are quite straightforward to generate (49), but a large number of Legendre terms might be required to accurately represent the given function.

In a similar practical vein, expressing the ξ-dependence as a linear expansion in known functions (here, simple powers) means that all information about this degree of freedom required in calculation may be summarized in a table of matrix elements of these functions over the (usually) accurately known free diatom wave functions. Expectation values of powers of ξ for a number of levels of H_2 and D_2 are listed in Table IV of Ref. (10), and others may be readily generated using standard methods (51).

An important conceptual advantage of the double linear expansion of Equation 7 is the fact that, within the framework of first-order perturbation theory, the different terms correlate with different types of diatom monomer transitions. In particular, assuming a harmonic oscillator description of the diatom vibration, k=1 terms are the primary source of direct $\Delta v=\pm1$ coupling, k=2 terms the main source of $\Delta v=\pm2$ coupling, or more generally, direct transitions associated with any given value of Δv are only driven by potential terms corresponding to $k \geq |\Delta v|$. Similarly, direct rotational coupling between diatom rotational channels j and j' can only come from potential terms corresponding to $\lambda \geq |j-j'|$. Of course, if the interchannel coupling is very strong (or very weak, see below), these first-order considerations will not be dominant; however, they can still provide considerable physical insight, even if only used as a reference point against which exact results may be compared.

One situation in which individual terms in Equation 7 almost have independent physical significance is in the description of the discrete spectra of the hydrogen-inert gas Van der Waals molecules. There, the characteristic frequency associated with the stretching of the Van der Waals bond is significantly smaller than that for diatom rotation and more than two orders of magnitude smaller than that for diatom vibration, and to a fairly good approximation, v and j are good quantum numbers. As a result, the energy levels of a complex formed from hydrogen in j=0 depend almost solely on the $\lambda=0$ contributions to the potential, or more generally, levels of complexes formed from hydrogen in any given j state will be largely insensitive to potential terms corresponding to $\lambda > 2j$. Since the experimental data contain separate sets of transitions between levels of complexes formed from: a) only j=0 diatoms, b) only j=1 diatoms, c) j=0 and j=2 diatoms, and d) j=1 and j=3 diatoms, the effects of potential terms corresponding to different orders of anisotropy (i.e., to different values of λ) may in principal be separated. Unfortunately, the data used to obtain the best existing surfaces for these systems (9,10) did not allow terms corresponding to $\lambda > 2$ to be resolved, but it is hoped that the better-resolved spectra now available (29) will change this situation.

Similarly to the above, the ξ-dependence of the potential energy surface affects the data in a qualitatively different manner than does the R- or θ-dependence, so its effect may also be resolved (5). For the case of H_2-Ar, this point is illustrated in Figure 2, which shows the types of transitions which occur between complexes formed from the j=0 diatoms. In this case, the hydrogen effectively behaves as if it is spherical, and the observed transitions involve the end-over-end rotational levels of the complex in its ground vibrational state (n=0). The differences between the frequencies of pairs of transitions such as the P(5) and R(4) lines shown here define the level spacings, and hence are the source of information on the radial (R) behaviour of the spherical ($\lambda=0$) part of the potential. At the same time, if the upper and lower curves were identical, the averages of the frequencies of such pairs of transitions would be exactly equal to the accurately known distance between the potential asymptotes (i.e., the free diatom level spacing). In practise, however, they are ca. 1 cm^{-1} smaller than that. This indicates that the potential curve for a complex formed from vibrationally excited H_2 is relatively deeper than that for one formed from the ground state diatom, a result which correlates with the fact that the diatom bond length increases with the degree of vibrational and/or rotational excitation. This is the source of our information about the ξ-dependence of these potential energy surfaces (5-7,9,10).

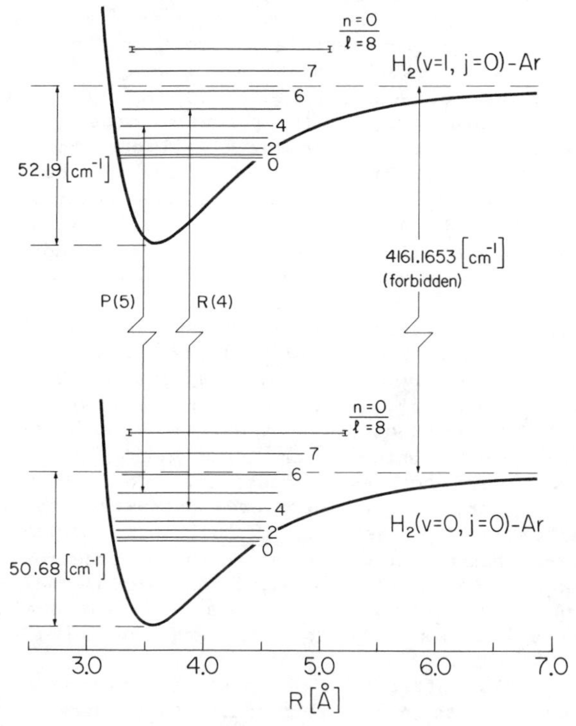

Figure 2.
Potential curves and energy levels for states involved in the $Q_1(0)$ transitions of H_2-Ar. Reproduced with permission from Ref.(5). Copyright 1974, American Institute of Physics.

Limiting Behaviour. As in other types of physical modelling, when devising potential energy surfaces it is important to apply external constraints on limiting behaviour whenever possible. In particular, while many physical observables are affected by the form and strength of the asymptotically-dominant inverse-power term in the intermolecular potential, this dependence is almost never sufficiently localized that the associated potential coefficients may be accurately determined from experiment. This is certainly true for the discete spectra and other available properties of most Van der Waals molecules. As a result, one should always attempt to obtain realistic theoretical or semi-empirical estimates of these coefficients, and incorporate them into the potential model. Fortunately, theoretical methods are now sufficiently well developed that it is often possible to generate realistic estimates of the spherically averaged part of such long range potential coefficients, and sometimes also of their θ- and ξ-dependence. Introduction of such constraints facilitates the determination of much more realistic overall potential energy surfaces than could otherwise be obtained (6,10).

Another type of limiting behaviour constraint, one which has as yet seen relatively little use, concerns the behaviour of an atom-diatom potential energy surface at small values of the diatom bond length r. To illustrate it, let us consider the specific example of H_2-Ar. From the viewpoint of the Ar atom, the H_2 molecule is a non-spherical charge distribution which looks rather like a slightly distorted He atom, and becomes increasingly like a He atom as the diatom bond length $r \to 0$. It is therefore appropriate to introduce a "collapsed diatom limit" constraint (9,10), which requires that in the r=0 (or $\xi=-1$) limit, the potential energy surface for H_2-Ar must become identical to the accurately-known (one dimensional) potential curve for He-Ar. In other words, in this limit the sum of all $\lambda>0$ contributions to Equation 7 go to zero, and that of the $\lambda=0$ part becomes the He-Ar potential.

Of course, this latter type of constraint may only be readily used for molecules whose ground electronic state correlates with the ground state of the atom obtained in the collapsed diatom limit. However, its use in such cases should yield a much better knowledge of the ξ-dependence of the associated potential energy surfaces than could otherwise have been possible (9,10).

Molecular Hydrogen-Inert Gas Potential Energy Surfaces

Since their overall three-dimensional potential energy surfaces are more accurately known than those for any other atom-diatom systems, the species of most interest here are molecular hydrogen with Ar, Kr and Xe. The current best surfaces for these systems (9,10) are represented by expansions of the form of Equation 7 with $\lambda=0$ and 2 and k=0, 1, 2 and 3, where the diatom stretching coordinate is defined as $\xi=(r-r_0)/r_0$ and the constant $r_0=0.7666348$ Å is the expectation value of r for H_2 in its ground vibration-rotation state. The associated radial strength functions $V_{\lambda k}(R)$ have the $BC_3(6,8)$ functional form

$$V(R) = A\exp(-\beta R) - [C_6/R^6 + C_8/R^8]D(R) \qquad (8)$$

where the damping function D(R) equals $\exp[-4(R_0/R-1)^3]$ for $R<R_0$ and

unity for $R>R_0$. Each $V_{\lambda k}(R)$ function is then defined by specifying the associated values of ε, R_e, β, and C_6, which in turn define

$$A = \{\varepsilon(8-R_eD_1) - 2D_0C_6/(R_e)^6\}\exp(\beta R_e)/(\beta R_e-8+R_eD_1) \quad (9)$$

$$C_8 = \{(6-R_eD_1-\beta R_e)C_6/(R_e)^6 + \varepsilon\beta R_e/D_0\}(R_e)^8/(\beta R_e -8+R_eD_1) \quad (10)$$

where $D_0 \equiv D(R=R_0)$ and $D_1 \equiv [\partial D(R=R_0)/\partial R]/D_0$, and the damping function cutoff distance R_0 is defined as R_e^{00} (i.e., R_e for $(\lambda,k)=(0,0)$).

These surfaces were obtained subject to the two types of limiting behaviour constraints mentioned above (9,10). The limiting large-R behaviour was imposed by using reliable semi-empirical dispersion coefficients and their $P_2(\cos\theta)$ anisotropies, calculated as functions of the diatom bond length r (9,52), to define the values of the C_6 coefficients for all (λ,k). As the experimental data could only discern the k=0 and 1 components of the potential (5-7, 9,10), the collapsed diatom limit constraint was imposed by defining the k=2 functions by the requirements that at $\xi=-1$ (i.e., at r=0), the sum of $(-1)^k$ times the $A^{\lambda k}$, $C_8^{\lambda k}$ and $C_6^{\lambda k}$ constants for $\lambda=2$ be identically zero, and for $\lambda=0$ yield an effective spherical potential with the same depth, equilibrium distance and C_6 constant as the known pair potential (53) between the corresponding inert gas

Table I. Parameters of $BC_3(6,8)$ potential energy surfaces for H_2 and D_2 interacting with Ar, Kr and Xe (9,10); note that $\beta^{\lambda k}=\beta^{00}$ for all λ and k.

λ	k	parameter	H_2,D_2-Ar	H_2,D_2-Kr	H_2,D_2-Xe
0	0	ε /cm^{-1}	50.87	58.73	64.78
		R_e/Å	3.5727	3.7181	3.9366
		β /Å^{-1}	3.610	3.399	3.484
		C_6/cm^{-1}Å^6	134500.	190300.	268200.
0	1	ε /cm^{-1}	34.2	39.36	53.72
		R_e/Å	3.769	3.847	4.0024
		C_6/cm^{-1}Å	115800.	167300.	247500.
0	2	ε /cm^{-1}	1.904	3.791	7.562
		R_e/Å	4.3575	4.3052	4.3167
		C_6/cm^{-1}Å^6	3051.	5565.	11710.
0	3	C_6/cm^{-1}Å^6	-25270.	-36890.	-55800.
2	0	ε /cm^{-1}	5.72	8.25	9.67
		R_e/Å	3.743	3.834	3.948
		C_6/cm^{-1}Å^6	13500.	19450.	28550.
2	1	ε /cm^{-1}	21.6	18.6	19.8
		R_e/Å	3.949	3.967	4.014
		C_6/m^{-1}Å^6	29600.	43250.	65120.
2	2	ε /cm^{-1}	13.914	7.48	5.790
		R_e/Å	4.027	4.131	4.221
		C_6/cm^{-1}Å^6	5705.	8624.	13860.
2	3	C_6/cm^{-1}Å^6	-10395.	-15176.	-22710.

atom (Ar, Kr or Xe) and He. Note that the k=3 contributions to these surfaces consist only of the C_6/R^6 term (i.e., $A=C_8=0$ for k=3).

For the three types of complexes mentioned above, the parameters defining these surfaces are listed in Table I (9,10). Using H_2-Ar as an example, a number of their more prominent features are illustrated in Figures 3-5. The first of these figures simply shows the $V_{\lambda k}(R)$ functions for this case; the manner in which the strength and relative positions of the various functions depend on λ and k is similar for all three systems.

Figure 4 presents a set of two-dimensional (R,θ) contour diagrams for H_2-Ar corresponding to various fixed values of the diatom stretching coordinate ξ. In this diagram, the H_2 molecule is assumed to lie at the origin with its axis horizontal; because of its symmetry, a single quadrant suffices to define the whole potential surface for that value of ξ. According to the definition of ξ, the upper right hand quadrant ($\xi=0$) corresponds to the diatom bond length being fixed at its average value for H_2 in its ground vibration-rotation state, while the lower left hand quadrant ($\xi=-1$) corresponds to the collapsed-diatom limit. The two remaining segments then correspond to the diatom bond length being fixed at the inner ($\xi=-0.26$) and outer ($\xi=0.33$) classical turning points for $H_2(v=1,j=2)$.

The most noteworthy feature of Figure 4 is simply the large differences in the angular behaviour associated with diatom bond lengths ranging across the classically-allowed region. Another view of this ξ-dependence of the potential anisotropy is provided by Figure 5, which shows how the effective potential anisotropy strength function $V_2(R,\xi)$ (the sum over k of the $\lambda=2$ terms in Equation 7) depend on the diatom stretching coordinate. The two extreme ξ values used there are again the classical inner and outer turning points for $H_2(v=1,j=2)$, while the other values are the expectation values of ξ for $H_2(0,1)$ and $H_2(1,2)$. The complete change in character of the anisotropy for different diatom bond lengths within the classically accessible region suggests that predissociation or inelasticity calculations which do not take account of the diatom stretching dependence of the potential energy surface may be quite unreliable, a prediction which is quantitatively confirmed below.

It is important to remember that the potential energy surfaces of Table I were obtained from fits to data for complexes formed from both H_2 and D_2 (9,10), and that they should be isotopically invarient. Isotope effects enter the description of the dynamics both through the value of μ and through the diatom vibrational averaging which gives rise to different diagonal and off-diagonal matrix elements of $\mathbf{U}$ in Equation 3. For heteronuclear isotopes such as HD, the potential must also be re-expanded about the (displaced) centre-of-mass of the diatom, but this is a straightforward procedure (49) which introduces odd anisotropy terms to its analytic representation, but does not change the potential.

Of course, the potential energy surfaces of Table I are by no means the last word on these systems. Even leaving aside the question of the inclusion of terms corresponding to larger values of λ and k, the choice of potential form and the assumption that the exponent parameter β is the same for all (λ,k), are somewhat arbitrary. Indeed, for molecular hydrogen with Ar, it has recently been

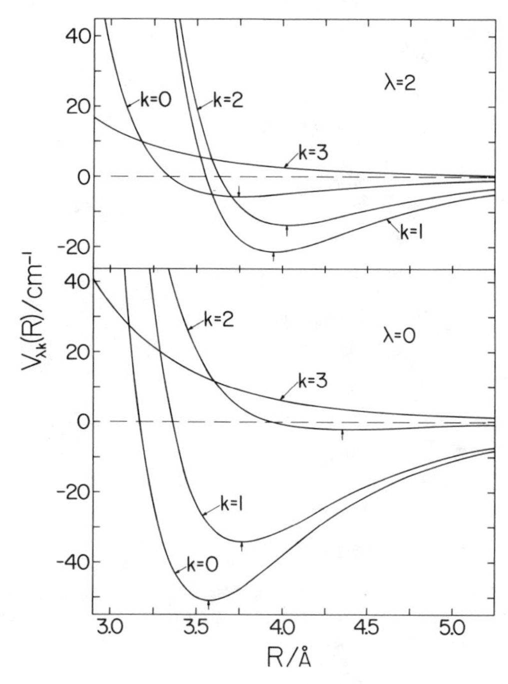

Figure 3.
Radial strength functions of the $BC_3(6,8)$ potential for H_2-Ar; vertical arrows denote the minima of the various curves. Reproduced with permission from Ref.(10). Copyright 1981, John Wiley and Sons Ltd.

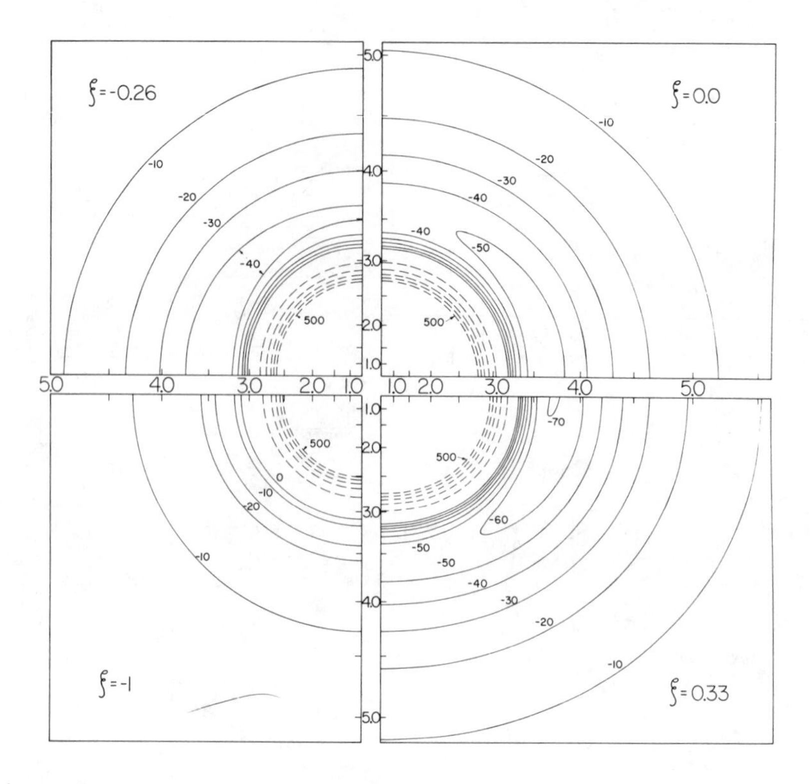

Figure 4.
Contour diagram for the $BC_3(6,8)$ H_2-Ar potential of Table I; contours marked in cm^{-1}. The H_2 molecule is assumed to lie horizontally at the origin, and each quadrant corresponds to a different fixed H_2 bond length.

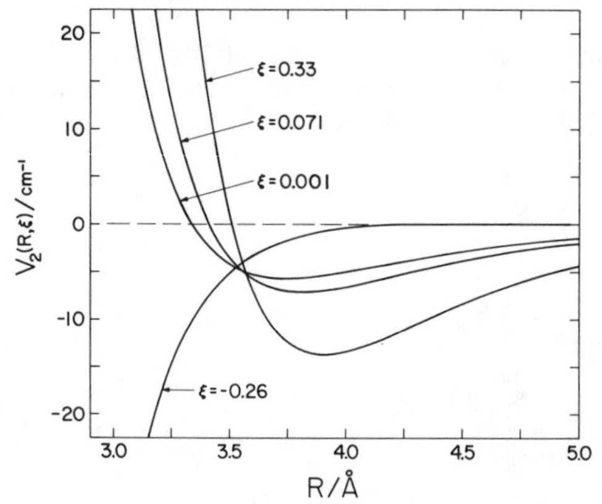

Figure 5.
Effective anisotropy strength functions for the H_2-Ar potential surface of Table I, corresponding to different (fixed) values of the diatom stretching coordinate ξ. Reproduced with permission from Ref.(10). Copyright 1981, John Wiley and Sons Ltd.

shown (54) that relaxing this last restriction and allowing the exponent parameter to be different for $\lambda=0$ and 2 yields a surface whose predicted inelastic rotational cross-sections are in better agreement with experiment, while retaining excellent agreement with the spectroscopic data. However, the surfaces of Table I were those used in the predissociation calculations for these species reported to date, and those for hydrogen with Kr and Xe remain the best currently available.

Predissociation By Internal Rotation: Vibrational Predissociation With $\Delta v=0$

Some of the most extensive studies of "vibrational" predissociation reported to date are those for the predissociation by internal rotation of the the $H_2(v=1,j=2)$-Ar and HD(1,2)-Ar complexes (17-23). While the resulting dissociation products may in principle include diatom fragments corresponding to $v=0$ and $j\leq 9$, the overwhelmingly dominant process (by ca. 7 orders of magnitude!) is "predissociation by internal rotation" to produce H_2 or HD with $v=1$ and $j\leq 1$. As prototype weak-coupling complexes, these are system for which approximate computational methods should be most reliable. Moreover, as one of the cases for which the potential energy surface which controls the dynamics is most well known, they are systems for which an understanding of sensitivity to features of the potential energy surface might allow comparisons with experiment to yield new information regarding this potential.

An important point to remember about vibrational predissociation is that a given type of complex usually has a large number of levels, all of which predissociate at different rates, and when there is more than one open channel, also with different branching ratios. For example, the metastable levels associated with the $n=0$ stretching vibration of the Van der Waals bond of $H_2(1,2)$-Ar are schematically illustrated by Figure 6; this diagram also shows how the potential anisotropy splits the end-over-end rotation (ℓ) levels into the observed states associated with different values of the total angular momentum quantum number J. Resonance energies and widths calculated for this system using both close-coupling (E^{cc} and Γ^{cc}) and various approximate methods (E^{γ} and Γ^{γ}, for γ=SFD, BFD and BL) are listed in Table II. Although they are merely different rotational sublevels associated with the same stretching state of the Van der Waals bond, and they all predissociate via the same mechanism, the exact widths (Γ^{cc}) of these levels are seen to differ by more than a factor of five. Note that states corresponding to odd $(j+\ell+J)$ parity are omitted from consideration here, as they can only predissociate to yield a $v=0$ fragment, a much slower process (22,25).

The question of whether this prototype weak-anisotropy system may be accurately described by one of the simple decoupling methods mentioned above is also examined in Table II. The first two approximate methods considered are the space-fixed (SFD) and body-fixed (BFD) versions of the distortion approximation of Levine et al.(40,41) while the third is their "best local" (BL) approximation, according to which the angular functions Φ_a^J are optimized at each value of R. It is immediately clear that the space-fixed angular basis functions are the more appropriate here (note, however that

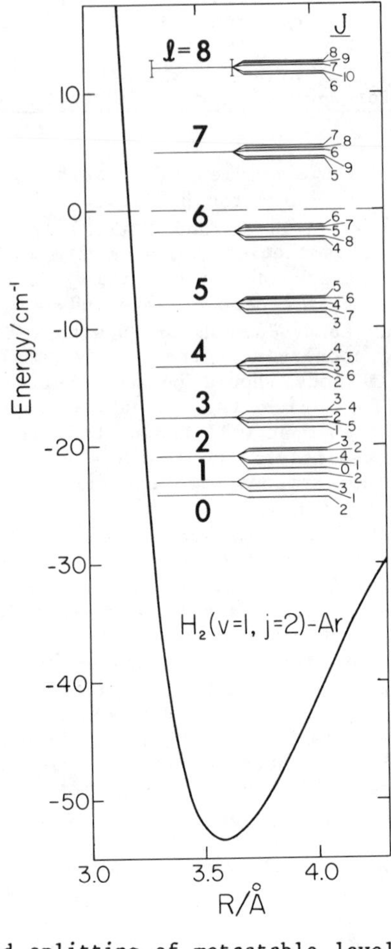

Figure 6.
Anisotropy-induced splitting of metastable levels of $H_2(1,2)$-Ar. Reproduced with permission from Ref.(5). Copyright 1974, American Institute of Physics.

Table II. Comparison of exact (CC) and approximate resonance energies E and widths Γ (in cm^{-1}) for n=0 levels of $H_2(1,2)$-Ar taken from Ref.(19). The zero of energy is the $H_2(1,2)$ excitation threshold, and $\Delta E^{\gamma}=(E^{\gamma}-E^{\gamma})$.

v	J	E	ΔE^{SFD}	ΔE^{BFD}	ΔE^{BL}	Γ^{cc}	Γ^{SFD}	Γ^{BFD}	Γ^{BL}
0	2	-23.23	0.21	3.45	-0.48	0.0394	0.032	0.0	0.002
1	1	-22.65	-0.02	2.41	-0.06	0.0530	0.052	0.0	0.036
	3	-22.24	0.11	5.78	-0.36	0.0414	0.039	0.0	0.009
2	0	-20.73	-0.07	-0.07	-0.00	0.1103	0.112	0.112	0.109
	2	-19.31	-0.16	1.26	0.42	0.0235	0.044	0.0	0.055
	4	-20.08	0.07	7.99	-0.27	0.0420	0.042	0.0	0.017
3	1	-17.24	-0.05	-2.46	0.03	0.0551	0.064	0.109	0.077
	3	-15.94	-0.07	1.17	0.29	0.0237	0.039	0.0	0.052
	5	-16.82	0.05	10.07	-0.21	0.0411	0.042	0.0	0.024
4	2	-12.83	-0.05	-4.69	0.06	0.0408	0.049	0.104	0.058
	4	-11.58	-0.04	1.08	0.21	0.0219	0.035	0.0	0.045
	6	-12.52	0.04	a	-0.16	0.0388	0.040	a	0.029
5	3	-7.49	-0.05	-6.79	0.05	0.0319	0.039	0.097	0.044
	5	-6.26	-0.02	1.00	0.15	0.0191	0.029	0.0	0.037
	7	-7.25	0.04	a	-0.12	0.0351	0.036	a	0.030
6	4	-1.31	-0.04	-8.72	0.04	0.0244	0.030	0.087	0.033
	6	-0.11	-0.02	a	0.10	0.0156	0.023	a	0.027
	8	-1.11	0.03	a	-0.08	0.0298	0.031	a	0.028
RMS deviation:			0.08	5.15	0.22				

[a] This level incorrectly predicted to lie above dissociation.

the opposite is true for intermediate coupling cases such as Ar-HCℓ (33)). However, while the SFD method is the best of the approximate procedures considered, the level energies it yields are typically in error by more than a full level width, and the root mean square relative error associated with its widths is ca. 35%. Thus, while the SFD method does qualitatively explain the bulk of the differences among the various Γ^{cc} values in Table II, its inadequacies are unfortunately too large to allow its predictions to be the basis of attempts to refine the potential energy surface. A more detailed discussion of the nature and weaknesses of the various approximate methods may be found in Ref.(19).

The results in Table II were calculated from the $BC_3(6,8)$ potential of Table I, subject to three approximations: (i) the effect of predissociation to v''=0 was assumed to be negligible, (ii) the dependence of the potential matrix of Equation 11 on j' was ignored, and (iii) coupling to closed channels corresponding to j'≥4 or v'≥2 was neglected (19). Of these, (iii) is believed to be most serious. Tests have shown that it can raise the level energies by as much as 0.03 cm^{-1}, approximately half of which is due to neglect of coupling to j'≥4 and half to neglect of v'≥2 (19,57). However, cancellation effects will make the effect on transition frequencies much less than the experimental uncertainties. The effect on widths is somewhat more significant though, as coupling to $j' \geqq 4$ can increase them by ca. 1% and coupling to $v' \geqq 2$ decrease them by as much as 4% (19,57). Moreover, while it had a negligible effect on the energies (< 0.001 cm^{-1}), approximation (ii) exaggerates the predicted widths by ca. 3%. However, these changes are still fairly small relative to the associated experimental uncertainties. Thus, any significant disagreements between experiment and the predicted E^{cc} and Γ^{cc} values of Table II must be due to deficiencies in the potential energy surface.

Effect of Diatom Stretching Dependence. The features of the potential energy surface most central to a discussion of its effect on the predissociation process are not the individual radial strength functions $V_{\lambda k}(R)$, but rather the vibrational matrix elements (integrated over the diatom bond length) of the full potential

$$\langle v,j|V(R,\xi,\theta)|v',j'\rangle = V_0(v,j;v',j'|R) + V_2(v,j;v',j'|R)\, P_2(\cos\theta) \qquad (11)$$

The functions $V_\lambda(v,j;v',j'|R)$ all have the same qualitative behaviour as the analogous (i.e., same λ) $V_{\lambda 0}(R)$ function of Figure 3. The depth $\bar{\varepsilon}$ and equilibrium distance $\bar{R}_e$ parameters characterizing the diagonal potential matrix elements $V_\lambda(v,j;v,j|R)$ for H_2-Ar yielded by the vibrational averaging for various levels of H_2, are listed in the first three columns of Table III. The final column in this table then corresponds to the surface implied by a "frozen equilibrium" (ξ=0) approximation, according to which only the leading (k=0) terms are retained in the potential expansion of Equation 3. The spherical (λ=0) components of the diagonal potential matrix elements are of course the vibrationally averaged supporting potentials such as those shown in Figures 2 and 6.

The last row of Table III lists the exact (close-coupling) width calculated for Δj=-2 predissociation of a $(n,\ell,j,J)=(0,2,2,0)$ level (e.g., the fifth lowest level in Figure 6) on the corresponding surface. Although (see Table II) each metastable level in general has a different width, the radial matrix elements which

Table III. Potential characteristics and predicted widths for Δj=-2 predissociation of the $(n,\ell,j,J)=(0,2,2,0)$ level of H_2-Ar on vibrationally averaged supporting potentials corresponding to various (v,j). Results from Ref.(19).

		(v,j) = (1,2)	(0,2)	(0,0)	"k=0"
λ = 0:	$\bar{\varepsilon}$ /cm^{-1}	52.31	50.81	50.68	50.87
	$\bar{R}_e$/Å	3.5985	3.5772	3.5762	3.5727
λ = 2:	$\bar{\varepsilon}$ /cm^{-1}	7.59	5.92	5.84	5.72
	$\bar{R}_e$/Å	3.843	3.768	3.763	3.743
Γ^{cc}/cm^{-1}		0.110	0.043	0.040	0.034

govern their predissociation rates are affected by changes in the potential in qualitatively the same way. Thus, the effect of the diatom stretching-dependence on the other levels considered in Table II should be proportional to its effect on the prototype level considered here.

While the depth of the effective isotropic potential increases by ca. 3% upon $v=0\rightarrow1$ vibrational excitation of the H_2 (see Table III and Figure 2), the analogous strength parameter $\bar{\varepsilon}$ for the potential anisotropy increase by 30% (see also Figure 5). However, the level widths should vary approximately as the square of the matrix elements of the effective anisotropy strength function $V_2(v,j;v,j\,|R)$, so this change alone does not explain the factor of three difference between the widths in the first two columns of Table III. An even more important consideration is the fact that the increase of 2% in the $\bar{R}_e$ value of $V_2(v,j;v,j\,|R)$ as v increases from 0 to 1 is more than three times as large as that for $V_0(v,j;v,j\,|R)$, so that for vibrationally excited hydrogen, the radial matrix elements coupling the open and closed channels will sample much more of the steep repulsive portion of this coupling function.

The results in the first two columns of Table III imply that $H_2(v=1,j=2)$-Ar complexes will predissociate almost 3 times as rapidly as $H_2(v=0,j=2)$-Ar. However, within a first-order treatment, rotational inelasticity depends on the same type of squared matrix element of $V_2(v,j;v'j'\,|R)$ as does the level width, except that the (isoenergetic) wavefunctions being coupled are both continuum functions lying above the rotational threshold. In terms of Figure 1, they would be continuum eigenfunctions of $V_1(R)$ and $V_2(R)$ at energies above the asymptote of $V_2(R)$. It seems reasonable to expect these overlap integrals to depend on the potential in much the same way as do those appearing in the predissociation problem, so the results in Table III imply that the rotational inelastic cross-sections of vibrationally excited H_2 will be much larger than those for the ground state diatom.

The last two columns in Table III correspond to non-physical situations, in that the effective potentials do not correspond to H_2 in j=2. However, comparison of the second and third columns serves to show that even the effect of diatom stretching due solely to centrifugal distortion can significantly affect the level widths. Similarly, the results in the last column show that the two-dimensional (R,θ) potential surface obtained on freezing the H_2 bond length at its expectation value for (v,j)=(0,0) has significantly different properties than the proper vibrationally averaged surface associated with this same state. This again emphasizes that potential energy surfaces which do not incorporate monomer stretching dependence in a realistic way may not in general be expected to yield predissociation lifetimes or inelastic cross-sections in good agreement with experiment.

Predictions, Comparisons, and Implications Regarding Potentials. Another point of considerable physical interest concerns the differences in the predissociation rates of H_2-Ar, H_2-Kr and H_2-Xe. For the same bellweather level considered in Table III, widths (Γ^{cc}) calculated on the proper vibrationally-averaged surfaces for these three species are compared in Table IV. Following the arguments

Table IV. Effective anisotropy strength parameters and predicted widths for the $(n,\ell,v,j,J)=(0,2,1,2,0)$ levels of H_2-Ar, Kr, and Xe implied by the potential energy surfaces of Table I. Results from Refs. (9 and 19).

	H_2-Ar	H_2-Kr	H_2-Xe
$\bar{\varepsilon}(\lambda=2)/cm^{-1}$	7.59	9.73	11.22
Γ^{cc} $/cm^{-1}$	0.110	0.073	0.039

used above, the trend seen here implies that the rotational inelasticity of molecular hydrogen colliding with inert gas atoms will decrease from Ar to Kr to Xe. This prediction is in agreement with the observation that inelastic rotational cross sections are smaller for D_2 colliding with Ar than for D_2 with Ne (58). However, this trend is opposite to that for both the effective potential anisotropy strength parameter $\bar{\varepsilon}(\lambda=2)$ (see Table IV), and for anisotropy-induced level splittings such as those shown in Figure 6.

The source of this apparent contradiction is simply the fact that the spectroscopically-observed level splittings depend on an expectation value of $V_2(v,j;v,j|R)$ weighted by the square of a bound state type wave function such as χ_b of Figure 1, while the predissociation level widths depend on the square of an overlap integral of the (very similar) anisotropy strength function $V_2(v,j;v',j'|R)$ with a product of continuum and bound-type wave functions similar to χ_c and χ_m of Figure 1. The latter are clearly relatively much more strongly dependent on the strength of the anisotropy at short distances where it has large positive values, while the former will depend mainly on its relatively small negative values at larger distances (see Figure 5). Thus, the apparent discrepancy mentioned above is readily explained by the fact that the anisotropy strength function in the hydrogen-inert gas pair potential both shifts to relatively smaller distances, and becomes stronger, with increasing size of the inert gas partner.

This apparent sensitivity to the potential anisotropy in different regions suggests that this aspect of these potential energy surfaces could be more accurately determined if the discrete spectra were analysed together with either predissociation level widths or inelastic rotational cross sections. For H_2-Ar this has recently been done by Buck et al.(54). The essence of the difference between their improved potential and that of Table I is that they relaxed the requirement that the exponent parameter β of Equation 8 have the same value for $\lambda=0$ and 2. This allowed the anisotropy strength function to be stronger at the short distances important to the inelastic cross sections (and predissociation level widths), while retaining essentially the same shallow well at larger distances. The width of the prototype $(n,\ell,v,j,J)=(0,2,1,2,0)$ H_2-Ar level of Table IV for this surface is 0.186 cm^{-1}, some 70% larger than that implied by the potential of Table I. Thus, our present best estimates of the rotational predissociation level widths for $H_2(1,2)$-Ar are that they are ca. 1.7 times larger than the Γ^{cc} values of Table II.

While no inelastic rotational cross-sections have as yet been reported for hydrogen colliding with Kr or Xe, McKellar (29) has measured a predissociation width of 0.11 cm^{-1} for the $(n,\ell,\overline{v},j,J)$= (0,2,1,2,0) level of H_2-Kr. As is apparently the case (see above) for H_2-Ar, for which such widths have not yet been resolved, this value is somewhat (ca. 50%) larger than the prediction implied by the potential of Table I (see Table IV). Thus, it seems reasonable to expect that an analogously refined potential for H_2-Kr could be obtained from a simultaneous analysis of this observed level width and the discrete infrared spectrum.

Comparisons with experiment are also possible for HD-Ar, for which McKellar has been able to measure predissociation-induced widths for most of the Δn=0 N- and T-branch ($\Delta\ell=\pm3$) lines associated with the $S_1(0)$ transitions $(v,j)=(0,0)\rightarrow(1,2)$ of the component diatom While the potential energy surface of HD-Ar is the same as that for H_2-Ar, it must be re-expanded in a Legendre expansion with its origin at the displaced centre of mass of the diatom. It is important to note that the averaging over the diatom stretching coordinate to obtain the potential matrix elements of Equation 11 should not be performed until after this new angular expansion is obtained, as significant errors (ca. 10% in the widths and 21% of Γ^{cc} in the energies (59)) are introduced if the vibrational averaging is done first (24,59). This is the source of most of the differences between the predictions for HD-Ar in Refs. (21) and (24).

Figure 7 shows a comparison of these experimental line widths (29) with predictions based on the potential of Table I (24); within the uncertainties in the former, the agreement seen there is exact. The predissociation in this case is largely due to the λ=1 contribution to the potential anisotropy arising from the coordinate transformation mentioned above, and it is much faster than that for H_2-Ar. The dominant contributions to this λ=1 anisotropy are the first derivatives of the λ=0 and 2 radial strength functions (60), which make approximately equal contributions to these widths (24). The agreement seen in Figure 7 therefore attests to the quality of the isotropic (λ=0) potential and to that of the λ=2 anisotropy at distances over the potential well. At the same time, it does not contradict the finding of Ref.(54) that the anisotropy strength function of Table I should be strengthened at short distances. The latter possibility has also been suggested by Kidd and Balint-Kurti (23) on the basis of their examination of the level widths associated with the $\Delta\ell=\pm1$ transitions in the $S_1(0)$ HD-Ar spectrum. The reason for the apparent discrepancy between their conclusion and the excellent agreement seen in Figure 7 may be that the other HD-Ar calculations (21,24) neglected the effects of coupling to closed channels with $v'\geq2$, which has been shown to narrow such resonances significantly (57).

Vibrational Predissociation With $\Delta v=-1$

Results and Predictions. Detailed close coupling calculations for "real" $\Delta v<0$ vibrational predissociation of weak-coupling systems such as the hydrogen-inert gas complexes are more difficult and computationally more expensive than those for predissociation by internal rotation. The computational expense arises simply from the very large increase in the number of channels which must be included in order to obtain converged results. The difficulty, on the other hand, arises from the fact that these resonances have very small widths, usually 10^{-7} cm^{-1}, which makes them very difficult to find.

Fortunately, across a narrow interval the direct scattering contribution to the S-matrix is a very smooth function of energy. As a result, it is possible to detect the tail of the Breit-Wigner lineshape function in the S-matrix eigenphase sum at energies as much as 10^5 widths away from the centre of a resonance. Once such a tail has been observed, it is a fairly routine matter to use fits to the lineshape function to iteratively converge on the precise resonance energy and width. To complement the above convergence procedure, realistic initial trial energies may be obtained by neglecting coupling to the closed channels and solving for the eigenvalues of the resulting bound-state problem using either scattering (61) or bound-state (19,39) type methods (62).

For the prototype level $(n,\ell,v',j',J)=(0,0,1,0,0)$ of H_2-, D_2- and HD-Ar, the results of close coupling calculations for the partial and total predissociation widths are listed in Table V. Tests showed that the angular basis sets used to obtain these results were fully converged (25); they included diatom rotation states up to $j'=8$ for $v'=1$ and up to $j''=10$ for $v''=0$. The absence of odd-j'' dissociation products for H_2- and D_2-Ar merely reflects the fact that their potentials have no odd-j' anisotropy terms.

Table V. Total and partial widths for vibrational predissociation of the $(n,\ell,v',j',J)=(0,0,1,0,0)$ levels of H_2-, D_2- and HD-Ar to yield $(v''=0,j'')$ fragments; results from Ref.(25).

	j''	H_2-Ar	D_2-Ar	HD-Ar
$\Gamma_{mj}/10^{-10}$ cm^{-1}	0	2.0	0.07	4.
	1	-	-	10.
	2	18.5	0.37	12.
	3	-	-	12.
	4	136.	1.25	18.
	5	-	-	20.
	6	45.	0.85	3.
	7	-	-	37.
	8	0.01	0.02	40.
$\Gamma_m(total)/10^{-10}$ cm^{-1}		202.	2.5	156.

The first thing which strikes one about these widths is their size, some seven orders of magnitude smaller than the rotational predissociation widths discussed above. Thus, it is clear that line broadening of this type will not be observable for these species, and hence cannot be looked on as a possible source of information on the $k>0$ contributions to the potential expansion (see Equation 7), which are responsible for this type of coupling. On the other hand, as in the case of rotational predissociation, the same type of information about the coupling functions contained in these widths is also found in vibrationally inelastic cross sections, both being sensitive to behaviour at smaller distances than are important for the line shifts in the discrete spectra.

Calculations analogous to those of Table V have not yet been reported for complexes of hydrogen with other inert gases. However, our experience with the case of predissociation by internal rotation and our understanding of the potential energy surfaces involved should allow us to make reliable predictions reqarding the trend from one system to the next.

In this case, the dominant contributions to the coupling will be proportional to the $V_{01}(R)$ and $V_{21}(R)$ radial strength functions. As can be seen from Table I, the strength parameters of the former increase and those of the latter remain approximately constant with increasing size of the inert gas partner. However, as for the anisotropy strength functions discussed above, both of these coupling functions shift to smaller distances, relative to the isotropic supporting potentials, as the inert gas partner becomes larger. This means that the corresponding off-diagonal matrix elements become relatively less dependent on the large values of the coupling function associated with its steep "repulsive" wall, which in turn implies that predissociation widths and inelastic vibrational cross-sections will both <u>decrease</u> with increasing size of the inert gas partner. Once again, this prediction contradicts the trends implied by the relative magnitudes of the coupling strength parameters (here, $\varepsilon^{\lambda=0,k=1}$) and level shift, and this apparent discrepancy is again readily understood in terms of the properties of the potential surface.

A second interesting point is the way in which the partial widths depend on the rotational quantum number j'' of the ($v''=0$) dissociation fragment. For H_2- and D_2-Ar, the potential energy surface has only $\lambda=0$ and 2 components, so the initial $(v',j')=(1,0)$ state may only couple directly to products corresponding to $j''=0$ and 2. Thus, the presence of <u>any</u> flux in channels corresponding to $j''>2$ can only be due to high order coupling. In particular, the (dominant) $j''=4$ fragments may only be produced by coupling which is at least second order, while flux into $j''=6$ requires third or higher order mixing. Moreover, the fact that this behaviour remains when the $v'=1$ angular basis set is truncated at $j'=0$ (see Table VI, below) indicates that it is coupling among the open channels which is most important.

The results for HD-Ar in Table V are qualitatively different from those for H_2- and D_2-Ar, because the centre of mass transformation introduces both odd and high-order potential anisotropies which allow direct coupling to high j'' states and allow the diatom fragment to be produced with odd as well as even j''. Even so, if one ignores the flux into the two highest-j'' open channels, these

HD-Ar results are quite similar to those for the homonuclear diatom isotopes, with partial widths which initially increase to a maximum near $j''=4$ and then die off. In both cases, it seems reasonable to interpret this behaviour as an initial growth in the influence of high-order coupling, driven by momentum gap minimization considerations, eventually being supressed by the very small propensities for such high-order events. The large HD-Ar flux into $j''=7$ and 8 would then be due to direct or low-order coupling due to the high-order potential anisotropy terms introduced by the potential transformation. Although the strength of these coupling functions decreases rapidly with increasing order (see Figure 8), momentum gap minimization considerations very strongly favour these (more) near-resonant processes.

Unfortunately, the importance of this high-order coupling effectively means that the predictions of Table V may have little absolute significance. This is because the best available potential surfaces for this system (10,54) have only a $P_2(\cos\theta)$ anisotropy, while the true interaction energy must also contain (at least small) terms corresponding to $\lambda=4$, 6, 8, ... etc. Even if the latter are very weak, the fact that they would permit direct (first-order) coupling to high j'' states might make their contributions to the overall flux into those states even more important than that of the high-order $P_2(\cos\theta)$ coupling mentioned above. Thus, until more is known about their high-order anisotropies and its stretching dependence, even for surfaces as sophisticated and accurate as those of Table I, predictions of vibrational inelasticity or vibrational predissociation widths may not fully agree with experiment.

Convergence Considerations and Approximate Methods. Another consideration which bears on the reliability of results such as those presented above is the question whether or not the calculations are really fully converged. For the prototype $(n,\ell,v',j',J)=(0,0,1,0,0)$ level of H_2-Ar, this question is examined in Tables VI and VII, which show how the partial and total widths depend on the size of the angular basis set used for the $v'=1$ and $v''=0$ levels, respectively. The calculations of Table VI used a fully-converged open

Table VI. Dependence of partial widths of the $(n,,v',j',J)=(0,0,1,0,0)$ level of H_2-Ar on the $v'=1$ rotational (j') basis set size. Close-coupling results from Ref.(25); energies in cm^{-1}.

		$\Gamma_{mj''}/10^{-10}$ cm^{-1}					
j'_{max}	$\Delta E_m(j'_{max})$	$j''=0$	2	4	6	8	total
0	23.0	12.0	96.	343.	67.	0.01	518.
2	359.7	2.8	39.	630.	718.	0.01	1391.
4	1133.3	2.0	19.	137.	94.	0.07	252.
6	2316.3	2.0	18.	136.	45.	0.01	202.
8	3869.6	2.0	18.	136.	45.	0.01	202.

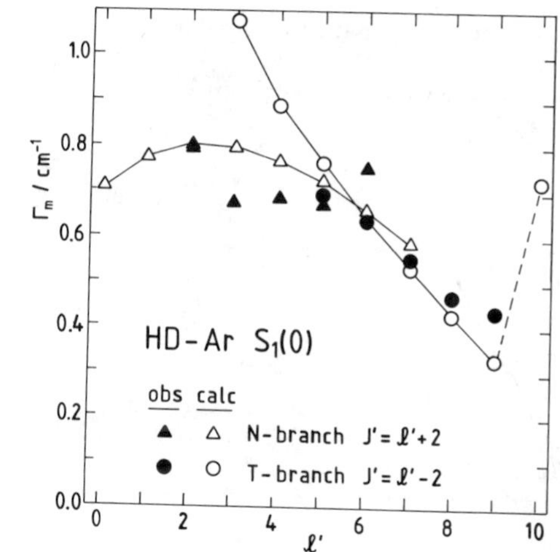

Figure 7.
Comparison with experiment (solid points) of calculated widths (open points) of rotationally predissociating levels of HD(1,2)-Ar. Reproduced with permission from Ref.(24). Copyright 1983, American Institute of Physics.

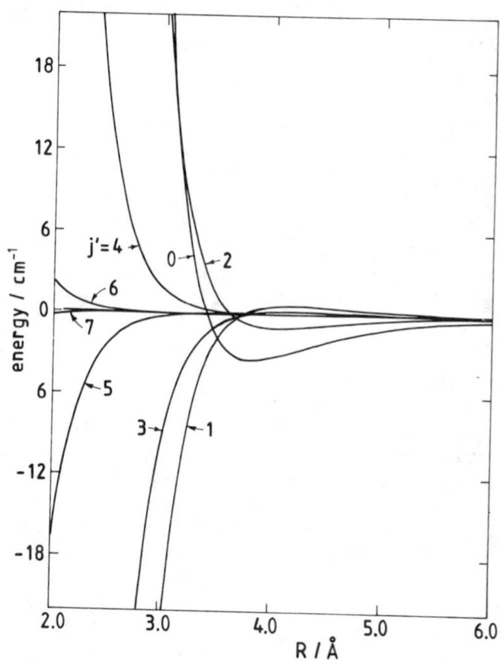

Figure 8.
Potential matrix elements between the (v'=1,j'=0) and (v"=0,j") channels for J=0 HD-Ar. Reproduced with permission from Ref.(25). Copyright 1983, American Chemical Society.

Table VII. Dependence of partial widths of the $(n,\ell,v',j',J)=(0,0,1,0,0)$ level of H_2-Ar on the v"=0 rotational (j") basis set size. Close-coupling results from Ref.(25); energies in cm^{-1}.

j''_{max}	$\Delta E_m(j''_{max})$	$\Gamma_{mj''}/10^{-10}\ cm^{-1}$ j" = 0	2	4	6	8	total
0	-4139.1	0.4	-	-	-	-	0.4
2	-3784.7	1.2	8.5	-	-	-	9.7
4	-2970.2	1.9	17.0	143.	-	-	162.
6	-1724.0	2.0	18.4	137.	45.	-	204.
8	- 86.8	2.0	18.5	137.	46.	.01	202.

channel (v"=0) angular basis set, while those summarized in Table VII used aconverged v'=1 angular basis set. As indicated in Figure 1, $\Delta E_m(j_{max})$ is the energy gap from the metastable level to the channel threshold for $j=j_{max}$. In both cases, one sees a sensitivity to the basis set size which, in view of the magnitude of the energy gaps and the fact that the potential has only a $\lambda=2$ anisotropy, at first seems somewhat suprising. However, in view of the importance of high-order coupling noted above, this sensitivity should have been expected.

One other convergence problem, which has only recently been pointed out (57), is associated with $\Delta v>0$ coupling to closed vibrational channels. In the calculations of Refs.(17-25) (and Tables V-VII), its effect was assumed to be negligible, and perturbation theory arguments indicate that it will affect level energies by only ca. 0.001 cm^{-1}. However, Kidd and Balint-Kurti (57) have shown that it can reduce the widths of rotationally predissociating levels by ca. 4% and those of pure vibrationally predissociating levels such as those considered here by as much as 30%! Thus, even as solutions to a model problem with only a $P_2(\cos\theta)$ anisotropy, the results of Table V are inexact because of this lack of vibrational basis set convergence. However, all of our qualitative conclusions about the importance of high-order coupling, potential dependence and related matters should still be valid.

Both the smallness of the total widths discussed here, and their variation from one isotope to the next, are qualitatively consistent with the trends predicted by the momentum gap arguments mentioned above (15,43-46). However, a perturbation theory treatment such as that which gave rise to the momentum gap arguments yields predictions in error by more than an order of magnitude (25). The reason for this, and for the extreme sensitivity to basis size seen in Tables VI and VII, is qualitatively explained by Figure 9.

Within the SFD approximation, the level width is given by the golden rule expression of Equation 6. For the same prototype H_2-Ar level considered above, Figure 9 shows the initial and final state

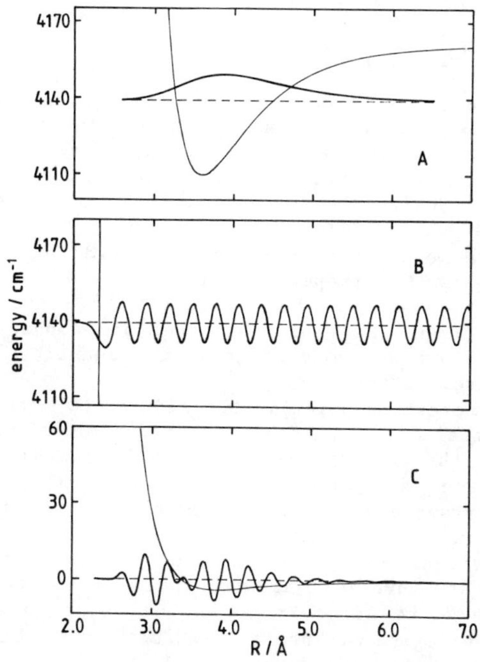

Figure 9.
Effective potential curves and wave functions for the closed (segment A) and open (segment B) channels, and the off-diagonal coupling function and golden rule integrand (segment C) for the (J=0) predissociation $H_2(1,0)$-Ar → $H_2(0,0)$+Ar.

coupling function and golden rule integrand (segment C). It is clear that the integrand is highly oscillatory, and it turns out that the value of the associated overlap integral is only ca. 10^{-4} of the area of a single loop of the integrand shown in Figure 9C. Thus, even a very small change in the character or amplitude of the initial or final state wave function can have a relatively large effect on the resulting matrix element, and hence on the predicted predissociation level width.

Concluding Remarks

One obvious conclusion from the above study is that calculated vibrational and rotational predissociation levels widths are extremely sensitive to the quality of the wave function used to represent the "initial" and "final" states, as well as to the coupling function itself. While this makes it difficult to use observations of this type to obtain new information about the potential energy surface, it also means that they should provide an extremely stringent constraint on the properties of a surface so determined. We have also seen that while they sometimes give useful qualitative information, none of the approximate methods considered herein are reliable enough that they may be used in a quantitative analysis of experimental level widths.

Another interesting point is the similarity between the type of information contained in vibrational predissociation data and that associated with the corresponding (above threshold) inelastic cross sections, and the fact that these phenomena are sensitive to the potential energy surface in a different region than are the discrete transition frequencies. Moreover, the present predictions that inelastic rotational and vibrational cross sections for molecular hydrogen will vary inversely as with the size of inert gas partner, decreasing from Ne to Ar to Kr to Xe, and that they will increase rapidly with the degree of internal excitation of the diatom, are of considerable practical importance.

LITERATURE CITED

1. Smalley, R.E.; Levy, D.H.; Wharton, L. J. Chem. Phys. 1976, 64, 3266-3276.
2. Levy, D.H. Adv. Chem. Phys. 1981, 47, 323-362.
3. Gough, T.E.; Miller, R.E.; Scoles, G. J. Chem. Phys. 1978, 69, 1588-1590.
4. Toennies, J.P.; Welz, W.; Wolf, G. J. Chem. Phys. 1979, 71, 614-642.
5. Le Roy, R.J.; Van Kranendonk, J. J. Chem. Phys. 1974, 61, 4750-4769.
6. Le Roy, R.J.; Carley, J.S.; Grabenstetter, J.E. Faraday Disc. Chem. Soc. 1977, 62, 169-178.
7. Carley, J.S. Faraday Disc. Chem. Soc. 1977, 62, 303.
8. Dunker, A.M.; Gordon, R.G. J. Chem. Phys. 1978, 68, 700-725.
9. Carley, J.S. Ph.D. Thesis, University of Waterloo, Waterloo, 1978.
10. Le Roy, R.J.; Carley, J.S. Adv. Chem. Phys. 1980, 42, 353-420.

11. Holmgren, S.L.; Waldman, M.; Klemperer, W. J. Chem. Phys. 1978, 69, 1661-1669.
12. Hutson, J.M.; Barton, A.E.; Langridge-Smith, P.R.R.; Howard, B.J. Chem. Phys. Lett. 1980, 73, 218-223.
13. Hutson, J.M.; Howard, B.J. Mol. Phys. 1981, 43, 493-516.
14. Hutson, J.M.; Howard, B.J. Mol. Phys. 1982, 45, 769-790 and 791-805.
15. Beswick, J.A.; Jortner, J. Adv. Chem. Phys. 1981, 47, 363-506.
16. Janda, K.C. "Predissociation of Polyatomic Van der Waals Molecules", Adv. Chem. Phys. 1984 (in press).
17. Beswick, J.A.; Requena, A. J. Chem. Phys. 1980, 72, 3018-3026.
18. Beswick, J.A.; Requena, A. J. Chem. Phys. 1980, 73, 4347-4352.
19. Le Roy, R.J.; Corey, G.C.; Hutson, J.M. Faraday Disc. Chem. Soc. 1982, 73, 339-355.
20. Chu, S.-I.; Datta, K.K. J. Chem. Phys. 1982, 76, 5307-5320.
21. Datta, K.K.; Chu, S.-I. Chem. Phys. Lett. 1983, 95, 38-45.
22. Balint-Kurti, G.G.; Kidd, I.F. Faraday Disc. Chem. Soc. 1982, 73, 133-136 and 404.
23. Kidd, I.F.; Balint-Kurti, G.G. Chem. Phys. Lett. 1983, 101, 419-423.
24. Hutson, J.M.; Le Roy, R.J. J. Chem. Phys. 1983, 78, 4040-4043.
25. Hutson, J.M.; Ashton, C.J.; Le Roy, R.J. J. Phys. Chem. 1983, 87, 2713-2720.
26. Le Roy, R.B.; Bernstein, R.B. J. Chem. Phys. 1971, 54, 5114-5126.
27. McKellar, A.R.W.; Welsh, H.L. J. Chem. Phys. 1971, 55, 595-609.
28. McKellar, A.R.W.; Welsh, H.L. Can. J. Phys. 1972, 50, 1458-1464.
29. McKellar, A.R.W. Faraday Disc. Chem. Soc. 1982, 73, 89-108.
30. Toennies, J.P.; Welz, W.; Wolf, G. J. Chem. Phys. 1976, 64, 5305-5307.
31. Arthurs, A.M.; Dalgarno, A. Proc. Roy. Soc. London 1960, A256, 540-551.
32. Child, M.S. "Molecular Collision Theory"; Academic Press: London, 1974.
5305-5307.
33. Ashton, C.J.; Child, M.S.; Hutson, J.M. J. Chem. Phys. 1983, 78, 4025-4039.
34. Bacic, Z.; Simons, J. Int. J. Quant. Chem. 1980, S14, 467-475.
35. Chu, S.-I. J. Chem. Phys. 1980, 72, 4772-4776.
36. Atabek, O.; Lefebvre, R. Chem. Phys. 1981, 55, 395-406.
37. Shapiro, M. J. Chem. Phys.19, 56, 2382-2591.
38. Beswick, J.A.; Shapiro, M. Chem. Phys. 1982, 64, 333-341.
39. Grabenstetter, J.E.; Le Roy, R.J. Chem. Phys. 1979, 42, 41-52.
40. Levine, R.D. J. Chem. Phys. 1967, 46, 331-345.
41. Levine, R.D.; Johnson, B.R.; Muckerman, J.T.; Bernstein, R.B. J. Chem. Phys. 1968, 49, 56-64.
42. Micha, D.A. Phys. Rev. 1967, 162, 88-97.
43. Beswick, J.A.; Jortner, J. Chem. Phys. Lett. 1977, 49, 13-18.
44. Beswick, J.A.; Jortner, J. J. Chem. Phys. 1978, 68, 2277-2297 and 2525.
45. Ewing, G.E. Chem. Phys. 1978, 29, 253-270.
46. Ewing, G.E. J. Chem. Phys. 1979, 71, 3143-3144.

47. Beswick, J.A.; Delgado-Barrio, G. J. Chem. Phys. 1980, 73, 3653-3659.
48. Segev, E.; Shapiro, M. J. Chem. Phys. 1983, 78, 4969-4984.
49. Liu, W.-K.; Grabenstetter, J.E.; Le Roy, R.J.; McCourt, F.R. J. Chem. Phys. 1978, 68, 5028-5031.
50. Pack, R.T. Chem. Phys. Lett. 1978, 55, 197-201.
51. Le Roy, R.J. "Further Improved Computer Program for Solving the Radial Schrodinger Equation for Bound and Quasibound (Orbiting Resonance) Levels" University of Waterloo Chemical Physics Research Report CP-230R, 1984.
52. Thakkar, A.; Carley, J.S.; Le Roy, R.J. unpublished work 1977.
53. Smith, K.M.; Rulis, A.M.; Scoles, G.; Aziz, R.A.; Nain, V. J. Chem. Phys. 1977, 67, 152-163.
54. Buck, U.; Meyer, H.; Le Roy, R.J. "Determining the Anisotropic Interaction Potential of D_2-Ar from Rotationally Inelastic Cross-Sections" J. Chem. Phys. 1984, 81 (in press).
55. Holmgren, S.L.; Waldman, M.; Klemperer, W. J. Chem. Phys. 1977, 67, 4414-4422.
56. Hutson, J.M.; Howard, B.J. Mol. Phys. 1980, 41, 1123-1141.
57. Kidd, I.F.; Balint-Kurti, G.G. "Theoretical Calculations of Photodissociation Cross Sections for the Ar-H_2 Van der Waals Complex" J. Chem. Phys. 1984 (in press).
58. Buck, U. Faraday Disc. Chem. Soc. 1982, 73, 187-203.
59. Hutson, J.M.; Le Roy, R.J. Appendix A of "Predissociation of HD-Ar Van der Waals Molecules by Internal Rotation", University of Waterloo Chemical Physics Research Report CP-196, 1982.
60. Kreek, H.; Le Roy, R.J. J. Chem. Phys. 1975, 63, 338-344.
61. Danby, G. J. Phys. B 1983, 16, 3393-3410.
62. Hutson, J.M. "Vibrational Predissociation and Infrared Spectrum the Ar-HCℓ Van der Waals Molecule", submitted to J. Chem. Phys.

RECEIVED June 11, 1984

14

Complex-Coordinate Coupled-Channel Methods for Predissociating Resonances in van der Waals Molecules

SHIH-I CHU

Department of Chemistry, University of Kansas, Lawrence, KS 66045

Complex-Coordinate Coupled-Channel (CCCC) methods are presented for the accurate and efficient treatment of the resonance energies and widths (lifetimes) of multichannel rotationally predissociating van der Waals (vdW) molecule resonances. Algorithms for dealing with the complex scaling of numerical and piecewise analytical potentials are also presented. The CCCC methods for vdW complexes are formulated in both the space-fixed (SF) and the body-fixed (BF) coordinates. The SFCCCC method is more appropriate for the treatment of weak-coupling complexes (such as $Ar-H_2$), whereas the BFCCCC method is better for strong-coupling complexes (such as Ar-HCl). These methods have been applied successfully to a number of vdW molecules, using reliable potential surfaces determined by experiments. In particular, the predicted widths for Ar-HD are in good agreement with the recent experimental data of McKellar.

It is well known that the resonance states characterized by complex energies correspond to poles of the resolvent operator $(E-\hat{H})^{-1}$ in the complex-energy plane on a non-physical higher Riemann sheet (1). Numerous techniques have been developed to compute these poles. These include close coupling, Feshbach, R-matrix, stabilization and other methods. In some methods, a complete scattering calculation is carried out, followed by a fitting of the results to a Breit-Wigner formula to extract the resonance parameters, while in others the pole of the resolvent is computed directly. In all of these methods the asymptotic form of the wave function plays an essential role in performing the calculation and/or extracting the scattering information. Recently, Aguilar and Combes (2), Balslev and Combes (3), and Simon (4), have suggested an elegant direct way to perform analytical continuation of the Hamiltonian into the complex plane. The method is based on the analytic properties of the Hamiltonian $\hat{H}(\vec{R})$ under the complex coordinate transformation (also known as the complex scaling, coordinate-rotation, and dilatation analytical

0097–6156/84/0263–0263$07.50/0

transformation), $\vec{R} \rightarrow \vec{R}e^{i\alpha}$, where α can be real or complex. As a result of the complex coordinate transformation (2-4), the eigenvalues corresponding to the bound states of $\hat{H}$ stay invariant, while the branch cuts associated with the continuous spectrum of $\hat{H}$ are rotated about their respective thresholds by an angle -2α (assuming $0 < \alpha < \pi/2$), exposing the complex resonance states in appropriate strips of the complex energy plane. A crucial point from the computational point of view is that, although the eigenfunctions associated with the resonance states normally diverge at infinity, they now become localized, i.e. square integrable. The square integrability of complex-coordinate resonance wave functions naturally led to the extension of well-established bound-state techniques to the determination of resonance energies (E_R) and widths (Γ) of metastable states. Recent applications of the complex coordinate method to various areas of atomic and molecular physics demonstrate the power and usefulness of such a direct technique (for recent reviews, see references 5 and 6).

In this paper we focus our interest on the extension of the complex coordinate method (CCM) to the study of multi-channel rotational predissociation resonances in van der Waals (vdW) molecules. Van der Waals molecules are weakly bound complexes of atoms and molecules which characteristically have small dissociation energies ($10\ cm^{-1}$ - $500\ cm^{-1}$) and large bond lengths and retain the individual properties of the molecular constituents within the vdW aggregate. Van der Waals molecules in metastable states can predissociate by two different mechanisms. These are tunneling through a potential energy barrier and conversion of internal excitation energy into relative translation of fragments. Vibrational or rotational predissociation occurs when one of the monomers in a vdW molecule is internally excited, and the internal vibrational or rotational energy exceeds the intermolecular (vdW) binding energy. In the case of rotational predissociation (RP), the anisotropy of the interaction potential permits this excess internal rotational energy to flow into the vdW bond, and the complex eventually breaks up. The metastable state has a finite lifetime τ (related to Γ by $\tau=\hbar/\Gamma$), which can be measured in principle through the observable broadening of spectroscopic lines. The predissociation line width data provide valuable information on the potential well depths and anisotropies of the intermolecular interactions, and the fragmentation of the vdW bond via the vibrational/rotational predissociation mechanism provides an illuminating example of intramolecular dynamics on a single electronic potential surface. There is therefore currently considerable experimental and theoretical interests in the study of the structure and dynamics of vdW molecules (for recent reviews, see references 7-9). In this paper we present a short review of the complex-coordinate coupled-channel (CCCC) methods recently developed in our laboratory for the accurate determination of the resonance energies and widths (lifetimes) of rotationally predissociating vdW molecules.

We note that in spite of the increasing popularity of the complex coordinate method (CCM), most atomic and molecular problems studied so far have involved the use of simple analytic potentials, so that the complex scaling transformation $\vec{R} \rightarrow \vec{R}e^{i\alpha}$ can be directly applied (5, 6). However, in the case of more complex molecular problems, such as the predissociation of vdW molecules, very often accurate potential energy surfaces are available only as tabulated numbers or as sophisticated piecewise analytic forms. There is no previous demonstration that the CCM can be applied successfully to realistic complicated systems. In the next section, we present a stable computational algorithm to alleviate the problems associated with numerical and piecewise potentials (10). Then we present the complex-coordinate coupled-channel (CCCC) formalism in both the space-fixed (SF) and body-fixed (BF) coordinates. The SFCCCC method is then applied to the determination of the resonance energies and widths (lifetimes) of Ar-H_2 and Ar-HD vdW complexes, using reliable anisotropic potential surfaces recently determined from experiments. Comparison of the SFCCCC results with other theoretical and experimental data are made. This is followed by a concluding remarks in the final section.

Complex Scaling Method for Numerical and Piecewise Potentials

Let $\{\phi_i(R)\}$ be an appropriate L^2 (square integrable) basis set for a specific problem under consideration. One of the main efforts in complex scaling calculations is to compute efficiently and reliably the matrix elements $<\phi_i(R)|V(Re^{i\alpha})|\phi_j(R)>$, of the complex-rotated potential $V(Re^{i\alpha})$ obtained by applying $R \to Re^{i\alpha}$ to the potential V(R). However, in the case that V(R) is only available in tabulated numercal form or in some piecewise analytical form, it is not obvious how one can compute these complex matrix elements directly and accurately. For example, a straightforward application of the complex coordinate transformation to cubic-spline interpolated complex numerical potentials indicates that dilatation analyticity is badly destroyed (10).

To bypass these numerical instabilities, we have developed a method (10) which takes advantage of several well-established transformation theorems and stable quadrature algorithms. The procedure consists of the following three steps: Step (i). The identity

$$<\phi_i(R)|V(Re^{i\alpha})|\phi_j(R)> = e^{-i\alpha}<\phi_i(Re^{-i\alpha})|V(R)|\phi_j(Re^{-i\alpha})> \quad (1)$$

is used to transform the complex rotated potential back to the real potential. Step (ii). The inner projection technique (11) is adopted such that

$$<\phi_i(Re^{-i\alpha})|V(R)|\phi_j(Re^{-i\alpha})> =$$

$$\sum_{mn}<\phi_i(Re^{-i\alpha})|\psi_m(R)> <\psi_m(R)|V(R)|\psi_n(R)> <\psi_n(R)|\phi_j(Re^{-i\alpha})>, \quad (2)$$

where $\{\psi_m(R)\}$ are real, L^2-orthonormal basis functions. The simplifying feature here is that the evaluation of the complex-scaled-Hamiltonian matrix elements is now replaced by the much simpler calculation of a complex overlap matrix and the matrix elements of real potentials. For several of the orthonormal basis sets we have used so far, such as harmonic oscillator, Laguerre, and Gaussian, the complex overlap matrix elements $<\phi_i(Re^{-i\alpha})|\psi_m(R)>$ can be derived in closed analytic forms and evaluated exactly (10). Thus the uncertainty associated with the complex rotated potentials can be completely removed. Step (iii). The matrix elements of the real potential V(R), in numerical or piecewise analytical form, $\langle\psi_m(R)|V(R)|\psi_n(R)\rangle$, can also be evaluated accurately by using the well-known transformation theory and quadrature scheme introduced by Harris, Engerholm, and Gwinn (HEG) (12).

The above-mentioned three-step transformation procedure allows us to expand the scope of the method of complex scaling to virtually any system. The method has been applied successfully to the study of predissociating resonances of several vdW molecules, including Ar-H_2, Ar-HD, Ar-N_2 and Ar-HCl etc., involving either piecewise or numerical potentials.

We note that an alternative way for the treatment of the problem associated with the numerical or piecewise potential is to adopt Simon's exterior complex scaling procedure (13). This idea has been used to facilitate the treatment of molecular electronic structure within the Born-Oppenheimer approach (14, 15). Complex scaling only beyond a certain distance R_o has the advantage that the interior region can be treated in the usual manner (without the use of complex coordinates) while the wave function can be made to vanish asymptotically in the exterior region. Extension of the exterior complex scaling method to the study of vdW molecule predissociation has been suggested recently (16, 17).

Complex-Coordinate Coupled-Channel Approach to Predissociation of Atom-Diatom Van Der Waals Complexes

The theory of atom-diatom van der Waals molecules may be formulated using either space-fixed or body-fixed coordinates. The coordinates and axis systems are shown in Figure 1. The primed axes refer to a space-fixed (SF) coordinate system while the unprimed axes refer to the body-fixed (BF) coordinate system. The BF axes $\{\vec{x}, \vec{y}, \vec{z}\}$ are defined such that $\vec{z}$ is parallel to $\vec{R}$, the vector of length R running from the diatom centre-of-mass to the atom, and $\vec{y}$ lies in the $\{\vec{R}, \vec{r}\}$ plane, where $\vec{r}$ is the diatom internuclear vector. In the SF frame, the dynamics of the system are described by R, r, and the polar angles specifying the orientation of $\hat{R}$ and $\hat{r}$. In the BF frame, the relevant coordinates are R, r, θ ($=\cos^{-1}(\hat{r}\cdot\hat{R})$), and the three Euler angles defining the orientation of $\{\vec{x}, \vec{y}, \vec{z}\}$ relative to $\{\vec{x}', \vec{y}', \vec{z}'\}$.

Hamiltonian and Complex-Coordinate Coupled-Channel Formulation in Space-Fixed Coordinates (18-21). The hamiltonian of a triatomic vdW molecule A...BC, after separating out the motion of the centre of mass, may be expressed in the SF frame as

$$H(\vec{R},\vec{r}) = (1/2\mu)[-\hbar^2 R^{-1}(\partial^2/\partial R^2)R + \vec{\ell}^2(\hat{R})/R^2] + H_d(\vec{r}) + V(R,r,\theta). \quad (3)$$

Here $\mu = m_A(m_B + m_C) / (m_A + m_B + m_C)$ is the reduced mass, $\vec{R}$ is the axis of the complex, $\vec{r}$ that of the diatom BC, $\vec{\ell}$ is the angular momentum of BC and A about each other, H_d is the vibration-rotation hamiltonian of the diatom, $V(R,r,\theta)$ is the interaction potential of A and BC, and $\theta = \cos^{-1}(\hat{r}\cdot\hat{R})$. The interaction potential for vdW

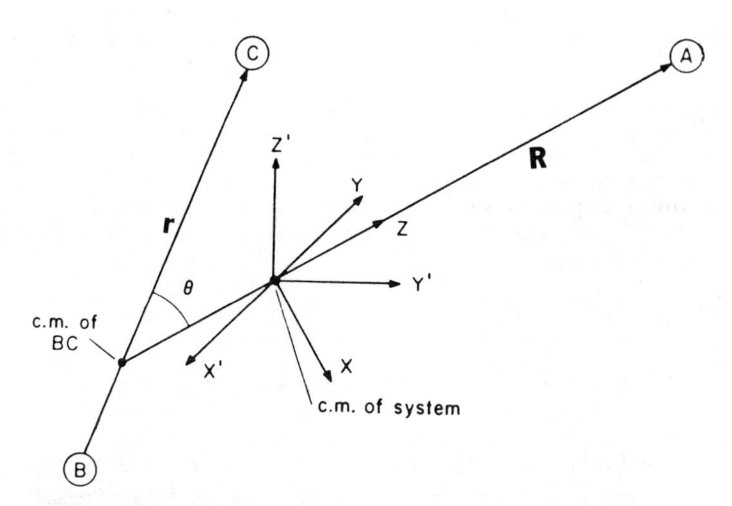

Figure 1. Center-of-mass coordinate systems for atom-diatom van der Waals molecules. The primed axes are used in the space-fixed (SF) formulation, the unprimed axes in the body-fixed (BF) formulation.

complexes may be conveniently expressed as an expansion in terms of Legendre polynominal as well as a power series expansion in the diatom stretching coordinate $\xi(r)$ (9):

$$V(R,r,\theta) = V(R,\xi,\theta) = \sum_{\lambda k}\sum \xi^k P_\lambda(\cos\theta)\, V_{\lambda k}(R), \tag{4}$$

where $\xi(r) = (r-r_o)/r_o$ and r_o is an isotope-independent constant. The eigenfunctions of $H_d(\vec{r})$ may be written as $\phi_{vj}(r)Y_{jm}(\hat{r})$, where v and j are respectively the vibrational and rotational quantum numbers of the diatom BC, Y_{jm} is a spherical harmonic, and $\phi_{vj}(r)$ satisfies the radial Schrödinger equation

$$\{-(\hbar^2/2\mu_{BC})[r^{-1}(\partial^2/\partial r^2)r - j(j+1)/r^2] + V_{BC}(r) - E_d(v,j)\}\phi_{vj}(r)=0. \tag{5}$$

In equation 5, $E_d(v,j)$ is a vibration-rotation eigenvalue of the free diatom, μ_{BC} is the reduced mass, and $V_{BC}(r)$ is the potential.

In the spaced-fixed total-angular-momentum representation, with quantum numbers J and M, where $\vec{J}=\vec{\ell}+\vec{j}$ and M denotes the quantum number for a component of $\vec{J}$, a convenient angular basis for the expansion of the wave function is the total-angular-momentum eigenfunction defined by

$$\mathcal{Y}_{J\ell j}^{M}(\hat{R},\hat{r}) = \sum_{m_\ell}\sum_{m_j} (\ell j m_\ell m_j|\ell j JM)Y_{\ell m_\ell}(\hat{R})Y_{jm_j}(\hat{r}), \tag{6}$$

where (....|....) is a Clebsch-Gordan coefficient. It is expedient to define a scheme for labeling eigenstates of the complex uniquely. In the zero-coupling limit, corresponding to an isotropic interaction potential, each state may be labelled by $|vj\ell JM\rangle$ and may be decomposed into a product of radial and angular functions,

$$|vj\ell JM\rangle = \psi^J_{vj\ell}(R)\phi_{vj}(r)\mathcal{Y}^M_{J\ell j}(\hat{r},\hat{R}). \tag{7}$$

When the couplings are turned on, only J, M and the parity $p = (-1)^{j+\ell+J}$ will be good quantum numbers in the perturbed state. However, for weak coupling complexes, such as $Ar\text{-}H_2$ and Ar-HD, v, j and ℓ remain nearly good quantum numbers. Thus the zero-coupling function $|vj\ell JM\rangle$ provides a good basis function for level energies and widths.

Figure 2 shows an example of the zero-coupling potential curves (labeled by j and ℓ) and zero-coupling-limit level energies for the angular channels contributing to the $J^p = 0^+$ states of the $Ar....N_2(j)$ vdW complex, where N_2 is treated as a rigid rotor (10). Note that the states above the $E_d(j=0)$ threshold are all metastable resonance states and can predissociate via the anisotropic coupling mechanism.

We now discuss the complex-coordinate coupled-channel (CCCC) formulation in the SF frame. According to the theory of complex-coordinate transformation, the energy (E_R) and the width (Γ) associated with a metastable state of the vdW molecule may be determined by the solution of the complex eigenvalue of a non-hermitian hamiltonian $H_\alpha(\vec{R}e^{i\alpha},\vec{r})$, obtained by applying the complex-coordinate

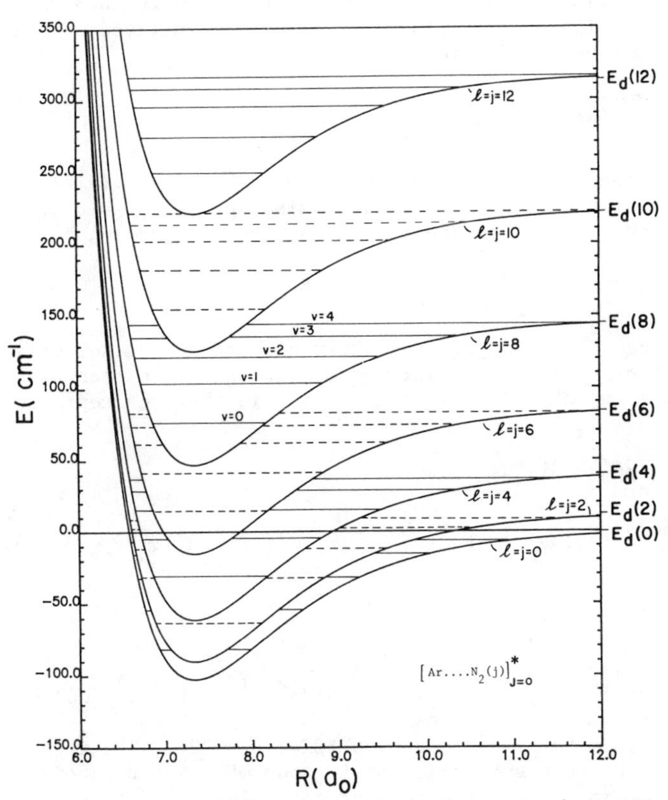

Figure 2. The isotropic-channel potential curves (specified by j and ℓ) of the $Ar...N_2$ system, where j is the rotational quantum number of N_2 and ℓ is the relative orbital angular momentum of Ar and N_2. $E_d(j)$ is the diatom level energies. Shown also here are the zero-coupling limit level energies for the angular channels contributing to J=M=0 states of the $Ar...N_2$ complex. Reproduced with permission from Ref. 10. Copyright 1982, North-Holland Physics Publishing.

transformation to the vdW bond (dissociating) coordinate, $R \to Re^{i\alpha}$. In the CCCC formalism, the total wave function of the coordinate-rotated hamiltonian $H_\alpha(\vec{R}e^{i\alpha}, \vec{r})$, for a given J, M and p, is expanded in terms of the complete set of the zero-coupling state functions $|vj\ell JM\rangle$ allowed by the symmetry. We further expand the radial function $\psi^J_{vj\ell}(R)$ in terms of orthonormalized square-integrable (L^2) basis functions $\{\chi_n(r)\}$,

$$\psi^J_{vj\ell}(R) = \sum_{n=1}^{N_\gamma} a_n(\gamma)\chi_n(R), \tag{8}$$

where γ specifies the channel quantum numbers, $\gamma = (vj\ell JM)$, N_γ is the size of the truncated radial basis, and $\langle\chi_n|\chi_m\rangle = \delta_{nm}$. For convenience, let us define the channel basis function to be

$$|\gamma n\rangle \equiv \chi_n(R)\phi_{vj}(r)\mathcal{Y}^M_{J\ell j}(\hat{r},\hat{R}) \tag{9}$$

and arrange the order of the matrix elements of $H_\alpha(\vec{R}e^{i\alpha},\vec{r})$ in such a way that n is allowed to vary from 1 to N_γ within each channel block $\gamma = (vj\ell JM)$. The matrix element in the $|\gamma n\rangle$ representation is then

$$\langle\gamma'n'|H_\alpha(\vec{R}e^{i\alpha},\vec{r})|\gamma n\rangle$$

$$= e^{-2i\alpha}(\hbar^2/2\mu)\langle\chi_{n'}|\ -R^{-1}(\partial^2/\partial R^2)R + \ell(\ell+1)/R^2|\chi_n\rangle\delta_{vv'}\delta_{jj'}\delta_{\ell\ell'}\delta_{JJ'}\delta_{MM'}$$

$$+ E_d(v,j)\delta_{vv'}\delta_{jj'}\delta_{\ell\ell'}\delta_{nn'}\delta_{JJ'}\delta_{MM'}$$

$$+ \sum_{\lambda k}\sum \langle\phi_{v'j'}|\xi^k|\phi_{vj}\rangle\cdot\langle\chi_{n'}|V_{\lambda k}(Re^{i\alpha})|\chi_n\rangle f_k(\ell'j',\ell j;J)\ \delta_{JJ'}\delta_{MM'} \quad , \tag{10}$$

where f_k is a Percival-Seaton coefficient (22). The resulting matrix of H_α is a symmetric complex one whose complex eigenvalues may be determined via the secular determinant

$$\mathrm{Det}|(H_\alpha)_{\gamma'n',\gamma n} - E\delta_{\gamma'\gamma}\delta_{n'n}| = 0. \tag{11}$$

The desired metastable states can be then identified by the stationary points (10, 18-21) of the α trajectories of complex eigenvalues.

Hamiltonian and Complex-Coordinate Coupled-Channel Formulation in Body-Fixed Coordinates (23). In the BF frame, the Hamiltonian is identical to equation 3 except that $\vec{R}$ and $\vec{r}$ are expressed relative to the unprimed axes of figure 1 and the angular momentum operator of the rotation of $\vec{R}$ (i.e. $\vec{\ell}$) is written as $\vec{\ell} = \vec{J}-\vec{j}$. The operator $(\vec{J}-\vec{j})^2$ may be expressed as

$$(\vec{J} - \vec{j})^2 = (\vec{J}^2 + \vec{j}^2 - 2\,J_z^{\,2}) - (J_+j_- + J_-j_+), \tag{12}$$

where $J_\pm = J_x \pm i\,J_y$ and $j_\pm = j_x \pm i\,j_y$. Making use of equation 12, the BF Hamiltonian becomes

$$H(\vec{R},\vec{r}) = [-(\hbar^2/2\mu)R^{-1}(\partial^2/\partial R^2)R+(\vec{J}^2 + \vec{j}^2 - 2J_z^2)/2\mu R^2] + H_d(\vec{r})$$

$$+ V(R,r,\theta) - (J_+j_- + J_-j_+)/2\mu R^2 \qquad (13)$$

Here the coupling between the different modes of vdW vibrational motions is induced by the anisotropic part of $V(\vec{R},\vec{r})$ and by the last term in equation 13.

In the BF frame, the projection of the rotational angular momentum $\vec{j}$ on the body-fixed $\vec{z}$ axis (i.e. $\vec{R}$) is denoted by the "tumbling" angular momentum quantum number Ω, so that the eigenfunctions of $\vec{j}^2$ are $Y_{j\Omega}(\hat{r})$. By conservation of angular momentum, the projection of the total angular momentum $\vec{J}$ on this BF $\vec{z}$-axis is also equal to $\Omega\hbar$. In the total angular momentum (J, M) representation (where M is the z component of J in a SF axis system), an appropriate angular basis for wave function expansion in BF coordinates is

$$\Theta_{j\Omega}^{JM}(\hat{R}, \hat{r}) = \left(\frac{2j+1}{4\pi}\right)^{1/2} D_{\Omega M}^{J}(\hat{R}, \hat{r})Y_{j\Omega}(\hat{r}) \ , \qquad (14)$$

where $D_{\Omega M}^{J}$ is the symmetric top wave function. We now define a scheme for the labeling of the eigenstates of the vdW complex. In the zero-coupling limit, each state may be labeled by $|JMj\Omega\rangle$ and may be decomposed into products of radial and angular functions

$$|JMj\Omega\rangle = \psi_{j\Omega J}(R)\phi_{vj}(r)\Theta_{j\Omega}^{JM}(\hat{R}, \hat{r}). \qquad (15)$$

When the coupling terms are turned on, only J and M are good quantum numbers. Nevertheless, the zero-coupling function $|JMj\Omega\rangle$ still forms a convenient basis function for our present study.

In the BFCCCC formalism, the total wave function of the complex-scaled hamiltonian $H_\alpha(\vec{R}e^{i\alpha}, \vec{r})$, for a given J and M, is expanded in terms of the zero-coupling frunctions $|JMj\Omega\rangle$ allowed by the symmetry. We further expand the radial function $\psi_{j\Omega J}(R)$ of $|JMj\Omega\rangle$ in terms of an orthonormalized L^2 basis function $\{\chi_n(R)\}$,

$$\psi_{j\Omega J}(R) = \sum_{n=1}^{N_\gamma} a_n(\gamma)\chi_n(R) \ , \qquad (16)$$

where γ specifies the channel quantum number, $\gamma = (JMj\Omega)$, $N\gamma$ is the size of the truncated radial basis, and $\langle\chi_n|\chi_m\rangle = \delta_{nm}$. For convenience, let us define the basis function

$$|\gamma n\rangle \equiv \chi_n(R) \ \phi_{vj}(r) \ \Theta_{j\Omega}^{JM}(\hat{R}, \hat{r}) \ , \qquad (17)$$

and arrange the order of the matrix elements of $H_\alpha(\vec{R}e^{i\alpha}, \vec{r})$ in such a way that n is allowed to vary from 1 to N_γ within each channel block $\gamma = (JMj\Omega)$. The matrix element in the $|\gamma n\rangle$ representation is then

$$\langle\gamma'n'|H_\alpha(\vec{R}e^{i\alpha},\vec{r})|\gamma n\rangle$$

$$= e^{-2i\alpha}(\hbar^2/2\mu)\langle\chi_{n'}|-d^2/dR^2+[J(J+1)+j(j+1)-2\Omega^2]/R^2|\chi_n\rangle\delta_{JJ'}\delta_{MM'}\delta_{vv'}\delta_{jj'}$$

$$\times\ \delta_{\Omega\Omega'} + \sum_{\lambda k}\sum\langle\phi_{v'j'}|\xi^k|\phi_{vj}\rangle\cdot\langle\chi_{n'}|V_{\lambda k}(Re^{i\alpha})|\chi_n\rangle d_k(j',j,\Omega)\ \delta_{JJ'}\delta_{MM'}\delta_{\Omega\Omega'}$$

$$-e^{-2i\alpha}(\hbar^2/2\mu)\langle\chi_{n'}|R^{-2}|\chi_n\rangle\lambda_J(\Omega,\Omega')\delta_{JJ'}\delta_{MM'}\delta_{jj'}\delta_{vv'}$$

$$+E_d(v,j)\ \delta_{vv'}\delta_{jj'}\delta_{\Omega\Omega'}\delta_{JJ'}\delta_{MM'}\delta_{nn'}\ , \qquad (18)$$

where

$$d_k(j',j,\Omega) = (-1)^\Omega\ [(2j+1)(2j'+1)]^{1/2}\begin{pmatrix} j' & k & j \\ 0 & 0 & 0\end{pmatrix}\begin{pmatrix} j' & k & j \\ -\Omega & 0 & \Omega\end{pmatrix}\ , \qquad (19)$$

and

$$\lambda_J(\Omega,\Omega') = [J(J+1) - \Omega(\Omega+1)]^{1/2}\cdot[j(j+1) - \Omega(\Omega+1)]^{1/2}\delta_{\Omega',\Omega+1}$$

$$+ [J(J+1) - \Omega(\Omega-1)]^{1/2}\cdot[j(j+1) -\Omega(\Omega-1)^{1/2}\delta_{\Omega',\Omega-1}\ , \qquad (20)$$

and $\begin{pmatrix} \cdot & \cdot & \cdot \\ \cdot & \cdot & \cdot\end{pmatrix}$ is a 3-j symbol. The resulting matrix of H_α is again a symmetric complex one.

Although the BFCCCC method described in this section is formally equivalent to the SFCCCC formulation, the SF theory appears more appropriate for weak-coupling complexes, whereas the BF theory is more natural for strong-coupling complexes. In addition, for strong-coupling vdW complexes which are only slightly asymmetric tops, such as Ar-HCl(24), the "tumbling" angular momentum quantum number Ω in the BF frame is nearly conserved, and the centrifugal decoupling (CD) approximation (namely, ignoring the $J_\pm j_\mp/2\mu R^2$ coupling terms in equation 13 or λ_J terms in equation 18) will be effective. This drastically reduces the number of coupled channels and provides a tremendous simplification and computational advantage in the BF frame. Extension of the BFCCCC-CD method to the study of rotational predissociation of the Ar-HCl vdW complex will be discussed elsewhere (16).

Rotational Predissociation of Ar-H_2 and Ar-HD Van Der Waals Complexes

The complex-coordinate coupled-channel (CCCC) formulations described in the last section have been applied to the determination of resonance energies and widths (lifetimes) of several rotationally predissociating vdW molecules, including Ar-H_2 (19), Ar-HD (20), Ar-N_2 (10), and Ar-HCl (16). In this section, we present the results for the Ar-H_2 and Ar-HD systems. The Ar-H_2 system is one of the most extensively studied vdW molecules, and several reliable anisotropic potential energy surfaces derived from experimental data are available. Although rotational predissociation (RP) line widths for Ar-H_2 systems have not been measured, McKellar (25) has recently successfully obtained the RP widths for the Ar-HD(v,j) vdW molecules from the (improved) IR-high resolution spectra. The direct comparison between the theoretical and experimental results allow a critical assessment of the quality of the theoretical methods as well as the reliability of the potential surfaces used in the calculations.

SFCCCC Calculations of $Ar-H_2$ Atom-Rigid Rotor vdW Complex (19). Calculations for the $Ar-H_2$ system were performed by invoking the space-fixed complex-coordinate coupled-channel (SFCCCC) formulation and the use of several realistic atom-rigid rotor anisotropic potentials (derived from experimental data) of the form,

$$V(R,\ r=r_e,\ \theta) = V_o(R) + V_2(R)P_2(\cos\theta). \tag{21}$$

Six different potential functions for V_o and V_2 were employed (19), including three Lennard-Jones LJ(12,6) types, LJ(I), LJ(II) and LJ(III), two Buckingham-Corner BC(6,8) types, BC(I) and BC(II), and the semiempirical potential of Tang and Toennies (TT). The LJ(I) potential (both V_o and V_2) is that obtained by LeRoy and van Kranendonk (26). In LJ(II), V_o is that of LeRoy and van Kranendonk (26), whereas V_2 is determined by Zandee and Reuss (27). In LJ(III), V_o is that obtained by Helbing et. al. (28), and V_2 is that of Zandee and Reuss (27). In the first BC potential, V_o is that obtained by LeRoy et. al. (29), whereas V_2 is determined by Zandee and Reuss (27). For BC(II) potential, both V_o and V_2 are determined by LeRoy and Carley (30) from the spectroscopic data. It is now generally believed that the Buckingham-Corner-type potentials, BC(I) and BC(II), and the TT potential (31) provide more realistic description of the $Ar-H_2$ system than the LJ potentials. Detailed description of the potential parameters has been given in reference 19. Figure 3 depicts two of the potential functions, BC(I) and TT, used. For all the six potentials, we consider the metastable state correlating with the isotropic channel $|j=2,\ \ell=2,\ J=M=0\rangle$. Because of the symmetry of H_2, only the following angular basis need to be considered: $(j=0,\ \ell=0)$, $(j=2,\ \ell=2)$, $(j=4,\ \ell=4)$,.... etc. For the $Ar-H_2$ system, we found that the resonance energy and width (lifetime) of the metastable state can be determined satisfactorily by including the only open channel $(j=0,\ \ell=0,\ J=0,\ M=0)$ and one closed channel $(j=2,\ \ell=2,\ J=0,\ M=0)$. Very high precision can be achieved by including one additional closed channel $(j=4,\ \ell=4,\ J=0,\ M=0)$ in the basis set.

The matrix structure in the $|\gamma n\rangle$ representation is of three-by-three block form (for the three channel case) as depicted in Figure 4. Within each diagonal block specified by the channel quantum number $\gamma=(j\ell JM)$, we used the orthonormal harmonic oscillator L^2-basis

$$\chi_n(R) = \left[\frac{\beta}{\pi^{1/2}2^n n!}\right]^{1/2} H_n(\beta x)e^{-\frac{1}{2}\beta^2x^2}\ ,\ n=1,2...,N_\gamma \tag{22}$$

to expand the radial wave function $\psi_{J\ell j}(r)$ defined in equation 8. In equation 22, H_n is a Hermite polynomial, $x=R-R_o$, and β is an adjustable non-linear parameter. The potential matrix elements for

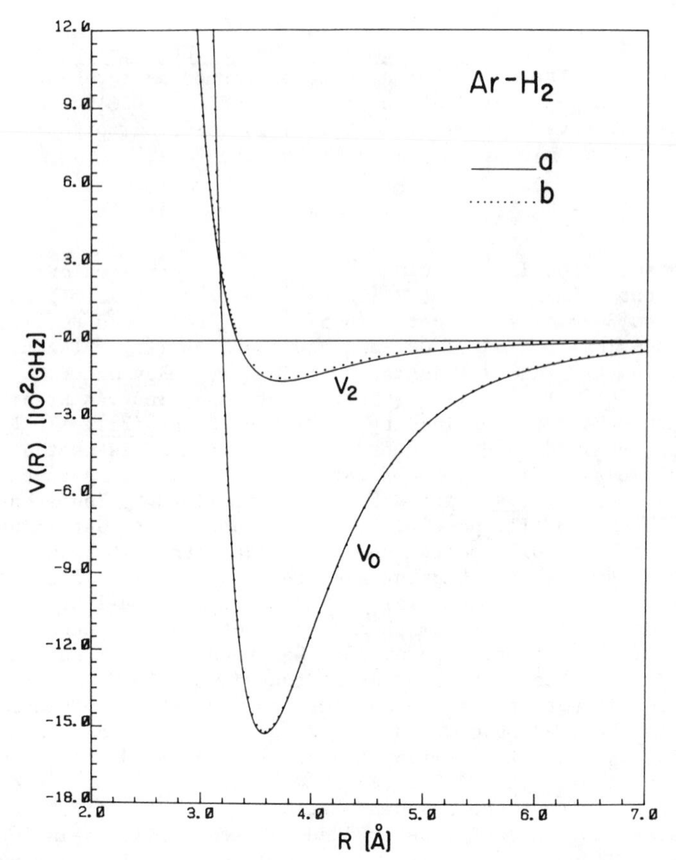

Figure 3. Potential surfaces $V_0(R)$ and $V_2(R)$ of BC(I) and TT potentials for the Ar-H_2 system. Reproduced with permission from Ref. 19. Copyright 1982, American Institute of Physics.

$\langle j=\ell=0, n \mid H \mid j'=\ell'=0, n' \rangle$	$\langle j=\ell=0, n \mid V_2 P_2 \mid j'=\ell'=2, n' \rangle$	**0**
$\langle j=\ell=2, n \mid V_2 P_2 \mid j'=\ell'=0, n' \rangle$	$\langle j=\ell=2, n \mid H \mid j'=\ell'=2, n' \rangle$	$\langle j=\ell=2, n \mid V_2 P_2 \mid j'=\ell'=4, n' \rangle$
0	$\langle j=\ell=4, n \mid V_2 P_2 \mid j'=\ell'=2, n' \rangle$	$\langle j=\ell=4, n \mid H \mid j'=\ell'=4, n' \rangle$

$(n, n' = 1, 2, \ldots\ldots, N)$

Figure 4. Matrix structure of the predissociation Hamiltonian H in the $|\gamma n\rangle$ representation. Here $\gamma=(j\ell, J=M=0)$ specifies the channel quantum numbers, n is the index number for the harmonic oscillator radial basis, $n=1,2,\ldots,N$, and V_2P_2 is the anisotropic potential. Reproduced with permission from Ref. 19. Copyright 1982, American Institute of Physics.

the LJ(12,6) potentials can be computed readily using the HEG quadrature (12). For the inhomogeneous, piecewise BC and TT potenials, we have employed the three-step procedure described earlier to calculate the complex scaled potential matrix elements. Once the potential matrix elements have been computed, the symmetric complex matrix as depicted in Figure 4 can be set up and the complex eigenvalue of the matastable state correlates with the isotropic channel $|j=\ell=2, J=M=0\rangle$ can be determined. To facilitate the location of the optimum α-trajectory, the parameters R_o and β can be adjusted in such a way that the spurious widths (19) associated with the unperturbed diagonal complex eigenvalues (i.e. the imaginary parts of the eigenvalues of the diagonal block $\langle j=\ell=2, n|H_\alpha|j'=\ell'=2, n'\rangle$ that are correlated with the predissociation resonances) be kept as small as possible. As a practical guide for multichannel problems, it is important that these (unperturbed) spurious widths are smaller in magnitude than the genuine (perturbed) predissociation widths of physical interest.

Typical examples of the α-trajectories are shown in Figures 5 and 6. It is seen that the resonance positions can be clearly identified by the sharp stationary points [where $d(E_R + i\Gamma/2)/d\alpha \approx 0$] in the α-trajectories. Table I shows a typical example of the convergence of the resonance positions with respect to the basis size N_γ (potential LJ(III)). We note that E_R (= real part of the complex eigenvalue) converges to within 10^{-3} cm^{-1} with $N_\gamma > 30$, and to within 10^{-4} cm^{-1} with $N_\gamma > 40$. The convergence of the half-width (Γ/2 or imaginary part of the complex eigenvalue) is somewhat slower: N_γ=35 is needed to converge Γ/2 to within 10^{-3} cm^{-1} and N_γ=45 to within 10^{-4} cm^{-1}.

TABLE I. Convergence study of the position of the resonance state (ℓ=j=2, J=M=0) of Ar....H_2 van der Waals molecule (potential LJ(III)) with respect to basis size N_γ. Shown here are calculations with two channel blocks (with the same basis size N_γ), and R_o = 9.0 a_o, β = 1.5 a_o^{-1} (a_o = Bohr radius) and rotational angle α = 0.07 radian.

N_γ	E_R (cm^{-1})	Γ/2 (cm^{-1})
20	347.2774	0.0327
30	347.1553	0.0105
35	347.1541	0.0090
40	347.1539	0.0087
45	347.1539	0.0085
50	347.1539	0.0083
60	347.1539	0.0083

Shown in Figure 7 is the comparison of the predicted resonance positions (3-channel-block converged results) for the six potentials considered. We notice that the results are sensitive to the potential energy surfaces used. In particular, the line widths predicted

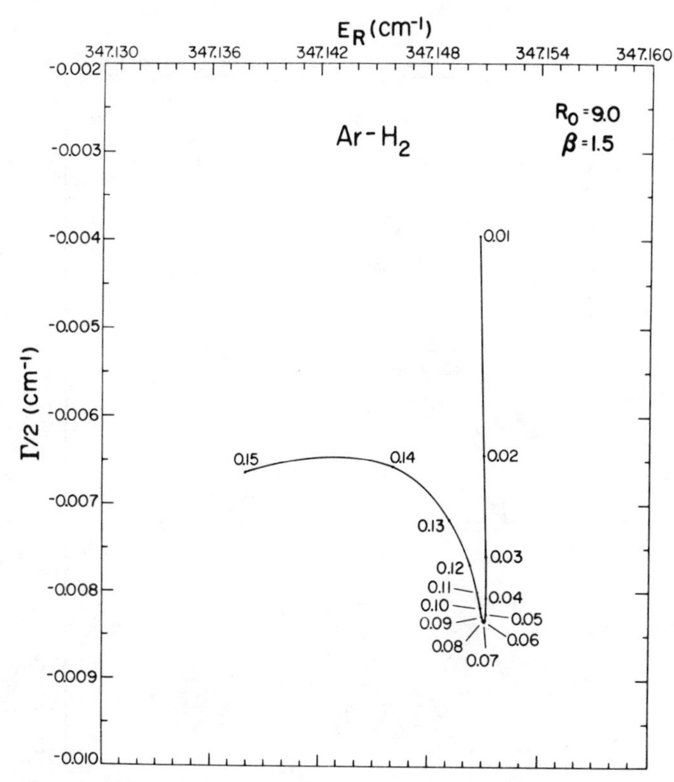

Figure 5. A typical SFCCCC α-trajectory for the complex eigenvalue associated with the rotational predissociation of the metastable level (j=ℓ=2, J=M=0) of the Ar...H_2 vdW molecule (with potential LJ(III)). The numbers on the dots shown in the figure indicates the rotational angles (in radians) used. Reproduced with permission from Ref. 19. Copyright 1982, American Institute of Physics.

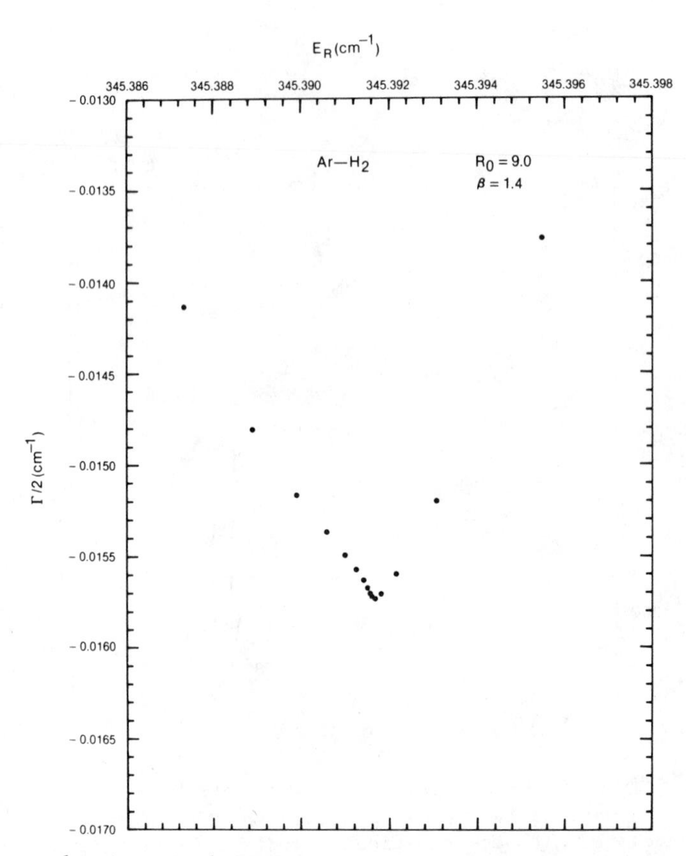

Figure 6. A typical SFCCCC α-trajectory for the $Ar...H_2$ metastable state ($j=\ell=2$, $J=M=0$) with potential BC(II). Reproduced with permission from Ref. 19. Copyright 1982, American Institute of Physics.

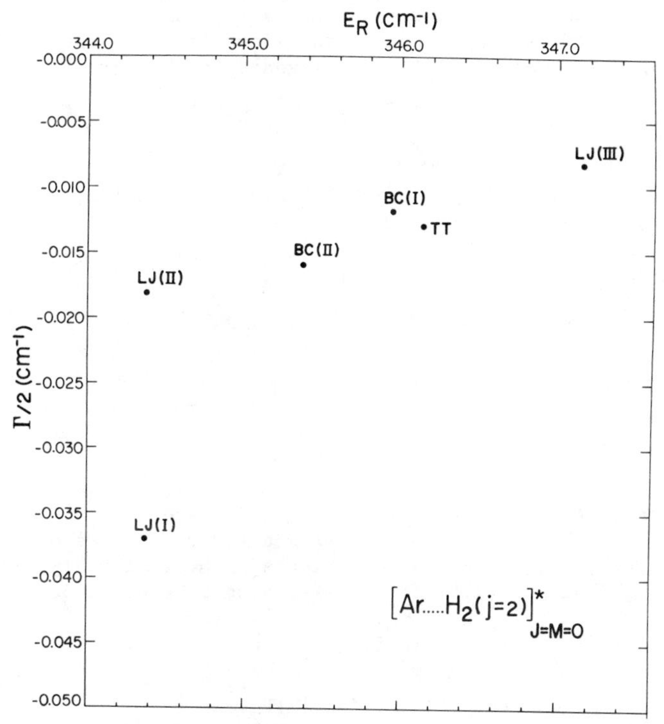

Figure 7. Comparison of the predicted resonance positions associated with the metastable level ($j=\ell=2$, $J=M=0$) of Ar...H_2 vdW molecule for the six potential surfaces considered: LJ(I)-LJ(III), BC(I), BC(II), and TT. Reproduced with permission from Ref. 19. Copyright 1982, American Institute of Physics.

for the (less accurate) LJ potentials vary by a factor as large as four. However, the agreement among the more recent potentials, namely, BC(I), BC(II) and TT, is much closer: the resonance energies (E_R) agree to within 1 cm^{-1} and the line widths (Γ) to within 30%.

<u>SFCCCC Calculations of $Ar-H_2$ Atom-Vibrating Rotor vdW Complex</u>. A more complete description of the rotational predissociation dynamics in vdW molecule should in principle also include the effect of diatom (H_2) bond stretching. Currently only one potential surface, the $BC_3(6,8)$ potential of Carley and LeRoy(9), contains detailed information about its dependence on the length of the diatom bond. The $BC_3(6,8)$ $Ar-H_2$ potential (9) is expanded in the form

$$V(R,\xi,\theta) = \sum_{k=0}^{3} \sum_{\lambda=0,2} \xi^k P_\lambda(\cos\theta) V_{\lambda k}(R) \; , \tag{23}$$

where $\xi = (r-r_o)/r_o$ with $r_o = 0.7666348$ Å. The radial strength function $V_{\lambda k}(R)$ has the form

$$V_{\lambda k}(R) = A^{\lambda k}\exp(-\beta_\lambda R) - D(R)[C_6{}^{\lambda k}/R^6 + C_8{}^{\lambda k}/R^8] \tag{24}$$

where

$$D(R) = \exp[-4(R_o/R-1)^3] \qquad \text{for } R<R_o = R_e^{oo}$$

$$= 1 \qquad \text{for } R\geq R_o . \tag{25}$$

The parameters ($A^{\lambda k}$, β_λ...etc.) are listed in reference 9. In practice, it is the vibrationally averaged (over ξ) forms of this surface $\langle\phi_{vj}(r)|V(R,\xi,\theta)|\phi_{v'j'}(r)\rangle$ which are physically significant. The resulting diagonal vibrationally averaged potentials are

$$\bar{V}(v,j|R,\theta) \equiv \langle\phi_{vj}(r)|V(R,\xi,\theta)|\phi_{vj}(r)\rangle$$

$$= \bar{V}_o(v,j|R) + \bar{V}_2(v,j|R)P_2(\cos\theta). \tag{26}$$

Here the vibrationally averaged anisotropy strength functions $\bar{V}_\lambda(v,j|R)$ are defined by (λ=0,2)

$$\bar{V}_\lambda(v,j|R) = \bar{A}^\lambda\exp(-\beta_\lambda R) - D(R)[\bar{C}_6^\lambda/R^6 + \bar{C}_8^\lambda/R^8] \; , \tag{27}$$

where

$$\bar{Q}^\lambda = \sum_{k=0}^{3} Q^{\lambda k} \langle\phi_{vj}(r)|\xi^k|\phi_{vj}(r)\rangle \tag{28}$$

for Q=A, C_8, and C_6. The parameters $\bar{A}^\lambda$, $\bar{C}_8^\lambda$, and $\bar{C}_6^\lambda$ defining the vibrationally averaged potentials for various types of $Ar-H_2(v,j)$ complexes are given in Table I of reference 32.

The calculations reported in the present section are new results concerned with the internal rotational predissociation of the ground (n=0) vibrational levels of the Ar-H_2(v=1, j=2) complex and were carried out using the $\bar{V}_\lambda$(v=1, j=2|R) anisotropy strength functions in reference 32. For most of the calculations, the coupling to closed channels corresponding to $j>j_{max}=2$ is found to be insignificant and ignored. As an example of the number of coupled channels involved, consider the even parity resonance state associated with (v=1, j=2, ℓ=4, J=4). The SFCCCC calculations in this case will involve one open channel (v=1, j=0, J=ℓ=4), and three closed channels, namely, (v=1, j=2, ℓ=2, J=4), (v=1, j=2, ℓ=4, J=4), and (v=1, j=2, ℓ=6, J=4). Figure 8 shows a typical α-trajectory whose stationary point yields the resonance energy. The converged resonance energies E_R and widths Γ of the n=0 metastable levels of Ar-H_2(v=1,j=2) determined from the SFCCCC calculations are tabulated in table II. Also listed here are the results of close coupling scattering (CCS) calculations reported by LeRoy et al. (32) using the same potential surface and approximations. The excellent agreement of the SFCCCC and CCS data, in most cases within 10^{-3} cm^{-1}, provides a mutual check on the reliability of both theoretical methods.

Table II. Comparison of SFCCCC with CCS resonance energies E_R(cm^{-1}) and widths Γ (cm^{-1}) for the Ar-H_2 (v=1, j=2) resonances. The zero of E_R is defined as the j=2 rotational energy threshold.

ℓ	J	E_R(SFCCCC)	Γ(SFCCCC)	E_R(CCS)	Γ(CCS)
2	0	-20.728	0.111	-20.727	0.110
1	1	-22.654	0.0532	-22.653	0.0530
3	1	-17.244	0.0558	-17.243	0.0551
0	2	-23.238	0.0394	-23.238	0.0394
2	2	-19.311	0.0238	-19.311	0.0235
4	2	-12.828	0.0424	-12.828	0.0408
1	3	-22.238	0.0416	-22.237	0.0414
3	3	-15.943	0.0240	-15.943	0.0237
5	3	-7.490	0.0322	-7.490	0.0319
2	4	-20.078	0.0422	-20.078	0.0420
4	4	-11.581	0.0222	-11.580	0.0219
6	4	-1.304	0.0252	-1.306	0.0244

SFCCCC Calculations of Ar-HD vdW Complex (20). The potential surface for Ar-HD can be obtained from the BC_3(6,8) potential of Ar-H_2 by performing the asymmetric isotope frame transformation (33) to a coordinate system based on the centre-of-mass of HD. This transformation introduces Legendre terms of odd order into the potential expansion, so that the diagonal vibrationally averaged Ar-HD potential, for example, has the form

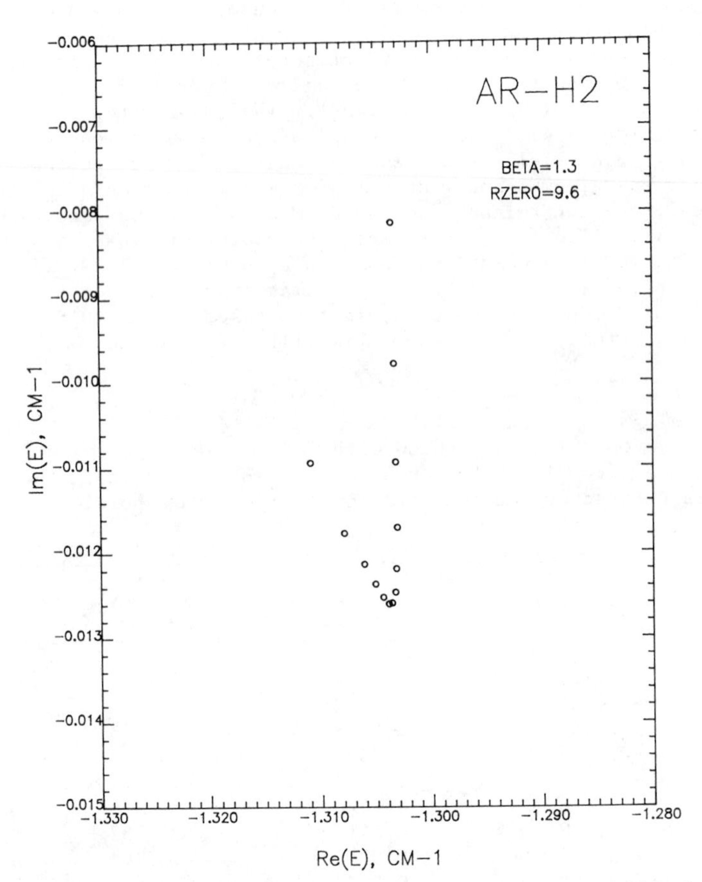

Figure 8. A typical SFCCCC α-trajectory for the Ar...H_2(v=1,j=2) vdW resonance with ℓ=6 and J=4. The potential surface used is the BC_3(6,8) surface of Carley and LeRoy.

$$\bar{V}(v,j|R,\theta) = \sum_{\lambda=0,1,2...}^{\lambda_{max}} \bar{V}_{\lambda}(v,j|R)P_{\lambda}(\cos\theta) \quad . \qquad (29)$$

The vibrationally averaged anisotropy strength radial functions $\bar{V}_{\lambda}(v,j|R)$ for the case of interest here (v=1, j=2) are plotted in Figure 9.

The Ar-HD(v=1, j=2) predissociating states have been observed experimentally (25) in the $S_1(0)$ band of the vdW complex, which corresponds to transitions between states of Ar-HD(v"=0, j"=0) and Ar-HD(v'=1, j'=2). The allowed transitions consist of four branches:

N-branch $J''=\ell''$ $\rightarrow$ $\ell'=\ell''-3$, $J'=J''-1$
P-branch $J''=\ell''$ $\rightarrow$ $\ell'=\ell''-1$, $J'=J''-1$, J'', $J''+1$
R-branch $J''=\ell''$ $\rightarrow$ $\ell'=\ell''+1$, $J'=J''-1$, J'', $J''+1$
T-branch $J''=\ell''$ $\rightarrow$ $\ell'=\ell''+3$, $J'=J''+1$.

Only the N- and T- branch spectral lines are well resolved and their widths can be determined experimentally.

The SFCCCC calculations were carried out for all N- and T-branch predissociating states observed experimentally. The basis set of HD includes all channels corresponding to v=1, j=0, 1, 2, and 3. Channels corresponding to HD(v=0,j) states can be safely ignored, as the vibrational predissociation (VP) rates from v=1 to v=0 states were found to be least three orders of magnitude smaller than the rotational predissociation (RP) rates in the v=1 states. Since the radial potentials $\bar{V}_{\lambda}(R)$ were generated in numerical rather than in analytic form, the radial coupling matrix elements $\langle\chi_{n'}(R)|\bar{V}_{\lambda}(Re^{i\alpha})|\chi_n(R)\rangle$, in harmonic oscillator basis, were computed using the three-step procedure aforementioned. Shown in Figure 10 is a typical SFCCCC α-trajectory whose stationary point is taken as the resonance position sought. The converged resonance energies E_R and widths Γ corresponding to N- and T- branch predissociating states of Ar-HD(v=1,j=2) were tabulated in reference 20. There again we found the SFCCCC widths are in excellent agreement with the recent independent calculations of LeRoy et al. (32, 34) using the CCS method. Shown in Figure 11 is the comparison of the SFCCCC calculated widths with the experimental data recently obtained by McKellar (25). The agreement appears gratifying, considering the sensitivity of the width calculations with respect to potential surface used. The discrepancy between the theoretical and experimental widths as well as the recent hyperfine spectroscopy of $Ar-H_2$ (35) suggest that further refinement of the BC_3 (6,8) anisotropic potential may be possible.

Conclusion

The results described in this paper for vdW molecule predissociation demonstrate that the CCCC formalisms provide accurate and efficient methods for the direct prediction of resonance energies and widths of metastable states. In work published elsewhere (36), the CCCC method has also been extended and applied to the first determination of the energies and widths of the autoionizing resonances of the hydrogen atom in intense magnetic fields. The utility and advantages of the CCCC methods may be summarized as follows: (1) It is an ab initio method (given a defined "exact" hamiltonian). (2) Only

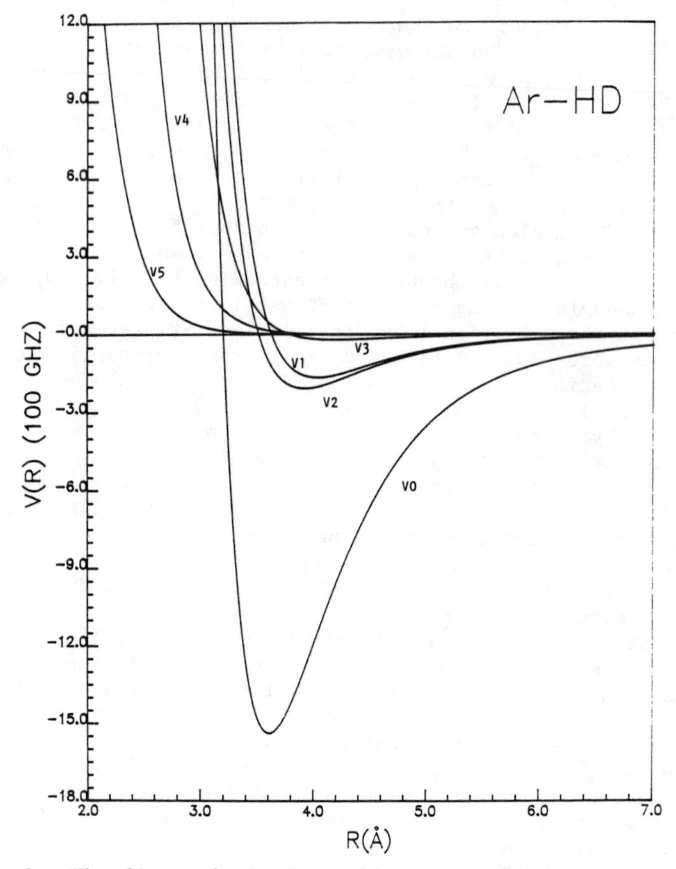

Figure 9. The diagonal vibrationally averaged anisotropy strength radial functions $\bar{V}_{\lambda}(v,j|R)$ for Ar-HD(v=1, j=2). Reproduced with permission from Ref. 20. Copyright 1983, North-Holland Physics Publishing.

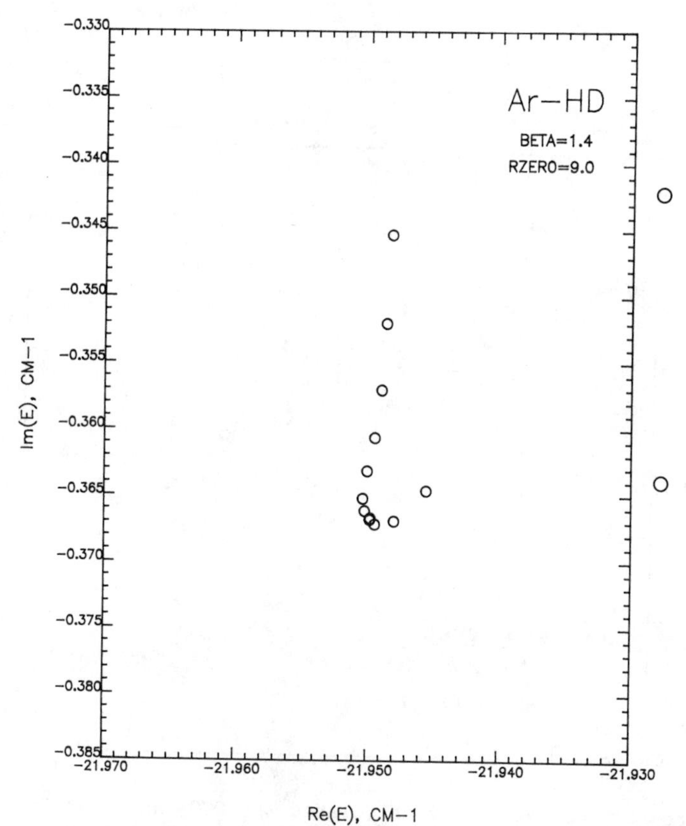

Figure 10. A typical SFCCCC α-trajectory for the complex eigenvalue associated with metastable (J=5, ℓ=3) state of the Ar-HD(v=1, j=2) system. Reproduced with permission from Ref. 20. Copyright 1983, North-Holland Physics Publishing.

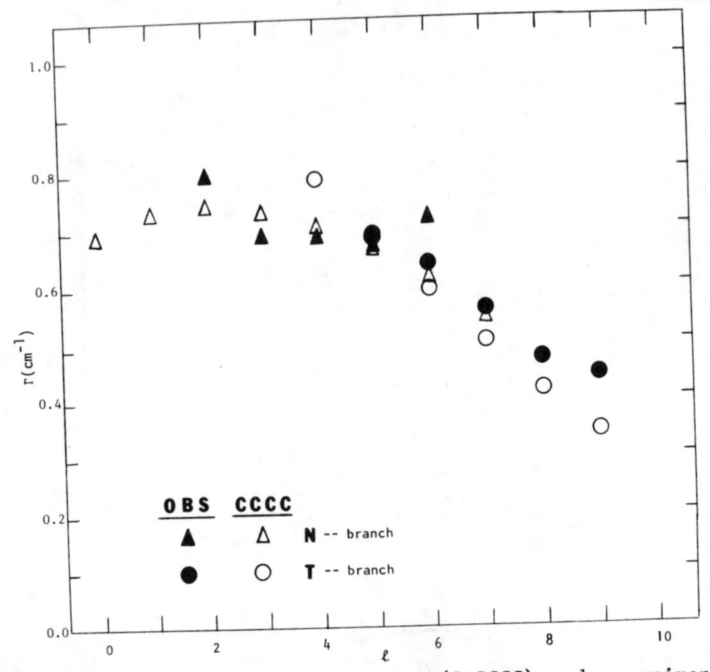

Figure 11. Comparison of calculated (SFCCCC) and experimental widths for rotationally predissociating states of Ar-HD(v=1,j=2). Reproduced with permission from Ref. 20. Copyright 1983, North-Holland Physics Publishing.

bound-state structure calculations are required and no asymptotic conditions need to be enforced. (3) The resonance energies are obtained directly from eigenvalue analysis of appropriate non-hermitian matrices, the imaginary parts of the complex eigenvalues being related directly to the lifetimes of the metastable states. (4) The CCCC method is applicable to many-channel problems involving multiply coupled open continua as well as close channels.

Combining with the non-hermitian Floquet theory (also called the complex quasi-energy formalism) (37, 38), the CCCC method can be extended to the study of photodissociation or multiphoton dissociation of vdW molecules. Work in this direction is in progress.

Acknowledgments

The author acknowledges fruitful collaborations with several of his coworkers, particularly, Dr. K.K. Datta and Dr. S.K. Bhattacharya. This work has been supported in part by the United States Department of Energy (Division of Chemical Sciences) and by the Alfred P. Sloan Foundation. The author is grateful to the United Telecom Computing Group (Kansas City) for generous support of the CRAY computer time.

Literature Cited

1. M.L. Goldberger and K. Watson, "Collision Theory", Wiley: New York, 1964.
2. J. Aguilar and J.M. Combes, Commun. Math. Phys. 22, 269 (1971).
3. E. Balslev and J.M. Combes, Commun, Math. Phys. 22, 280 (1971).
4. B. Simon, Ann. Math. 97, 247 (1973).
5. W.P. Reinhardt, Ann. Rev. Phys. Chem. 33, 223 (1982).
6. B.R. Junker, Adv. At. Mol. Phys. 18, 208 (1982).
7. D.H. Levy, Advan. Chem. Phys. 47, 323 (1981).
8. J.A. Beswick and J. Jortner, Advan. Chem. Phys. 47, 363 (1981).
9. R.J. LeRoy and J.C. Carley, Advan. Chem. Phys. 42, 353 (1980).
10. K.K. Datta and S.I. Chu, Chem. Phys. Lett. 87, 357 (1982).
11. B. Schneider, Chem. Phys. Lett. 31, 237 (1975).
12. D.O. Harris, G.G. Engerholm and W.D. Gwinn, J. Chem. Phys. 43, 1515 (1965).
13. B. Simon, Phys. Lett. 71A, 211 (1979).
14. J.D. Morgan and B. Simon, J. Phys. B14, L167 (1981).
15. C.W. McCurdy, Phys. Rev. A21, 464 (1980).
16. K.K. Datta and S.I. Chu (in preparation).
17. R. Lefebvre, "Siegert Quantization, Complex Rotation and Molecular Resonances", (preprint).
18. S.I. Chu, J. Chem. Phys. 72, 4772 (1980).
19. S.I. Chu and K.K. Datta, J. Chem. Phys. 76, 5307 (1982).
20. K.K. Datta and S.I. Chu, Chem. Phys. Lett. 95, 38 (1983).
21. Z. Bacic and J. Simons, Int. J. Quantum Chem. 14, 467 (1980).
22. I.C. Percival and M.J. Seaton, Proc. Cambridge Phil. Soc. 53, 654 (1957).
23. S.I. Chu, Chem. Phys. Lett. 88, 213 (1982).
24. S.L. Holmgren, M. Weldman and W. Klemperer, J. Chem. Phys. 69, 1661 (1978).

25. A.R.W. McKellar, Faraday Discuss. Chem. Soc. 73, 89 (1982).
26. R.J. LeRoy and J. van Kranondonk, J. Chem. Phys. 61, 4750 (1974)
27. L. Zandee and J. Reuss, Chem. Phys. 26, 345 (1977).
28. R. Helbing, W. Gaide, and H. Pauly, Z. Phys. 208, 215 (1968).
29. R.J. LeRoy, J.S. Carley, and J.E. Grabenstetter, Faraday Discuss. Chem. Soc. 62, 169 (1977).
30. J.S. Carley, Ph D. Thesis, University of Waterloo, Waterloo, Ontario, 1978.
31. K.T. Tang and J.P. Toennis, J. Chem. Phys. 74, 1148 (1981).
32. R.J. LeRoy, G.C. Corey, and J.M. Hutson, Faraday Discuss. Chem. Soc. 73, 339 (1982).
33. S.I. Chu, J. Chem. Phys. 62, 4089 (1975).
34. J.M. Hutson and R.J. LeRoy, J. Chem. Phys. 78, 4040 (1983).
35. M. Waaijer, Ph.D. thesis, Katholieke University, Nijmegen, The Netherlands, 1981.
36. S.K. Bhattacharya and S.I. Chu, J. Phys. B16, L471 (1983).
37. S.I. Chu, J. Chem. Phys. 75, 2215 (1981) and references therein.
38. S.I. Chu, C. Laughlin and K.K. Datta, Chem. Phys. Lett. 98, 476 (1983).

RECEIVED June 11, 1984

Vibrationally Excited States of Polyatomic van der Waals Molecules

Lifetimes and Decay Mechanisms

W. RONALD GENTRY

Department of Chemistry, University of Minnesota, Minneapolis, MN 55455

The existing data on the infrared photodissociation of polyatomic van der Waals molecules are reviewed from a dynamical perspective. Correlations are examined between the lifetimes derived from the homogeneous linewidths for photodissociation and the structural and dynamical parameters of the various molecular systems. Two conclusions emerge: that the linewidths are determined principally by microscopic dynamics rather than statistics, and that the dynamics are dominated by vibrational-rotational coupling instead of vibrational-translational or vibrational-vibrational coupling. On this basis, the hypothesis is offered that the linewidth-derived lifetimes correspond not to the vibrational predissociation rates, but to the rates of vibrational relaxation within the metastable complexes by coupling of the initial vibrational state to those van der Waals modes which might best be described as internal rotations and/or librations.

If we describe a collisional resonance by the (deceptively) simple expression

$$A + B \rightarrow AB^* \qquad (1)$$

where AB* is long-lived compared to the A + B interaction time in a "direct" collision, then it is apparent that information about the resonance state should in principle be obtainable from studies of the microscopic reverse process -- the dissociation of AB*. Since collisional resonances in heavy-particle systems are difficult to study directly, the exploitation of time-reversal symmetry is an especially appealing idea for examples in which A and B are both molecules. The case which we will consider here is the infrared photodissociation of a van der Waals (vdW) molecule:

0097-6156/84/0263-0289$06.00/0

$$A\cdot B \xrightarrow[1]{h\nu} A\cdot B^* \xrightarrow[2]{} A + B \quad , \qquad (2)$$

where A·B and A·B* represent, respectively, the weakly-bound vdW molecule before and after absorption of a single infrared photon. The reverse of step 2 corresponds to a vibrational Feshbach resonance, which has never been observed experimentally for heavy-particle systems. The forward process, on the other hand, can be observed quite easily. One generally needs a supersonic molecular beam in order to achieve low enough temperatures to form A·B without having macroscopic condensation of the sample, and a detector which can distinguish A·B from A, B or larger clusters A_nB_m. The depletion of the A·B signal is then observed when infrared radiation from a laser is allowed to intersect the molecular beam.

The first photodissociation experiments with vdW molecules were actually done in the visible and UV spectral regions by Levy and coworkers (1). These beautiful experiments reveal the vibrational predissociation dynamics within an electronically excited state, and are therefore subject to the possible complication that the dynamics following vertical excitation of the system may be influenced by structural differences between the ground and excited electronic states. Another complication is that internal conversion may compete with fluorescence. However, important advantages of vibronic excitation are (1) that many initial vibrational states are easily accessible, (2) that product states may be identified from the dispersed fluorescence and (3) that the fluorescence lifetime provides an internal "clock" against which the rates of energy redistribution and dissociation may be compared (2). Because of the significant differences between the vibrational and vibronic excitation experiments the present discussion will be limited to the former.

Reports of infrared photodissociation experiments with vdW molecules first appeared in 1978, from the laboratory of Gough, Miller and Scoles (3), and then from ours (4,5). In our own case, the observation of photodissociation was almost accidental. We were trying to excite ethylene with a CO_2 laser in preparation for experiments on collisional transfer of vibrational energy. Unable to detect any ethylene monomer excitation, presumably because the very narrow absorption of the cold ethylene falls between lines of the CO_2 laser spectrum, we set our mass spectrometer detector to one of the dimer masses. Immediately, we saw that the laser caused almost total obliteration of the dimer signal. We then discovered that dissociation occured with every line available from our laser, over the entire 900 cm^{-1} - 1100 cm^{-1} frequency range. What seemed at the time to be nearly an incredibly large linewidth has since proven to be rather typical of polyatomic vdW molecules, as data has emerged from several laboratories on a wide variety of molecular systems (6).

The data on infrared photodissociation of vdW molecules provides, in principle, information on several topics of great current interest in addition to collisional resonances. One of these is the structure of weakly bound systems. Another is the rate of "intramolecular" energy flow from the vibration which is initially localized within A or B into the vdW coordinates which express the location and orientation of A and B with respect to each other. The situation is analogous to that of a large polyatomic in which a localized high-frequency vibration is coupled to a "bath" of low-frequency modes having a high density of states. Still another related topic is that of unimolecular reaction dynamics, in which vdW molecules would appear to be particularly simple model systems.

In this paper I will try to summarize briefly the existing data on the infrared photodissociation of polyatomic vdW molecules, and to discuss the extent to which it is now possible to provide answers to several questions which have been raised. These are:

1) What interpretation should be given to the homogeneous linewidths and their corresponding uncertainty-principle lifetimes?
2) Is unimolecular decay (dissociation) governed by statistics or by microscopic dynamics?
3) Do the photodissociation data provide information which is directly relevant to the understanding of collisional resonances?

I have not attempted to be comprehensive, but have selected those data for discussion which seem to shed light most directly on the above questions. The interested reader is referred to a review by Janda (6) for a more extensive bibliography. Although this discussion will deal only with vdW molecules in which at least one of the constituents is a polyatomic molecule, it is worth pointing out that the answers to the above questions may be quite different for atom-diatom and diatom-diatom systems. Convincing evidence has been provided recently from both experiment (7,8) and theory (9,10) that the dynamics of such "simple" systems can occur on a time scale many orders of magnitude longer than has yet been observed for a polyatomic system.

For most systems, the only data which is available is in the form of photodissociation spectra, i.e., the photodissociation yields as functions of laser frequency. In a few cases, dissociation-product speed and angle distributions have been measured as well. Table I gives the vdW molecule peak frequencies ν_0, shifts $\Delta\nu_0$ relative to the frequencies of the isolated molecules, and lifetimes τ derived from the linewidths, for systems in which the evidence for homogeneous broadening is reasonably convincing (11-20). There seems to be a systematic difference between the apparently homogeneous line widths measured in pulsed laser and cw laser experiments, for reasons which are not understood at present. Although these differences are not important for the discussion which follows, the values of τ are tabulated in separate columns for the two kinds of experiments, to facilitate comparisons. Table II summarizes the product energy disposition data for those cases in which the translational energy distributions have been estimated (12,16,18,21,22). The data in Table II are all from our laboratory, although very similar results for ethylene dimers were obtained by Bomse, Cross and Valentini (22). Listed are the mean product kinetic energies $\langle E \rangle$, the root-mean-square momenta P_{rms} and the lifetimes τ. The reader is cautioned in the use of both tables that error limits and other caveats vary from one system to another

Table I. Summary of frequency and linewidth/lifetime data for the infrared photodissociation of polyatomic vdW molecules.

System	Mode	Type	$\nu_o(cm^{-1})$	$\Delta\nu_o(cm^{-1})$	τ(ps) cw	τ(ps) Pulsed	Ref.
$C_2H_4 \cdot C_2H_4$	9 (b_{2u})	CH_2 asym. str.	3099.8	-5.2	1.0		11
$C_2H_4 \cdot C_2H_4$	1 (a_g)	CH_2 sym. str.	3013.4	-12.6	1.6		11
$C_2H_4 \cdot C_2H_4$	11 (b_{3u})	CH_2 sym. str.	2982.4	-6.6	1.0		11
$C_2H_4 \cdot C_2H_4$	2+12 (b_{3u})		3072.6	-26.4	1.1		11
$C_2H_4 \cdot C_2H_4$	7 (b_{1u})	CH_2 wag	952.8	+3.6	0.5	0.3	12,13,14
$C_2H_4 \cdot C_2D_4$	7 (b_{1u})	CH_2 wag	953.8	+4.6		0.4	12
$C_2H_4 \cdot C_2D_4$	12 (b_{3u})	CH_2 def.	1076.6	-1.3		1.5	12
$C_2D_4 \cdot C_2D_4$	12 (b_{3u})	CH_2 def.	1075.1	-2.8		1.1	12
$C_2H_4 \cdot Ne$	7 (b_{1u})	CH_2 wag	949.1	-0.1	11		6
$C_2H_4 \cdot Ar$	7 (b_{1u})	CH_2 wag	953.9	+4.7	1.8		6
$C_2H_4 \cdot HF$	7 (b_{1u})	CH_2 wag	974.8	+25.6	1.6		15
$C_2H_4 \cdot HCl$	7 (b_{1u})	CH_2 wag	964.3	+15.1	3.3		15
$C_2H_4 \cdot NO$	7 (b_{1u})	CH_2 wag	951.5	+2.3	0.7		15
$C_2H_4 \cdot C_2F_4$	7 (b_{1u})	CH_2 wag	954.7	+5.5	0.9		13

$OCS \cdot Ar$	$2\nu_2$	bend	1046.0	-1.0		5	16
$OCS \cdot OCS$	$2\nu_2$	bend	1044.7	-2.3		1.5	16
$OCS \cdot CD_4$	$2\nu_2$	bend	1045.4	-1.6		0.9	17
$OCS \cdot C_2H_6$	$2\nu_2$	bend	1044.6	-2.4		1.3	17
$OCS \cdot C_4H_{10}$	$2\nu_2$	bend	1043.7	-3.3		1.2	17
$OCS \cdot C_6H_{14}$	$2\nu_2$	bend	1044.0	-3.0		1.2	17
$CH_2OH \cdot CH_3OH$	ν_8	C-O str.	1045.2	+11.7		0.4	18
$SF_6 \cdot SF_6$	ν_3	asym. str.			3.5		19,20

Table II. Summary of product kinetic energy and linewidth/lifetime data.

System	⟨E⟩(cm^{-1})	P_{rms}(10^{-20} g cm/s)	τ(ps)	Ref.
$(C_2H_4)_2$	71	81	0.3-0.5	12
$(OCS)_2$	ε	41	1.5	16
$(CH_3OH)_2$	2	13	0.4	18
$(SiH_4)_2$	110	108	≲1	21

and sometimes involve issues too complicated to discuss in the space available here.

Since the experimental results seem reasonably lucid while the interpretations at this point are less so, it seems best to separate the empirical and the theoretical issues in the discussion which follows. Thus, the following section will deal empirically with the lifetimes τ derived directly from the experimental linewidths, while the discussion of the meaning and implications of these lifetimes is deferred to the final section.

Linewidth/Lifetime Correlations with Structural and Dynamical Parameters

Although there are only 22 cases listed in Table I, the variety is great enough to permit several questions to be answered with regard to the correlation of the linewidth-derived lifetime of the initial state with various structural and dynamical features. First, it is important to observe that the total range of lifetimes for these systems is remarkably small -- about 0.4 to 4 ps for molecule-molecule systems and 1.8 to 11 ps for atom-molecule systems. With this in mind, let us examine several parameters which one might expect to be correlated with the lifetime.

Photon Energy and Vibrational Mode. The extensive data on ethylene dimers from our laboratory (12) and those of Janda (13) and Miller (11) span about a factor of three in photon energy and six different vibrational modes (including the 2 + 12 combination). This is certainly the most thoroughly investigated system to date. If we examine just the $(C_2H_4)_2$ isotopic species, we see that the higher-energy states have lifetimes in the range 1.0 to 1.6 ps, while the lower-energy ν_7 state has a lifetime of 0.3-0.5 ps. Comparing the ν_7 excitation of C_2H_4 in $C_2H_4 \cdot C_2D_4$ with the ν_{12} excitation of C_2D_4, one finds that the lifetime of the ν_{12} state is 4 times that of the ν_7 state, even though its energy is 13% larger. At first glance, this might appear to be a strong inverse correlation of the lifetime with the excitation energy. However, one should also notice that the short-lived ν_7 mode excitation is the in-phase, out-of-plane motion of all four H-atoms, while all of the longer-lived states correspond to in-plane deformations. The appropriate correlations may therefore be with the geometry of the mode rather than with its energy. In any case, there is no evidence to suggest that the lifetime decreases with an increase in the energy deposited in the vdW molecule.

Vibrational Density of States or Number of Modes. We have just noted that the lifetimes for decay of different modes in the same vdW molecule can differ by a factor of four. We now wish to compare that range with the range of lifetimes for different vdW molecules, in which the constituent covalent molecules vary in complexity. It is natural to look for such correlations, because ultimately the photon energy which remains after breaking the vdW bond must be partitioned among the internal and translational degrees of freedom of the separating products. Comparing all of the atom-molecule systems in Table I to all of the molecule-molecule systems, it is evident that the atom-molecule systems tend to be slightly longer-lived. Within the molecule-molecule systems, however, there seem to be no obvious correlations between lifetimes and the vibrational density of states or degrees of freedom either in the molecule which is excited or in the other molecule in the vdW pair. Consider the ν_7 excitations of C_2H_4 bound to HF, HCl and NO. In none of these cases is the lowest excited vibrational state of the diatom accessible with the available energy. The lifetimes vary by almost a factor of 5. Comparing systems having C_2H_4 bound to C_2H_4, C_2D_4 and C_2F_4, the lifetimes go up only slightly in this progression, as the number of vibrational states energetically accessible to the dissociation products goes from 0 in $(C_2H_4)_2$, to 4 in $C_2H_4 \cdot C_2D_4$, to about 24 in $C_2H_4 \cdot C_2F_4$.

An even better example is found in the systems in which OCS excited in the first overtone of the ν_2 bend is bound to hydrocarbons ranging from CD_4 to C_6H_{14}. Although the vibrational density of states and number of degrees of freedom in the hydrocarbon vary enormously in this series, the lifetimes all fall in the range from 0.9 to 1.3 ps.

Bond Energy. Thinking of the vdW molecule dissociation as a unimolecular reaction, and the rate of reaction as equal to the rate at which energy equal to the vdW bond dissociation energy accumulates in the vdW coordinates, one is led to expect the reaction rate to depend on the amount of energy in the vdW molecule in excess of that which is required for dissociation. We have already seen that the state lifetimes in ethylene dimers do not decrease when the excitation energy is raised from about 1000 cm^{-1} to about 3000 cm^{-1}. However, both excitations are much greater than the vdW bond energy of perhaps 200 cm^{-1} so that one might imagine the rate of reaction to be unusually large in both cases.

An extremely different case for comparison is that of $(CH_3OH)_2$. CH_3OH has, of course, the same number of heavy and light atoms as C_2H_4, but the CH_3OH dimer is hydrogen-bonded, with a dissociation energy almost equal to the ν_8 excitation energy supplied by the laser. In fact, the evidence is good that the dimer bond energy is slightly greater than the photon energy in this experiment and that the only dimers which are dissociated are those which retain enough of their heat of formation to reach the dissociation limit by absorption of a single photon (18,23). The excess energy is difficult to estimate quantitatively, although it is certainly very small. Based on the ratio of translational to internal energies which have been determined for other systems (*vide infra*) a reasonable guess for the excess energy might be 20 cm^{-1}, or about a factor of 150 less than that of the ν_9 excitation of $(C_2H_4)_2$. The linewidth-derived lifetimes, however, are about the same for these two systems.

Product Momentum and Kinetic Energy. Finally, we consider correlations between the lifetimes for several systems and the momenta and kinetic energies of the dissociation products. There are only a few cases in which this information is available, and they are listed in Table II. The product kinetic energies are all small compared to the photon energy of about 1000 cm^{-1}. In the case of $(CH_3OH)_2$, this could be due entirely to the large hydrogen-bond energy. In each of the other systems, however, the vdW bond energy must be considerably smaller than the photon energy (even though the vdW bond energies are not known quantitatively). The small product kinetic energy in these cases implies that most of the energy which is available to the products goes into internal rather than translational motion. The estimated fractions going into translation are about 10% for $(C_2H_4)_2$, about 1% for $(OCS)_2$, and about 14% for $(SiH_4)_2$. Only in the case of $(OCS)_2$ are any excited vibrational states energetically accessible in the products. Since OCS is excited in the $2\nu_2$ overtone, the dissociation could possibly correspond to a $\Delta v = -1$ transition which leaves a single quantum of excitation in ν_2, with the balance of the internal energy going into rotational degrees of freedom. In the other three systems, all of the energy not going into translation must go into rotation of the products. Based on this very limited set of data, one might postulate an inverse correlation between the product kinetic energy and the anisotropy of the intermolecular interaction within the cluster. The SiH_4 interaction would appear to be the most nearly spherical, and the OCS interaction the least spherical, considering the location of the center of mass and the dipole moment of OCS, both of which lead to nonzero first-order moments in a Legendre expansion of the potential. Since the partitioning of product energy is presumably determined by the relative strengths of vibrational-rotational and vibrational-translational coupling, such a structural correlation is reasonable.

The lifetimes τ, however, do not appear to be correlated with the product kinetic energy or momentum. Over a range of about a factor of 70 in kinetic energy and about a factor of 8 in momentum, the lifetimes show no systematic variation.

Interpretations and Conjectures

Linewidths and Unimolecular Decay Rates. Now let us consider the questions posed in the Introduction, in the light cast by these data. The first two questions, having to do with the interpretation of the lifetimes τ and the dynamics of unimolecular decay, are closely related. One possible interpretation is that τ is the predissociation lifetime of the vdW molecule. It is important to realize that a direct measurement of the excited vdW molecule population as a function of time following excitation has not been made in any of the cases cited. These are all frequency-domain experiments. A time-domain measurement is certainly possible in principle, and will probably be done in the near future. In the meantime, we must make do with circumstantial evidence.

Most of us are preconditioned by the conventional wisdom developed over the last 30 years or so to think of unimolecular decay rates as being dominated by statistical factors rather than by microscopic features of the dynamics. This is the basis of the RRK and RRKM theories which have been generally successful in interpreting thermal rate data for unimolecular reactions (24). Adams has shown recently that it is possible for an RRKM calculation based on a reasonable parameterization of the system to reproduce the linewidth-derived lifetime for a vdW molecule photodissociation (25). However, if the linewidth-derived lifetimes τ actually are the inverse predissociation rates, then the above data show clearly that the fundamental statistical assumptions of such an approach are qualitatively incorrect for these vdW systems. Consider the following statistical predictions:

1) Within a given system, the rate of reaction will increase with an increase in the total energy. The lifetime of $(C_2H_4)_2$ is no shorter with excitation at 3000 cm^{-1} than with excitation at 1000 cm^{-1}. If one allows comparisons between different systems, the nearly equivalent lifetimes for $(C_2H_4)_2$ and $(CH_3OH)_2$ indicate clearly that the dominant factor in determining τ cannot be the amount of energy in excess of the dissociation limit.

2) At a given energy of excitation, the reaction rate will not depend on the vibrational mode in which the energy is placed initially. The data on ν_7 excitation of C_2H_4 versus ν_{12} excitation of C_2D_4 in the mixed dimer $C_2H_4 \cdot C_2D_4$ refute this prediction convincingly. The lifetimes differ by a factor of 4.

3) At a given total energy, the rate of reaction will decrease with an increase in the number of degrees of freedom in which the energy may be distributed. The OCS·hydrocarbon data show that the vibrational complexity of the hydrocarbon does not influence the lifetime.

Since the lifetimes τ are evidently determined by microscopic dynamics rather than by statistics, one must conclude either that statistical theories do not apply to the dissociation of polyatomic vdW molecules or that the lifetimes τ are not the predissociation lifetimes (or, possibly, that both statements are true). The first choice -- abandoning a statistical description -- should not be too upsetting, for two reasons. First, there have been very few tests of statistical theories on a microscopic level, and it is easy to imagine that even if a statistical model were microscopically incorrect it might be valid (on the average) for macroscopic observations. Second, vdW molecules do represent a special class of systems, in which the dissociation energies are very small, and which contain vibrational frequencies in the vdW coordinates which are unusually low. It is possible that a statistical description not valid for vdW molecule dissociation might still be valid for dissociation of covalent bound species.

Having expressed the appropriate caveats, I will now argue that the preponderance of available circumstantial evidence favors the hypothesis that the lifetimes τ actually correspond to the rates of processes occurring within the vdW molecule which are greater than the rates of dissociation.

First, it must be pointed out that broad homogeneous lines can be observed for systems in which dissociation does not occur. An example is napthalene, in which vibronic linewidths up to 4 cm^{-1} for the jet-cooled molecule were observed by Smalley and coworkers, who attributed the linewidth to very rapid intramolecular relaxation (26,27). Another example is the CH overtone excitation of benzene, studied by Berry and coworkers (22,29), which displays spectroscopic linewidths of order 100 cm^{-1}, corresponding to state lifetimes less than 100 fs (30).

Second, whether or not one prefers a statistical description of dissociation, the very small range of lifetimes is very difficult to accept, given the large variations in bond energy, vibrational frequency, mode geometry, and molecular structure within these systems. The available dynamical models for vibrational predissociation (31), including a curve-crossing model applied by Ewing to $(C_2H_4)_2$ (32), all predict much longer predissociation lifetimes than those corresponding to the observed linewidths and appear to be highly sensitive to structural details.

Third, the isotropic product angular distributions which have been demonstrated by laser polarization-modulation experiments (12,16) seem to suggest lifetimes greater than a rotational period of the complex. In the case of $(C_2H_4)_2$, the rotational period is about two orders of magnitude longer than τ (12).

Fourth, the vibronic excitation experiments of Brumbaugh *et al* (33) reveal fluorescence from Ar·s-tetrazine complexes on the nanosecond time scale of the fluorescence. The vibrational predissociation lifetimes which were deduced for several vibrational modes fell in the range of a few tens of nanoseconds. Although the vibronic excitation experiments differ in many ways from the infrared experiments, it is difficult to imagine why the vibrational predissociation lifetimes for the upper electronic state of this complex should be 10^3 times as long as any of the systems listed in Table I, given the insensitivity of those lifetimes to virtually all structural parameters.

Finally, we also get some information from the product kinetic energy distributions in the center of mass (c.m.) coordinate system. The lowest product kinetic energies which were observed experimentally for $(OCS)_2$ and $(CH_3OH)_2$ were 1.6 cm^{-1} and 1.3 cm^{-1} respectively (16,18). Multiplying the corresponding relative speeds by the respective lifetimes τ, we obtain distances of 0.5Å and 0.2Å by which the OCS and CH_3OH products will separate within the time τ. The calculated distances will be even smaller if one allows for the fact that the molecules are initially at rest in the c.m. system and cannot be accelerated to their final relative speeds instantaneously. At least in these cases, for which the translational motion is so small during the linewidth-derived lifetime, it seems especially unlikely that a direct population measurement will indicate that dissociation has occurred in the time τ.

How then are we to interpret these results? The most important clue, I believe, is in the large amounts of energy going into product rotation in every case yet examined. This suggests strongly that the dominant feature of the decay dynamics is the coupling of vibration in the mode excited initially into rotation of the covalently bound molecules. It is helpful at this point to consider how one may best describe the coordinates of the system at various points in the photodissociation process. Four cases which are useful conceptually are summarized in Table III, which gives the numbers of translational (T), rotational (R) and vibrational (V) coordinates for a system of two nonlinear polyatomics A and B having n_A and n_B atoms, respectively. The rotational degrees of freedom are further subdivided into rotational motions of the A and B substituents $R_{A,B}$, and orbital motions R_{orb}. (Similar tables are easily constructed for systems in which A or B is a linear molecule or an atom.) Two independent molecules each have three translations and three rotations, and the total number of vibrations is the sum of $3n_A$-6 for A and $3n_B$-6 for B. However, for close approaches of the free molecules, it is convenient to use collision coordinates, in which the three translations in the center of mass coordinate system are transformed into the one relative separation (r) and the two angles (θ,ϕ) of a polar coordinate system. For the A·B complex, one limiting case is obtained by treating the complex as a single "rigid" molecule having n_A+n_B atoms. In this limit there are the three trivial translations of the system c.m., three orbital rotations of A and B about the c.m., and $3(n_A+n_B)-6$ vibrations. One may think of the 6 extra vibrational degrees of freedom in A·B which are not present in separated A and B as "vdW" vibrations. However, quite a different description is appropriate in the limit of a "floppy" A·B complex, in which both A and B are free to rotate independently of each other, giving a total of 6 internal rotational degrees of freedom. In this limit there is only one "vdW" vibration, in the separation coordinate r, and two orbital rotational coordinates, θ and ϕ, as in the scattering description. Even if the vdW complex A·B is rigid initially, the internal A and B rotations will rapidly become more free as the separation r increases during dissociation. In going from the rigid limit to the floppy limit the vdW coordinates are transformed from 6 vibrations (or librations) into one vibration and 5 internal rotations. (The sixth internal rotation comes from R_{orb} by uncoupling the rotations of A and B about their separation axis.)

Table III. Ways of partitioning the degrees of freedom in a system consisting of two nonlinear polyatomics A and B.

	A·B Complex		Free A + Free B	
	"Rigid" limit	"Floppy" limit	Collision Coordinates	Independent Coordinates
T*	3	3	4	6
$R_{A,B}$	0	6	6	6
R_{orb}	3	2	2	0
V	$3(n_A+n_B)-6$	$3(n_A+n_B)-11$	$3(n_A+n_B)-12$	$3(n_A+n_B)-12$

*T includes the three translations of the system center of mass, so that the total number of degrees of freedom is $3(n_A+n_B)$.

We now come to the central point of the interpretation offered here. Thinking of the coupling between a covalent mode of A or B and the vdW coordinates, one can see that decay of the initially excited vibrational state can occur by coupling to any of the "bath" states in these vdW coordinates, but only coupling to the translational coordinate r can result in dissociation. There would appear to be no reason to expect the decay of the initial state to be dominated by vibrational coupling to the separation coordinate r rather than to those 5 vdW coordinates which represent internal rotations and/or librations. In fact, the large rotational-to-translational energy ratio in the dissociation products suggests quite the opposite. It could be that the dynamics revealed by the vdW molecule photodissociation spectroscopy are more closely related to line-broadening mechanisms in liquids than to vibrational predissociation. As Flygare has pointed out (34), Lorentzian broadening of vibrational lines in a liquid can be connected to an exponential decay of the rotational dipole autocorrelation function, with a time constant of the same order as the values of τ in Table I. Therefore, even ignoring the possible influence of other vibrational states within the covalent molecules, we clearly must recognize the possibility that it will be necessary to distinguish between at least two intramolecular relaxation rates. Instead of relying on the simplest kinetic scheme for homogeneous line-broadening, in which the ground state is connected by photon absorption to the excited state, which in turn is strongly and irreversibly coupled to the continuum, we may have to insert an intermediate set of states between the initial excited state and the continuum.

A revealing spectral and dynamical analysis of this coupling scheme has been carried out by Stannard and Gelbart (35), who illustrated the results for a small molecule (H_2O) and a large molecule (benzene). A slightly simplified (36) version of their scheme, adapted for the purposes of the present discussion, is given below.

$$|s\rangle \xrightarrow{\Gamma} \{|\ell\rangle\} \xrightarrow{\gamma} \{|q\rangle\} \qquad (3)$$

Initial localized vibration	internal rotations and/or librations	translational quasicontinuum

Here $|s\rangle$ represents the vibrational state excited initially, which is localized in the internal coordinates of one or both covalent molecules. This state is coupled "strongly" to the set of states $\{|\ell\rangle\}$. The $|\ell\rangle$ states are librational in character at low levels of excitation and can become nearly free rotations at higher energies. They are in turn coupled "weakly" to states $\{|q\rangle\}$, which represent the low-frequency motions of the c.m. of one covalent molecule with respect to the c.m. of the other. The spectrum generated by this coupling scheme contains a set of peaks with energy half-widths γ, having spacings ε of the order of the $|\ell\rangle$ spacings, and formed into an overall envelope of halfwidth Γ. The corresponding time evolution of $|\langle s|e^{-(i/\hbar)Ht}|s\rangle|^2$ shows an initial nearly exponential decay with rate constant $\Gamma/\hbar$ and "recurrences" separated by roughly $\varepsilon/\hbar$. The damped recurrences follow an exponential envelope with the rate constant $\gamma/\hbar$. The hypothesis suggested here is that the Lorentzian lineshapes observed in the photodissociation experiments under consideration reflect the rate $\Gamma/\hbar$ instead of the rate $\gamma/\hbar$ which can be identified with dissociation.

As Stannard and Gelbart illustrate for the case of benzene excited in a local C-H stretch overtone, this kinetic scheme can lead to observation in a spectroscopic experiment of only the fast decay Γ, even if the experiment is done on a long time scale. The condition which leads to suppression of information on the slow decay is that $\gamma >> \varepsilon$, which causes the amplitudes of the recurrences to be damped. In our adaptation, this corresponds to decay of the internal rotational-librational states $\{|\ell\rangle\}$ with a rate $\gamma/\hbar$ which is fast enough so that the $|\ell\rangle$ states are overlapped with each other by lifetime broadening. Given the large rotational density of states which one expects at these levels of excitation, it does not seem unreasonable that this criterion might be satisfied. For example, Ewing has estimated conservatively the number of internal angular momentum states which might be involved in the photodissociation of $(C_2H_4)_2$ as 10^3. The actual number may be considerably larger. It is clearly possible, at least in principle, for the condition $\Gamma >> \gamma >> \varepsilon$ to be satisfied under conditions which lead in a spectroscopic experiment to a large Lorentzian linewidth ($\Gamma \sim 10$ cm^{-1}, $\hbar/\Gamma \sim 0.5$ ps) while the predissociation lifetime is much longer (say, $\gamma \sim 10^{-1}$cm^{-1}, $\hbar/\gamma \sim 50$ ps).

The hypothesis outlined above is appealing for several reasons. Perhaps most significantly, it helps to explain the small range of linewidths for these systems. While the number of vibrational degrees of freedom and the vdW bond energies vary over large ranges, the number of internal rotational-librational degrees of freedom is the same for all vdW molecules composed of two nonlinear polyatomics. Given the large product rotational excitations which are apparently ubiquitous, it is much easier to accept vibrational-rotational coupling rates which are uniform to within an order of magnitude than it is to accept vibrational predissociation rates which are nearly independent of the molecular system. Perhaps the longer lifetimes for atom-polyatomic than for polyatomic-polyatomic vdW molecules reflect simply that the number of internal rotational degrees of freedom is 3 in the atom-molecule case instead of 6 (or 2 instead of 4 for linear molecules). Also, this hypothesis helps to explain linewidths which depend on the geometry of the excited vibrational mode more than on its energy, as in the case of $(C_2H_4)_2$. If the staggered-parallel structure which has been postulated is correct for this system, it is easy to imagine how the coupling of the ν_7 out-of-plane vibration into internal rotations will be very rapid, while all in-plane excitations will relax at a uniformly slower rate. Finally, in a kinetic scheme which involves two distinct relaxation rates (or possibly more!) the outcome of a spectral measurement will depend on its time scale. It is possible that the differences in linewidths observed in experiments with pulsed and cw lasers might be attributable to such effects.

Relevance to Collisional Resonances. Unfortunately, both the experimental evidence and the interpretation given above suggest that spectroscopic experiments on the infrared photodissociation of polyatomic vdW molecules will not be of much help in understanding collisional resonances. In the first place, the uniformity of the linewidths for drastically different molecular systems and vibrational modes indicates that there is not much information on either the structures or the microscopic dynamics of these systems contained in the linewidth data, regardless of whether the linewidths are identified with internal relaxation or with predissociation. Second, the large product rotational energies imply, by microscopic reversibility, that the excited states represented by A·B* in Equation 2 can be accessed with significant probability only in collisions of very highly rotationally excited states of A and B at low kinetic energies. Furthermore, the A and B molecules in the vdW system have more or less well-defined initial orientations with respect to each other, which implies a selected rotational phase relationship, at least for short times following photoabsorption. These phases cannot be selected experimentally in a collision experiment, even if the necessary rotational states of A and B are available. Thus, the resonances contained in the photodissociation dynamics are likely to make only small contributions to the scattering amplitudes, and only for a very small subset of the collisions involved in any real scattering experiment. Finally, the idea of relating vdW molecule predissociation to collisional resonances seems attractive conceptually only if the resonance state can be identified with a well-defined state of the vdW system, such as the state which is excited initially by photoabsorption. If, however, the reverse of the predissociation process produces a large number of intermediate states, such as $\{|\ell\rangle\}$ in (3), each potentially having its own separate lifetime, then the description of these resonances will be extremely complicated if it is possible at all.

The only consolations I have to add is that the relationship between infrared photodissociation and collisional resonances will certainly be much simpler for atom-diatom vdW molecules, and that the experimental data for such systems will probably be available soon.

Acknowledgments

This research was supported by the National Science Foundation under grant no. CHE-8205769. The author has enjoyed the collaboration of excellent coworkers in this general area of research, including M.A. Hoffbauer, K. Liu, M.J. Howard, S. Burdenski, R.D. Johnson and C.F. Giese, and he appreciates helpful comments on this manuscript from Dr. Kopin Liu and Prof. Donald G. Truhlar.

Literature Cited

1. Smalley, R.E.; Levy, D.H.; Wharton, L. J. Chem. Phys. 1976, 64, 3266.
2. Levy, D.H. Adv. Chem. Phys. 1981, 47, 323.
3. Gough, T.E.; Miller, R.E.; Scoles, G. J. Chem. Phys. 1978, 69, 1588.
4. Hoffbauer, M.A.; Gentry, W.R.; Giese, C.F. In "Laser-Induced Processes in Molecules"; Kompa, K. and Smith, S.D., Eds; Springer Series in Chemical Physics; Vol. 6, 1978, p. 252.

5. Gentry, W.R. In "Electronic and Atomic Collisions, Invited Papers and Progress Reports"; Oda, N. and Takayanagi, K., Eds.; North Holland, Amsterdam; 1980, p. 807.
6. The latest review is in press: Janda, K. Adv. Chem. Phys. 1984.
7. McKeller, A.R.W. Faraday Disc. Chem. Soc. 1982, 73, 89.
8. Pine, A.S.; Lafferty, W.J. J. Chem. Phys. 1983, 78, 2154.
9. Hutson, J.H.; LeRoy, R.J. J. Chem. Phys. 1983, 78, 4040.
10. Hutson, J.H.; Ashton, C.J.; LeRoy, R.J. J. Phys. Chem. 1983, 87, 2713.
11. Fischer, G.; Miller, R.E.; Watts, R.O. Chem. Phys. 1983, 80, 147.
12. Hoffbauer, M.A.; Liu, K.; Giese, C.F.; Gentry, W.R. J. Chem. Phys. 1983, 78, 5567.
13. Casassa, M.P.; Bomse, D.S.; Janda, K.C. J. Chem. Phys. 1981, 74, 5044.
14. Geraedts, J.; Snels, M.; Stolte, S.; Reuss, J. Chem. Phys. (in press).
15. Casassa, M.P.; Western, C.M.; Celli, F.G.; Brinza, D.E.; Janda, J. Chem. Phys. 1983, 79, 3227.
16. Hoffbauer, M.A.; Liu, K.; Giese, C.F.; Gentry, W.R. J. Phys. Chem. 1983, 87, 2096.
17. Hoffbauer, M.A.; Giese, C.F.; Gentry, W.R. J. Chem. Phys. 1983, 79, 192.
18. Hoffbauer, M.A.; Giese, C.F.; Gentry, W.R. J. Phys. Chem. 1984, 88, 181.
19. Geraedts, J.; Setiadi, S.; Stolte, S.; Reuss, J. Chem. Phys. Lett. 1981, 78, 277.
20. Geraedts, J.; Stolte, S.; Reuss, J. Z Phys. 1982, A304, 167.
21. Johnson, R.D.; Gentry, W.R., in preparation.
22. Bomse, D.S.; Cross, J.B.; Valentini, J.J. J. Chem. Phys. 1983, 78, 7175.
23. A similar case is $(NH_3)_2$. See Howard, M.J.; Burdenski, S.; Giese, C.F.; Gentry, W.R. J. Chem. Phys. (in press).
24. For a recent review and bibliography, see Truhlar, D.G.; Hase, W.L.; Hynes, J.T. J. Phys. Chem. 1983, 87, 2664.
25. Adams, J.E. J. Chem. Phys. 1983, 78, 1275.
26. Powers, D.E.; Hopkins, J.B.; Smalley, R.E. J. Chem. Phys. 1981, 74, 5971.
27. Smalley, R.E. J. Phys. Chem. 1982, 86, 3504.
28. Bray, R.G.; Berry, M.J. J. Chem. Phys. 1979, 71, 4909.
29. Reddy, K.V.; Heller, D.F.; Berry, M.J. J. Chem. Phys. 1982, 76, 2814.
30. Hutchinson, J.S.; Reinhardt, W.P.; Hynes, J.T. J. Chem. Phys. 1983, 79, 4247.
31. For a review, see Beswick, J.A.; Jortner, J. Adv. Chem. Phys. 1981, 47, 363.
32. Ewing, G.E. Chem. Phys. 1981, 63, 411.
33. Brumbaugh, D.V.; Kenny, J.E.; Levy, D.H. J. Chem. Phys. 1983, 78, 3415.
34. Flygare, W.H. "Molecular Structure and Dynamics"; Prentice-Hall, 1978; pp. 437-444.
35. Stannard, P.R.; Gelbart, W.M. J. Phys. Chem. 1981, 85, 3592.
36. Direct coupling of $|s\rangle$ to $\{|q\rangle\}$, which also contributes to γ, has been omitted from the scheme for the purposes of this argument.

RECEIVED June 19, 1984

Photodissociation of van der Waals Molecules
Do Angular Momentum Constraints Determine Decay Rates?

M. P. CASASSA[1], COLIN M. WESTERN[2], and KENNETH C. JANDA

Arthur Amos Noyes Laboratory of Chemical Physics, California Institute of Technology, Pasadena, CA 91125

Experimental data pertinent to the vibrational predissociation mechanism of two types of van der Waals complex are presented and discussed. First, variations in the infrared band shape for excitation of the ethylene out-of-plane wag, ν_7, in the series of molecules $(C_2H_4)_2$, C_2H_4:HF, C_2H_4:Ne are discussed in terms of structure and relaxation mechanisms. Second, rotationally resolved laser excited fluorescence spectra for $NeBr_2$ and $NeCl_2$ are presented. There is a strong dependence of decay rate on molecular structure. Relaxation lifetimes vary from less than 10^{-12} s for C_2H_4 dimer to greater than 10^{-5} s for $NeCl_2$. Trends observed are qualitatively predicted by consideration of linear and angular momentum gap arguments. Other possible interpretations are discussed.

The prototypical van der Waals molecule infrared photodissociation experiment was originally described by Klemperer in 1974 (1). He proposed that infrared active constituents of weakly bonded clusters formed in molecular beams be excited with a laser. Since typical vibrational quanta are larger than typical van der Waals bond energies, subsequent intramolecular vibrational energy redistribution would lead to fragmentation that could be detected as beam loss with a mass spectrometer.

The information from such experiments, of which there are now many examples(2,3), is important to our understanding of both dynamics and structure. First, since the dynamics of the initially excited state are reflected in spectral line widths, factors which control the rate of vibrational energy flow within molecules can be studied by recording spectra as a function of molecular structure. Fast randomization of this energy is a postulate of statistical unimolecular rate theories. On the

[1]Current address: National Bureau of Standards, Gaithersburg, MD 20899
[2]Current address: Department of Chemistry, University of Bristol, U.K.

0097-6156/84/0263-0305$06.00/0

other hand, the potential success of laser selective chemistry depends on the existence of metastable vibrational energy distributions. Structural information about the nature and effects of the weak bonding interaction are obtained from spectral shifts, intensities and, in favorable cases, rotational structure in the photodissociation spectra. Rotational structure, vibrational frequencies and band intensities are all influenced by the intramolecular potential.

Experimental study of the vibrational predissociation of van der Waals molecules was pioneered by Levy and coworkers in studies of complexes containing the I_2 molecule(4). Detection of infrared photodissociation was first reported in 1978 by Gough *et al.* in a study of N_2O clusters (5). Shortly afterward Hoffbauer *et al.* reported infrared photodissociation of ethylene clusters using mass spectrometer detection (6). The ethylene example was quickly studied by a variety of investigators because its homogeneously broadened line shape is wide enough to be accurately recorded with a line tuned CO_2 laser(7-10). Relaxation lifetimes obtained from infrared linewidths have been found to range from less than 10^{-12} s for $(C_2H_4)_2$ to longer than 10^{-7} s for $(HF)_2$ (11). Even in simple triatomic molecules a wide range of relaxation rates has been observed. HeI_2 (B state, $v > 10$) relaxes in 10 to 300 ps (4) while $NeCl_2$ (X state, $v = 1,2$) does not relax until after 10^{-5} s (12).

Theoretical models for understanding these phenomena have been developed by several groups. In particular, Ewing has attempted to determine propensity rules for relaxation rates (13). At first the rules concentrated on product momenta. Recently it has become clear that the ability of the metastable complex to excite product angular momentum channels can also be important. There is a growing body of information, not discussed in this paper, on the propensity for intramolecular V-V transfer to govern relaxation rates. The goal of these studies, then, is to isolate as much as possible the effects of the intramolecular potential, vibrational symmetry and product channel availability on the overall rates. Naturally, all of these effects are correlated with each other. By judicious choice of examples, however, the relative importance of individual effects can be demonstrated.

In our laboratory we have measured infrared photodissociation spectra of a series of bimolecular complexes containing ethylene. The aim has been to vary the interaction potential and the nature of the decay channels to observe how these influence the photodissociation spectra. Molecules studied include complexes of ethylene with rare gases, hydrogen halides, and nonhydrogen-bonding polyatomic molecules. In each case, vibrational predissociation was observed following excitation of the C_2H_4 ν_7 mode, an in-phase out-of-plane wag. These three types of complexes have dramatically different relaxation rates which we interpret as being due to angular momentum constraints. We also report laser excited fluorescence experiments on $NeBr_2$ and $NeCl_2$ which indicate that linear momentum constraints dominate the order of magnitude of the decay rate for these molecules.

The ethylene cluster experiments and interpretation are discussed below as follows. Section II summarizes important aspects of the experiment. Section III gives details on the spectral analysis with respect to inhomogeneous versus homogenous broadening as well as saturation effects. Section IV discusses structural information obtained from the spectra. Section V discusses the vibrational predissociation mechanism for these complexes and compares the results to other systems.

Experimental Results

The apparatus used to obtain infrared photodissociation spectra consists of four vacuum chambers to provide sufficient differential pumping to isolate a mass spectrometer from a continuous, high-pressure supersonic molecular beam source. By directing the CO_2 laser radiation collinear to the molecular beam, each molecule is irradiated on the order of 10^{-3} s. The long irradiation time allows infrared transitions to be saturated at modest laser powers. It is then relatively easy to detect changes in molecular beam intensity as a function of laser frequency. The geometry also provides a uniform distribution of laser intensity over the molecular beam. This allows us to calculate absolute transition moments with a high degree of precision.

Two experimental parameters are critical to properly interpret the experiments. First, the molecular formula of the cluster producing a spectrum must be known. Second, the laser intensity along the molecular beam must be determined. Determination of the cluster formula is aided by mass selective detection. For polyatomic van der Waals molecules, however, fragmentation of the cluster upon ionization is often extensive. For electron impact (7) or photoionization (14) of the dimer of ethylene, for instance, the parent ion is a minor product. Instead $C_3H_5^+$ (loss of CH_3) and $C_4H_7^+$ (loss of H) dominate the mass spectrum. For larger clusters the problem is even more severe. The practical problems caused by fragmentation are illustrated in Figure 1. Each of the illustrated spectra were determined by monitoring the disappearance of the $HF:C_2H_4^+$ mass peak as a function of CO_2 laser frequency. The mole ratio of expansion gases and total pressures are given to the right of the spectra. Although the top spectrum is obtained with a dilute mixture of only 0.1% HF and 0.2% C_2H_4 it is dominated by the photodissociation of large clusters from the molecular beam. Only when the expansion mixture is diluted to 0.02% HF and 0.1% C_2H_4 is the spectrum of the pure bimolecular complex $HF:C_2H_4$ observed. The case of $HF:C_2H_4$ is actually a favorable one because complexation induces a large frequency shift which clearly indicates spectral purity. In many cases no such shift occurs and determination of spectral purity is suspect.

To ensure that the absolute laser intensity coincident with the molecular beam is measured, a pyroelectric detector was installed in the beam flight chamber on a translation stage. Since the laser beam is apertured by the same collimators as the molecular beam, and since the spatial intensity profile of both beams is uniform in the flight chamber, the power reading gives an accurate measure of the laser intensity actually experienced by the van der Waals molecules.

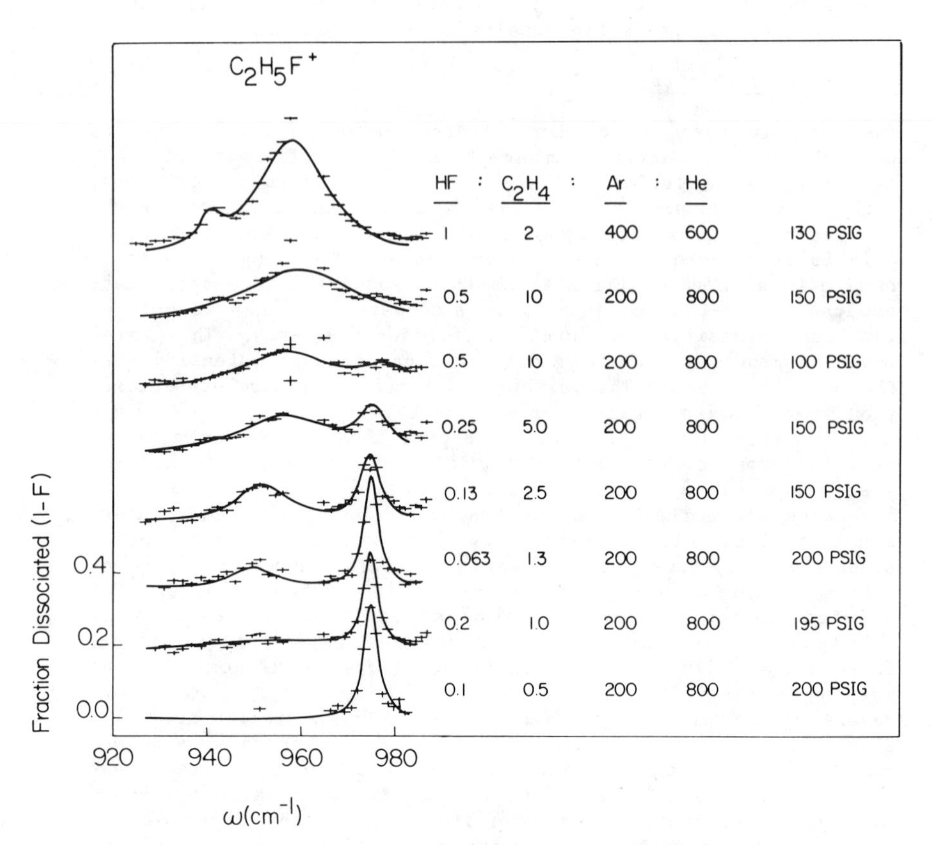

Figure 1. Photodissociation spectra of $(C_2H_4)_m(HF)_n$ clusters observed at m/e = 48 ($C_2H_5F^+$) for a variety of expansion conditions. Reproduced with permission from Ref. 16. Copyright 1983, American Institute of Physics.

To obtain spectra of very dilute molecular beams with low laser intensities requires extensive signal averaging. Using ion counting techniques the molecular beam intensity is measured with the laser on and off over many thousands of cycles of a mechanical chopper to obtain the absolute fraction of molecular loss. Signal to noise is limited only by the counting statistics.

Figure 2 shows three photodissociation spectra obtained by the above techniques. The spectra represent the range of results in studying a variety of ethylene complexes. The points are the measured data. Solid lines are computer-generated bands obtained from models discussed below. Other ethylene clusters studied are listed in Table I. Each of these spectra were excited near the 949 cm^{-1} ν_7 frequency of monomeric ethylene. Ethylene has no other infrared active modes within the range of the CO_2 laser. The spectrum observed for $(C_2H_4)_2$ is typical of C_2H_4 bound to nonhydrogen bonding polyatomics: It is quite broad and its center is slightly blueshifted from the monomer ν_7 fundamental. Spectra of C_2H_4 bound to HF and HCl are strongly blue shifted and significantly narrower, while still exhibiting a single band. Spectra of C_2H_4 bound to Ar and Ne are structured, with especially sharp features in the case of $Ne:C_2H_4$.

Results of the band profile analysis described below support the assertion that the vibrational motion in each cluster closely resembles that of the unperturbed ν_7 mode of ethylene. This might be surprising since the spectra in Figure 2 differ dramatically in appearance. The differences are all attributable to different types of bonding interaction and different decay rates.

Table I. Ethylene Cluster Photodissociation Parameters

Cluster	$\omega_o(cm^{-1})$	$\gamma(cm^{-1})$	τ(ps)	$<\mu^2>(10^{-3}D^2)$	$V_2(cm^{-1})$	Ref.
$Ne:C_2H_4$	949.1	≤0.5	≥10.0	-	12.5	15
$Ar:C_2H_4$	953.9	≤3.0	≥ 1.7	-	30.0	15
$C_2H_4:HF$	974.4	1.6	3.3	77.	-	17
$C_2H_4:HCl$	964.1	1.6	3.3	52.	-	17
$C_2H_4:NO$	951.5	7.4	0.7	51.	-	7
$C_2H_4:C_2F_4$	954.7	6.0	0.9	46.	-	7
$(C_2H_4)_2$	952.3	12.0	0.4	96.	-	7
C_2H_4	949.0	-	-	35.5	-	18

Band Profiles

The band profile analysis we have used includes inhomogeneous broadening, lifetime broadening and saturation effects. Its application permits extraction of both dynamical and structural information. Some of the observed spectra exhibit inhomogenous structure while others appear to be homogeneously broadened. In particular, the intensity and symmetry of the $(C_2H_4)_2$ profile strongly suggest homogeneous broadening. Homogeneous widths determined by band profile analysis are strictly phenomenological. The widths give an unambiguous measure of relaxation rates

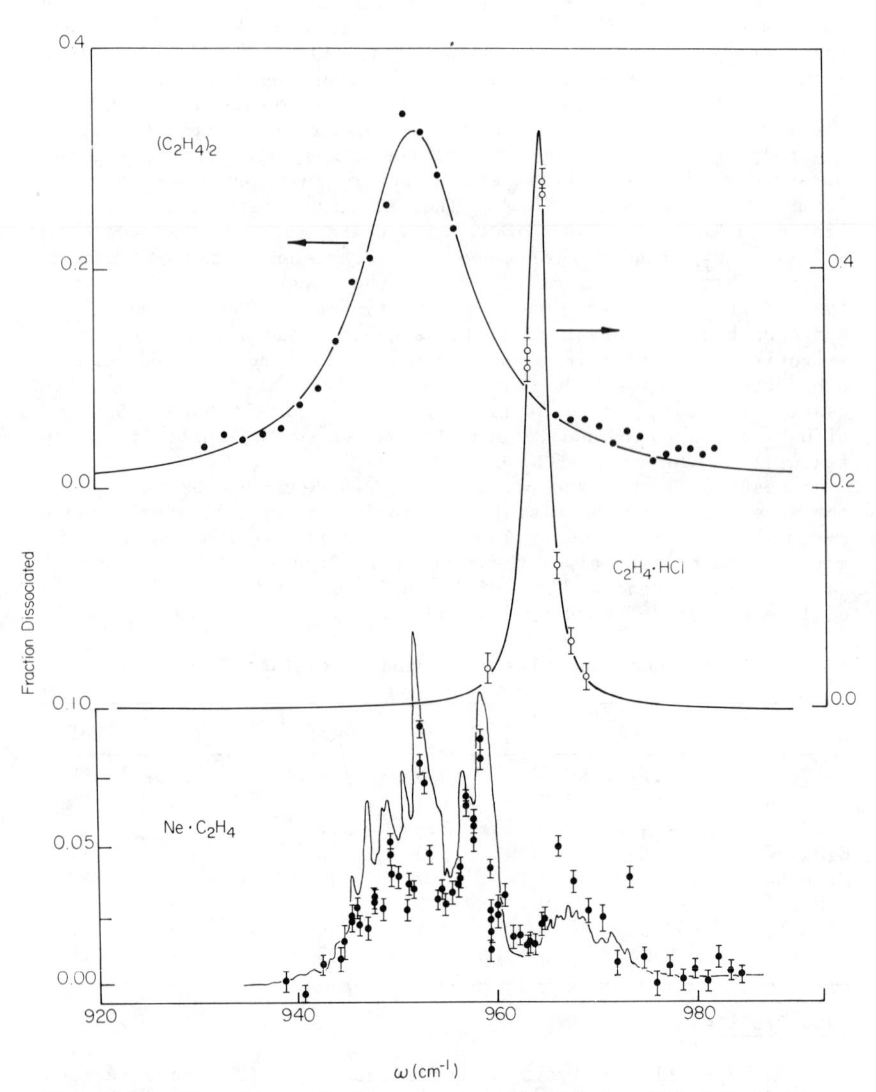

Figure 2. Photodissociation spectra of $(C_2H_4)_2$, $HCl:C_2H_4$ and $Ne:C_2H_4$. The points represent data and the solid lines are theoretical curves generated using parameters from Table I and the models discussed in the text. Note that in each case absolute, not relative, intensities are compared.

of the initially excited motions, but their relationship to actual dissociation mechanisms must be inferred. A discussion of linebrodening and dissociation mechanisms is given in Section V.

The photodissociation lineshape of a multilevel system in which an upper group of levels is dissociative is given by

$$F(t)=\sum_{\ell} f_{\ell}(0)\exp[-\Sigma\Omega^2_{u\ell}\gamma_u t(4(\omega_u - \omega_{\ell} - \omega)^2 + \gamma_u)^{-1}] \quad (1)$$

F(t) is the fraction of all of the molecules which remain intact after irradiation for time t at light frequency ω. The ℓ subscripts label lower bound levels with energy ω_{ℓ} and the u subscripts label metastable upper levels with energy ω_u. $f_{\ell}(0)$ is the Boltzmann population number for level ℓ at t = 0. Upper levels decay at rates given by γ_u. Transition moments and laser intensity are included in the Rabi frequencies $\Omega_{u\ell}$. Equation (1) is valid if $\gamma_u > \Omega_{u\ell}$ and if γ_u leads ultimately to dissociation. It is important to note that the γ_u need not be the actual predissociation rates (17).

In Eq.(1), the distribution of initial states and the distribution of transition moments from each initial state to the group of dissociative states give rise to inhomogeneous broadening. Levels included in the sum may be as completely specified as desired. For our analyses, the sums have explicitly included terms for individual J, M rotational sublevels. For $Ne:C_2H_4$ and $Ar:C_2H_4$ the sums were expanded to include internal rotation levels. Homogeneous broadening for an individual transition is entailed in the γ_u, which are the widths of the Lorenzian terms in Eq.(1). These can dominate the inhomogeneous effects to give rise to spectra which are homogeneous in an empirical sense. Saturation effects arise in Eq.(1) if the exponential arguments are large ($\geq$ 1). Saturation in this context means that significant population is removed from lower sublevels. Since the $\Omega_{u\ell}$ include a range of transition moments, some transitions will saturate more quickly than others. The result is that different transitions can exhibit different widths even if the γ_u are equal. This type of saturation effect, which is really a sort of laser-induced inhomogeneity, can occur at moderate laser fluences (>10 mJ/cm^2).

The most important parameters determined for C_2H_4 clusters are listed in Table 1. These are the natural widths, band origins, corresponding lifetimes, and vibrational transition moments. For $Ar:C_2H_4$ and $Ne:C_2H_4$ we have also determined barriers to internal rotation.

van der Waals Interactions

In analyzing the photodissociation spectra, line frequencies and transition moments based on an assumed molecular structure, as well as decay rates, are convoluted through Eq.(1). The parameters are adjusted to give the best agreement with observations. Band positions and infrared intensities also reflect the van der Waals interactions. Summarized in this section are the important conclusions regarding the structures and interactions in the van der Waals molecules we have studied.

C_2H_4 Dimer. The $(C_2H_4)_2$ spectrum appears to be a single band and is the broadest spectrum observed. Its intensity and symmetry suggest that it is dominated by homogeneous broadening. Inhomogeneous widths expected for any reasonable structure at low temperature are small compared to the observed width. The band center is slightly blue shifted from the monomer ν_7 band origin and the intensity is larger than would be expected for two non-interacting ethylene monomers. Since a single band is observed, $(C_2H_4)_2$ may be best thought of as a rigid molecule. This is based on comparison with the hindered internal rotors Ne:C_2H_4 and Ar;C_2H_4 described below.

The breadth of the $(C_2H_4)_2$ spectrum makes it insensitive to some important aspects of the dimer geometry. In particular, the separation of the centers of mass of the ethylene planes cannot be determined. Data from which structural information can be gleaned are the blue shift and the laser power dependence of the band profile. Based on measurements performed in our laboratory, at moderately low fluence (4mJ/cm^2), and similar measurements reported by Gentry and coworkers (8) at higher fluence (40mJ/cm^2), we can say that the skew angle which the dimer axis makes with the vibrational transition moment is between 25° and 90° (17). This means that the transition occurs as either a hybrid or perpendicular band. This conclusion is based on fits of Eq.(1) to the available data with transition moments and selection rules determined by the band type. The fits are sensitive to band type because inhomogeneous saturation effects, resulting from a distribution of transition moments in Eq.(1), influence the profile at higher fluence.

The dipole-dipole interaction between the transition moments of the ethylene subunits can lead to either a red or a blue frequency shift, depending on their relative orientation (19). If the planes of the subunits are parallel, which is reasonable based on calculations (20) and on the observed crystal structure (21), then the observed blue shift corresponds to a skew angle of 63° for a center of mass separation of 4.1 Å. This is in good agreement with the discussion in the preceeding paragraph. Other weak interactions are induction and dispersion forces, which would lead to a red shift, and repulsive valence forces which would cause a blue shift. Using reasonable values, these forces compensate and lead to slight shifts. The dipole-dipole interaction is expected to be chiefly responsible for the observed blue shift in $(C_2H_4)_2$.

Induction forces, on the other hand, may be responsible for the enchanced ν_7 intensity. The squared transition moment expected for the dimer with two independent ethylene partners is twice that of the monomer. Inspection of Table 1 shows that the observed probability is somewhat higher than this. Vibration in each molecule induces oscillations in the polarizable charge distribution of the partner, leading to an enhanced dipole moment derivative. Numerical results for C_2H_4 bound to a polarizable atom show that this enhancement is of the same magnitude as that observed (17).

C_2H_4:HCl and C_2H_4:HF. The geometrical structures of C_2H_4:HCl and IR spectra of these clusters have also been observed in rare gas matrices (24). The gas phase ν_7 spectra are single bands which are significantly narrower and more blue shifted than that of $(C_2H_4)_2$. Their intensities are enhanced compared to C_2H_4.

Structures determined from microwave spectra place the diatomic symmetrically above the ethylene plane with the hydrogen directed towards the ethylene π-electron cloud (22,23). The fact that ν_7 occurs as single bands indicates that these structures are rigid. Otherwise it would be possible to excite internal rotations as in the cases of Ne:C_2H_4 and Ar:C_2H_4. The ν_7 transition then must occur as a simple parallel band. Parameters which can be adjusted in Eq.(1) to best reproduce observations are the temperature, natural width, and vibrational transition moment. Like $(C_2H_4)_2$, the C_2H_4:HCl spectrum is dominated by homogeneous broadening. On the other hand, the homogeneous and inhomogeneous widths of C_2H_4:HF are comparable because of a relatively large end-over-end rotational constant.

The large blue shifts are attributable to an electrostatic interaction between partners rather than stiffening of the C-H bonds upon formation of the weak bond. This is supported by the observation that only out-of-plane modes of C_2H_4 exhibit shifts in the infrared spectra of matrix isolated clusters (24). Simple numerical calculations, with the molecules approximated as collections of point charges, give blue shifts of the same magnitude as those observed (16). Similarly, the intensity enhancements are largely attributable to interaction of the ν_7 oscillation and the polarizable but otherwise passive, van der Waals partner. Note that the oscillating induced dipole moments listed in Table 1 are quite uniform, with the exception of C_2H_4:HF, with an average squared transition moment per ethylene subunit equal to 48.3 x $10^{-3} D^2$. The larger enhancement for C_2H_4:HF may reflect a change in the charge distribution in the C-H bonds upon cluster formation, although not so large as to effect the frequency shifts.

Ne:C_2H_4 and Ar:C_2H_4. Spectra of Ne:C_2H_4 and Ar:C_2H_4 have the most complex structure observed in any ethylene cluster. These clusters are evidently quite loose compared to other clusters in Table 1. They are the only clusters wherein it is possible to excite internal rotations in combination with ν_7. By analogy with Ar:C_2H_4 (25), the two ethylene carbons and the Ne or Ar atoms are expected to form a "T". In this geometry one can envision that the internal rotation of ethylene about the C=C axis is less sterically hindered than other internal motions. If rotation about this axis is nearly free, this motion can be excited in combination with ν_7 with large intensity. This is analogous to rotational subbands superimposed on a vibrational band. Indeed the spectra of Ne:C_2H_4 and Ar:C_2H_4 resemble the ν_7 perpendicular band of ethylene monomer at low temperatures.

A model spectrum which uses the above features is compared to the data in Figure 2. The model includes only the internal rotation about the C=C axis and end-over-end rotation of the entire complex. Other van der Waals modes--stretching and end-over-end libration of C_2H_4 -- would be best treated as simple vibrations. Energy levels and rotational transition moments were calculated by diagonalizing an assymmetric rotor Hamiltonian with a hindered internal rotation. The double-welled hindering potential was of the form:

$$V(\theta) = \frac{1}{2} V_2(1-\cos 2\theta) \quad (2)$$

where θ is the angle the van der Waals bond axis makes with a line normal to the ethylene plane. Basis functions were products of symmetric top wavefunctions for the end-over-end motion of the entire complex and the internal rotor. The latter were simple sine and cosine functions. The overall vibrational transition moment was assumed to be that of ethylene monomer. Energy levels and transition moments were convoluted with Eq.(1) to generate the $Ne:C_2H_4$ spectrum shown in Figure 2. The appearance of the observed and calculated spectra are in good agreement. Note that absolute intensities are compared.

Barriers determined for hindered internal rotation of C_2H_4 about the C=C axis in $Ne:C_2H_4$ and $Ar:C_2H_4$ are 12.5 and 30.0 cm^{-1}, respectively. Assuming that the molecule is "T" shaped, the best agreement between calculated and observed spectra is obtained if the potential minima are in the ethylene plane. However, since calculated spectra are only weakly sensitive to the equilibrium position of the rare gas, the actual equilibrium geometry has not been determined. Similarly, the calculated spectra were insensitive to the van der Waals bond lengths. These were assumed to be those calculated using pair-wise Lennard-Jones potentials. The slight blue shift of the ν_7 band origin of the rare-gas ethylene clusters from that of ethylene is attributed to a repulsive interaction of the wagging hydrogens and the rare gas atom.

The low barriers to internal rotation in $Ne:C_2H_4$ and $Ar:C_2H_4$ give rise to the internal rotation subbands with splittings of the order of 10 cm^{-1}. No corresponding bands have been observed for the other van der Waals molecules listed in Table 1. Excitation of van der Waals librations in non-rare gas clusters in combination with ν_7 would be well shifted from the ν_7 origin. Intensities would be expected to be low, corresponding to vibrational combination bands. Note that the terms "loose" and "rigid" have been implicitly defined to reflect our ability to excite internal rotation at low temperatures (~5°K). It is conceivable that at higher temperatures these classifications would change.

Photodissociation Dynamics

The Ethylene Complexes. Natural linewidths in van der Waals molecule spectra correspond to decay rates of the initially excited state. The range of lifetimes for the ethylene clusters is from 0.44 ps for $(C_2H_4)_2$ to > 10.0 ps for $Ne:C_2H_4$ Presumably because the ethylene vibrational motion is a large amplitude, out-of-plane bending motion these lifetimes represent the low end of the range observed for van der Waals molecules to date. At the high end of the range are HF dimer (11) and $NeCl_2$ (12) which are excited in stretching motions and have lifetimes greater than

10^{-7} s and 10^{-5} s, respectively. Can this range of results be explained in terms of the dissociation mechanism?

Conceivable broadening mechanisms are direct predissociation, evolution of other metastable states which ultimately dissociate, or pure dephasing. First we present a qualitative discussion of a direct predissociative mechanism which is consonant with observed trends. If the broadening is due to direct decay into continuum states, then trends observed should resemble expectations based on golden-rule or similar expressions. As such, decay rates depend on the efficiency of various channels and the number of such channels. The former give rise to propensity rules which have been developed by Ewing (13).

The principle propensity rule is that channels which minimize translational momenta of the fragments are most efficient. The poor efficiency of V-T mechanisms has been observed in $(C_2H_4)_2$ photodissociation experiments where product velocities have been measured (8,9). V-V mechanisms are more efficient, but in the case of most of the clusters listed in Table 1, the V-V channels are energetically inaccessible. In all cases, the V-T,R channels are open. Of these, the most efficient will be channels which minimize both the linear and angular momentum of the fragments. One can visualize, and Ewing has demonstrated (26), that anisotropic binding potentials enhance the coupling of vibration to rotation. The strength of the interaction would also depend upon the geometry of the excited complex. These rules apply to individual channels. If there are many open channels, the overall rate is faster.

In the case of $(C_2H_4)_2$ the van der Waals bond energy, D_0, is approximately 350 cm^{-1} (20). Excess energy, ΔE, that must be incorporated into fragment motion following ν_7 photodissociation is approximately 600 cm^{-1}. There are no open V-V channels. The staggered, planes-parallel structure would seem optimal for coupling ν_7 motion to fragment rotation. There are many nearly resonant V-R channels. Two examples which give the range of rotational quantum number (J,K) values energetically accessible are 1) both fragments with (J,K) = (17,0); and 2) the fragment state (8,8) + (8,7). Alternatively, anisotropy in the potential enhances rates for non-resonant V-R channels. For example, Ewing has calculated a rate of 10^{10} s^{-1} for the 2(4,4) channel, which incorporates only 160 cm^{-1} into rotation. It is encouraging to note that thousands of such channels are open, leading to subpicosecond lifetimes.

C_2H_4:HCl has no open V-V channels since $D_0 \approx 500$ cm^{-1} and ΔE 460 cm^{-1}. The "T" shaped geometry may be less favorable for V-R coupling than the $(C_2H_4)_2$ geometry, but the potential must be strongly anisotropic. There are some nearly resonant V-R channels. An example gives the fragments: C_2H_4(J=5, K=5) + HCl(J=5) which incorporates 440 cm^{-1} into rotation. Overall, fewer dissociative channels are accessible and most have larger kinetic energy release than for $(C_2H_4)_2$ dissociation.

Ne:C_2H_4 is a loose complex compared to those above. This suggests that the interaction potential is more isotropic and coupling of vibration to rotation is weaker. Since the bond is very weak ($D_0 \approx 70$ cm^{-1} based on pairwise Lennard-Jones potentials), the V-R,T mechanism requires large kinetic energy release in the fragments. A V-V,R,T channel is accessible since the ν_{10} torsional mode of C_2H_4 occurs near 826 cm^{-1}. In either case, ethylene rotational angular momentum can only be compensated by orbital angular momentum. Since the interaction is weak and there are few efficient open channels, Ne:C_2H_4 could be expected to decay more slowly than either $(C_2H_4)_2$ or C_2H_4:HCl.

Thus, the observed trend in lifetimes -- $(C_2H_4)_2 \ll C_2H_4$:HCl $\ll$Ne:C_2H_4 -- can be explained in terms of the direct predissociation model. In this picture the rates are controlled by a complex interplay of important effects: the potential which couples the energy out of the initially excited mode, the amount of linear and angular momentum which must be incorporated into fragment motion, and the overall number of product channels available. This inferred mechanism is subject to further tests. These include measurements of product velocity and rotational distributions. Real time measurements of either the population in the initially pumped level or the appearance of product fragments would be of great significance.

Elsewhere in this volume Gentry discusses arguments in favor of an indirect photodissociation mechanism for these and other clusters. In such a mechanism the excitation energy would decay first to the van der Waals vibrational modes followed by a slower dissociation step. The direct and indirect mechanisms are very similar in that the energy acceptor modes for the two mechanisms are the same for both pictures. The point of disagreement (if there is one) is whether the intermediate state is more naturally thought of as those of the cluster or those of the products. Our intuition is that such states are so short lived that their identification as states is of limited use in describing the dynamics of the molecule.

The indirect mechanism hypothesis is based largely on the observation that linewidths for a variety of systems are within an order of magnitude of each other. The data in Figure 2 shows that there is, however, a clear dependence of linewidth on cluster structure. In any case, we are not convinced that an indirect photodissociation mechanism would lead to a uniform set of linewidths any more than would a direct mechanism. Clearly, more work will be required to understand which of the two pictures is more accurate.

Two mechanisms which can be discounted at this time are pure dephasing and resonant energy transfer to other covalent modes of the molecule. In addition to coupling strengths both of these mechanisms depend on the "bath" density of states of modes not directly involved as dissociation channels. The dependence of pure dephasing rates on occupation numbers is such that the low temperatures used in these experiments rule out such a relaxation mechanism. The mechanism whereby energy is redistributed among bound modes is not viable for the ethylene complexes based on the dependence of rate on molecular structure (15).

Molecules on Excited Electronic Surfaces. Data similar to that discussed here has been obtained for a variety of molecules on excited electronic state surfaces. Examples include rare gases bound to glyoxal (27), benzene (28), tetrazene (29) and stilbene

(30). The examples indicate that mixing of vibrational levels in the photodissociation process is highly selective and that the amount of energy dumped into the rotational and translational modes is minimal.

By constrast the careful analysis of the data of Levy et al. on HeI_2 vibrational predissociation indicates that the mechanism is a nearly pure V-T process (2,31). This can be rationalized on the basis of a small energy gap and poor kinematics for excitation of rotation.

To examine the applicability of simple momentum gap arguments to these simple triatomic systems we have recently examined the free-jet laser excited fluorescence spectra of $NeCl_2$ (11,32) and $NeBr_2$. A sample spectrum for $NeBr_2$ is shown in Figure 3. It is seen that both states involved in the transition (Br_2 B-X, v=10-0) are long lived enough to yield rotationally resolved frequency and linewidth data. A plot of the lifetime versus energy gap for several vibrational levels of $NeBr_2$ and NeI_2 (33) is shown in Figure 4. The energy gaps as a function of v for these two molecules actually overlap because although the vibrational frequency of Br_2 is 30% larger than that of I_2, the anharmonicity of Br_2 is more than double that of I_2.

What is observed in Figure 4 is that the lifetime of the metastable complex is more strongly dependent upon energy gap for NeI_2 than for $NeBr_2$. One might expect the potential energy surfaces for the two molecules to be quite similar. Perhaps the momentum gap constraints are weaker for $NeBr_2$ because the rotational degree of freedom can accept more energy with less angular momentum for the lighter molecule.

Summary

There is a growing body of evidence that energy gap laws are useful in rationalizing the relative rates of vibrational predissociation of various van der Waals molecules. As molecular complexity increases, so do the possible number of product channels. It remains to be seen if the dynamics of a molecule like ethylene dimer can be quantitatively understood.

Acknowledgments

The infrared spectroscopy of ethylene complexes was supported by the United States Department of Energy. The visible laser excited fluorescence work was supported by the National Science Foundation. The data for Figures 3 and 4 were obtained by Fritz Thommen and Dwight Evard using a single frequency laser borrowed from the San Francisco Laser Center supported by the National Science Foundation under Grant No. CHE79-16250. Kenneth C. Janda would like to acknowledge fellowships from the Alfred P. Sloan Foundation and the Camille and Henry Dreyfus Foundation. This is contribution no. 7029 from the Arthur Amos Noyes laboratory of Chemical Physics.

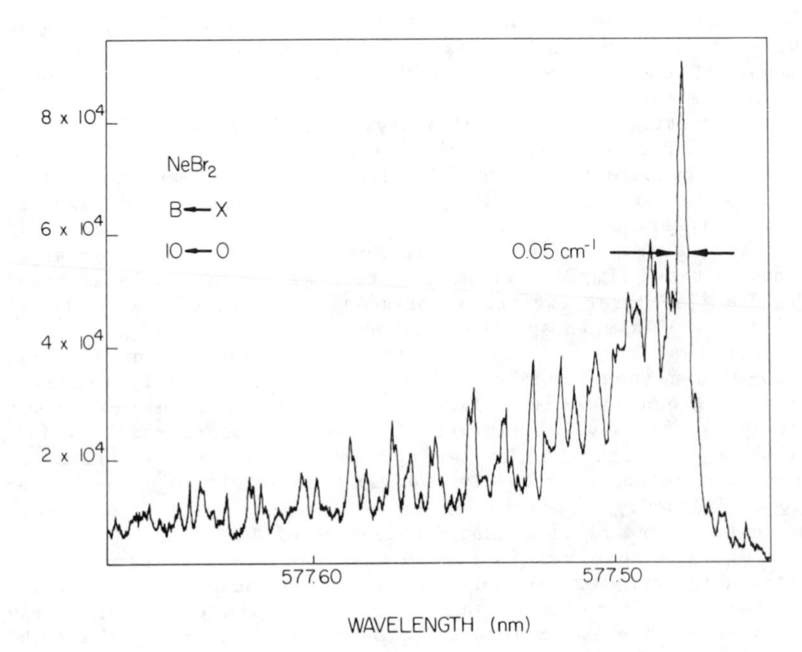

Figure 3. A rotationally resolved laser excited fluorescence spectrum of $NeBr_2$. The spectral widths are slightly lifetime broadened by the decay of the v = 10 level of the excited electronic state.

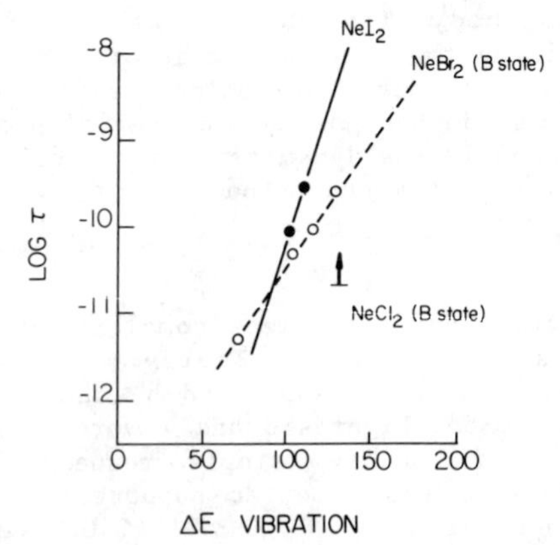

Figure 4. Comparison of the lifetime as a function of v versus the vibrational energy for several levels of $NeBr_2$ and NeI_2. For $NeBr_2$ the energy levels are v = 10,14,15,24. For NeI_2 the levels are v = 10,14. (NeI_2 data is from Ref. 33.)

Literature Cited

1. Klemperer, W. Ber. Bunsenges Phys. Chem. 1974, 78, 128.
2. Beswick, J. A.; Jortner, J. Adv. Chem. Phys. 1981, 47, 363.
3. Janda, K. C. Adv. Chem. Phys., in press.
4. Levy, J. H. Adv. Chem. Phys. 1981, 47, 323.
5. Gough, T. E.; Miller, R. E.; Scoles, G. J. Chem. Phys. 1978, 69, 1588.
6. Hoffbauer, M. A.; Gentry, W. R.; Giese, C. F. In "Laser-Induced Processes in Molecules"; Kompa, K. L.; Smith, S. D., Ed.; Springer-Verlag: Berlin, 1979.
7. Casassa, M. P.; Bomse, D. S.; Janda, K. C. J. Chem. Phys. 1981, 74, 5044.
8. Hoffbauer, M. A.; Liu, K; Giese, C. F.; Gentry, W. R. J. Chem. Phys. 1983, 78, 5567.
9. Bomse, D. S.; Cross, J. B.; Valentini, J. J. J. Chem. Phys. 1983, 78, 7175.
10. Geraedts, J; Snels, M; Stolte, S.; Reuss, J. Chem. Phys., submitted for publication.
11. Pine, A. S.; Lafferty, W. J. J. Chem. Phys. 1983, 78, 2154.
12. Brinza, D. E.; Swartz, B. A.; Western, C. M.; Janda, K. C. J. Chem. Phys. 1983, 79, 1541.
13. Ewing, G. E. Faraday Discuss. Chem. Soc. 1982, 73, 325.
14. Ono, Y.; Lin, S. H.; Tzeng, W. B.; Ng, C. Y. J. Chem. Phys. 1984 , 80, 1482.
15. Western, C. M.; Casassa, M. P.; Janda, K. C. J. Chem. Phys. 1984, in press.
16. Casassa, M. P.; Western, C. M.; Celii, F. G.; Brinza, D. E.; Janda, K. C. J. Chem. Phys. 1983, 79, 3227.
17. Casassa, M. P.; Western, C. M.; Janda, K. C. J. Chem. Phys., submitted for publication.
18. Nakanaga, T.; Kondo, S.; Saeki, S. J. Chem. Phys. 1979, 70, 2471.
19. Garaedts, J.; Waayer, M.; Stolte, S.; Reuss, J. Faraday Discuss. Chem. Soc. 1982, 73, 375.
20. Avoird, A. van der; Wermer, P.; Mulder, F.; Berns, R. Topics in Current Chemistry 1980, 93, 1.
21. van Hes, G. J. H.; Vos, A. Acta Cryst. 1977, B33, 1653.
22. Aldrich, P. D.; Legon, A. C.; Flygare, W. H. J. Chem. Phys. 1981, 75, 2126.
23. Shea, J. A.; Flygare, W. H. J. Chem. Phys. 1982, 76, 4857.
24. Andrews, L.; Johnson, J. L.; Kelsall, B. J. Chem. Phys. 1982, 76, 5767.
25. DeLeone, R. L.; Muentor, J. S. J. Chem. Phys. 1980, 72, 6020.
26. Ewing, G. E. Chem. Phys. 1981, 63, 411.
27. Halberstadt, N.; Soep, B. Chem. Phys. Lett. 1982, 87, 109.
28. Stephenson, T. A.; Rice, S. S. J. Chem. Phys., submitted.
29. Haynam, C. A.; Levy, D. H. J. Chem. Phys. 1983, 78, 2091.
30. Zwier, T. S.; Carrasquillo, E.; Levy, D. J. Chem. Phys. 1983, 78, 5493.
31. Segev, E.; Shapiro, M. J. Chem. Phys. 1983, 78, 4969.
32. Brinza, D. E.; Western, C. M.; Evard, D. D.; Thommen, F.; Swartz, B. A.; Janda, K. C. J. Phys. Chem. 1984, 88, in press.
33. Kenny, J. E.; Johnson, K. E.; Sharfin, W.; Levy, D. J. Chem. Phys. 1980, 72, 1109.

RECEIVED June 11, 1984

UNIMOLECULAR DYNAMICS

17

Classical, Semiclassical, and Quantum Dynamics of Long-Lived Highly Excited Vibrational States of Triatoms

R. M. HEDGES, JR., R. T. SKODJE, F. BORONDO, and W. P. REINHARDT

Department of Chemistry, University of Colorado and Joint Institute for Laboratory Astrophysics, National Bureau of Standards and University of Colorado, Boulder, CO 80309

Triatoms with doubly vibrationally excited predissociating states of exceptionally long lifetime are theoretically investigated using several techniques. For a two-degrees-of-freedom model of H_2O fully converged quantum estimates of resonance lifetimes are made, confirming the correspondence principle expectation that exceptionally long lived states exist and display non-RRKM behavior in the sense that lifetimes do not always decrease with increasing energy above the dissociation limit. Using the quantum results as a benchmark, the validity of a Golden-Rule-type formula is demonstrated, and the formula is then applied for physically realistic values of frequency, anharmonicity, and well depth: States with lifetimes of up to 0.1 sec are found. The paper ends with presentation of preliminary adiabatic semiclassical estimates of resonance energies for HOD in two and three degrees-of-freedom models.

The possibility that highly excited molecules can display strongly non-statistical behavior has been an important theme of experimental and theoretical research during the past decade. To name only two examples motivating such interest, the predissociation of very weakly coupled van der Waals molecules provides a possibility of long lived high energy (relative to dissociation) systems; and, the possibility of infrared laser directed chemistry depends of at least short term localization of energy, defying statistical randomization. The questions to be addressed here may be simply stated: How do dissociation times for covalently bonded systems depend on excitation mechanism? Namely, are there specific modes of highly excited polyatomic molecules that live longer than isoenergetic ones? Are there sequences of modes where lifetimes increase with increasing energy?

Classical mechanical studies of such questions date from the work of Bunker (1) and co-workers. Our attention was drawn to such problems by the work of Wolf and Hase (2,3) who showed that model systems of two and three degrees of freedom displayed large apparently trapped volumes of classical phase space well above the clas-

0097-6156/84/0263-0323$06.00/0

sical dissociation limit. Such volumes of non-dissociating phase space gave rise to strongly nonstatistical (non-RRKM) classical dissociation kinetics (3). Hase (4) subsequently identified such non-dissociating classical trajectories as being quasi-periodic, and thus quantizable using the underlying invariant tori (5). Noid and Koszykowski (6) and Hase (4) carried out such primitive quantizations, obtaining semiclassical estimates of the energies of the highly excited states. The question remained as to the lifetimes of the states.

An analogous series of classical calculations has been carried out by Uzer, Hynes and Reinhardt (7), in a study of dissociation following mode-specific classical excitation, modeling the experiments of Rizzo et al. (8) on the laser dissociation of HOOH via pumping of the high OH vibrational overtones. These workers (7) found that trajectories with classical initial conditions corresponding to initial high excitation in the OH stretch seem not to dissociate, energy being "trapped" in a Fermi resonance coupling the OH stretch to the HOO bend in this six-degrees-of-freedom problem. This suggests non-RRKM lifetimes of dissociation for the mode-specific excitation used in the experiments. Unfortunately, these experiments (8) give no indication of actual dissociation time scales, but only that the OH is produced in a distribution of excited states consistent with statistical. As repeatedly pointed out by Hase (9), this type of time independent observation does not preclude a non-RRKM distribution of dissociation lifetimes. The classical work on HOOH again requires that a technique be developed for determining the lifetimes of the quantum resonances suggested by the correspondence principle. As the trapped volumes of phase space often appear stable, techniques based on classical stability analysis of trapped but unstable periodic orbits (10) yield no estimates of lifetimes, and a quantum or semiclassical method must be employed.

Realistic quantum studies of highly excited systems with many vibrational degrees of freedom are currently prohibitive if all couplings and degrees of freedom are to be accounted for. Several two-degrees-of-freedom systems have been investigated at "large" values of Plancks constant: Waite and Miller (11,12) have looked at the tunneling predissociation in the cubic Henon-Heiles problem; Numrich and Kay (13) and Christoffel and Bowman (14) have carried out calculations of the resonance widths of an harmonic oscillator coupled to a Morse oscillator; Hedges and Reinhardt (15,16) have investigated Feshbach type resonances in ABA triatomics modeled as two mass-coupled Morse oscillators. Large values of Planck's constant are used as fewer basis states, or channels, are needed to describe the system, allowing fully converged quantum estimates of widths to be made. However, such highly quantum models need not correlate with classical dynamics (5,17-19) and may, indeed, miss many features of the classical-quantum correspondence. This is well illustrated by the reanalysis of the Henon-Heiles problem at smaller $\hbar$, by Bai et al. (20), who come to quite different conclusions to those of Waite and Miller (11,12), who found no mode specificity.

In the first two parts of the present paper the classical and quantum work of Refs. (15,16) is reviewed, and extended (21,22), via a Golden-Rule-type ansatz (23) to give resonance widths for the doubly excited states of ABA triatoms for $\hbar = 1$. In agreement with correspondence principle expectations, states of extraordinarily long lifetime are routinely found, the longest of the present study being 0.1 seconds. In the third part, preliminary numerical studies of a "new" adiabatic semiclassical quantization technique (24,25) are presented (26) and applied to determinations of resonance positions for two- and three-degrees-of-freedom models of HOD (27).

Classical Dynamics Above Dissociation: ABA Local Mode Systems with Two Degrees of Freedom

A useful model system, representing the interaction of two "local mode" oscillators as might occur in a light-heavy-light system, such as water, is described by the Hamiltonian

$$H = \frac{p_1^2}{2\mu_{12}} + \frac{p_2^2}{2\mu_{12}} - \frac{p_1 p_2}{m_2}$$

$$+ D\left(1 - \exp(-(r_1 - r_1^o))\right)^2 + D\left(1 - \exp(-(r_2 - r_2^o))\right)^2 \qquad (1)$$

where the coupling between the Morse type local mode oscillators is simply the momentum coupling through the central mass. The Hamiltonian of Equation (1) can represent a collinear triatom or a bent system, such as H_2O, for appropriate choices of the effective masses μ_{12}, m_2. Figure 1 shows a Poincare surface of section (defined by zeroes of the variable $Q_2 = r_1 - r_2$) for classical dynamics determined by the Hamiltonian above — using parameters appropriate for H_2O — for an energy equal to the classical dissociation energy. It is evident that the dynamics are governed by the presence of invariant tori even at the dissociation energy. As the energy is increased, the area occupied by tori on the composite surface of section slowly decreases: Figure 2 indicates the situation at 5/4 of the dissociation energy. Tori are still easily visible at 7/4 D, although the area on the surface of section occupied by tori is decreased (15,16). The presence of invariant tori at, and above, the classical dissociation limit indicates that whole volumes of the classical phase space are non-dissociating for the model Hamiltonian. This suggests the presence of long-lived quantum resonances, which if the correspondence principle holds, should be recognizable well above dissociation.

Quantum Dynamics of ABA Systems

Complex Coordinate Studies. The Hamiltonian of Equation (1) is an analytic function of positions and momenta. We may thus expect that the analytically continued Hamiltonian

$$H(\theta) = e^{-2i\theta}\left[\frac{p_1^2}{2\mu_{12}} + \frac{p_2^2}{2\mu_{12}} - \frac{p_1 p_2}{m_2}\right]$$

$$+ D[1 - \exp-((r_1 e^{i\theta} - r_1^o))]^2 + D[1 - \exp(-(r_2 e^{i\theta} - r_2^o))]^2 \qquad (2)$$

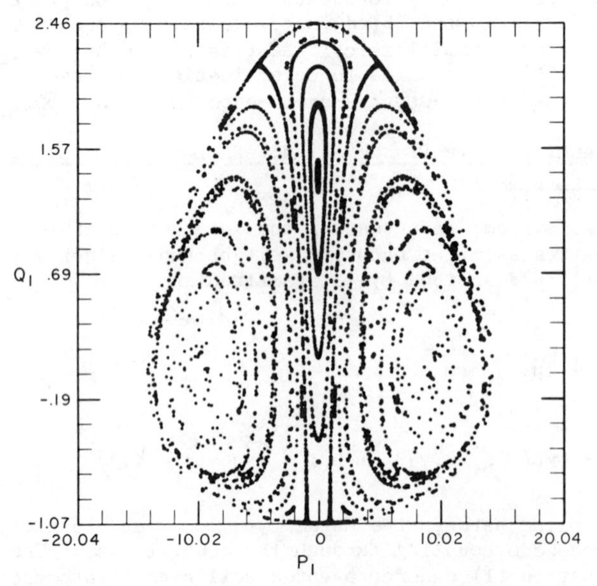

Figure 1. Composite of several Poincaré surfaces of section for the mass ratio $\mu_{12}:m_2 \cong 1:64$ (appropriate to H_2O), at the classical dissociation energy, D. Invariant tori dominate the phase space structure. Reproduced from ref. (16), with permission. Copyright 1983, American Institute of Physics.

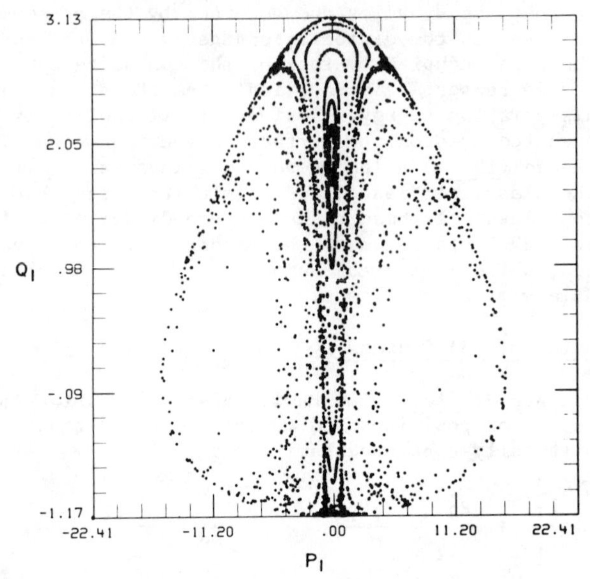

Figure 2. As in Fig. 1, but E = 5/4 D. The fraction of phase directly coupled to dissociative channels grows as a function of energy above D, but trapped invariant tori remain, even at quite high energy. Reproduced from ref. (16), with permission. Copyright 1983, American Institute of Physics.

defined by the scale transformation $r \rightarrow r \exp(i\theta)$, will have the usual properties of dilatation analytic (or "complex-scaled") Hamiltonians (see the reviews of References (28-30)). Namely: (1) the real bound state eigenvalues are independent of the transformation; (2) complex resonance eigenvalues are "exposed" as cuts of the corresponding resolvent are rotated off the real axis; and (3) the eigenfunctions corresponding to the resonance eigenvalues are square integrable, and thus may be approximated by usual linear variational methods. Hedges and Reinhardt (15,16) have carried out a series of such variational calculations using a variational spline basis contracted by prediagonalization of the one-degree-of-freedom Morse problem. Use of this non-analytic basis resulted from a compromise between accuracy and computational flexibility. The basis allowed easy spanning of the required coordinate space near the tops of the anharmonic wells, but gave only "asymptotic" convergence (such as that discussed in (31)) for narrow resonances: imaginary parts of eigenvalues less than 10^{-5} in absolute magnitude were unreliable.

As these calculations have been described in detail elsewhere (15,16), we present results only for the mass ratio 1:64:1, which corresponds to H_2O with a bond angle of 104.5°. Figure 3 shows the quantum eigenvalues, plotted as lifetime (decreasing upwards) versus energy in units of the dissociation energy, D. The well depth in these calculations corresponds to ten bound states per local bond mode, or, put another way, to an effective value of $\hbar = 2.4$ atomic units, where $\hbar = 1$ would correspond to experimental reality. Nevertheless, the figure indicates the existence of very long lived states (up to 10^6 harmonic vibrational periods) even for this excessively quantum-like model. Several comments are in order: the classical bound states in the continuum of Figures 1 and 2 correspond to doubly excited vibrations with approximately equal energies in each bond mode, with the asymmetric stretch normal mode type classical motion (see (32) for a discussion of the difference between classical local and normal mode behavior) being the most stable. The quantum results mirror this in that the doubly excited states with bond mode quantum numbers (n,n) display high stability above dissociation. For example, the longest lived state in Figure 3 is the (4,4), where (0,0) denotes the local mode ground state. Such doubly excited states are familiar in nuclear physics as compound state, or Feshbach, resonances, and in atomic spectroscopy as autoionizing, or Auger, states of atoms and molecules (33). Just as in the atomic case (33,34), sequences of states with quantum numbers of the form (n,n+m) do not necessarily have shorter lifetimes as a function of increasing m, in contradistinction to statistical expectations. This follows from the increase in period of the local bond modes as dissociation is neared, and the detuning of any near frequency resonances as m increases in a sequence of (n,n+m) states.

Approximate Quantum Studies: The Zeroth Order Golden Rule. Decreasing the value of $\hbar$ below ~2.4 is prohibitive if all complex eigenvalues are to be simultaneously determined, as has been done in the above studies. It is of course possible to determine individual complex eigenvalues of much larger complex symmetric matrices using inverse iteration (35), or, larger matrices might be treated using an altogether different type of algorithm (i.e., that of Ref. (36)) rather than the complex symmetric version of Givens used here. However, it seems preferable to use the $\hbar = 2.4$ calculations as benchmarks to test simple approximate methods, rather than to develop direct methods for working with the very large complex matrices which would arise in systems of more than two degrees of freedom. Rosen (23) in a very early study of coupled Morse systems suggested use of a simplified Golden Rule, which we now explore. Given doubly excited states, $|ij\rangle$, of an uncoupled Morse system lying above dissociation, decay is induced by the perturbation

$$V = -\frac{p_1 p_2}{m_2} \tag{3}$$

with the partial width into the bound (ℓ), free (k) channel $|\ell k\rangle$ being given by uncoupled Golden Rule expression

$$\Gamma_{(ij)} = \frac{2\pi}{\hbar m_2^2} \sum |\langle i|p_1|\ell\rangle\langle j|p_2|k\rangle|^2 \quad . \tag{4}$$

The total width is given by summing over all channels such that

$$E_{ij} = E_i + E_j = E_\ell + \frac{k^2}{2\mu_{12}} \quad . \tag{5}$$

Tables I and II give comparisons of the results of using the zero-order Golden rule, with the results of the fully converged complex coordinate computations of Refs. (15,16), as discussed above. With a small number of exceptions, it is seen that for a wide range of energies and coupling constants, the order of magnitude and trends of the widths are correctly given (21,22). Exceptions to this generally pleasing state of affairs occur when strong configuration interaction between doubly excited dissociating states takes place, leading to mixing of states with strongly different zero-order lifetimes. An example is the (3,3) resonance of Table II, where a three-order-of-magnitude discrepancy occurs. However, examination of the tables indicates that such major discrepancies are rare, and that the Zeroth-Order Golden Rule gives a qualitatively reasonable overview.

Zeroth-Order Golden Rule in the $\hbar = 1.0$ Case. Emboldened by the success of the simple Golden Rule as applied above, a series of results for couplings, well depths and masses appropriate to a two-degrees-of-freedom model of water, with a fixed bond angle of 104.5 are given in Figure 4. The Golden Rule, in accord with the correspondence principle, predicts (21,22) states of exceptional lifetime. In discussion of such results, two caveats must be made: (1) bend and rotational degrees of freedom have been omitted; and (2) possible mixing between the zero-order doubly excited states and the continuum, and more importantly, mixing between doubly excited states of differing lifetimes have been omitted. Nevertheless, as states of lifetimes of up to 0.1 sec are seen, unless the above effects entirely quench the multiply excited resonances, states with lifetimes in the micro- and milli-second range would seem to be a real possibility.

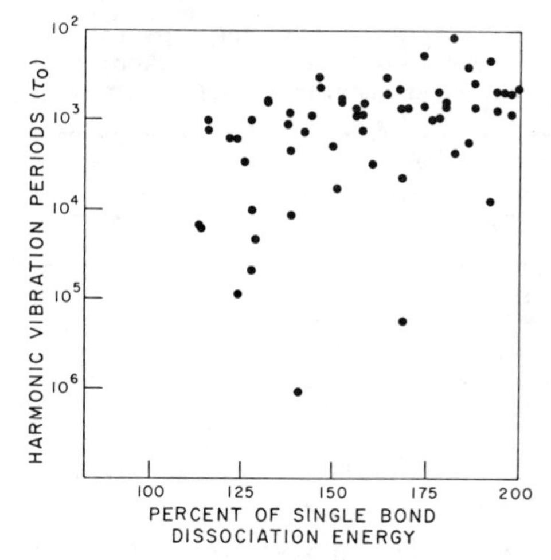

Figure 3. Complex eigenvalues for the 1:64:1 effective mass ratio for $\hbar \sim 2.4$ a.u. The longest lived state is $\sim 2 \times 10^6$ harmonic periods. Resonance lifetimes generally decrease as a function of: (1) energy above dissociation, (2) increasing coupling strength. However, there are certainly notable exceptions to these general trends. Reproduced from ref. (16), with permission. Copyright 1983, American Institute of Physics.

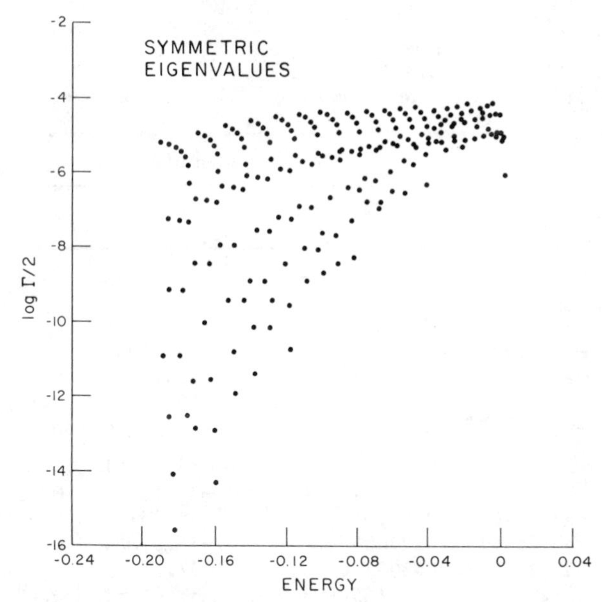

Figure 4. Golden Rule (Eq. (4)) results for lifetimes of a symmetric doubly excited states of H_2O. Plotted is log $\Gamma/2$ versus energy. Dissociation is at -0.19 in these (atomic) units. A $\Gamma/2$ of 10^{-16} implies a lifetime of ~0.1 sec.

Table I. Comparison for ABA mass ratio 1:12:1, of Golden Rule (Eq. (4)) positions and half-widths, with converged quantum results from Ref. (21).

Quantum State	Golden Rule (21)		Complex Coordinate (16,21)	
	Re(E)	Γ/2	Re(E)	Γ/2
2,4+	-42.91916	0.14029(-02)	-43.49182	0.69752(-02)
2,4-	-42.62260	0.11601(-02)	-42.73194	0.14159(-01)
1,6+	-42.01084	0.43667(-01)	-41.95854	0.37801(-01)
1,6-	-42.00713	0.43520(-01)	-41.90371	0.31363(-01)
3,3+	-41.68755	0.50959(-04)	( )	$<10^{-5}$
1,7+	-39.08114	0.31045(-01)	-39.19913	0.36024(-01)
1,7-	-39.07970	0.30986(-01)	-39.16935	0.34898(-01)
2,5+	-37.71722	0.12069(-02)	-38.01276	0.25396(-02)
2,5-	-37.63408	0.11299(-02)	-37.73714	0.53206(-02)
1,8+	-37.23546	0.17295(-01)	-37.19680	0.13527(-01)
1,8-	-37.23491	0.17272(-01)	-37.17793	0.11802(-01)
1,9+	-36.47333	0.29454(-02)	-36.47543	0.36676(-01)
1,9-	-36.47325	0.29420(-02)	-36.47517	0.36748(-01)
3,4+	-36.20733	0.79566(-02)	-36.00693	0.40807(-02)

Table II. As in Table I, but for mass ratio 1:64:1, appropriate to a two degrees of freedom model of H_2O.

Quantum State	Golden Rule (21)		Complex Coordinate (16,21)	
	Re(E)	Γ/2	Re(E)	Γ/2
2,4+	-44.23880	0.53243(-04)	-44.27512	0.61016(-04)
2,4-	-44.18184	0.44665(-04)	-44.18403	0.41544(-04)
3,3+	-43.19470	0.18044(-05)	-43.17689	0.11910(-02)
1,6+	-43.17317	0.18376(-02)	-43.17030	0.17206(-02)
1,6-	-43.17252	0.18325(-02)	-43.15637	0.48716(-03)
1,7+	-40.10424	0.13634(-02)	-40.10979	0.13578(-02)
1,7-	-40.10399	0.13614(-02)	-40.10910	0.13520(-02)
2,5+	-39.11810	0.46773(-04)	-39.12641	0.51053(-04)
2,5-	-39.10259	0.44122(-04)	-39.10397	0.45635(-04)
1,8+	-38.05107	0.84151(-03)	-38.05109	0.85632(-03)
1,8-	-38.05097	0.84069(-03)	-38.05089	0.77729(-03)
3,4+	-37.21938	0.30047(-05)	-37.21247	$<10^{-5}$
3,4-	-37.01355	0.29165(-03)	-37.01431	0.49795(-02)
1,9+	-37.01353	0.29141(-03)	-37.01387	0.50122(-02)

Adiabatic Generation of Semiclassical Resonance Eigenvalues

As mentioned previously, Noid and Koszykowski (6) and Hase (4) have obtained semiclassical estimates of the positions of resonances using surface-of-section techniques to quantize the invariant tori. However, in many cases this is not possible; the tori may not exist, being replaced by vague tori (37,38), or chaos. Even the mild situation of classical nonlinear resonance prevents standard trajectory based semiclassical techniques from being applied. This latter is well illustrated by a result of Hase (4), reproduced in Figure 5; rather than a smooth Poincare section through a single shank of a torus, we are confronted by an island chain. Smoothing techniques could be used here, as discussed elsewhere (18,38) but a simpler technique is available.

Solov'ev (24) and, independently Johnson (25), have suggested that one should take Ehrenfest (39) seriously. Given a Hamiltonian of the form

$$H(t) = H^{o} + \varepsilon(t)V \quad , \tag{6}$$

if H^{o} can be unambiguously semiclassically quantized, adiabatic turning on of the perturbation will preserve the quantization condition (i.e., actions are invariant under adiabatic variation of H(t)) while changing the energy. All one needs to do, then, is begin with a quantizing trajectory for H^{o}, turn on $\varepsilon(t)V$ slowly enough, and read off the final energy of the trajectory, which is now ideally on the adiabatically generated torus which meets the EBK quantization conditions! (Note that quantization of H^{o} is, itself, non-trivial if H^{o} is degenerate, precisely the analog of the usual problem of quantum degenerate perturbation theory.)

One can immediately ask, what if there is not such an EBK torus for the coupled system? The empirical result is that the method may well work anyway! Table III shows results (26) obtained by adiabatic quantization of the Hase (4) HCC two-degrees-of-freedom problem. Away from the 5:2 resonances (see Fig. 5) the adiabatic and Hase results are in accord, but the adiabatic method also successfully quantizes the resonance zones. Another illustration is given in Table IV where a two-degrees-of-freedom model of HOD is adiabatically quantized above the classical dissociation threshold (27). In this case much of the coupled phase space is actually chaotic at the higher energies, yet consistent results are obtained from the adiabatic quantization technique. Figures 6 and 7 indicate the sensitivity of the final energy to initial phase on the unperturbed torus for two- and three-degrees-of-freedom models of HOD, where the final "tori" do not exist as chaos has set in. Figure 8 indicates the complexity of the situation for a more strongly nonlinear problem: a section of an adiabatically generated torus (26) is shown for a high lying eigenstate of the Henon-Heiles problem. This — possibly fractal — torus nevertheless gave an excellent semiclassical eigenvalue.

The results of this section show that semiclassical estimates of resonance energies may be obtained — using the Solov'ev-Johnson ansatz — in regions of phase space where invariant tori do not exist. This technique may well turn out to be the semiclassical method of choice for many-degrees-of-freedom systems; even if tori exist, visualization of the caustic structure is difficult,

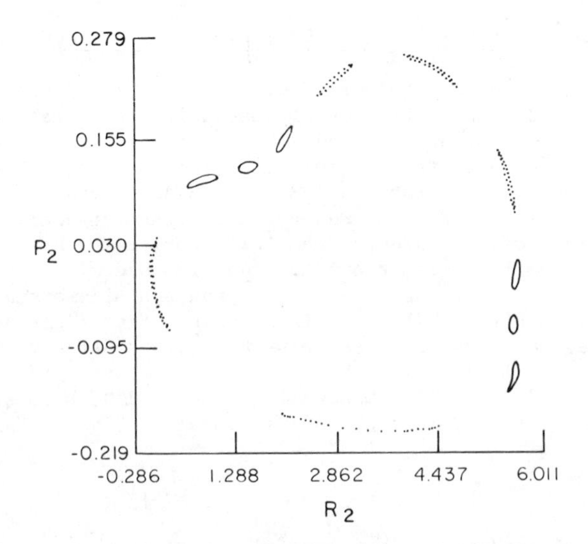

Figure 5. Island chain in the Hase (4) model problem. The presence of such classical resonances prevented Hase from carrying out EBK quantization for this model of H-C-C dynamics above dissociation. See Table III. Reproduced from ref. (4). Copyright 1983, American Chemical Society.

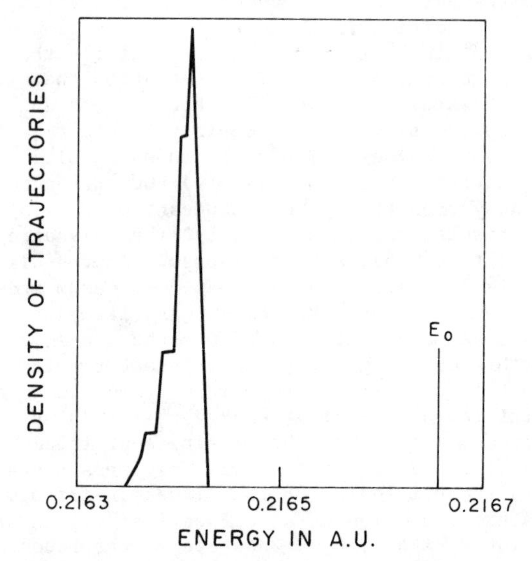

Figure 6. The density of adiabatically switched trajectories is plotted vs. the final energy for HOD in a two-degrees-of-freedom model initially in the (8,8) resonance state. The results are obtained from 83 randomly selected trajectories using a switching time of 30,000 a.u. The position E_0 is the energy of the (8,8) state obtained using the Hamiltonian H_0.

Table III. Adiabatically obtained semiclassical resonance energies for the H-C-C system of Ref. (4). Results of BR (Borondo, Reinhardt (26)) are compared with EBK results of Hase (4). As discussed in the text, the adiabatic method quantizes the classical resonances, where, as indicated by "...", Hase was unable to obtain results due to nonlinear resonance.

n_{CH}	n_{CC}	Hase	BR
1	0	14.24	14.25
1	1	17.94	17.95
1	2	21.57	21.58
1	3	...	25.14
1	4	28.63	28.64
1	5	...	32.08
1	6	35.43	35.45
1	7	...	38.76
1	8	41.98	42.00
1	9	...	45.18
1	10	48.29	48.30
1	11	51.35	51.36
1	12	54.35	54.34
1	13	57.28	57.27

Table IV. Adiabatically quantized semiclassical results for energies of doubly excited states of HOD, in a two degrees of freedom model, as obtained by Skodje and Reinhardt (27) using the technique of Refs. (24), (25). All of these states are above the classical dissociation threshold. ΔE is the uncertainty in the adiabatic quantization, as illustrated in Fig. 6. E_0 is the "unperturbed" energy, and E the result of the quantization procedure in this weakly coupled system.

n_1	n_2	E_0	E	ΔE
6	6	0.17328	0.17319	±0.000001
6	7	0.18327	0.18319	±0.000001
6	8	0.19286	0.19278	±0.00001
6	9	0.20204	0.20195	±0.000005
6	10	0.21081	0.21083	±0.00001
6	11	0.21918	0.21908	±0.00002
6	12	0.22714	0.22703	±0.00002
6	13	0.23470	0.23457	±0.00002

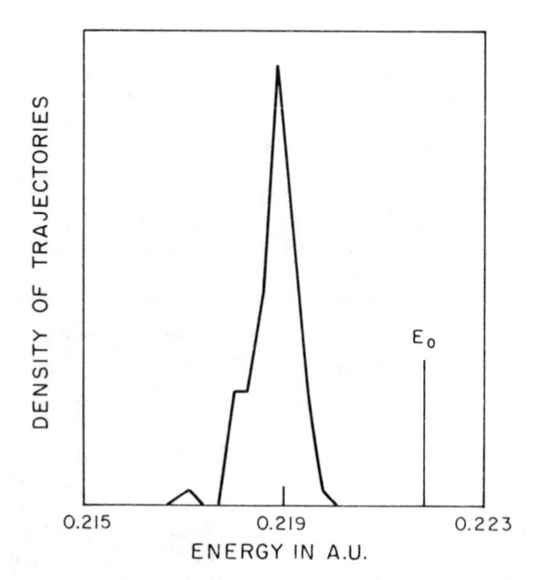

Figure 7. The density of adiabatically switched trajectories is plotted vs. the final energy for a three-degrees-of-freedom (stretch-stretch-bend) HOD initially in the (8,8,0) resonance state. The results are obtained from 80 randomly selected trajectories using the switching time of 70,750 a.u. for the stretch-stretch coupling and 100,000 a.u. for the bend stretch coupling. The position E_0 is the energy of the (8,8,0) state obtained using the Hamiltonian H_0.

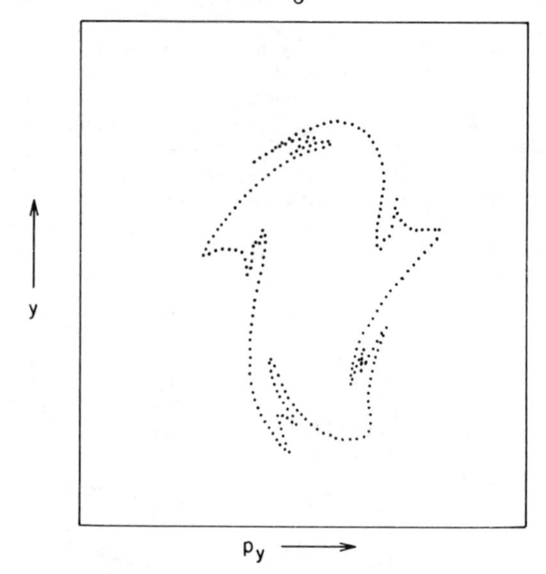

Figure 8. Convoluted adiabatically generated (26) torus for the Henon-Heiles problem in a region of classical chaos.

prohibiting straightforward applications of EBK theory. Yet, the problem of lifetimes will remain: As yet there is no trajectory-based method for direct determination of lifetimes from tori — or vague tori — trapped in volumes of phase space above dissociation.

Discussion

The present classical and quantum studies of simple co-valently bonded systems suggests that long lived resonances, analogous to Auger states of atomic and molecular electronic dynamics, may well exist, have surprisingly long lifetimes, and have strongly nonstatistical properties. Recent experiments of Kennedy and Carrington (40) have suggested highly excited states of H_3^+ with microsecond lifetimes. Their preliminary results do lend support to the possibility that systems with many degrees of freedom may well have excited states of very long lifetime, at energies well above dissociation. What will be the chemistry which follows?

The support of the National Science Foundation through grants CHE83-10122 and PHY82-00805 to the University of Colorado is gratefully acknowledged. F. Borondo was supported, in part, by a grant from the Comité Conjunto Hispano-Norteamericano para Asuntos Educativos y Culturales.

Literature Cited

1. Bunker, D. L.; Hase, W. L. J. Chem. Phys. 1973, 59, 4621.
2. Wolf, R. J.; Hase, W. L. J. Chem. Phys. 1980, 72, 316.
3. Wolf, R. J.; Hase, W. L. J. Chem. Phys. 1980, 73, 3779.
4. Hase, W. L. J. Phys. Chem. 1983, 86, 2873.
5. Noid, D. W.; Koszykowski, M. L.; Marcus, R. A. Ann. Rev. Phys. Chem. 1980, 32, 267.
6. Noid, D. W.; Koszykowski, M. L. Chem. Phys. Lett. 1980, 73, 114.
7. Uzer, A. T.; Hynes, J. T.; Reinhardt, W. P., work in progress.
8. Rizzo, T. R.; Hayden, C. C.; Crim, F. F. Faraday Disc. Chem. Soc. 1984, 75, 223.
9. Hase, W. L. In "Potential Energy Surfaces and Dynamics Calculations"; Truhlar, D. G., Ed.; Plenum: New York, 1981; p. 1.
10. Pollak, E. J. Chem. Phys. 1982, 76, 5843.
11. Waite, B. A.; Miller, W. H. J. Chem. Phys. 1980, 73, 3713.
12. Waite, B. A.; Miller, W. H. J. Chem. Phys. 1981, 74, 3910.
13. Numrich, R. W.; Kay, K. G. J. Chem. Phys. 1979, 71, 5352.
14. Christoffel, K. M.; Bowman, J. M. J. Chem. Phys. 1983, 71, 5352.
15. Hedges, R. M., Jr.; Reinhardt, W. P. Chem. Phys. Lett. 1982, 91, 241.
16. Hedges, R. M., Jr.; Reinhardt, W. P. J. Chem. Phys. 1983, 78, 3964.
17. Kay, K. G. J. Chem. Phys. 1979, 71, 1819.
18. Reinhardt, W. P. J. Phys. Chem. 1982, 86, 2158.
19. Reinhardt, W. P. In "Mathematical Analysis of Physical Systems"; Mickens, R., Ed.; Van Nostrand-Reinhold: New York, 1984.
20. Bai, Y. Y.; Hose, G.; McCurdy, C. W.; Taylor, H. S. Chem. Phys. Lett., 1983, 99, 342.

21. Hedges, R. M., Jr. Ph.D. Thesis, University of Colorado, Boulder, 1983.
22. Hedges, R. M., Jr.; Reinhardt, W. P. J. Chem. Phys., to be submitted.
23. Rosen, N. J. Chem. Phys. 1933, 1, 319.
24. Solov'ev, E. A. Sov. Phys.-JETP, 1978, 48, 635.
25. Johnson, B. R., unpublished work.
26. Borondo, F.; Reinhardt, W. P. J. Chem. Phys., to be submitted.
27. Skodje, R. T.; Reinhardt, W. P., unpublished work.
28. Reinhardt, W. P. Ann. Rev. Phys. Chem. 1982, 33, 223.
29. Junker, B. R. Adv. At. Mol. Phys. 1983, 18, 208.
30. Ho, Y. K. Physics Reports 1983, 99, 1.
31. Johnson, B. R.; Reinhardt, W. P. Phys. Rev. A 1983, 28, 1930.
32. Jaffé, C.; Brumer, P. J. Chem. Phys. 1980, 73, 5646.
33. Madden, R. P.; Codling, K. Astrophys. J. 1965, 141, 364.
34. Rau, A. R. P. J. Phys. B: At. Mol. Phys. 1983, 16, L699.
35. Maquet, A.; Holt, C. R.; Reinhardt, W. P. Phys. Rev. A 1983, 27, 2946.
36. Nauts, A.; Wyatt, R. E. Phys. Rev. Lett. 1983, 51, 2238.
37. Shirts, R. B.; Reinhardt, W. P. J. Chem. Phys. 1982, 77, 5204.
38. Reinhardt, W. P.; Jaffé, C. In "Quantum Mechanics in Mathematics, Chemistry, and Physics"; Gustafson, K.; Reinhardt, W. P., Eds.; Plenum: New York, 1981; p. 167.
39. Ehrenfest, P. Phil. Mag. 1917, 33, 500.
40. Kennedy, R. Ph.D. Thesis, The University of Southampton, (U.K.), 1983.

RECEIVED June 11, 1984

18

The Intramolecular Dynamics of Highly Excited Carbonyl Sulfide (OCS)

MICHAEL J. DAVIS and ALBERT F. WAGNER

Chemistry Division, Argonne National Laboratory, Argonne, IL 60439

The intramolecular dynamics of highly excited OCS has been studied within the framework of classical mechanics. One study involves the energy relaxation of OCS restricted to planar geometry. A second study investigates the unimolecular dissociation of OCS. It has been found that the relaxation of OCS is slow and appears to occur on two time scales. In addition, the relaxation still persists at 45 ps. Reasons for such an effect, as well as diagnostic techniques, are discussed. Since the causes for such effects persist above dissociation they lead to nonstatistical unimolecular dissociation, which also shows a high degree of mode specificity.

In this paper we present new results on the dynamics of highly energetic OCS. This work was motivated by the study of Carter and Brumer (1) on this same system. They discovered that the calculation of energy relaxation in OCS at high energy was not statistical while at the same time standard tests showed that individual trajectories had a pronounced statistical character. Such observations bear on the fundamental assumptions in RRKM theory (2), where some difficulty has been encountered in modeling the reaction dynamics of small molecules (3). In this current study we have tried both to calculate in more detail the relaxation dynamics of highly excited OCS (including energies in the dissociative regime) and to diagnose nonstatistical behavior using techniques developed in nonlinear dynamics (4).

Collinear Dynamics of OCS

Our calculated relaxation and dissociation dynamics of planar OCS are best understood by first examining these phenomena in the collinear world where OCS is not allowed to bend (5). The reduced degrees of freedom allow for a detailed study of statistical behavior by means of surfaces of section as well as the validation of other techniques that can be readily extended to more degrees of freedom.

0097-6156/84/0263-0337$06.00/0

Figures 1a-f show a series of surface of section (4) plots for collinear OCS generated from a single trajectory. The total energy is 20000 cm^{-1} (2.48 eV) where the dissociation energy (to O+CS) is 21900 cm^{-1} (2.72 eV). These are plots of normal coordinate and conjugate normal momentum for the normal mode which is predominately CS stretch and were generated in the usual manner. That is, during the course of a trajectory whenever the normal coordinate of the CO normal mode reached 0.0 and the normal momentum of that normal mode was positive the position and momentum of the CS normal mode were plotted. The plots in Figure 1 represent different time slices of the total surface of section with the times shown in the figure caption. The manner of selecting the time slices will be discussed shortly.

Although the density of points in each plot in Figure 1 is related to the length of its time slice, this does not explain the difference in appearance of each plot. For example, plots a, c, and e show similar behavior with relatively thick bands of P,Q values where either P or Q is large and the signs of P and Q are opposite. Also there is a conspicuous absence, most clearly seen in Figure 1c, of large positive values of Q and small negative values of P. In contrast, Figure 1d emphasizes both this region of phase space and those regions where P and Q are correlated with the same sign, Figure 1f correlates a narrow band of relatively small P,Q values, and Figure 1b correlates a very narrow band of the most extreme P,Q values of the same sign. Figure 1 presents a picture which seems to indicate that the phase space of highly excited collinear OCS is divided into different stochastic regions, though it is guaranteed that such regions will eventually be evenly occupied (6). The region presented in plots a, c, and e shows a more delocalized stochastic behavior than do plots b, d and f, which show remnants of quasiperiodic behavior, that with decreasing energy become even more localized into island structure associated with specific classical resonances (7). In fact, plots d and f show motion which is trapped around islands of quasiperiodic behavior. Such motion has been referred to as "vague torus" (8), since quasiperiodic trajectories move on a torus (4).

Exponential Separation in Collinear OCS

The time slices for Figure 1 are in fact the lifetimes of the trajectory in each of the regions of divided phase space. They were generated by the method of exponentially separating trajectories which can be briefly motivated as follows. It has been found numerous times (4) in studies of dynamical systems that stochastic motion is evident from the behavior of trajectories which are started initially very close. In the case of trajectories which are stochastic, the separation distance between these adjacent trajectories increases on the average exponentially with time. In the case of regular or quasiperiodic trajectories the separation distance between initial trajectories increases on the average linearly with time, a condition which can be proven (9). Though stochastic motion which is trapped near quasiperiodic motion resembles quasiperiodic motion, it is still stochastic, meaning initially close trajectories move apart exponentially. However, close analysis of such motion reveals that the trapped motion has a rate of exponential separation which is less than the untrapped.

In Figure 2, the bottom plot shows the rate of exponential separation for the trajectory used to generate Figure 1. Included on this plot are letters which correspond to the plots shown in Figure 1. The exponential separation was resolved using a technique developed by Benettin et al. (10) and is described more fully in reference 5. One can easily observe that when a trajectory moves in a more localized manner the rate of exponential separation decreases noticeably. The slopes for b, d, and f are much less than those for a, c, and e.

The top plot in Figure 2 shows a 23-line fit to the bottom plot. The manner in which such fits are made is described in reference 5. Such fits are important, since we can associate with each line segment a specific type of dynamical motion which can repeat itself several times in the course of a given trajectory. For simplicity, several of the plots in Figure 1 were generated by combining adjacent segments resolved in Figure 2. The motion pictured in Figure 1 is actually more complicated than it appears to be there. For the purposes of fitting the energy relaxation, a large number of line segments are needed to fit curves such as the one pictured in Figure 2 (5).

It has been shown several times in recent years that divided phase space leads to long time decays of correlation functions and complicates any attempt at a simple statistical description of relaxation or dissociation (11-19). As expected then, energy relaxation for OCS cannot be described by a single exponential decay (as will be explicitly shown for planar OCS). Although energy relaxation is too complicated to be described in detail here (see reference 5), several qualitative ideas are apparent from Figures 1 and 2. First, the relaxation time must exceed the average lifetime of trajectories in any one region of phase space. For example, the region occupied in Figure 1f has smaller values of P and Q than any other region and corresponds, unlike all the other regions in Figure 1, to an approximately equal sharing of energy in the CO and CS modes. Until enough time has elapsed for trajectories on the average to hop in and out of this region on a random basis, initial excess energy in one mode of motion will not relax to its final distribution. Second, like many systems with largely undivided phase space that exhibit uniform exponential separation (4), we expect the relaxation time in each region of divided phase space to be inversely proportional to the rate of exponential separation in that region (20). However the lifetime of a trajectory in each region is not related, in any direct way, to the exponential separation rate. In the particular trajectory displayed in Figures 1 and 2, the lifetime is in general an order of magnitude larger than the inverse of the exponential separation rate.

Energy Relaxation in Planar OCS

Given that phase space for highly excited collinear OCS is divided, it is not surprising that energy relaxation would be complicated in planar OCS, where the molecule is allowed to bend and rotate in a plane but no tumbling rotations are allowed. This expectation is borne out in Figure 3 which shows an energy relaxation plot for OCS

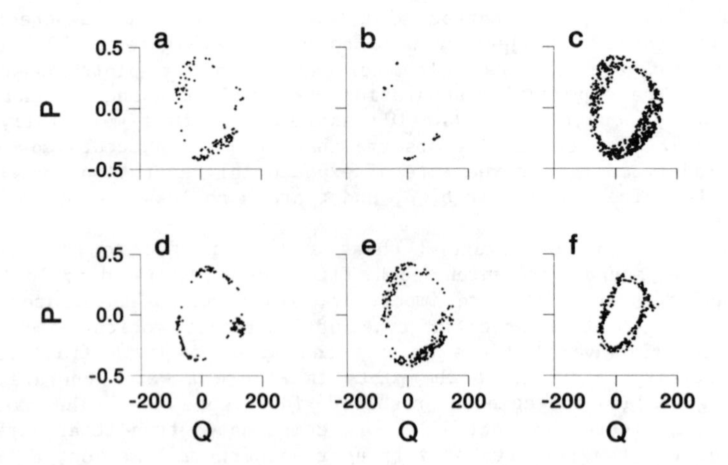

Figure 1. A series of time slices of a surface of section for a collinear OCS trajectory. The time ranges in ps of each plot are: a) 0.0 - 4.2, b) 4.2 - 7.0, c) 7.0 - 24.9, d) 24.9 - 30.0, e) 30.0 - 38.5, f) 38.5 - 45.3.

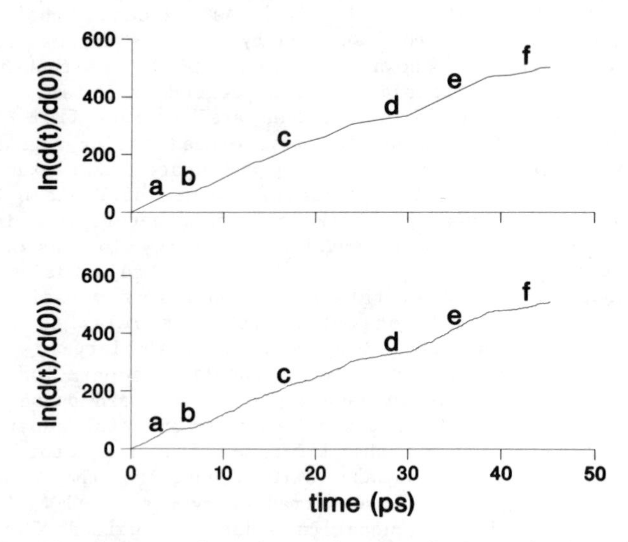

Figure 2. The bottom plot shows the ln of the relative distance between the collinear trajectory shown in Figure 1 and a collinear trajectory started initially very close to it. The top plot shows a 23 line fit to the bottom. The letters on these plots correspond to the plots in Figure 1.

initially excited in the normal mode that is predominately a CO stretch in the small displacement limit. The curves show the energy in each of the three vibrational normal modes averaged over an ensemble of 50 trajectories. This ensemble was chosen using a method developed by Hase and co-workers (21). Each trajectory in the ensemble was chosen so that both the normal mode which is predominately CS stretch (about 80% CS character) as well as the bend normal mode have zero point energy. The normal modes were resolved in the usual way (22) and the energy in each normal mode is defined as $E = \frac{1}{2}(P^2 + \omega^2 Q^2)$, where ω is the frequency of the normal mode and Q and P are the normal coordinate and momentum, respectively. The rest of the energy of the trajectory is in the CO normal mode. Adjustments are made to each trajectory, so that each has the correct angular momentum. In this case total angular momentum is zero and the total energy is as in Figures 1 and 2. This case is one of those studied by Carter and Brumer (1) who followed the relaxation over a much shorter time (2 ps) and used a different technique for generating the initial conditions of the trajectories.

In Figure 3 it appears that the energy relaxation occurs on two time scales. First the system relaxes at a fairly rapid rate for a time up to approximately 10 ps. This is followed by a much more gradual relaxation which is still occurring at the end of our calculation at 45 ps. While at first glance this final extremely slow relaxation may seem like an artifact of this calculation, we have observed such behavior in calculations on many different ensembles (23). In addition, a least squares fit to the last 35 ps shows that the energy in the CO mode is going down and the total error in this fit over this time range is less than the change in energy.

Figure 3 also includes a small line at the end of each plot which indicates the phase space average of the energy in each normal mode at a total energy of 20000 cm^{-1}. This shows that even after 45 ps the energy averages have not reached their statistical expectations which are the microcanonical averages shown on each figure. For example, the CO stretch energy average is approximately 0.2 eV higher than the phase space average, while the CS and bend normal modes are closer but slightly low in the former case and slightly high in the latter. The time scale of relaxation shown in Figure 3 can be put into perspective if one considers the lifetime of planar OCS at dissociative energies using RRKM theory (2). We have estimated such a lifetime to be approximately 10 ps at an energy 100 cm^{-1} above dissociation, a time at which the energy in the CO stretch is even further from microcanonical (0.4 eV).

The relaxation in Figure 3 is clearly nonstatistical. It does not occur with a single exponential and it happens on too long a time scale. Furthermore, this nonstatistical behavior has definite implications if it persists above dissociation (and it does, as will be shown later) since it implies mode specificity of unimolecular dissociation. This is due to the fact that energy tends to stay trapped in the mode of initial excitation, but dissociation occurs only when excess energy piles up in the weakest bond (the CS bond).

While the collinear phase space of OCS has been shown to be divided, it is of measure zero relative to the planar phase space. To verify that planar OCS has divided phase space would be very difficult with a surface of section approach as in Figure 1 (23).

The inability to make such an analysis is probably why so few studies have probed thoroughly a system of more than two degrees of freedom which possesses a divided phase space (16,24). However, it is not much harder than we demonstrated in collinear OCS to diagnose divided phase space using exponentially separating trajectories as in Figure 2.

Individual Trajectory Dynamics in Planar OCS

Figure 4 presents the exponential separation of one of the trajectories in the ensemble of trajectories used to generate the energy relaxation in Figure 3. Unlike two-dimensional systems where there is only one direction in which exponential separation is possible, in planar OCS there are two linearly independent directions where such behavior can occur (10). The top curve in Figure 4 shows the maximum exponential separation and is analogous to Figure 2 for collinear OCS. The bottom curve in Figure 4 shows exponential separation along the second linearly independent direction which is always perpendicular to the direction of maximum exponential separation pictured in the top plot. The resolution of exponential separation in two linearly independent directions is accomplished by the method of Benettin et al. (10).

The two curves shown in Figure 4 are both necessary to fully characterize the dynamics of planar OCS. First, notice how the slope of both curves changes depending upon whether the trajectory is in region a, b, or c. This indicates, as in the collinear case, that as the trajectory moves through phase space, it enters different stochastic regions with different characteristic values of the rate of exponential separation. Second, notice that in region b the bottom plot has a large reduction in slope. This is analogous to what was observed in collinear OCS, when a trajectory was trapped near a torus. However, unlike collinear OCS, in this same region the maximum rate of separation in the top plot has its highest value. Such behavior is indicative of a situation studied several times recently (10,25-27), because it represents behavior which is difficult to observe using previous techniques which were developed to probe stochastic behavior in systems with two degrees of freedom (4). In systems with two degrees of freedom, the problem in deciding whether a system is stochastic or not is equivalent to observing if an additional constant of the motion besides total energy exists locally. In systems with more than two degrees of freedom, during the course of evolution a trajectory can enter regions of phase space where there appears to be more than one constant of the motion, but less than the maximum allowed (N for N degrees of freedom). Such is the case pictured in region b. The trajectory is trapped near a region where one constant of motion has been lost, but a constant still exists in addition to the total energy. Behavior such as this must be taken into account if a full analysis of the dynamics of planar OCS can be accomplished and this is further examined in reference 23.

In collinear OCS, we could demonstrate the relationship between energy localization and exponential separation by comparing a surface of section for a trajectory with its rate of exponential separation (Figures 1 and 2). For planar OCS with one more degree of freedom, the surface of section is no longer a good means to

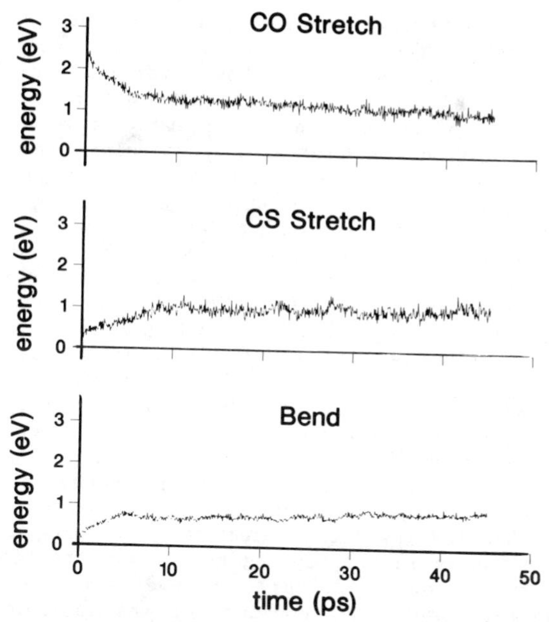

Figure 3. The ensemble average of the three normal mode energies vs. time. Total ensemble size is 50 and the planar trajectory initial conditions had most of the energy in the CO normal mode. Total energy of each trajectory is 20000 cm^{-1} and total angular momentum is zero.

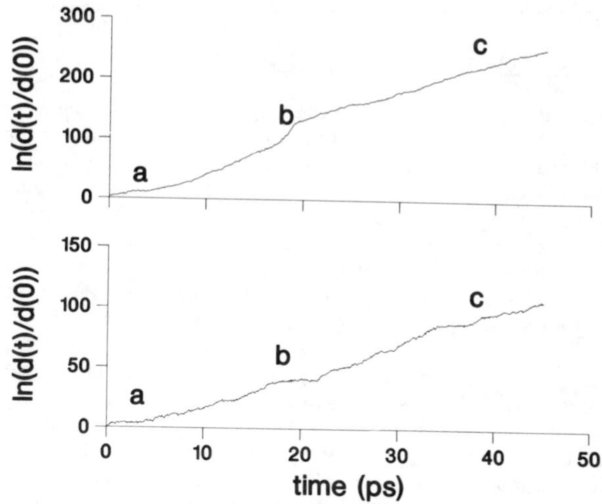

Figure 4. The top curve shows rate of maximum exponential separation for a single trajectory. The bottom curve shows the rate along a direction which is always perpendicular to the direction of maximum rate. This trajectory was chosen from the ensemble used to create Figure 3.

analyze this relationship. However, the trajectory whose exponential separation is plotted in Figure 4 does demonstrate an analogous relationship. This trajectory was chosen for presentation because it shows energy localization by merely monitoring the normal mode energies. Such a localization does not have to be apparent in a normal mode description, but for this particular trajectory it is and the time evolution of the resulting normal mode energies is shown in Figure 5. The top plot shows the CO energy, the middle shows the CS energy, and the bottom the bend energy.

The letters on the top of Figure 5 indicate the same regions as on Figure 4. Two immediate and not very surprising observations can be made concerning these plots. First, the normal mode energies are a poor description of energy flow in this highly excited trajectory. The second observation is that there are definite patterns to the energy flow. The regions a, b, and c represent, respectively, motion where (from left to right) the energy is trapped predominately in CO stretch, then CS stretch, and finally bend. The behavior observed in region a is common for most trajectories in the ensemble and leads to the initial relaxation observed over the first 10 ps of Figure 3. The behavior observed in region b indicates there is a stochastic flow between the two stretching normal modes, but there is no energy in the bending mode. For the time this trajectory spends in region b (~2.5 ps), the energy in the bending mode is approximately an additional constant of the motion, causing the behavior observed in Figure 4. Region c indicates a very long period during which most of the energy is trapped in the bending degree of freedom. As discussed for collinear OCS, if the average trajectory lifetimes in localized regions of phase space are long (for this one trajectory, 25 ps in region c), total energy relaxation lifetimes must be even longer, as in fact it is for Figure 3.

Constants of the Motion and Exponential Separation

A few final comments should be made concerning the behavior observed in Figures 4 and 5. The following is expected to be true (28). In systems of three or more degrees of freedom, all stochastic regions are connected, except for regions of measure zero - the collinear geometry is an example. This means that regions which appear to contain constants of the motion which are greater than 1 and less than N (for N degrees of freedom) will always be connected to those regions where there is only one constant of the motion, total energy. Trajectories which start in regions where it appears there are more than one constant but less than N will eventually wander into regions where there is only one, though this make take an extremely long time, for example, via Arnold diffusion (4,28). The motion pictured in region b of Figures 4 and 5 can never be asymptotically stable motion, except when symmetry is present (28). However, such trapping can occur on very long time scales and cause the same sort of long time delays in energy relaxation we have observed in the collinear dynamics, despite the fact that there is such a strong exchange of energy between two of the degrees of freedom.

Region b of Figures 4 and 5 provides a good example of the effect described in the previous paragraph because it provides a case which is easily observed using normal mode energies. In fact,

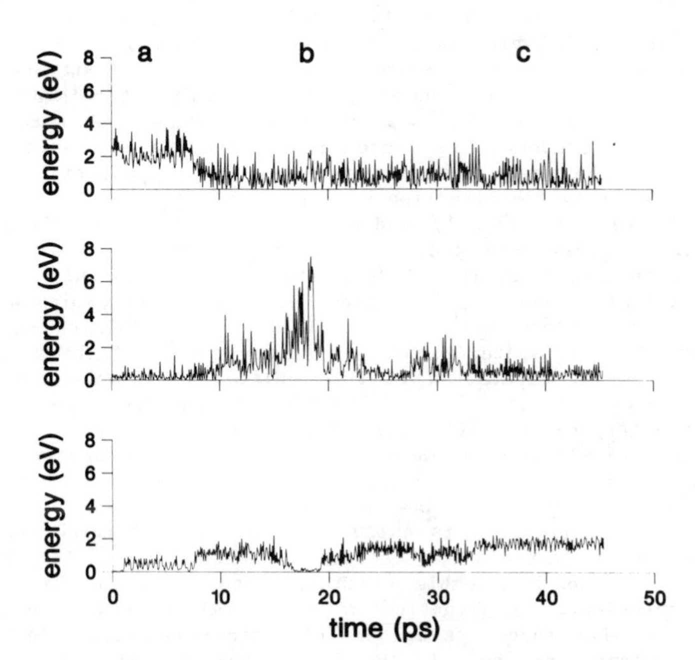

Figure 5. The time development of the normal mode energies for the trajectory used to generate Figure 4. The top plot shows CO energy, the middle CS energy, and the bottom bend energy. The letters at the top indicate different dynamical motions and correspond to the letters on Figure 4.

one could argue that the additional information provided by the bottom plot in Figure 4 is not essential from a qualitative point of view since the same information is available by just monitoring the energy in the bend (quantitatively such information is necessary to accurately define a lifetime). However, many regions of phase space may exist where similar behavior occurs but is not easily observed using normal mode energies. One could envision situations where there is a strong exchange of energy between all normal modes, yet an additional constant of the motion would exist or appear to exist for significant times. The bottom plot in Figure 4 shows such behavior at 35 ps. There is a definite decrease in the slope at this point, a behavior which needs further investigation.

As a final point, we emphasize that the complexity observed in Figures 4 and 5 cannot become apparent by only determining the maximum rate of exponential separation in systems with more than two degrees of freedom. The presence of temporary constants of motion less than the maximum number requires that exponential separation along more than one direction be calculated (10).

Figures 3-5 clearly indicate the presence of divided phase space, but they were generated by trajectories with most of their energy initially in the CO stretch. Similar calculations (along with energy relaxation) have been performed on ensembles with other initial conditions (23). As a further indication of the divided nature of phase space, Figure 6 displays the ln of the average rate of exponential separation of three different ensembles of 50 trajectories. The curve labeled "CO Str." is the same ensemble used to generate Figure 3, while the curves labeled "CS Str." and "Bend" were generated from ensembles with most of their energy in the CS stretch and the bend, respectively. Figure 6 shows that while on average the ensembles show exponential separation, the rate of exponential separation is quite different in all three cases. The difference is not due to different numbers of quasiperiodic trajectories in each ensemble (such trajectories would have zero rate of separation). Examination of each trajectory in the ensembles indicates that there are no quasiperiodic trajectories in the CS or Bend ensembles and four in the CO. A sampling of 100 trajectories with random partitioning of the energy in different normal modes shows only two quasiperiodic trajectories. Thus from Figure 6, it is clear that the average rate of exponential behavior is quite different depending upon which region of phase space the ensemble initially samples.

Dissociation of Planar OCS

The effects outlined in our previous discussion persist above dissociation for planar OCS. Figure 7 shows a plot which is analogous to Figure 6. These curves show the ensemble average of the exponential separation for three different sets of trajectories started along the respective normal modes. The total energy of the system is 22,400 cm^{-1}, or 500 cm^{-1} above dissociation and the ensembles include 50 trajectories each. Figure 7 once again shows the divided phase space nature of OCS, though the curves in Figure 7 are much closer than the ones in Figure 6. In addition, though a more thorough sampling of the surface needs to be done, it appears that

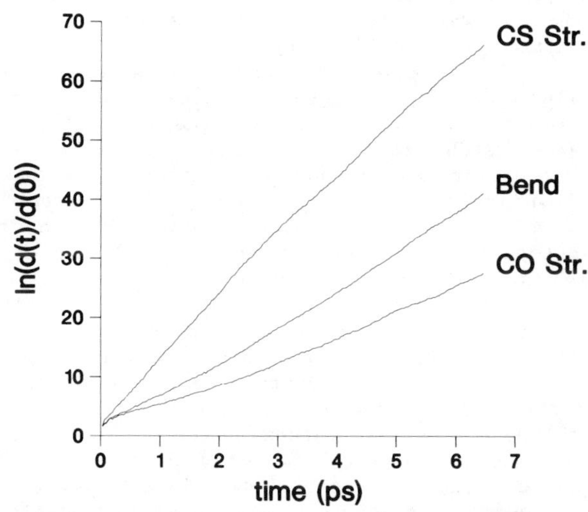

Figure 6. The ensemble average of the rate of exponential separation for three different ensembles of 50 planar trajectories. The energy and angular momentum are the same as in Figure 3.

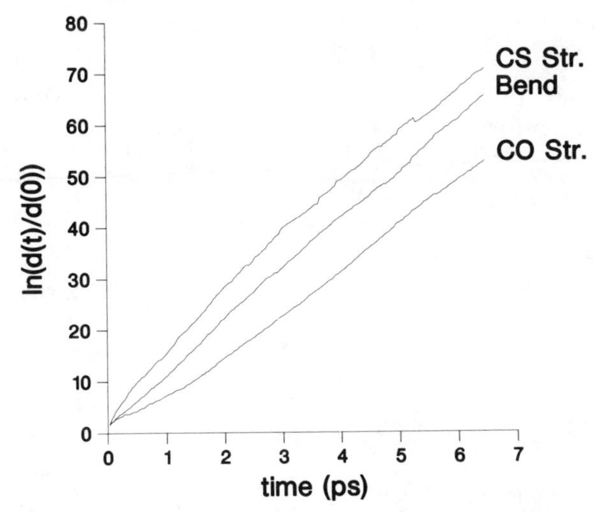

Figure 7. The average rate of exponential separation for three different ensembles of 50 planar trajectories. The total energy is 22,400 cm^{-1} and total angular momentum is zero.

there is much less trapping and quasiperiodic behavior at this energy than there was at 20000 cm^{-1}.

As we noted previously, behavior such as that observed in Figure 7 can lead to interesting nonstatistical effects. Since the CS bond is the one which can break, any tendency which causes the dynamics to linger in the vicinity of high excitation in the CS stretch normal mode will cause excessive dissociation for ensembles which initially possess most of their energy in the CS normal mode. Any tendency to remain trapped away from the CS stretch, for example near a pure CO stretch, may greatly decrease the dissociation rate. Such trapping behavior has already been discussed in this paper, so the curves shown in Figure 8 are not surprising, except perhaps in the degree by which they differ. These curves show the fraction of initial trajectories which remain bound as a function of time. The ensemble size is once again 50 and these are the same ensembles which were used to generate Figure 7. The initial excitations are labeled on each plot and the method used to create each ensemble was described above. Once again total angular momentum is zero.

The rates of dissociation observed in Figure 8 can be compared to an RRKM rate we estimate to be approximately 0.1 ps^{-1}, but more importantly the nonstatistical nature of the unimolecular decomposition can be discerned from the dramatic differences in the rate depending upon the initial excitation. Results such as the ones presented in Figure 8 point to a strong mode specificity in the dissociation of OCS. We believe that such classical behavior points to analogous quantum mechanical behavior involving resonances with dramatically different lifetimes, for example short lifetimes for states with large CS character and long lifetimes for those which are predominantly CO in character. At present, further work is being done on the classical dissociation of planar OCS (29).

Acknowledgments

Work performed under the auspices of the Office of Basic Energy Sciences, Division of Chemical Sciences, U. S. Department of Energy, under Contract W-31-109-Eng-38.

It is a pleasure to thank George Schatz and Nelson De Leon for many helpful discussions.

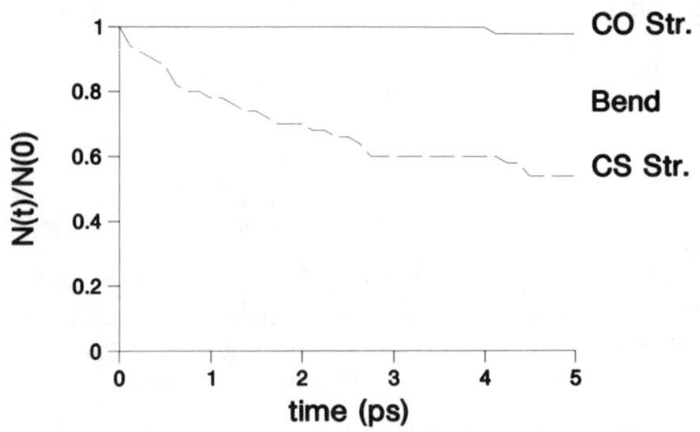

Figure 8. This figure was generated using the same three ensembles used in Figure 7 and shows the fraction of trajectories which stay bound vs. time.

Literature Cited

1. Carter, D.; Brumer, P. J. Chem. Phys. 1982, 77, 4208-4221.
2. Robinson, P. J.; Holbrook, K. A. "Unimolecular Reactions"; Wiley: New York, 1972.
3. Harding, L. B.; Wagner, A. F.; Bowman, J. M.; Schatz, G. C.; Christoffel, C. J. Phys. Chem. 1982, 86, 4312-4327.
4. Lichtenberg, A. J.; Lieberman, M. A. "Regular and Stochastic Motion"; Springer-Verlag: New York, 1983.
5. Davis, M. J.; Wagner, A. F., to be submitted.
6. Arnold, V. I.; Avez, A. "Ergodic Problems of Classical Mechanics"; Springer-Verlag: New York, 1968.
7. Walker, G. H.; Ford, J. Phys. Rev. 1969, 188, 416-432.
8. Shirts, R. B.; Reinhardt, W. P. J. Chem. Phys. 1982, 77, 5204-5217; and references cited therein.
9. Casati, G.; Chirikov, B. V.; Ford, J. Phys. Lett. A 1980, 77, 91-94.
10. Benettin, G.; Galgani, L.; Giorgilli, A.; Strelcyn, J. M. Meccanica, March, 1980, 21-30.
11. Chirikov, B. V. Phys. Rept. 1979, 52, 263-379; particularly section 5.5 and references cited there.
12. De Leon, N.; Berne, B. J. J. Chem. Phys. 1981, 75, 3495-3510.
13. Chirikov, B. V. In "Dynamical Systems and Chaos"; Garrido, L., Ed.; Springer-Verlag: New York, 1983; pp. 29-46.
14. Meiss, J. D.; Cary, J. R.; Grebogi, C.; Crawford, J. D.; Kaufman, A. N.; Abarbanel, H. D. I. Physica D 1983, 6, 375-384
15. Karney, C. F. F. Physica D 1983, 8, 360-380.
16. Casartelli, M. Nuovo Cimento B 1983, 76, 97-107.
17. Vivaldi, F.; Casati, G.; Guarneri, I. Phys. Rev. Lett. 1983, 51, 727-730.
18. Hase, W. L.; Duchovic, R. J.; Swamy, K. N.; Wolf, R. J. J. Chem. Phys. 1984, 80, 714-719; and references cited therein.
19. MacKay, R. S.; Meiss, J. D.; Percival, I. C. Phys. Rev. Lett. 1984, 52, 697-700.
20. Hamilton, I.; Brumer, P. J. Chem. Phys. 1983, 78, 2682-2690.
21. Sloane, C. S.; Hase, W. L. J. Chem. Phys. 1977, 66, 1523-1533.
22. Wilson, E. B.; Decius, J. C.; Cross, P. C. "Molecular Vibrations", McGraw-Hill: New York, 1955.
23. Davis, M. J.; Wagner, A. F., to be submitted.
24. Pfeiffer, R.; Brickmann J., preprint.
25. Contopoulos, G.; Galgani, L.; Giorgilli, A. Phys. Rev. A 1978, 18, 1183-1189.
26. Benettin, G.; Ferrari, G.; Galgani, L.; Giorgilli, A. Nuovo Cimento B 1983, 72, 137-147.
27. Stine, J. R.; Noid, D. W. J. Phys. Chem. 1983, 87, 3038-3042.
28. Tennyson, J. L.; Lieberman M. A.; Lichtenberg A. J. In "Nonlinear Dynamics and the Beam-Beam Interaction"; Month, M.; Herrera, J.C., Eds.; AIP Conference Proceedings No. 57, American Institute of Physics: New York, 1979: pp. 272-301.
29. Davis, M. J.; Wagner, A. F.; Brumer, P., to be submitted.

RECEIVED June 11, 1984

BIMOLECULAR REACTIVE SYSTEMS

19

Vibrationally Bonded Molecules

The Road from Resonances to a New Type of Chemical Bond

JOACHIM RÖMELT[1,3] and ELI POLLAK[2,4]

[1] Department of Theoretical Chemistry, University of Oxford, 1, South Parks Road, Oxford 0X1 3TG, U.K.

[2] Department of Chemistry and Materials and Molecular Research Division of the Lawrence Berkeley Laboratory, University of California, Berkeley, CA 94720

Bimolecular collinear reactions roughly may be divided into two categories: those occurring on attractive potential surfaces and those occurring on repulsive ones. For the former, the collision process is thought to proceed via a complex mechanism, the long living complex being a result of an attractive well between reactants and products. For a collision occurring on a 'repulsive' potential energy surface the process is assumed to be 'direct' since there are no attractive forces that can hold the colliding partners together. In this case the potential energy surface typically has a saddle point between reactants and products.

Surprisingly, early numerical computations for quantal scattering on saddle point surfaces exhibited spikes in the reaction probabilities as a function of the energy (1-4). Levine and Wu (5) showed that these spikes may be interpreted as resonances as they are associated with long (in comparison to vibrational periods) time delays. Subsequent numerical studies indicated that these phenomena are typical of direct reactions (6-10). It was suggested that the binding force forming the 'complex' was dynamic. Indeed it turned out that the corresponding vibrationally adiabatic potential energy surfaces - constructed using natural collision coordinates - exhibited wells in the interaction region (5,11). Furthermore Wyatt et al. (12) proved that if the adiabatic wells were eliminated then the resonances disappeared. Kuppermann and coworkers (10,13-15) provided additional evidence for the adiabatic trapping mechanism by constructing the vibrationally adiabatic potential energy surfaces using a radial coordinate system [see also Babamov and Marcus (16) and Launay and Le Dourneuf (17)]. Again, these surfaces exhibited wells in the interaction region and the exact resonance energies correlated qualitatively (typically within ± 1 kcal/mol) with virtual

[3] Permanent address: Lehrstuhl für Theoretische Chemie, Universitat Bonn, Wegelerstrasse 12, D-5300 Bonn 1, FRG.

[4] Permanent address: Department of Chemical Physics, Weizmann Institute of Science, Rehovot 76100, Israel.

0097-6156/84/0263-0353$06.50/0

or bound states of the adiabatic wells. Similar quality results, using a natural collision coordinate methodology were obtained by Garrett and Truhlar (18). Thus, by 1982, resonances in collinear reactive scattering on repulsive energy surfaces were qualitatively understood. Adiabatic wells are formed in vibrationally adiabatic potentials and are related to weakening of the binding forces in the interaction region (after all a bond is being broken). These wells may support long lived complexes, resonances. Despite the success in qualitative understanding of resonances a quantitative theory of resonance energies did not exist. With respect to the resonance widths and strengths appearing in different transitions (19) even a qualitative model was lacking.

Another fundamental question raised was whether there is a classical analog to the reactive scattering resonances. After all, quantal shape resonances have classical analogs - bound trajectories that cannot exit through the centrifugal barrier. Levine and Wu (5b) claimed to have found 'long lived' classical trajectories but they did not find bound quasiperiodic ones. Stine and Marcus (20), in a unique calculation, were able to show that the lowest lying resonance in the H_3 system could be accounted for, quantitatively, using a particular set of classical trajectories traversing the interaction region an increasing number of times. This calculation, however, was not only computationally tedious, but Duff and Truhlar (21) demonstrated that this semiclassical picture was not generally applicable.

Consider the adiabatic picture. If the adiabatic approximation is exact then the semiclassical analog of a resonance would be a quasiperiodic motion on an invariant torus. In fact resonances found by Chapman and Hayes (22) on surfaces with wells have been correlated successfully by Noid and Koszykowski (23) with classical quasiperiodic motion. But, a close inspection of the H_3 system by Costley and Pechukas (24) for quasiperiodic motion was unrevealing. In order to gain further insight into the classical analogue Pollak and Child (25) studied the properties of a specific set of periodic orbits - termed resonant orbits (RPO's) - that if quantized with integer action conditions, correlated well (with a typical accuracy of 0.5 kcal/mol) with exact quantal resonance energies. These orbits, however, were unstable for the H_3 and FH_2 systems, and as noted by Gutzwiller (26) it is dangerous to analyze quantal spectra in terms of isolated periodic orbits. Furthermore the integer action quantization law employed was ad hoc and there was no rigorous theoretical justification. Finally the connection of RPO's with the adiabatic picture was very loose and so not completely satisfactory. Again lots of details were available, but a complete picture was lacking.

To clarify these questions we have studied the quantal and semiclassical theory of reactive resonances. In section II the Diagonal corrected Vibrational Adiabatic Hyperspherical (DIVAH) model (27,28) is reviewed. This theory was the first to provide quantitative predictions (with a typical accuracy of 0.1 kcal/mole) of collinear quantal resonance energies. Furthermore the DIVAH model led us to the new and very surprising phenomenon of vibrationally bonded molecules (29-41). The connection and interrelation with semiclassical RPO theory (25,30,42-47) which predicted and interpreted these results, is presented in section III. Having noticed the possibility of vibrational bonding for collinear systems it was of fundamental interest to extend and to improve this concept to 3D systems. This is discussed in section IV. Finally the question of experimental evidence for vibrationally bonded molecules is raised in section V.

DIVAH model for collinear reactive scattering resonances

For a collinear atom-diatom reactive collision A+BC → AB+C in hyperspherical coordinates (8,9,13,48) the Hamiltonian representing angular (internal, vibrational) and radial (translational) motion is (with $\hbar=1$)

$$H(r,\varphi) = -\frac{1}{2m}\left(\frac{\partial^2}{\partial r^2} + \frac{1}{r}\frac{\partial}{\partial r} + \frac{1}{r^2}\frac{\partial^2}{\partial \varphi^2}\right) + V(r,\varphi) \tag{1}$$

The solution of the corresponding Schroedinger equation can be expanded in terms of eigenfunctions $\Phi_n(r,\varphi)$ for the angular motion at a fixed value r

$$\psi_{(r,\varphi)} = r^{-1/2} \sum_{n=0}^{\infty} \Phi_n(r,\varphi)\ \chi_n(r) \tag{2}$$

where the eigenfunctions $\Phi_n(r,\varphi)$ represent a complete and orthonormal set of adiabatic basis functions. Substitution of the expansion into the Schroedinger equation and integration over the angular variable φ gives a set of coupled differential equations in the radial coordinate (16). In a Born Oppenheimer-type approximation we assume an uncoupling of angular and radial motion and thus neglect all non-diagonal coupling elements. The resulting one-dimensional radial equation

$$\left(-\frac{1}{2m}\frac{\partial^2}{\partial r^2} + U_n(r) - E\right)\chi(r) = 0 \tag{3}$$

provides an effective potential (17,27,28)

$$U_n(r) = \varepsilon_n(r) - \frac{1}{8mr^2} - \frac{1}{2m}\left\langle \Phi_n(r,\varphi)\left|\frac{\partial^2}{\partial r^2}\right.\Phi_n(r,\varphi)\right\rangle \tag{4}$$

which includes the adiabatic energies $\varepsilon_n(r)$, a centrifugal term $-\frac{1}{8mr^2}$, and the diagonal adiabatic correction. The latter can be approximated via a finite difference scheme (27,28) and finally the effective potential simplifies to

$$U_n(r) = \varepsilon_n(r) - \frac{1}{8mr^2} - \frac{1}{m(\Delta r)^2}\left(1 - \langle \Phi_n(r+\Delta r,\varphi) \mid \Phi_n(r,\varphi)\rangle\right) \tag{5}$$

Hence, in the DIVAH theory the reactive scattering process is reduced to an elastic scattering problem on the effective DIVAH potential $U_n(r)$. The one-dimensional radial equation can now be solved for the bound as well as the metastable states whose lifetime τ may be determined by $\tau = 2\hbar\, d\eta(E)/dE$. Here $\eta(E)$ is the phase shift. Both bound and metastable states are related to resonance energies of the corresponding reaction process. Two categories of resonances are expected: shape-type resonances as metastable states on $U_n(r)$ occurring in $P^R_{n,n}(E)$ reaction probabilities and Feshbach-type resonances - due to the interaction of the one-dimensional states with the continua of lower lying angular states.

The signature of these resonances may appear in diagonal ($P^R_{n,n}(E)$) and off-diagonal ($P^R_{n,n'}(E)$) reaction probabilities.

In Fig. 1 the effective potentials $U_n(r)$ for the four lowest angular states of the H_3 system (gerade (g) and ungerade (u) symmetry) are compared to the corresponding adiabatic eigenenergy curves $\varepsilon_n(r)$. For the excited angular states we find wells in the adiabatic potentials confirming the idea of adiabatic trapping. However, the diagonal adiabatic correction which reflects the dynamical coupling to the other angular states and which is always repulsive in nature (27), modulates the shape of the effective potential quite a bit. Only when taking into account both terms and calculating the bound and metastable states on the effective potential $U_n(r)$ does one obtain excellent agreement with the corresponding resonance energies in the $P^R_{n,n'}(E)$ reaction probability curves. A comparison with accurate computations is provided in Table I. Furthermore the interplay of both of these contributions to the effective potential is responsible for the change of the resonance energy locations relative to the corresponding threshold energies as one goes from n=1 to higher quantum numbers. As can be seen from Fig. 1, for n=1,2 the adiabatic correction shifts the resonances over the threshold energy. With increasing quantum number n this effect is finally overcome by a more and more bonding eigenenergy curve $\varepsilon_n(r)$ turning the one-dimensional metastable states into one-dimensional bound states with respect to their threshold energy.

Table I. Resonance energies for the collinear $H+H_2$ reaction: a comparison of one-dimensional model predictions in comparison to exact quantum mechanical results on the Porter-Karplus II surface (49).

n	E_n [a] (eV)	Resonance energies (eV): From exact quantum-mechanical calculation of P^R_{oo} (9,48) [b]	DIVAH model	Simple hyper-spherical model (14)
0	-4.474	-	-	-
1	-3.953	-3.870	-3.872	-3.930
2	-3.464	-3.443	-3.446	-
3	-3.007	-3.022	-3.028	-

a) E_n = threshold energy for vibrational state n, relative to dissociation into three free atoms.

b) P^R_{oo} is the ground-state-to-ground-state reaction probability.

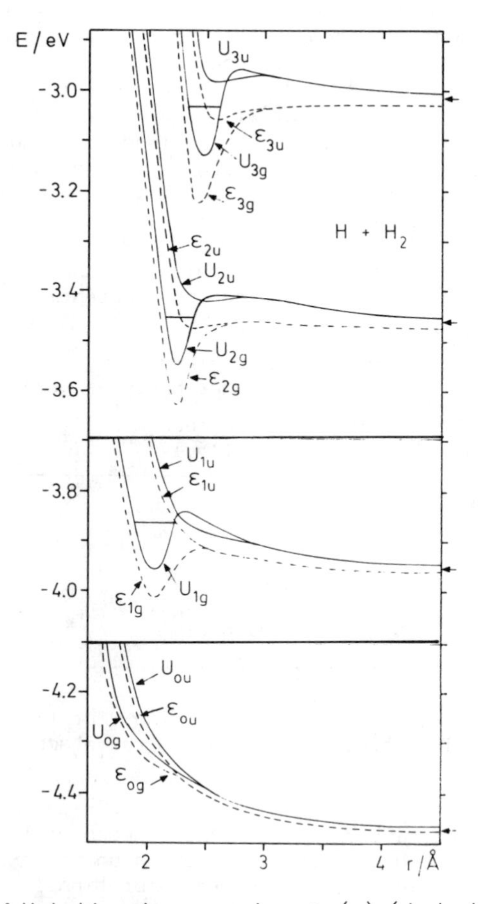

Figure 1. Adiabatic eigenenergies $\varepsilon_n(r)$ (dashed lines) and effective potentials $U_n(r)$ (full lines) for the collinear $H+H_2$ reaction (27) on the Porter-Karplus II surface (49) for n=0,1,2,3. The energies are relative to the dissociation energy into three free atoms.

The $F+H_2$ system (on the Muckerman V surface (50)) is taken as representative of the class of asymmetric reactions. In asymmetric reactions, the asymmetric nature of the electronic potential energy surface may introduce a new feature in form of avoided crossings (cf. Fig. 2) in the adiabatic eigenenergy surfaces. In this case the adiabatic diagonal correction includes two different types of contributions: the first coming from the near degeneracy of two adiabatic energies $\varepsilon_n(r)$ and $\varepsilon_{n+1}(r)$ at the region of the avoided crossing, and the second resulting from the potential coupling of the angular states as discussed for the H_3 system. It can be shown (17,27) that if the crossing is strongly localized, as it is in the $F+H_2$ system, a potential spike occurs due to the curve crossing effect (51) (cf. Fig. 2). But as long as it does not disturb the energetic structure of the bound states in the adiabatic wells, the model is still satisfactory (cf. Table II) for comparison to accurate quantal calculations.

Table II. Resonance and threshold energies for the collinear $F+H_2$ FH+H reaction on the Muckermann V potential energy surface (50).

	E [a] (eV)	Resonance Energies (eV): Exact quantum-mechanical calculation (15,17)	Resonance Energies (eV): DIVAH model
$F+H_2(v=0)$	-4.480	-	-
FH(v=3)+H	-4.463	-4.465	-4.460
FH(v=4)+H	-4.035	-4.052	-4.064
$F+H_2(v=1)$	-3.966	-3.95 ± 0.01 [b]	-3.959

[a] E = threshold energy for the corresponding vibrational state, relative to dissociation into three free atoms.

[b] Estimated from Fig. 3 of Ref. (17).

The adiabatic correction is not always necessarily large. Consider the Cl+HCl → ClH+Cl exchange reaction on the extended LEPS surface of Ref. (28). The coupling of the vibrational states is small, the adiabatic correction minor and the reaction proceeds nearly adiabatically. This statement can be easily verified from the reaction probabilities displayed in Fig. 3. The off-diagonal probabilities are small compared to the diagonal ones. Also the signature of resonance states in the n-th adiabatic surface $U_n(r)$ is most noticeable in the $P^R_{nn}(E)$ or $P^R_{n,n\pm1}(E)$ reaction probabilities but dies off very rapidly with increasing Δn. It has been shown that this adiabaticity of the reaction depends primarily on the masses of the atoms involved (15,16,28,52). The adiabatic approximation becomes better with decreasing skew angle, i.e. for Cl+HCl the effective potentials $U_n(r)$ and the adiabatic energies $\varepsilon_n(r)$

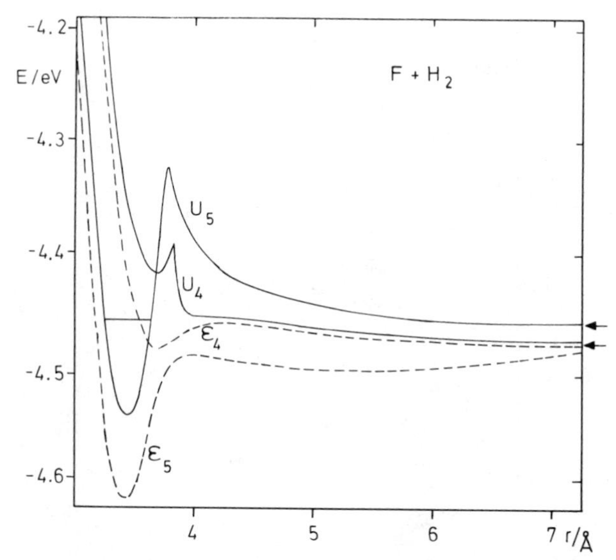

Figure 2. Adiabatic eigenenergies $\varepsilon_n(r)$ (dashed lines) and effective potentials $U_n(r)$ (full lines) of the collinear $F+H_2 \rightarrow FH+H$ reaction for $F+H_2(v=0)$ ($\hat{=}\ \varepsilon_4, U_4$) and $FH(v=3) + H$ ($\hat{=}\ \varepsilon_5, U_5$) (27). The energies are relative to the dissociation energy into three free atoms.

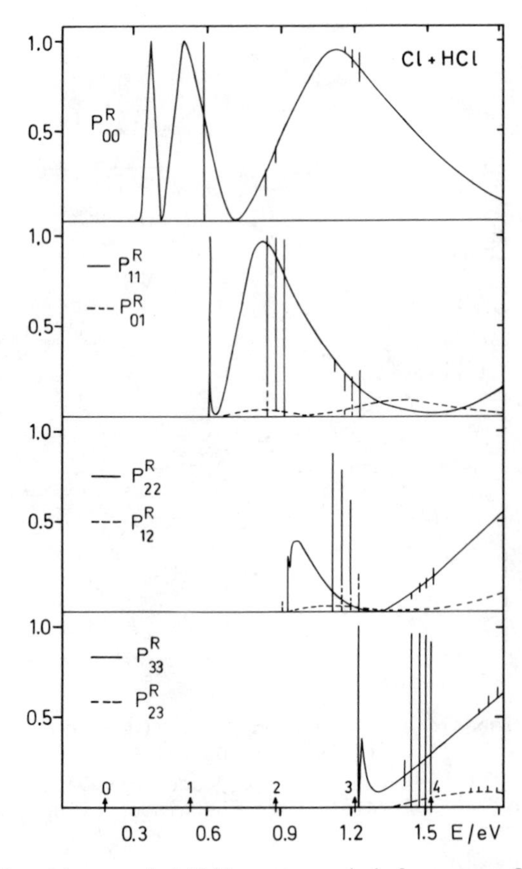

Figure 3. Reaction probability versus total energy for the collinear Cl+HCl → ClH+Cl reaction. The total energy is measured from the minimum of the asymptotic HCl potential well. The arrows and numbers on the abscissae indicate the vibrational energies and associated quantum numbers of the isolated HCl molecule (28).

almost coincide. From Table III, which includes a comparison to accurate quantal calculations for the same potential energy surface, it can be seen that the predictions are exact to within 10^{-3} eV (28). Most of these very narrow resonances (cf. Fig. 3) were first predicted by the DIVAH model and have subsequently been verified by two-dimensional quantal calculations. The resonance structure strongly resembles a spectrum of the transition complex. It is a very appealing idea to invert this 'spectral' information to obtain the vibrational structure of the transition complex. With regard to the width of resonances, it is well known that the widths of Feshbach resonances are related to the coupling potential describing the interaction of the bound or metastable state with the continuum of a lower lying angular state. As discussed above this coupling decreases with the skew angle of the system. Consequently, the Feshbach-type resonances become sharper and sharper with decreasing skew angle.

Table III. Approximate and exact resonance energies for the collinear Cl+HCl exchange reaction. The exact resonance energy is for the transition indicated.

$n^{a)}$	Parity[b)]	E_n(eV)[c)]	$E_{v' \leftarrow v}$(eV)[d)]	v'	$\leftarrow v$	Type of resonance
0	g					
0	u					
1	g	0.58835	0.58848	0	0	Feshbach
1	g	0.61991	0.61960	1	1	Shape
1	u					
2	g	0.84459	0.84492	1	1	Feshbach
2	g	0.88312	0.88359	1	1	Feshbach
2	g	0.91576	0.91609	1	1	Feshbach
2	u					
3	g	1.12229	1.12272	2	2	Feshbach
3	g	1.16072	1.16158	2	2	Feshbach
3	g	1.19597	1.19691	2	2	Feshbach
3	g	1.22551	1.22592	2	2	Feshbach
3	u	1.23193	1.23194	3	3	Shape

a) Quantum number of the effective vibrationally adiabatic potential.

b) Parity (u or g) of the effective vibrationally adiabatic potential.

c) Approximate one-dimensional resonance energy obtained from the solution of equation (3).

d) Exact resonance energy.

The most exciting result of these investigations, however, was the realization that if DIVAH theory predicted a bound state below the zero point energy of reactants and products, then this state would represent a truly bound molecule since a non-adiabatic transition to lower lying vibrational states is not possible (30). In fact, the bound state computations of Meyer showed that collinear IHI does have four truly bound states on a minimum-free potential energy surface (31). Furthermore, the exact collinear energies agreed excellently with the DIVAH predictions (cf. Table IV), and therefore vibrational bonding can be regarded as an extreme case of resonances in a 'direct' scattering process. We had thus predicted the existence of a new type of chemical bond which is formed because of an effective vibrationally adiabatic well - hence the name vibrational bonding. The mass combination in the IHI system (heavy-light-heavy) resembles very much the H_2^+ molecule. The electron is located mainly between the two protons, attracting both of them and thus overcoming the repulsion between the two protons and binding the system. The fast electronic motion provides an effective (adiabatic) potential for the slowly moving nuclei. Analogously, in vibrational bonding the bond is a result of the fast nuclear motion of a light-particle group of nuclei which provides an effective binding potential for the rest of the system.

Table IV. Bound-state energy levels (kJ/mol) of collinear IHI[a].

	Exact solution of mathematical-two-dimensional problem	Vibrationally adiabatic approximation[c]
E_{0g}	-299.175	-299.173
E_{1g}	-297.846	-297.844
E_{2g}	-296.609	-296.605
E_{3g}	-295.512[b]	-295.506

a) Surface A of Ref. (8,10). The levels lie between the saddle point at -303.78 kJ/mol and the HI zero point energy at -294.75 kJ/mol, relative to dissociation into three free atoms.

b) Converged within ~ 0.005 kJ/mol.

c) Note that the excellent agreement can only be obtained after insertion of the diagonal correction. The magnitude of this term is < 0.01 kJ/mol.

The potential energy surfaces (LEPS A from Refs. (8,10)), the effective potentials U_n (n=0; gerade and ungerade symmetry) and the bound states of the collinear IHI and IDI are shown in Fig. 4. As argued for the Cl+HCl system the reaction is very adiabatic, and thus the effective potentials and the adiabatic eigenenergies $\varepsilon_n(r)$ coincide within 10^{-4} eV. It should be noticed that isotopic substitution of the hydrogen atom by its heavier isotope deuterium tends to destabilize the vibrational bonded molecule. This inverted isotope effect is mainly due to a smaller variation of the IDI zero point energies, indicating that the heavier deuterium

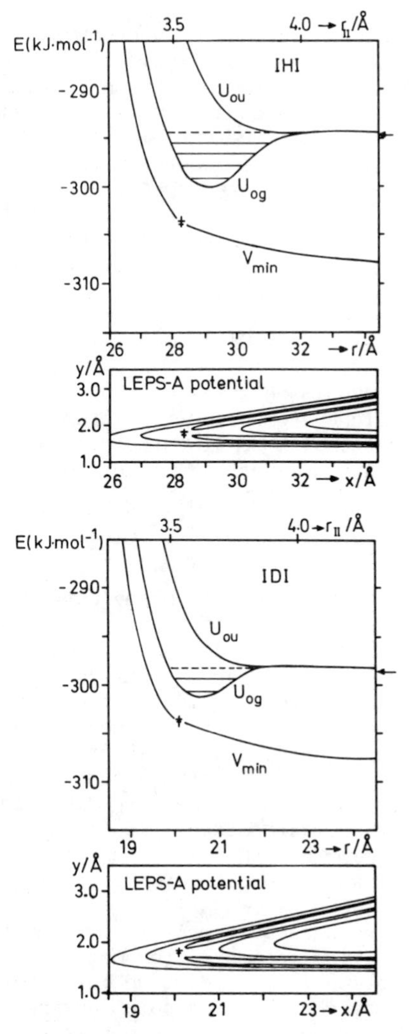

Figure 4. Vibrational bonding in the collinear IHI and IDI systems. The LEPS A potential energy surface (8,10) in mass-scaled radial coordinates $[x = (m_{A,BC}/m_{BC})^{1/2} r_{A,BC}, y = r_{BC}$ are shown in the lower panels. Contours are at -268.4, -294.7 and 304.0 kJ/mol for IDI. The position of the saddle point is indicated by ≠. The adiabatic gerade and ungerade surfaces as well as the static minimum potential energy are shown in the upper panels as a function of the radial coordinate r. Bound (resonance) states are given as solid (dashed) lines. Energies are relative to dissociation into three free atoms. Also shown is the relative I-I internuclear distance r_{II}.

is less effective in overcoming the repulsive forces. In all cases the binding energy is of the order of 1-10 kJ/mol, thus being roughly comparable to the binding energy of van der Waals molecules (38,39). We have seen that the DIVAH model is ideally suited for light atom transfer systems. Here, the radial motion corresponds to the slow heavy atom motion while the angular motion corresponds to the asymmetric fast light atom movement. Thus there is a good separation of time scales and an adiabatic approximation in radial coordinates is appropriate. Resonances of course appear also in heavy atom transfer systems for which, at least at low total energies one expects an adiabatic reaction path hamiltonian approach to be quite reasonable as shown by Truhlar and coworkers (18,53,54). Aquilanti and coworkers (55,56) have studied the applicability of adiabatic radial coordinate theories for heavy atom transfer reactions. Schor and Manz (57) also find that DIVAH model is successful for heavy atom transfer systems.

Similarly, it should be stressed that vibrational bonding is not limited to light atom transfer systems. Atabek and Lefebvre (35) have given an example for vibrational bonding in non-reactive systems. Pollak (36) and Meyer (37) have studied vibrational bonding in heavy atom transfer reactions.

Resonant Periodic Orbits

In principle, adiabatic potential energy curves may also be constructed semiclassically. If r is the radial (hyperspherical) coordinate and φ the (hyperspherical) angle, then for each fixed r one may evaluate an action integral for the angular motion. Quantizing this action would then yield the semiclassical estimate for the adiabatic potential. If the adiabatic approximation were exact, then a classical trajectory initiated at a barrier or well of the n-th adiabatic surface (with no radial momentum but with angular action corresponding to the n-th vibrational state) would stay at the well or barrier forever: it would be a periodic orbit (44,58). We have shown that this statement may be turned around: a periodic orbit is an adiabatic barrier or well of some adiabatic potential energy surface.

Consider then an adiabatic well in the hyperspherical coordinate system. Classically, the motion of the periodic orbit at the well would be an oscillation from a point on the inner equipotential curve in the reactant channel to a point on the same equipotential curve in the product channel. This is qualitatively the motion of what are termed "resonant periodic orbits" (RPO's). For example the RPO's of the IHI system are given in Fig. 5. Thus, finding adiabatic wells in the radial coordinate system corresponds to finding RPO's and quantizing their action. Note that in Fig. 5 we have also plotted all the periodic orbit dividing surfaces (PODS) of the system, except for the symmetric stretch. By definition, a PODS is a periodic orbit that starts and ends on different equipotentials. Thus the symmetric stretch PODS would be an adiabatic well for an adiabatic surface in reaction path coordinates. However, the PODS in the entrance and exit channels shown in Fig. 5 may be considered as adiabatic barrieres in either the radial or reaction path coordinate systems. Here, the barrier in radial coordinates, has quantally a tunneling path between the entrance and exit channels.

Thus the PODS are the classically allowed part of the angular motion.

The tunneling does cause though a more serious difficulty. The RPO's move across a double well potential so that to obtain a good semiclassical estimate of the quantal energies one would have to incorporate tunneling below the barrier and reflection contributions above the barrier (56). This difficulty can be circumvented by using a primitive semiclassical approach (45,46). Consider first the case of a symmetric exchange reaction. For each fixed value of r, the angular potential is a symmetric double well (or, at small r, a single well potential). The quantal states will be either gerade or ungerade with respect to reflection about the symmmetric stretch. As r goes to infinity, the barrier separating the two wells becomes very large so that for states well below the three-body dissociation energy, $U_n^g(r)=U_n^u(r)$. For large r the doublet n and n+1 vibrational states are well approximated semiclassically by an (n+1/2)h action condition in either well or overall by a (2n+1)h action condition over the double well. As r decreases, the doublet splits, $U_n^g < U_n^u$, as a result of tunneling and above-barrier reflection. However, one may define an average eigenvalue $U_{\bar{n}}(r)$ as

$$U_{\bar{n}}(r) = \frac{1}{2}\left(U_n^g(r) + U_n^u(r)\right) \tag{6}$$

and $U_{\bar{n}}(r)$ does have a well defined primitive semiclassical analog: The action over the double well is quantized by a (2n+1)h action condition. In other words, the resonant orbit with (2n+1)h action will give the adiabatic well energy and location of the $U_{\bar{n}}(r)$ adiabatic potential energy curve. For example, in Fig. 6 we compare the energies and locations of quantized RPO's of the IHI system with the quantal (averaged) adiabatic surfaces.

A similar method may also be devised for asymmetric reactions. Instead of considering the general case, we will deal with a specific system - the collinear FHH reaction; generalization to any asymmetric system is straightforward. As $r \to \infty$, one finds for the angular coordinate an asymmetric double well potential. Since the barrier separating the two wells is very large, the quantal states are well localized in either the entrance or exit channels; hence, we label these quantal states by an entrance channel index j or an exit channel index k. U_j and U_k are well approximated by (j+1/2)h or (k+1/2)h action conditions in the entrance or exit channel wells, respectively. If for a certain pair of indexes $U_j \simeq U_k$ then, just as in the symmetric case, the two states will mix as r decreases, because of tunneling or above-barrier reflection. In the FHH system, the lowest resonance occurs on an adiabatic potential energy curve which may be correlated with the asymptotic H_2(j=0) and HF(k=3) vibrational states (cf. Fig. 2). Using the primitive semiclassical prescription, we may approximate this well by looking for an RPO with $\{(j+1/2)h + (k+1/2)h\}$ = 4h action. Similarly, the next collinear resonance occurs on a well that mixes the asymptotic H_2(j=1) and HF(k=4) vibrational states and so may be approximated by an RPO with 6h action. Thus, in general, the adiabatic well responsible for a resonance may be approximated by an RPO with (j+k+1)h action, where j and k are determined by the correlation of the well with

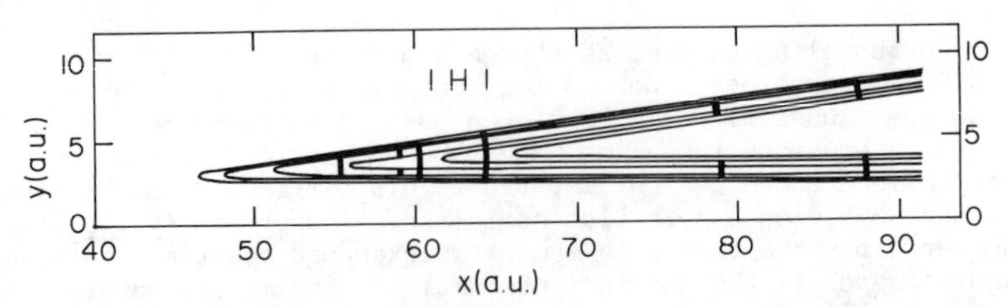

Figure 5. RPO's of the collinear IHI system on the LEPS A surface of Ref. (8,10). Equipotential contours are at 0.1, 0.5 and 0.9 eV relative to the bottom of the asymptotic HI well. The mass scaled coordinates x,y are defined in Fig. 4. The RPO's are the heavy lines denoting trajectories that have their two turning points on the same equipotential lines.

Reproduced with permission from Ref. 45.

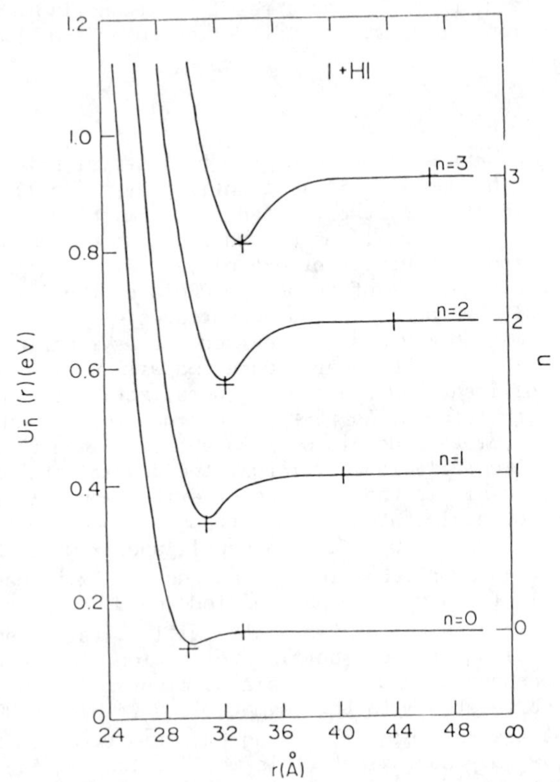

Figure 6. Average adiabatic surfaces for the IHI system. The crosses denote the energies and locations of the quantized orbits. The solid lines show $U_{\bar{n}}(r)$ determined from quantal computations. The coordinate on the right hand side denotes the asymptotic vibrational energies of HI.

Reproduced with permission from Ref. 45.

asymptotic reactant and product states of nearly the same energy. Given the adiabatic wells, one may use them to study the properties of resonances formed because of them. As we have shown elsewhere (60), given a periodic orbit one can evaluate the adiabatic frequency ω_A for motion perpendicular to the well or barrier formed by that orbit. We have found (44), in all systems we analyzed, that the adiabatic frequencies of the RPO's are real, confirming that RPO's actually are adiabatic wells.

One of the main advantages of dealing with periodic orbits is that they are uniquely determined by the electronic potential energy surface and the masses of the system. This is not true for the quantal adiabatic surfaces. As shown above, the quantal adiabatic approximation became quantitatively reliable only with the advent of DIVAH theory - that is one must incorporate diagonal adiabatic corrections within the adiabatic approximation. By studying the properties of RPO's we have shown that the adiabatic diagonal corrections are basically curvature corrections that are automatically incorporated in the RPO's (45).

In addition to the adiabatic frequency ω_A for the motion perpendicular to the RPO one can also determine the stability frequency ω_{st} or what is known as a characteristic eigenvalue (61) for the motion perpendicular to the RPO, by linearisation of the equations of motion about the periodic orbit. Pollak has shown (60) that the adiabatic frequency is just the first-order Magnus approximation to the stability frequency. This implies that when the adiabatic assumption is good then $\omega_A \simeq \omega_{st}$. For example, in the H_3 system, the frequency Ω of the RPO is $\sim$ 2000 cm^{-1} while $\omega_A \simeq 2200$ cm^{-1}, that is $\omega_A > \Omega$. Evidently the adiabatic approximation cannot be too good, in fact the RPO's are unstable, $|\omega_{st}| \approx 1300$ cm^{-1}. Note that in this case, qualitatively, the deviation of ω_{st} from ω_A is providing information on the extent of non-adiabatic transitions. In Ref. (42) we have shown that if the RPO is unstable then the quantal resonance width may be evaluated from the imaginary stability frequency. In an A + BC system, as the light atom B gets lighter and the atoms A,C get heavier, one expects the adiabatic approximation about the RPO to get better. This statement is in complete agreement with the corresponding findings of the quantal computations. In fact, for the IHI system, $\omega_A \approx \omega_{st}$ at all energies, which implies that the RPO is stable in IHI. For light atom transfer systems where the antisymmetric light-atom motion is much faster than the symmetric heavy-atom motion, the RPO is stable at almost all energies below the three-body dissociation limit (29,42). Classically we find a large family of bound quasiperiodic orbits although the potential energy surface employed was minimum free. These orbits were the first example of vibrational bonding.

Three-dimensional Treatment of Resonances and Vibrational Bonding

So far we have described the phenomenon of resonances and vibrational bonding for collinear systems. Of course, the real world is three-dimensional and hence it was of interest to extend these concepts to 3D systems. For the semiclassical theory Pollak and Wyatt (43,46) have shown that RPO's are useful for determining properties of resonances in 3D, too. The underlying idea is an adiabatic periodic reduction method (43). A general triatomic molecule undergoes rotational, bending and stretching motion. Usually the overall rotation is much slower than the vibrational motion because of the relatively large moment of inertia. Furthermore, it is often the case that bending motion is much slower than the antisymmetric stretching motion. Thus one may use a Born-Oppenheimer-type separation, in which one first freezes the rotational and bending angles, then finds the RPO at each set of fixed angles, and finally quantizes it. Each RPO corresponds automatically to a nearly frozen symmetric stretch motion. Thus we obtain an adiabatic well $U_n(\gamma)$ dependent on the bend angle γ. Averaging the Hamiltonian over the (fast) antisymmetric stretching motion of the RPO provides an effective Hamiltonian for the bending motion of the adiabatic well with stretch vibrational quantum number n. The quantized bend states are obtained by semiclassical quantization of the effective bend Hamiltonian. Finally one averages the Hamiltonian over the quantized bend motion to obtain the effective rotational resonance Hamiltonian.

To illustrate this method we consider in some detail the resonances of the 3D FHH system on the Muckerman V surface (50). As noted in the previous section, collinearly the resonance may be identified with an RPO having 4h action. Defining the bend angle γ via the Natanson-Smith-Radau (62,63) coordinate system, we find the 4h action RPO at fixed γ. This provides the curve $E_1(\gamma)$ (shown in Fig. 7) where $E_1(\gamma)$ is the energy of the γ dependent RPO. The kinetic energy for the bending motion has a γ dependent mass like coefficient $B_x^1(\gamma)$ (cf. Fig. 7) which may be found by averaging the total Hamiltonian over the motion of the γ dependent RPO. The effective Hamiltonian for the bending motion of the resonance is

$$h_1(\gamma) = B_x^1(\gamma)\, P_\gamma^2 + E_1(\gamma) \tag{7}$$

To find the lowest bend level one may quantize the bend motion semiclassically. This provides the bend energy level, shown as a dashed line on the lower right hand side of Fig. 7. Finally one averages the rotational constant $B_x^1(\gamma)$ over the bend motion to obtain the rotational constant B_1 for overall rotational motion. The energy dependence of the resonance on total angular momentum is simply $\hbar^2 B_1 J(J+1)$. The dependence of the resonance energy on J has been computed by Redmon and Wyatt (64) using a j_z conserving approximation, and by Walker and Hayes (65) using the approximate BCRLM methodology. Both of these studies agree well with the prediction of the semiclassical RPO theory as shown in Fig. 8.

We have also used the periodic reduction method to predict with good accuracy the 3D structure of vibrationally bonded molecules. It should be stressed though, that in principle it is not necessary to use periodic reduction. As shown in Fig. 9 the RPO's of the IHI system are stable also in 3D, one can find bound quasiperiodic orbits and quantize them semiclassically directly without resorting to periodic reduction.

A similar treatment of the 3D IHI system was given on a quantum mechanical level (34). For each value of the radial coordinate r we evaluated within a harmonic approximation the zero point bending

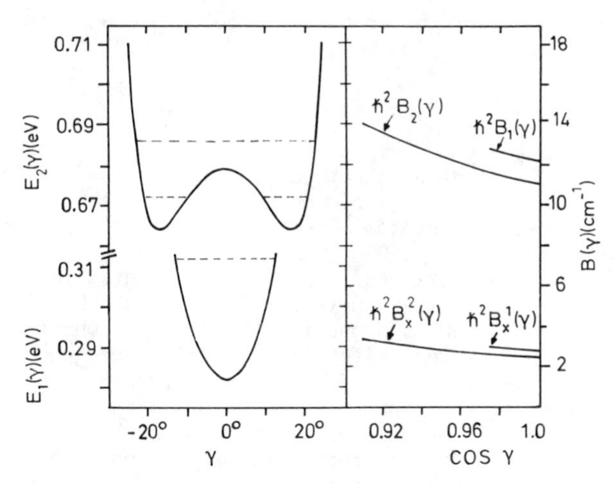

Figure 7. Adiabatic angle dependent potential energy curves $E_n(\gamma)$ and force constants $B_n(\gamma)$, $B_x^n(\gamma)$ for resonances in 3D FH_2 on the Muckerman V (50) potential energy surface. n=1,2 denotes the first and second vibrational resonance levels of the FH_2 system. The dashed lines denote the quantized bend levels. Note the bend level substructure for the n=2 case. For further explanation see text. **Reproduced with permission from Ref. 46.**

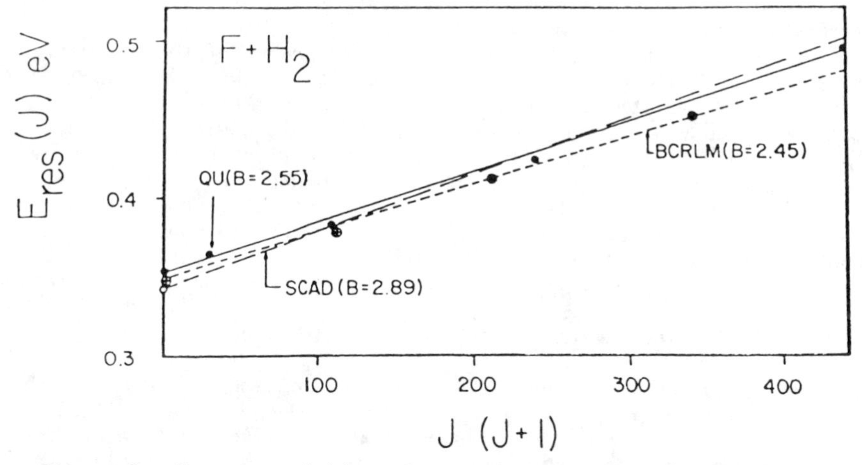

Figure 8: Energy variation of resonance energies for the 3D $F+H_2$ reaction. SCAD denotes the semiclassical adiabatic results based on the adiabatic reduction of RPO's. QU denotes J_z conserving results (46) and BCRLM are results from Ref. (63). The rotational constants B, in cm^{-1}, that emerge from the various treatments are indicated in parenthesis for each curve. **Reproduced with permission from Ref. 46.**

energy $\varepsilon_b^o(r) = h_{\omega b}$. Thus the ground state adiabatic potential in 3D includes a bend level contribution; i.e. in analogy to the collinear approximation

$$U_o^{3D}(r) = \varepsilon_{og}(r) + h_{\omega b}(r) \tag{8}$$

The different contribution are displayed in Fig. 10. Clary and Connor (32,40) computed the exact quantum mechanical, fully coupled 3D energy at zero total angular momentum J. All these studies (semi-classical, approximate quantum mechanical and exact quantum mechanical) found that for the LEPS A potential surface, addition of bending energy to the collinear bound-state energies leaves only one bound 3D, J=0 IHI state, while for IDI no truly bound state exists in 3D. These results prove that the phenomenon of vibrational bonding is not an artifical effect of the collinear geometry, although bending seems to weaken the bond considerably. Again, we find an anomalous isotope effect: substitution of hydrogen by deuterium causes destruction of the molecule instead of stabilisation. Because the LEPS A surface is only a crude estimate of the true electronic potential energy surface, further investigations have been undertaken using the more refined DIM-3C surface of Last (66), and the occurring interplay of the van der Waals interaction and vibrational bonding has been studied (38,39). From all of these investigations the following results emerged:
The dissociation energy for IHI → I+HI (or HLH → H+LH in general) has a contribution from vibrational bonding which apparently does not depend very much on the details of the electronic potential surface, but is mainly determined by the kinematic properties of the system. This contribution, due to variations of the antisymmetric stretching and the bending zero point energies, stabilizes IHI by approximately 6 kJ/mol. Normal mode frequencies of the adiabatic molecule are quite different from frequencies of the asymptotic diatoms. The IHI stretching frequency should be smaller than the corresponding vibrational frequency of the I_2 molecule. The bending frequency of IHI is larger than the symmetric stretching frequency, and probably only a few bending levels are bound. The antisymmetric stretch of IHI is predicted to possess the largest frequency. One observes destabilisation upon deuteration (the vibrational bonding contribution decreases to about 3 kJ/mol), and this isotope effect should be observable in temperature dependent-spectroscopic studies.

Experimental Evidence for Vibrational Bonded Molecules

The ultimate question regarding vibrational bonding is - can it be observed experimentally? At this point we do not have a definite answer, but the following discussion should convince the reader that it is worthwhile to pursue the problem. The spectroscopic predictions and estimates given at the end of the previous section show that in general a vibrationally bonded molecule has properties strikingly different from those of van der Waals molecules or hydrogen bonded molecules (while having about the same binding energy). Just these properties should allow an experimental verification.

For example, in the past decade Pimentel (67), Noble (67,69), Ault (69) and others (70) have studied in some detail the properties

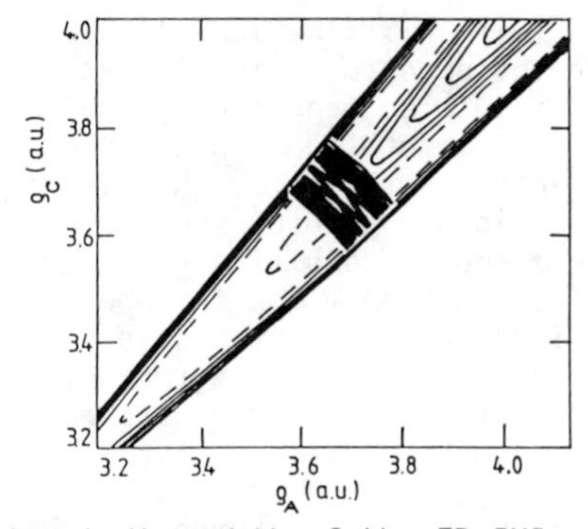

Figure 9. A quasiperiodic orbit of the 3D IHI system projected on the collinear plane. Potential contours are given at 0.1 eV intervals, the dashed contour is at 0.1 eV relative to the bottom of the asymptotic IHI well. The coordinates ρ_A, ρ_C are mass-scaled; for their definition, see Ref. (33).

Reproduced with permission from Ref. 33. Copyright 1983, North Holland.

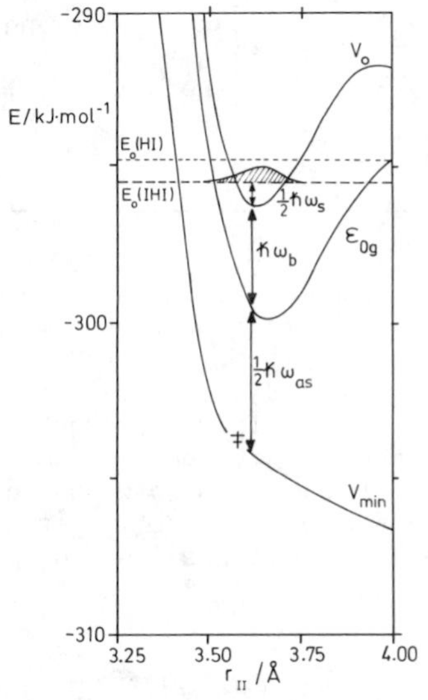

Figure 10. Potential energy for the symmetric IHI and IDI stretching mode U_0 as a sum of the repulsive collinear minimum potential energy V_{min} plus the zero-point energy of the anti-symmetric stretching mode plus that of the degenerate bending mode. The resulting IHI energy E_0 (IHI), as well as the HI ground state energy E_0 (IH) are given as a dashed and dotted line, respectively. The position of the saddle point is indicated by ≠. The asymmetric stretch, symmetric stretch and bend level zero point energies are denoted as $1/2\,\hbar\omega_{as}$, $1/2\,\hbar\,\omega_s$ and $\hbar\omega_b$ respectively.

Reproduced with permission from Ref. 34. Copyright 1983, North Holland.

of molecules like YHX^-(with X,Y = F,Cl,Br,I) isolated in matrices. In Table V we compare the spectroscopically determined normal mode frequencies of IHI^- and IDI^- as measured by Noble and the theoretical predictions based on our computations on the LEPS A surface. Of course the potential energy surface for IHI^- is different from the IHI surface assumed in our studies. But as noted many properties of such molecules are rather insensitive to the details of the potential energy surface. Not only the isotope effect of the experimental investigation is reproduced (cf. Table V), but even the absolute frequencies are in reasonable agreement. All this suggests that it is worthwhile to take a closer experimental look at the IHI (and IDI) systems in matrices.

Table V. Spectroscopic constants of three-dimensional IHI and IHI^-.

System	Reference	$\nu_1(cm^{-1})$	$\nu_2(cm^{-1})$	$\nu_3(cm^{-1})$
IHI^-	Noble (68)	120	?	682
IHI	Manz et al. (38)	137	280	878
IDI^-	Noble (68)	124	?	470
IDI	Manz et al. (38)	140	211	594
HI	Herzberg (71)			2308
DI	Herzberg (71)			1631

Another possibility is given by gas-phase studies using supersonic beams. Vibrationally bonded molecules are bound as tightly as van der Waals molecules; there is no reason why they should not be observable with the same techniques (72,73). So far the case that has been analyzed in detail has one bond breakup and a formation of a single bond; this represents a process in which basically only one frequency becomes weaker in the interaction region. In more complex systems more than one force constant may become weaker and the effect could be larger.

To conclude: At present vibrational bonding provides a challenge to experimentalists - can they verify its existence?

Acknowledgments

This work is a result of a close collaboration of the authors with Drs. J. Manz and R. Meyer. We would like to thank them for many stimulating discussions. E.P. would like to thank Prof. W.H. Miller and J.R. is indebted to Dr. M.S. Child for their kind hospitality during the writing of this review. We would like to thank Prof. D.G. Truhlar, Dr. S.K. Gray and Dr. E.L. Sibert for their critical comments on this manuscript. The kind support of Prof. S.D. Peyerimhoff is also gratefully acknowledged. This work was partially supported by the Director, Office of Energy Research, Office of Basic Energy Sciences, Chemical Sciences Division of the U.S. Department (Contract No.: DE-AC03-76SF00098) and by the Fonds der Deutschen Chemischen Industrie.

Literature Cited

1. Truhlar, D.G.; Kuppermann, A. J. Chem. Phys. 1970, 52, 3841; 1972, 56, 2232
2. Wu, S.F.; Johnson, B.R.; Levine, R.D. Mol. Phys. 1973, 25, 839
3. Schatz, G.C.; Bowman, J.M.; Kuppermann, A. J. Chem. Phys. 1973, 58, 4023; J. Chem. Phys. 1975, 63, 674, 685
4. Connor, J.N.L.; Jakubetz, W.; Manz, J. Mol. Phys. 1975, 29, 347
5. Wu, S.F.; Levine, R.D. Mol. Phys. 1971, 22, 881; Chem. Phys. Letters 1971, 11, 557
6. Baer, M. Mol. Phys. 1974, 27, 1429
7. Chapman, E.M.; Hayes, E.F. J. Chem. Phys. 1977, 66. 2554
8. Manz, J.; Römelt, J. Chem. Phys. Letters 1980, 76, 337
9. Kuppermann, A.; Kaye, J.A.; Dwyer, J.P. Chem. Phys. Letters 1980, 74, 257
10. Kaye, J.A.; Kuppermann, A. Chem. Phys. Letters 1981, 77, 573
11. Truhlar, D.G. J. Chem. Phys. 1970, 53, 2041
12. Latham, S.L.; McNutt, J.F.; Wyatt, R.E.; Redmon, M.J.; J. Chem. Phys. 1978, 69, 3746
13. Dwyer, J.P., Ph.D. thesis, 1978, California Inst. of Technology
14. Kuppermann, A. "Potential energy surfaces and dynamics calculations"; Truhlar, D.G.; Plenum Press, New York, 1981, p. 375
15. Babamov, V.K.; Kuppermann, A. J. Chem. Phys. 1982, 77, 1891
16. Babamov, V.K.; Marcus, R.A. J. Chem. Phys. 1981, 74, 1790
17. Launay, J.M.; LeDourneuf, M. J. Phys. B 1982, 15, L455
18. Garrett, B.C.; Truhlar, D.G. J. Phys. Chem. 1982, 86, 1136; 1983, 87, 4554 (E)
19. Manz, J.; Pollak E.; Römelt, J. Chem. Phys. Letters 1982, 29, 26
20. Stine, J.R.; Marcus, R.A. Chem. Phys. Letters 1975, 29, 575
21. Duff, J.W.; Truhlar, D.G. Chem. Phys. Letters 1976, 40, 251
22. Chapman, E.M.; Hayes, E.F. J. Chem. Phys. 1975, 62, 4400; 1976, 65, 1032
23. Noid, D.W.; Koszykowski, M.L. Chem. Phys. Letters 1980, 73, 114
24. Costley, J.; Pechukas, P. J. Chem. Phys. 1982, 77, 4957
25. Pollak, E.; Child, M.S. Chem. Phys. 1981, 60, 23
26. Gutzwiller, M. J. Math. Phys. 1971, 12, 343
27. Römelt, J. Chem. Phys. 1983, 79, 197
28. Bondi, D.K.; Connor, J.N.L.; Manz, J.; Römelt, J. Mol. Phys. 1983, 50, 467
29. Pollak, E. Chem. Phys. 1983, 78, 1228
30. Pollak, E. In "15th Jerusalem Symposium on Quantum Chemistry and Biochemistry: Intramolecular Dynamics" Jortner, J.; Pullmann, B., Eds.; Reidel Publ. Comp., Dordrecht, Holland, 1982, p. 1
31. Manz, J.; Meyer, R.; Pollak, E., Römelt, J. Chem. Phys. Letters 1982, 93, 184
32. Clary, D.C.; Connor, J.N.L. Chem. Phys. Letters 1983, 94, 81
33. Pollak, E. Chem. Phys. Letters 1983, 94, 85
34. Manz, J., Meyer, R., Römelt, J. Chem. Phys. Letters 1983, 96, 607
35. Atabek, O.; Lefebvre, R. Chem. Phys. Letters, 1983, 98, 559
36. Pollak, E. Chem. Phys. Letters 1983, 102, 319
37. Meyer, R. Chem. Phys. Letters 1983, 103, 63
38. Manz, J.; Meyer, R.; Pollak, E.; Römelt, J.; Schor, H.H.R. Chem. Phys. 1984, 83, 333.
39. Manz, J.; Meyer, R.; Schor, H.H.R. J. Chem. Phys. 1984, 80, 1562.
40. Clary, D.C.; Connor, J.N.L. J. Phys. Chem. in press

41. Pollak, E., Comment on At. Mol. Phys., in press
42. Pollak, E. J. Chem. Phys. 1982, 76, 5843
43. Pollak, E.; Wyatt, R.E. J. Chem. Phys. 1982, 77, 7689; 1983, 78, 4464
44. Pollak, E. In "Theory of Chemical Reaction Dynamics"; Baer, M., Ed.; CRC Press, New York, 1984, to be published
45. Pollak, E.; Römelt, J. J. Chem. Phys. in press
46. Pollak, E.; Wyatt, R.E. J. Chem. Phys. in press
47. Marston, C.C.; Wyatt, R.E. J. Chem. Phys. in press
48. Hauke, G.; Manz, J.; Römelt, J. J. Chem. Phys. 1980, 73, 5040 Römelt, J. Chem. Phys. Letters 1980, 74, 263
49. Porter, R.N.; Karplus, M. J. Chem. Phys. 1964, 40, 1105
50. Muckerman, J.T. In "Theoretical Chemistry: Theory of Scattering: Papers in Honor of Henry Eyring; Henderson, D., Ed.; Academic, New York, 1981, Vol. 6A, p. 1
51. Note that the adiabatic potentials displayed in Fig. 2 are different from those of Launay and LeDourneuf (17). These authors choose a different origin for their radial coordinate system.
52. Hiller, C.; Manz, J.; Miller, W.H.; Römelt, J. J. Chem. Phys. 1983, 78, 3850
53. Skodje, R.T.; Schwenke, D.W.; Truhlar, D.G.; Garrett, B.C. J. Phys. Chem., in press
54. Skodje, R.T.; Schwenke, D.W.; Truhlar, D.G.; Garrett, B.C. J. Chem. Phys. in press
55. Aquilanti, V.; Grossi, G.; Lagana, A. Chem. Phys. Letters 1982, 93, 174, 179
56. Aquilanti, V.; Cavelli, S.; Grossi; Lagana, A. J. Mol. Structure 1983, 93, 319
57. Manz, J.; Schor, H.H.R.; Chem. Phys. Letters in press
58. Pollak, E. J. Chem. Phys. 1981, 74, 5586; 1981, 75, 4435
59. Pollak, E.; Child, M.S.; Pechukas, P. J. Chem. Phys. 1980, 72, 1669
60. Pollak, E. Chem. Phys. 1981, 61, 305
61. Arnold, V.I. Mathematical Methods of Classical Mechanics Springer Verlag, New York, 1980, Chap. 5
62. Adamov, M.N.; Natanson, G.A. Vestn. Leningr. Univ. 1973, 4, 28
63. Smith, F.T. Phys. Rev. Lett. 1980, 45, 1157
64. Redmon, M.J.; Wyatt, R.E. Chem. Phys. Lett. 1979, 63, 209
65. Walker, R.B.; Hayes, E.F. J. Chem. Phys. 1984, 80, 246
66. Last, I. Chem. Phys. 1983, 69, 193
67. Noble, P.N.; Pimentel, G.C. J. Chem. Phys. 1968, 49, 3165
68. Noble, P.N. J. Chem. Phys. 1972, 56, 2088
69. Ault, B.S. Acc. Chem. Res. 1982, 15, 103, and Refs. therein
70. Milligan, D.E.; Jacox, M.E. J. Chem. Phys. 1971, 55, 2550; 1975, 63, 2466
71. Herzberg, G. "Spectra of Diatomic Molecules"; van Nostrand, Princeton, 1950
72. Levy, D.H. Adv. Chem. Phys. 1981, 47, 323
73. Beswick, J.A.; Jortner, J. Adv. Chem. Phys. 1981, 47, 363

RECEIVED June 11, 1984

20

Bimolecular Reactive Collisions

Adiabatic and Nonadiabatic Methods for Energies, Lifetimes, and Branching Probabilities

BRUCE C. GARRETT[1], DAVID W. SCHWENKE[2], REX T. SKODJE[2], DEVARAJAN THIRUMALAI[2], TODD C. THOMPSON[2], and DONALD G. TRUHLAR[2]

[1]Chemical Dynamics Corporation, Columbus, OH 43220
[2]Department of Chemistry, University of Minnesota, Minneapolis, MN 55455

Several approximate methods for calculating resonance energies and widths for atom-diatom reactive collisions are discussed. In particular, we present resonance energy calculations by semiclassical and quantal vibrationally adiabatic models based on minimum-energy and small-curvature paths, by the semiclassical SCF method, by quantal SCF and configuration-mixing methods, and by close coupling calculations. We also present total width calculations based on analytic continuation by polynomials and Padé approximants of configuration-mixing stabilization graphs, and we present total width and partial-width calculations based on close coupling calculations and on the Feshbach formalism in reaction-path coordinates with a small-curvature tunneling approximation for adiabatic decay and a reaction-path-curvature coupling operator for nonadiabatic decay. The model calculations are judged by their agreement with the accurate close coupling calculations, and we also compare the resonance energies and total widths to values obtained semiclassically from resonant periodic orbits. To illustrate the methods we consider the collinear reactions H + FH → HF + H and D + FD → DF + D on the low-barrier model potential of Muckerman, Schatz, and Kuppermann and the collinear and three-dimensional H + H_2 reactions on Porter-Karplus surface number 2. Finally we use an accurate potential energy surface for the three-dimensional H + H_2 reaction to predict the energies of several series of observable resonances for a real system.

Resonances in reactive collisions were first observed in quantum mechanical scattering calculations for the collinear H + H_2 reaction (1-9; for a review of early calculations on this system see reference 10 and for a recent review of the quantum mechanical treatment of reactive resonances see reference 11). Resonances

0097-6156/84/0263-0375$07.50/0

have subsequently been found in many converged quantum scattering calculations for this and other collinear reactions on realistic potential energy surfaces (8,11-61) and for two coplanar reactions (61,62) and in close coupling and coupled states calculations for two three-dimensional reactions (61,63-66; further results from the study of reference 64 are given in 67) as well as many approximate three-dimensional calculations (see, for example, references in 68,69, and other chapters in this volume). Resonances are observed in quantum scattering calculations as oscillations in the reaction cross sections (or probabilities) as functions of energy (10,11,69,70), and they can also be identified from eigenphase sums (71-74), Argand diagrams (75-78), lifetime analyses (79-86), wavefunction analyses (87,88), and direct calculation of poles of the reaction amplitude density on the real axis (89) or of poles of the resolvent (90,91) or the scattering matrix (86,92-97) in the complex energy plane.

Resonances are defined formally as poles of the scattering matrix in the complex energy or momentum plane (98-100). The pole location in the complex energy plane may be written as

$$\varepsilon_{res} = E_{res} - i\Gamma/2. \quad (1)$$

At such energies the Schroedinger equation has a solution with purely outgoing waves at large distances from the origin; this solution is a pure resonance state (101). Depending on the context either ε_{res} or E_{res} is called the resonance energy and Γ is called the width. The width notation is appropriate because if an incoming wavepacket has an energy spread large compared to Γ, a part of this wavepacket of width Γ will be delayed in the interaction region due to the resonance (98). The effect of a resonance on scattering attributes is easy to describe only for the case of an isolated, narrow resonance (INR); i.e., a resonance which is separated from other resonances by much more than the sum of their widths and which is narrow enough that the background ("direct") scattering does not vary appreciably over its width. (For discussions of overlapping resonances see references 102-104.) An INR decays by first-order kinetics with a rate constant equal to $\Gamma/\hbar$ (99,100,105); its contribution to a collisional delay time is $4\hbar/\Gamma$ at the energy E_{res} and is $2\hbar/\Gamma$ averaged over the resonance width (79,85). If Γ is small then the wave function for real energies close to E_{res}, corresponding to scattering energies or energies of predissociating complexes formed by absorption of electromagnetic radiation, may be very similar in the interaction region to the resonant state at the nearby complex energy ε_{res}. Since resonance states tend to be more localized than typical continuum states, they are easier to interpret and classify, and since the resonant state affects the scattering processes connecting all initial and final channels for a finite range of energies, identification and characterization of resonances provides a unifying feature for a variety of observable dynamical phenomena. In particular, resonances may be classified in terms of system quantum numbers just as usefully as bound states are so classified.

For scattering transition probabilities and photon-absorption oscillator-strength distributions, an INR contributes an energy-dependent feature with full width at half maximum (FWHM) of Γ. In the absence of background contributions, this feature is Lorentzian and is centered at E_{res}; otherwise it may interfere constructively and/or destructively with the background to yield a more complicated shape function for the energy dependence. Although an INR exhibits the same FWHM for all initial and final states, it does not have the same amplitude for all processes. The relative amplitude for a given transition $\alpha \rightarrow \alpha'$ is controlled by the residue of the α',α element of the scattering matrix at the resonant pole. In the vicinity of its pole this element may be written

$$S_{\alpha'\alpha} = S^{d}_{\alpha'\alpha} - \frac{i\gamma_{\alpha'}\gamma_{\alpha}}{E - \varepsilon_{res}}, \qquad (2)$$

where E is the total energy, $S^{d}_{\alpha'\alpha}$ is the direct part, which is regular, and (99,100,105,106)

$$\sum_{\alpha} |\gamma_{\alpha}|^{2} = \Gamma. \qquad (3)$$

It is convenient to define the quantities

$$\Gamma_{\alpha} = |\gamma_{\alpha}|^{2} \qquad (4)$$

which are called partial widths and the quantities

$$\chi_{\alpha} = \gamma_{\alpha}\, \Gamma^{-\frac{1}{2}} \qquad (5)$$

and

$$P_{\alpha} = |\chi_{\alpha}|^{2} \qquad (6)$$

which, respectively, are components of normalized scattering matrix eigenvectors and branching fractions or probabilities for the resonance to decay into state α. The latter interpretation is evident by noting that in the absence of direct scattering the resonance leads to a factorizable Lorentzian transition probability of the form

$$|S^{r}_{\alpha'\alpha}|^{2} = \frac{\Gamma^{2}}{(E-E_{res})^{2} + \Gamma^{2}/4}\, P_{\alpha'}P_{\alpha},\ \alpha' \neq \alpha, \qquad (7)$$

and noting

$$\sum_{\alpha} P_{\alpha} = 1. \qquad (8)$$

Clearly a complete description of an INR requires specification of E_{res}, Γ, and all γ_{α} or χ_{α}.

If a resonance is not narrow, then the partial widths may still be defined in terms of the residue at the pole, but Equation 4 is no longer valid (92,93,107). Branching probabilities are now defined by

$$\frac{P_\alpha}{P_{\alpha'}} = \frac{|\gamma_\alpha|^2}{|\gamma_{\alpha'}|^2} \tag{9}$$

plus the normalization condition of Equation 8.

Reactive resonances are interpreted as long-lived, quasibound complexes in the interaction region of the potential energy surface. The lowest-energy resonance of the collinear $H + H_2$ reaction was first interpreted in semiclassical terms as the result of the interference between direct and long-lived quasiclassical trajectories (5,108); in the resonance picture, the interference is between direct, short-lived, background scattering and long-lived trajectories representing the resonance. Wu and Levine (2) interpreted this resonance as an adiabatically trapped quasibound state. In the adiabatic model for a collinear atom-diatom reaction, vibrationally adiabatic potential curves $V_a(n,s)$ are constructed by adding local vibrational stretching energies $\varepsilon_{str}(n,s)$ to the Born-Oppenheimer potential $V_{MEP}(s)$ as a function of the distance s along the minimum energy path (MEP),

$$V_a(n,s) = V_{MEP}(s) + \varepsilon_{str}(n,s) \tag{10}$$

where n is the vibrational state for the motion perpendicular to the reaction coordinate. The lowest-energy resonance of H_2 was interpreted as a quasibound state in the local well of the first excited adiabatic potential curve. This view was confirmed by analysis of the scattering wavefunction which showed that it had over 90% vibrationally excited character when projected onto wavefunctions of the symmetric stretch vibration (109).

The physical picture of a resonance as a long-lived, quasibound state has lead to the development of several approximate methods for calculating the locations and widths of resonances. The adiabatic model (see for example references 110-115) in reaction-path coordinates, also called natural collision coordinates, has been very successful for predicting both the locations and adiabatic partial widths for reactive resonances (46,56,116-119). Using Feshbach resonance theory (102,103), nonadiabatic partial widths can also be computed within the same framework (56,57,119, see also 120). In this method the nonadiabatic partial width is given by a golden-rule formula in which the operator is a reaction-path-curvature coupling term. The adiabatic approximation has also been used in other coordinate systems to locate reactive resonances, for example, the use of hyperspherical coordinates has been very successful for predicting resonance energies, especially for collinear heavy-light-heavy systems (11,47,121-126). In quantum mechanical stabilization calculations (127-130), bound-state linear variational methods are used instead of scattering calculations to calculate self-consistent-field (SCF) or configuration-interaction (CI) wavefunctions for the quasibound state; this method has recently been applied to bimolecular reactive resonances (131-135). Resonances are identified as the lowest-energy roots of the secular equation that are stable with respect to variation of a basis set parameter, e.g., the number of primitive basis functions or a scaling parameter. Total widths are obtained by analytically continuing approximate fits of the energy root, or two or more of them, as functions of the scaling parameter (133,135-138). [Although it is not considered here, the complex coordinate method provides a means of obtaining the complex resonance energy ε_{res} directly; i.e., without analytic continuation, using bound-state linear variational techniques (see for example, 91)]. The semiclassical SCF procedure (139) is essentially a generalization to multidimensional systems of the primitive WKB approximation (140) for bound state eigenvalues of one-dimensional potentials. This method has also been applied to locating reactive resonances (141-143) and evaluating the widths (142). Resonant periodic orbits (RPOs) are the basis of yet another pseudo-bound-state method for predicting resonances; RPOs are classical trajectories trapped in the interaction region of the potential energy surface with integer values of the classical action (144-147). These trajectories have been used to compute resonance energies, and widths are estimated from the instability of the RPOs (145).

To date, most of the work on reactive resonances has dealt with methods for calculating the resonance energy E_{res} and the total width Γ. However, a great deal can be learned from examination of the separate contributions of different decay channels to the total width. For example, as discussed above, branching ratios for decay into different channels are obtained from ratios of the partial widths or residue factors for those decay mechanisms. In the terminology of Taylor et al. (148) Feshbach internal-excitation resonances in adiabatic state n are denoted core-excited type I resonances if the resonance energy is below the threshold energy E_n^{thr} for this state and are called core-excited type II resonances if the resonance energy is above E_n^{thr} where the threshold energy is defined by

$$E_n^{thr} = \min[V_a(n,s=-\infty),V_a(n,s=+\infty)] \tag{11}$$

Core-excited type I resonances are sometimes called Feshbach resonances; they have only nonadiabatic contributions to the total width, but core-excited type II resonances have partial widths both for nonadiabatic decay in which n changes and for adiabatic decay by tunneling through the adiabatic barriers without changing n. Quasibound states of ground-state adiabatic potential curves are single-particle shape resonances, and they decay only adiabatically if they lie below the first excited-state threshold energy. Vibrational states of the ground-state adiabatic potential curve which lie below the ground-state threshold are truly bound states, and this phenomenon has been termed vibrational bonding (149-151). For larger polyatomic complexes more decay channels exist, e.g., decay can also occur nonadiabatically by intramolecular vibration-to-vibration energy transfer. Limited, but significant, progress has been made in developing methods to obtain partial widths from complex wavefunction calculations, e.g., by the Siegert method (107,152,153) or the complex-coordinate method (154,155). In the present article we summarize our recent work (56,57) using the Feshbach golden rule to calculate partial widths from real-valued wavefunctions.

In the present work we are interested in assessing the accuracy of several approximate methods for locating resonance energies and computing total widths and adiabatic and nonadiabatic partial widths. After a summary of the methods considered here we review and compare the results of these methods for the collinear H + FH → HF + H and D + FD → DF + D reactions on the low-barrier model potential surface of Muckerman, Schatz and Kuppermann (34). This system was chosen for review here because there are reliable quantum mechanical calculations of the resonance energies and widths (34,46,57) as well as several studies by approximate methods (46,57,117,118,131-133,135,141). Also, these systems display all three types of resonances, shape and type I and type II core-excited. We also review calculations on the lowest energy resonance of the collinear H + H_2 reaction on the Porter-Karplus surface number 2 (PK2,156). This resonance has also been thoroughly studied both by converged close coupling calculations (for example 4,61) and approximate methods (for example, for the methods discussed in this paper, see references (11,56,116-118,121,122,124,131-133,135,144-146). With a few exceptions, nearly all calculations of resonance energies and widths have been for collinear systems. We are also interested in the ease with which the approximate methods are extended to atom-diatomic reactions in three dimensions and to reactions involving more than three atoms. Therefore, we present new calculations of resonance energies for the three-dimensional H + H_2 reaction on the PK2 surface and on the accurate potential energy surface of Liu, Siegbahn, Truhlar, and Horowitz (LSTH, 157-159). Because of its current experimental interest (160-163), we also report calculations for the three-dimensional H + D_2 reaction on the LSTH surface.

Methods

Quantum mechanical scattering. In conventional quantum mechanical scattering calculations, resonances show up as oscillations in the reactive cross sections (or reaction probabilities) or an increase of π in the eigenphase sum. [Narrow resonances are difficult to locate because of the need to take very small energy steps to find them. The difficulty in using the eigenphase sum is that usually it is only known modulo π and unless the eigenphase sum is tracked through a resonance with sufficiently small energy steps, the resonance is missed. The definition of an absolute eigenphase sum (74) removes this difficulty and simplifies the task of locating narrow resonances.] Once the resonance is located, the width may be obtained by fitting the eigenphase sum in the vicinity of the resonance to a generalized Breit-Wigner form (72,73). The partial widths may be extracted from an analysis of the individual scattering matrix elements (46,56,164). The method used for the results presented here is given in the next paragraph.

Converged close coupling calculations were carried out using the R matrix propagation method of Light and Walker (165). These calculations yield the scattering matrix S as a function of energy E. There are several possible procedures one could use to extract resonance energies, total widths, and partial widths from close coupling calculations. For example, Macek and Burke (166) made 16-25-parameter nonlinear least-squares fits of reactance matrices to a resonant form analogous to Equation 6 to extract partial widths from 3- and 4-channel electron scattering calculations, and Fels and Hazi (167) used a nonlinear fitting procedure based on the variation of the eigenphases in the vicinity of a resonance and a transformation between partial widths corresponding to eigenchannels and those corresponding to physical channels to extract partial widths for a 2-channel model problem. A procedure for directly extracting the partial widths corresponding to physical channels for problems with an arbitrary number of channels was proposed by Ashton et al. (164), and a slightly modified version of this procedure was used in our previous work (56,57). In our work we first fit the eigenphase sum $\Delta(E)$ in the vicinity of $E = E_{res}$ to the INR form (72,73)

$$\Delta(E) = \Delta_0(E) + \arctan \frac{\Gamma}{2(E_{res} - E)} \tag{12}$$

where $\Delta_0(E)$ is a low-order real polynomial in E representing the background; this yields E_{res} and Γ. In step 2 each element of the upper triangular part of the scattering matrix is fit in the vicinity of $E = E_{res}$ to

$$S_{\alpha'\alpha} = S^d_{\alpha'\alpha} - \frac{i\, C_{\alpha'\alpha}}{E - \varepsilon_{res}} \tag{13}$$

where $S^d_{\alpha'\alpha}$ is a low-order complex polynomial in E representing the background and $C_{\alpha'\alpha}$ is an additional complex fitting parameter. For step 3, Ashton et al. suggested minimizing the function

$$D_3 = \left\{ \sum_{\alpha} \sum_{\alpha' \leq \alpha} \left[\frac{|C_{\alpha'\alpha}| - (\Gamma_{\alpha'}\Gamma_\alpha)^{1/2}}{|C_{\alpha'\alpha}|} \right]^2 \right\}^{1/2} \tag{14}$$

with respect to the set of Γ_α subject to the constraint

$$\sum_{\alpha} \Gamma_\alpha = \Gamma \tag{15}$$

in order to obtain a "best" set of estimates of the partial widths consistent with the INR constraint of Equation 3. [This is the procedure we used in references (56,57) except we included a spurious factor of one half in Equation 2 and hence also in Equation 14; this affects the quantitative values of the partial widths, but because of the constraint of Equation 15, the errors are small and they do not affect the qualitative trends. The partial widths given in this chapter are new values calculated without the spurious factor of one half and are based on D_4 presented below rather than D_3.]

Equating Equations 2 and 13 and using Equation 4 yields

$$|C_{\alpha'\alpha}| = \Gamma_{\alpha'}^{\frac{1}{2}} \Gamma_{\alpha}^{\frac{1}{2}} \qquad (16)$$

For the three collinear reactive resonances for which a partial width analysis is discussed in the present article, Equation 16 yields partial widths that sum to only about 50-90% of the total width obtained from Equation 12. This indicates a breakdown of the INR condition. Obtaining the best INR representation by using Equations 14 and 15 systematically changes the branching probabilities away from the values computed using Equations 4,9, and 16. We found that the INR constraint could be enforced without distorting the branching probabilities by replacing Equation 14 by

$$D_4 = \left\{ \sum_{\alpha} \sum_{\alpha' \leq \alpha} \left[|C_{\alpha'\alpha}| - (\Gamma_{\alpha'}\Gamma_{\alpha})^{\frac{1}{2}} \right]^2 \right\}^{\frac{1}{2}}, \qquad (17)$$

and Equation 17 (with the INR constraint of Equations 3 and 4) was used for all the results presented here.

When the INR approximation breaks down it may be better to use Equation 16 than to force satisfaction of the sum rule of Equation 3, but space does not permit further discussion of this point in this chapter. We do note though that the Feshbach formalism discussed below and used to compute approximate partial widths is consistent with the sum rule (102,103).

Vibrationally adiabatic models. In the adiabatic approximation the dynamics of a multidimensional system is reduced to motion in one mathematical degree of freedom governed by an effective Hamiltonian. Calculations of the resonance energies and widths for this one-dimensional model can be performed either semiclassically or quantum mechanically.

The reaction probabilities that determine some of the E_{res} and the adiabatic partial widths are nonzero only because of tunneling. We have considered two semiclassical methods to calculate the resonance energy and tunneling probabilities, namely the primitive WKB method (140), which simply quantizes the phase integral for motion in the well of the adiabatic potential, and a uniform semiclassical method (168,169), which also uses phase integrals in the classically allowed regions, but combines them with phase integrals for classically forbidden motion in the barrier region of the adiabatic potential curve. In the present work we use the primitive WKB method for energies of core-excited type I resonances, and we use the uniform expression for energies of shape and core-excited type II resonances. To obtain the adiabatic partial widths, it is necessary to use a uniform semiclassical expression (168,169). The calculation of the nonadiabatic partial widths is discussed below.

Different adiabatic models are possible depending upon the method used to incorporate reaction-path curvature effects in the kinetic energy term of the effective Hamiltonian (120). Several approximations have been used: in the minimum-energy-path (MEP) approximation the system is assumed to follow the minimum energy path [this has also been called the zero-curvature approximation (5)]; in the Marcus-Coltrin-path approximation the system is assumed to follow the path of outer turning points for the vibrational motion normal to the MEP (116,120,170); in the dynamical path (DP) approximation the system is assumed to follow a path on which the internal centrifugal forces are balanced by the potential energy surface (117); and in the small-curvature (SC) approximation the system is assumed to follow the MEP in classically allowed regions and to follow a path defined by the small-curvature tunneling approximation (120,171) in tunneling regions. In previous work (46,56,116,117) it was found that the methods which include the effects of reaction-path curvature generally give a better description of the adiabatic partial width. The difference between resonance energies calculated using these methods were small; typically the agreement was better than a few tenths of a kcal/mol. The relative agreement was not as good for the adiabatic partial widths obtained from these methods; differences were as large as a factor of two. In the following comparison, we limit our discussion to the SC approximation.

Three different approximations to the local vibrational energy $\varepsilon_{str}(n,s)$ have been considered: in Morse approximations I and II (172) the potential along the cut perpendicular to the reaction coordinate is fit to a Morse potential and the eigenvalues are given by an analytic formula (173); and in the WKB approximation the primitive semiclassical quantization condition is used to find $\varepsilon_{str}(n,s)$ for the actual potential along the cut perpendicular to the MEP. The WKB method was found to give a better description of the adiabatic potential curve near the adiabatic barrier maxima; however, the adiabatic potential curves are not described well by this method in regions of large reaction-path curvature because of a breakdown of the natural collision coordinates (118). For the quantum mechanical calculations only the Morse I approximation was used. When calculating E_{res} and Γ using the semiclassical methods, the Morse I approximation was found to give more reliable estimates of the resonance energies for the H + FH and D + FD reactions, whereas the Morse II approximation was better for $H + H_2$. For the collinear systems considered here, we will tabulate and discuss only those semiclassical resonance energies obtained using the adiabatic barriers computed using Morse approximation I, and for the three-dimensional cases considered here we only tabulate and discuss semiclassical results obtained by the Morse approximation II. The adiabatic partial widths have an exponential dependence upon the phase integrals in the tunneling regions of the adiabatic potential curves. Therefore, the calculated adiabatic partial widths have been found to be very sensitive to the energy at which they are computed and to the shape of the adiabatic barrier in the tunneling region (56,57,118). The most reliable estimates are obtained when the phase integrals are computed at the accurate resonance energy using the adiabatic potential curves obtained from the WKB approximation (118).

In the quantum mechanical calculations the resonance energy and widths are obtained by calculating adiabatic reaction probabilities and fitting these to a Lorentzian form. We have made calculations (46,116) in which reaction-path curvature is neglected (denoted the MEPVA method) and in which it is included using the Marcus-Coltrin path (denoted the MCPVA method). The differences between resonance energies calculated using these two methods were small (less than 0.1 kcal/mol for the cases studied) but the relative differences between adiabatic partial widths were larger (almost a factor of 2 in one case). Although the MCPVA results are presumably more accurate, we report here only the results of MEPVA calculations since they are available for more of the systems studied here.

Nonadiabatic Feshbach calculations. Using the reaction-path Hamiltonian and invoking an adiabatic separation of the reaction coordinate from all other coordinates, resonance energies and adiabatic partial widths are obtained by neglecting all off-diagonal terms of the Hamiltonian. The most important off-diagonal, nonadiabatic terms of the Hamiltonian are matrix elements of the kinetic energy term which includes reaction-path curvature. The diagonal elements of the reaction-path curvature operator are included, at least approximately, in the adiabatic calculations and the off-diagonal elements give rise to the nonadiabatic partial widths. The nonadiabatic partial width for decomposition of the resonance state into channel α is given by (56,57)

$$\Gamma^{N}_{nv;\alpha} = 2\pi \left[\langle \Psi^{res}_{nv} | H_{QP} | \Psi_{\alpha}(E_{res}) \rangle \right]^2 \quad (18)$$

where the resonance state is characterized by quantum numbers n and v for the vibrational and reaction coordinate motions, respectively. The wavefunction for the resonance state is approximated by

$$\Psi^{res}_{nv} = f_{nv}(s)\, \phi_n(u,s) \quad (19)$$

where ϕ_n is the adiabatic basis function for motion perpendicular to the reaction coordinate and $f_{nv}(s)$ is the bound-state wavefunction for the trapped motion in the adiabatic well. The wavefunction of the continuum state for channel α at the resonance energy E_{res} is approximated by

$$\Psi_\alpha(E_{res}) = f_\alpha(s)\, \phi_{n_\alpha}(u,s) \tag{20}$$

where $\phi_{n_\alpha}(u,s)$ is the adiabatic basis function for state n_α and $f_\alpha(s)$ is the continuum wavefunction for motion along the reaction coordinate in the adiabatic potential for state n_α, subject to the usual scattering boundary conditions. The coupling operator is approximated by

$$H_{QP} = \tfrac{1}{2}\,[H^{(I)} + h.c.] \tag{21}$$

where

$$H^{(I)} = -\frac{\hbar^2}{2\mu}\,[1+2\kappa(s)u]\,\frac{\partial^2}{\partial s^2} \tag{22}$$

h.c. is the hermitean conjugate of $H^{(I)}$, and $\kappa(s)$ is the curvature of the reaction coordinate. In previous work (57), the effect of anharmonicity and the neglect of parts of the coupling matrix elements on the nonadiabatic partial widths have been studied. In the present paper we report only those partial widths based on harmonic oscillator wavefunctions for the bound vibrational motion perpendicular to the reaction coordinate and including the effect of the reaction-path coupling operator on both vibrational and reaction coordinate wavefunctions.

Quantum mechanical stabilization calculations. Quantum mechanical stabilization calculations provide a very convenient and efficient way to estimate the resonance characteristics by bound-state techniques. The first step of the method as we apply it involves calculating the eigenvalues of the Hamiltonion in a square-integrable basis as a function of a suitable basis-set scaling parameter. This basis may be restricted to a product of linear combinations of single-mode functions, which yields SCF stabilization, or it may be general, which is called configuration-mixing or configuration-interaction (CI) stabilization. The resonant eigenvalue $\varepsilon(\alpha)$ is then fit to a suitable polynomial (133) or appropriate Padé approximant (135) or $\varepsilon(\alpha)$ is represented as a polynomial root (174) and polynomial coefficients are fit (138). The stability condition is then employed to find the complex solution α_r of

$$d\varepsilon/d\alpha\,\big|_{\alpha=\alpha_r} = 0. \tag{23}$$

Finally the resonance position E_{res} and its width Γ may be approximated by

$$\varepsilon(\alpha_r) = E_{res} - i\Gamma/2 \tag{24}$$

with $\varepsilon(\alpha_r)$ computed from the fit. An important advantage of the stabilization method is that one can profitably use well developed bound-state methods to calculate $\varepsilon(\alpha)$ efficiently. In addition, one can treat systems involving many degrees of freedom more easily than by quantum mechanical scattering calculations, which may become prohibitively expensive. If scattering calculations are indeed too expensive to carry out for a given system, then stabilization results may be invaluable for qualitative interpretative purposes even if they are only semiquantitative in accuracy. It is useful to point out that the stabilization method does not rely on any fit to the shape of the cross section or scattering matrix as a function of energy.

Semiclassical SCF calculations. The use of the semiclassical SCF procedure to treat reactive resonances (141) is a straightforward extension of the work of Gerber and Ratner (139). Numerical complications do arise, however, and convergence of the SCF procedure can be much more difficult in the semiclassical method than in the quantum mechanical one or even impossible. A similarity to the quantum mechanical SCF procedure is that both give a clear picture of where the probability density for the resonance is localized (132,134,141).

Numerical comparison of methods for collinear reactions.

Resonance energies. In Table I, the resonance energies for three collinear atom-diatom reactions are compiled. These include three resonances each for the H + FH and D + FD reactions and the lowest-energy resonance for $H + H_2$ on the PK2 surface. All energies are relative to the minimum of the asymptotic reactant well.

In Table I, n and v are quantum numbers in natural collision coordinates: n is the quantum number of the vibrational motion perpendicular to the reaction coordinate and v is the quantum number for the bound reaction-path motion in the well of the adiabatic potential. These identifications were originally based on the vibrationally adiabatic approximation (46,117) and were confirmed by quantal stabilization calculations (131). For the systems included in Table I, reaction-path curvature is only small to moderate and we expect the adiabatic model in natural collision coordinates to give a good description of the resonance energies. Using a quantum mechanical treatment in the adiabatic model (MEPVA) gives slightly better results than using the semiclassical model (SCSA). The results of the adiabatic approximation in hyperspherical coordinates are only available for one of the three systems reported here. Although the earlier hyperspherical results (11) are not nearly as good as the SCSA or MEPVA ones for the $H + H_2$ system, including diagonal corrections in the effective Hamiltonian greatly improved the results (124). Also, we expect the hyperspherical coordinates to give a better description for systems with large reaction-path curvature. In the stabilization calculations no

Table I. Resonance Energies (kcal/mol) for Three Collinear Atom-Diatom Reactions

System	n	v	accurate quantal	MEPVA	SCSA[a]	Stabilization[b] SCF	Stabilization[b] CI	SCSCF[c]	hyper-spher.	RPOs[d]
H + FH	0	0	6.29[e]	6.34[f]	6.50	6.74	6.31	6.42	...[g]	...
	1	0	15.35[e]	...	15.22	14.57	15.41	14.25	...	...
	1	1	17.78[e]	...	17.56	16.82	18.43	17.11	...	...
D + FD	0	0	4.86[e]	4.88[f]	4.99	...	...	4.92	...	...
	1	0	11.25[e]	...	11.44	...	...	10.51	...	...
	1	1	13.20[e]	...	12.94	...	...	12.54	...	...
$H + H_2$ (PK2)	1	0	20.1[h]	20.4[i]	20.7	18.08	20.33	...	18.8[j] 20.17[k]	19.99

[a] reference 117, [b] reference 131, [c] reference 141, [d] reference 144, [e] reference 57, [f] reference 46, [g]... indicates calculations not performed, [h] reference 56, [i] reference 116, [j] reference 11, [k] reference 124.

coordinate-dependent approximations are made, so this type of calculation should be valid for systems with small to large reaction-path curvature. The SCF description does not do as well for these systems as the adiabatic models; however, using configuration interaction the predictions of the resonance energies are greatly improved. The CI results are better than the MEPVA and SCSA results for the lowest-energy resonance of each system. The resonance energies predicted by the semiclassical SCF method are comparable in accuracy to the quantal SCF results. The resonant-periodic-orbit calculations appear very accurate for the $H + H_2$ resonance.

Partial widths. In Table II a compilation of adiabatic and nonadiabatic partial widths is given for the same resonances.

Table II. Partial Widths (kcal/mol) for Three Collinear Atom-Diatom Reaction

			adiabatic[a]			nonadiabatic[b]	
System	n	v	accurate quantal	MEPVA	SCSA[c]	accurate quantal	Feshbach
H + FH	0	0	0.0049[d]	0.0053[e]	0.0045	0.0[d]	0.0[d]
	1	0	0.0[d]	0.0[e]	0.0	0.479[d]	0.389[d]
	1	1	0.12	...[f]	0.107	0.15	0.184[d]
D + FD	0	0	0.0014[d]	0.0014[e]	0.0014	0.0[d]	0.0[d]
	1	0	0.0[d]	0.0[e]	0.0	0.309[d]	0.238[d]
	1	1	0.03	...	0.0435	0.10	0.127[d]
$H + H_2$ (PK2)	1	0	0.16	0.04[g]	0.16	0.37	0.48[h]

[a] sum of all partial widths for decay channels with same n as resonant state, [b] sum of all partial widths for decay channels with n different from resonant state, [c] reference 118, [d] reference 57, [e] reference 46, [f]... indicates calculations not performed, [g] reference 116, [h] reference 56.

For the lowest-energy resonances of H + FH and D + FD, the MEPVA and SCSA results give very similar-quality results for the adiabatic partial widths; however, for $H + H_2$ the MEPVA results underestimate the adiabatic partial width by a factor of 4 and the SCSA results are accurate to the number of significant figures quoted. This difference in the model calculations can be accounted for by the difference in the adiabatic potential used. In the MEPVA calculations the Morse I approximation is used in constructing the adiabatic potential curve, whereas for the SCSA calculations the WKB approximation is used to calculate $\varepsilon_{str}(n,s)$ in Equation 10. As discussed elsewhere (118), we expect the ground-state adiabatic barriers to be adequately described by the Morse I approximation, but for excited-state barriers a more accurate method for treating anharmonicities in the stretching vibration is needed. The Feshbach golden-rule formula is presently the only approximate method for calculating nonadiabatic partial widths. The typical errors are on the order of 0.03-0.09 kcal/mol, which is larger than the errors in the adiabatic partial widths as predicted by the SCSA method.

For the H + H_2 case, we can compare the partial width calculations to predictions made by Pollak (146) using an RPO analysis. His calculations indicated that the adiabatic partial width is "negligible" compared to the nonadiabatic one. Actually though, the accurate quantal results show an adiabatic/nonadiabatic ratio of 0.4, and the SCSA/Feshbach ratio is 0.3.

The useful accuracy obtained with the Feshbach golden rule formalism is encouraging because this approach is more general than the bimolecular reactive resonance calculations considered here. The same model used here for resonance energies and partial widths was applied to reactive tunneling in reference 120. More recently Carrington *et al.* (119) have treated the isomerization of vinylidene using a method that is similar in spirit to the one we applied to bimolecular resonances. One improvement made by these authors is to diagonalize the complex resonance energy operator. In their calculation the Hamiltonian matrix elements are quantized semiclassically using the Heisenberg correspondence principle and the derivative operator is not allowed to act on the vibrational wavefunction. (We found that the action of the derivative coupling operator on the vibrational wave function is very significant for collinear H + H_2 on the PK2 surface, for which it changes the predicted nonadiabatic partial width by a factor of 1.75.) In another application of the golden rule, Geiger *et al.* (175) have considered vibrationally nonadiabatic decay of the unimolecular decomposition resonances of C-O-H, as accessed in H + CO collisions. Since we have demonstrated that the intermode coupling responsible for vibrationally nonadiabatic decay in bimolecular reactive collision resonances of systems with single-saddlepoint potential energy surfaces and a system with a metastable well between two saddlepoints may be modelled quantitatively by the lowest-order reaction-path curvature operator, it would be interesting to see if this kind of treatment can also be applied successfully to similar intermode coupling effects that occur (176-178) in unimolecular decay resonances of systems with potential energy surfaces that exhibit stable wells. Although we (see also 179,180) have used the Feshbach approach to develop a formalism for the calculation of partial widths in the framework of natural collision coordinates, the Feshbach approach can also be applied in other coordinate systems. For example, Liedtke *et al.* applied the Feshbach method using Jacobi coordinates (26), although they did not calculate widths. Hyperspherical coordinates were mentioned in the introduction, and they often provide a useful separability in cases where reaction-path coordinates fail; furthermore, the coupling operator in hyperspherical coordinates is simpler than the reaction-path curvature operator of natural collision coordinates. Thus, golden-rule calculations in hyperspherical coordinates would be very interesting.

Total resonance widths. The total resonance widths are compiled in Table III. The reaction-path-Hamiltonian (RPH) method denotes using the SCSA method for adiabatic partial widths and the Feshbach golden-rule method for nonadiabatic partial widths and summing these to obtain the total width.

Table III. Resonance Widths (kcal/mol) for Three Collinear Atom-Diatom Reactions

System	n	v	accurate quantal	RPH[a]	Stabilization cubic[b]	Stabilization Padé[c]	RPO
H + FH	0	0	0.0049[d]	0.0045[d]	0.0059	0.0040	...[e]
	1	0	0.479[d]	0.389[d]	0.37	0.36	...
	1	1	0.27[d]	0.291[d]	0.67	0.25	...
D + FD	0	0	0.0014[d]	0.0014[d]	...	...	...
	1	0	0.309[d]	0.309[d]	...	...	...
	1	1	0.13[d]	0.171[d]	...	...	...
$H + H_2$ (PK2)	1	0	0.53[f]	0.62[f]	0.48	0.48	1.06[g]

[a] Reaction-path Hamiltonian model, see text for description. [b] reference 133, [c] reference 135, [d] reference 57, [e] ... indicates calculations not performed, [f] reference 56, [g] references 145,146

The good agreement seen in the separate contributions to the RPH methods is reflected in the good agreement for the total widths. Total widths can also be extracted from the stabilization calculations. The simple polynomial fit of the resonance energy as a function of scaling parameter is not as accurate as the Padé approximant method. The Padé approximant method gives total widths which are of about the same accuracy as the RPH model for the H + FH resonances, but the Padé method is more accurate for the $H + H_2$ resonance. Resonant periodic orbits have also been used to calculate a width for the lowest-energy resonance for $H + H_2$ (145,146); the result is too large by a factor of 2.

Resonances in three-dimensional atom-diatom reactions

$H + H_2$. For $H + H_2$, a reactive resonance has been observed in quantum mechanical scattering calculations for total angular momentum J=0 (61,64). For $F + H_2$, more details of reactive resonances in three dimensions have been uncovered. In plots of the quantum mechanical opacity function versus energy and total angular momentum J, peak values of the reaction probability are observed to shift to higher energies as J is increased (63). This "resonance ridge" indicates the dependence of the resonance energy upon J. Resonances in three-dimensional reactions have a much richer spectroscopy than those in collinear reactions; in addition to having a dependence upon the state of the bound stretching degrees of freedom in the interaction region, the resonance energies also have a dependence upon the quantum numbers for bending degree of freedom and the total angular momentum. In the present section, we present calculations for the H + H_2 reaction of resonance energies for various quantum numbers for both the PK2 and LSTH surfaces using the adiabatic model in natural collision coordinates.

For three-dimensional atom-diatom reactions with collinear minimum-energy-paths, the adiabatic potential curve is given by

$$V_a(\underline{n},J,s) = V_{MEP}(s) + \varepsilon_{str}(n_{str},s) + \varepsilon_b(n_b,s) + \varepsilon_b(n_b',s) + \hbar^2 J(J+1)/[2I(s)] \tag{25}$$

where $\varepsilon_b(n_b,s)$ is the energy level for the bending degree of freedom, I(s) is the moment of inertia as a function of the distance s along the MEP, and $\underline{n}$ is the collection of vibrational quantum numbers (n_{str},n_b,n_b'). In the adiabatic model, the resonances are interpreted as quasibound states in wells of these one-dimensional potential curves. The resonance energies will be labeled by the quantum number $\underline{n}$, by J, and by the quantum number v for motion along the reaction path in the adiabatic well. As described previously for collinear reactions, the resonance energies are computed semiclassically, using a primitive WKB method for core-excited type I resonances and a uniform expression otherwise. Also, the small-curvature (SC) method is used to incorporate the effects of reaction-path curvature. Unlike the collinear case, we used the Morse approximation II for fitting the local stretching vibrational potential instead of Morse I approximation since the Morse II approximation gave more accurate estimates of the resonance energies for the collinear H + H_2 resonances. The bending vibrational energies are evaluated by fitting the bending potential to a harmonic-quartic potential and computing the energy eigenvalues by a perturbation-variation method (181,182).

The results for the H + H_2 reaction on the PK2 and LSTH surfaces are shown in Tables IV and V, respectively. In both cases, we report all the sets of n_{str}, n_b, and n_b' with n_{str}= 1 and 2 for which we found resonance energies less than the maximum in the adiabatic barrier. Rather than give the resonance energies as a function of J, the resonance energies for the three lowest J values were fit to the form

$$E_{res}(\underline{n},J) = E_{res}(\underline{n},J{=}0) + B_{\underline{n}}J(J+1) + D_{\underline{n}}[J(J+1)]^2 \tag{26}$$

We note that, although the spacing between the resonances differing only in J is very small, the INR approximation need not be invalid for this reason because its validity only requires narrow resonances well separated from others with the same total angular momentum.

We compare the results in Table IV with the approximate resonant periodic orbit (RPO) calculations of Pollak and Wyatt (147) and with accurate quantal calculations (61). In the RPO calculations, the bending degrees of freedom are included using

Table IV. Spectroscopic Properties of Three-Dimensional H + H_2 Reactive Resonances on the PK2 Surface[a].

n_{str}	n_b	n_b'	v	SCSA[b] $E_{res}(\underline{n},J=0)$	SCSA[b] $B_{\underline{n}}$	SCSA[b] $D_{\underline{n}}$	RPO[c] $E_{res}(\underline{n},J=0)$	RPO[c] $B_{\underline{n}}$	Accurate[d] $E_{res}(\underline{n},J=0)$
1	0	0	0	22.68	0.0281	4.9(-6)	22.01	0.0236	22.5
	1	0	0	25.53	0.0288	-4.3(-6)	...[e]	...	
2	0	0	0	29.16	0.0285	-2.4(-7)	30.90	0.0201	
			1	33.01	0.0251	7.0(-6)	...	...	
	1	0	0	31.97	0.0282	-7.6(-7)	32.43	0.0208	
	1	1	0	34.77	0.0280	-8.9(-7)	33.96	0.0215	
	2	0	0	34.97	0.0279	-8.3(-7)	34.25	0.0215	
	2	1	0	37.74	0.0278	1.7(-7)	35.78	0.0221	
	2	2	0	40.70	0.0279	1.0(-6)	37.60	0.0228	
	3	0	0	38.10	0.0278	-2.1(-8)	...	...	
	3	1	0	40.87	0.0279	1.1(-6)	...	...	
	3	2	0	43.86	0.0282	1.4(-6)	...	...	

[a] Rotational constants $B_{\underline{n}}$ and $D_{\underline{n}}$ are explained in Equation 26. Energies in kcal/mol relative to the bottom of the asymptotic reactant vibrational well. [b] The SCSA method uses Morse approximation II to evaluate $\varepsilon_{str}(n_{str},s)$, [c] reference 147, [d] reference 61, [e] ... indicates that this method predicts no resonance would occur for these quantum numbers.

Table V. Spectroscopic Properties of the Three-Dimensional $H + H_2$ Reactive Resonances on the LSTH surface [a]

n_{str}	n_b	n_b'	v	SCSA[b] $E_{res}(\underline{n},J=0)$	SCSA[b] $B_{\underline{n}}$	SCSA[b] $D_{\underline{n}}$	Accurate[c] $E_{res}(\underline{n},J=0)$
1	0	0	0	22.58	0.0262	2.5(-6)	22.7
	1	0	0	25.18	0.0241	6.1(-5)	
2	0	0	0	28.65	0.0264	-4.0(-6)	
			1	31.88	0.0229	4.7(-6)	
	1	0	0	31.20	0.0261	-7.1(-7)	
	1	1	0	33.76	0.0259	-6.8(-7)	
	2	0	0	33.94	0.0259	-7.0(-7)	
	2	1	0	36.47	0.0258	2.9(-7)	
	2	2	0	39.18	0.0260	1.2(-6)	
	3	0	0	36.79	0.0258	5.1(-7)	
	3	1	0	39.33	0.0260	1.2(-6)	
	3	2	0	42.07	0.0262	-3.1(-7)	

[a] Rotational constants $B_{\underline{n}}$ and $D_{\underline{n}}$ are explained in Equation 26. Energies in kcal/mol relative to the bottom of the asymptotic reactant vibrational well.
[b] The SCSA method uses Morse approximation II to evaluate $\varepsilon_{str}(n_{str},s)$,
[c] reference 64

a sudden-like approximation and the rotational constant B_n is evaluated using an adiabatic reduction scheme. Therefore, it is interesting to compare these results with the vibrationally adiabatic ones which are based upon different approximations. The only accurate quantal results available for three-dimensional $H + H_2$ are for $n_{str}=1$, $n_b=n_b'=0$, and $J=0$. The adiabatic model (SCSA) is in better agreement with the accurate results than the RPO calculations, but both are very close and the two approximate calculations differ by only 0.7 kcal/mol. The two approximate calculations show similar trends in the dependence of the resonance energies on the bending state but also show some marked differences. For $n_{str}=1$, the SCSA calculations show that the $n_b=1$, $n_b'=0$, $J=0$ adiabatic curve should hold one resonance level, which is calculated to be about 0.05 kcal/mol below the maximum of the adiabatic curve. The RPO results do not predict a resonance for this state. For $n_{str}=2$, the SCSA and RPO predictions of the lowest resonance energy (for $n_b=n_b'=0$) differ by about 1.7 kcal/mol. This disagreement reflects the increased difficulty of predicting accurate resonance energies as the quantum numbers increase. Also of interest is the dependence of the resonance energy on the bending quantum numbers for this state. The SCSA method predicts that the adiabatic curves up to $n_b=3$ and $n_b'=2$ will hold resonance energy levels, whereas the RPO method predicts resonance levels only through $n_b=n_b'=2$. Also, the resonance energies go up faster with increasing numbers of bending quanta for the SCSA method. The trends in the dependence of the resonance energies upon J is fairly clear. The rotational constants B_n for the SCSA model are consistently higher than those for the RPO method. The values obtained from the SCSA model are all just slightly lower than the value of the rotational constant evaluated at the saddle point, $\hbar^2/[2I(s=0)] = 0.0295$ kcal/mol.

The agreement between the SCSA and accurate quantal results is as good for the LSTH surface (Table V) as for the PK2 surface. It is also interesting to compare the results on the two surfaces. Although the surfaces have substantially different bending potentials, the resonance energies are surprisingly similar for the two systems. However, the rotational constants B_n are consistently lower on the LSTH surface.

$H + D_2$. The adiabatic model (SCSA) was also applied to the $H + D_2$ reaction using the accurate potential energy surface (LSTH). For this reaction we found the adiabatic potential curves for $n_{str}=1$ and 2 would not support any resonance energy levels. Therefore, we predict that no low-energy reactive resonances will be observed for this system.

Concluding remarks.

Several approximate models for computing resonance energies and widths have been compared, with numerical illustrations for bimolecular reactive resonances for which an adiabatic separation in reaction-path coordinates is reasonable. For systems in which the reaction-path curvature is not too great, the adiabatic model in natural collision coordinates provides a good zero-order description of resonance states and reasonably accurate resonance energies and adiabatic partial widths, and it is most easily extended to reaction in three dimensions. When the reaction-path curvature is too large for natural collision coordinates to be useful, one can retain some of these advantages by making the adiabatic approximation in hyperspherical coordinates. We also consider three ways to go beyond the adiabatic approximation: (i) SCF stabilization calculations, in which the motion along two or more coordinates is assumed separable in an average sense, but neither coordinate is assumed to be adiabatic with respect to the other; (ii) configuration-mixing stabilization calculations, in which no separability is assumed at all; and (iii) Feshbach-theory calculations in which nonadiabatic coupling is added perturbatively to an adiabatic zero-order description. The stabilization calculations are reasonably accurate and are not limited to systems with small-to-moderate reaction-path curvature, but they provide only total widths and are slightly more complicated than adiabatic-based methods to extend to three-dimensional reactions. The Feshbach golden-rule calculations in an adiabatic basis have the advantage of providing partial widths, and the initial successes of this approach are very encouraging.

Acknowledgments

The authors are grateful to Bill McCurdy, Eli Pollak, John Taylor and Bob Walker for preprints and helpful discussions. The work at the University of Minnesota was supported in part by the National Science Foundation under grant no. CHE83-17944. The work at Chemical Dynamics Corporation was supported by the U. S. Army through the Army Research Office under contract no. DAAG-29-81-C-0015.

Literature Cited

1. Truhlar, D. G.; Kuppermann, A. J. Chem. Phys. 1970, 52, 3841.
2. Wu, S.-F.; Levine, R. D. Chem. Phys. Lett. 1971, 11, 557.
3. Wu, S.-F.; Levine, R. D. Mol. Phys. 1971, 22, 881.
4. Diestler, D. J. J. Chem. Phys. 1971, 54, 4547.
5. Truhlar, D. G.; Kuppermann, A. J. Chem. Phys. 1972, 56, 2232.
6. Johnson, B. R. Chem. Phys. Lett. 1972, 13, 172.
7. Wu, S.-F; Johnson, B. R.; Levine, R. D. Mol. Phys. 1973, 25, 609.
8. Wu, S.-F.; Johnson, B. R.; R. D. Levine Mol. Phys. 1973, 25, 839.
9. Schatz, G. C.; Kuppermann, A. J. Chem. Phys. 1973, 59, 964.
10. Truhlar, D. G.; Wyatt, R. E. Annu. Rev. Phys. Chem. 1976, 27, 1.
11. Kuppermann, A. In "Potential Energy Surfaces and Dynamics Calculations"; Truhlar, D. G., Ed., Plenum: New York, 1981; p. 375.
12. Schatz, G. C.; Bowman, J. M.; Kuppermann, A. J. Chem. Phys. 1973, 58, 4023.
13. Baer, M. Mol. Phys. 1974, 27, 1429.
14. Baer, M.; Halavee, U.; Persky, A. J. Chem. Phys. 1974, 61, 5122.
15. Kouri, D. J.; Baer M. Chem. Phys. Lett. 1974, 24, 37.

16. Chapman, F. M., Jr.; Hayes, E. F. J. Chem. Phys. 1975, 62, 4400.
17. Schatz, G. C.; Bowman, J. M.; Kuppermann, A. J. Chem. Phys. 1975, 63, 674.
18. Schatz, G. C.; Bowman, J. M.; Kuppermann, A. J. Chem. Phys. 1975, 63, 685.
19. Adams, J. T. Chem. Phys. Lett. 1975, 33, 275.
20. Connor, J. N. L.; Jakubetz, W.; Manz, J. Mol. Phys. 1975, 29 347.
21. Rosenthal, A.; Gordon, R. G. J. Chem. Phys., 1976, 64, 1641.
22. Chapman, F. M., Jr.; Hayes, E. F. J. Chem. Phys. 1976, 65, 1032.
23. Bowman, J. M.; Leasure, S. C.; Kuppermann, A. Chem. Phys. Lett. 1976, 43, 374.
24. Light, J. C.; Walker, R. B. J. Chem. Phys. 1976, 65, 4272.
25. Chapman, F. M., Jr.; Hayes, E. F. J. Chem. Phys. 1977, 66, 2554.
26. Liedtke, R. C.; Knirk, D. L.; Hayes, E. F. Int. J. Quantum Chem. Symp. 1977, 11, 337.
27. Wyatt, R. E.; McNutt, J. F.; Latham, S. L.; Redmon, M. J. Faraday Disc. Chem. Soc. 1977, 62, 322.
28. Wyatt, R. E. In "State-to-State Chemistry"; Brooks, P.R.; Hayes, E. F., Eds., American Chemical Society: Washington, 1977; p. 185.
29. Connor, J. N. L.; Jakubetz, W.; Manz, J. Mol. Phys. 1978, 35, 1301.
30. Gray, J. C.; Truhlar, D. G.; Clemens, L.; Duff, J. W.; Chapman, F. M.; Morrell, G. O.; Hayes, E. F. J. Chem. Phys. 1978, 69, 240.
31. Latham, S. L.; McNutt, J. F.; Wyatt, R. E.; Redmon, M. J. J. Chem. Phys. 1978, 69, 3746.
32. Askar, A.; Cakmak, A. S.; Rabitz, H. A. Chem. Phys. 1978, 33, 267.
33. Connor, J. N. L.; Jakubetz, W.; Manz, J. Mol. Phys. 1980, 39, 799.
34. Schatz, G. C.; Kuppermann, A. J. Chem Phys. 1980, 72, 2737.
35. Garrett, B. C.; Truhlar, D. G.; Grev, R. S.: Walker, R. B. J. Chem. Phys. 1980, 73, 237.
36. Garrett, B. C.; Truhlar, D. G.; Grev, R. S.; Magnuson, A. W.; Connor, J. N. L. J. Chem. Phys. 1980,73, 1721.
37. Hauke, G.; Manz, J.; Römelt, J. J. Chem. Phys. 1980, 73, 5040.
38. Kuppermann, A.; Kaye, J. A.; Dwyer, J. P. Chem. Phys. Lett. 1980, 74, 257.
39. Römelt, J. Chem. Phys. Lett. 1980, 74, 263.
40. Manz, J.; Römelt, J. Chem. Phys. Lett. 1980, 76, 337.
41. Manz, J.; Römelt, J. Chem. Phys. Lett. 1981, 77, 172.
42. Kaye, J. A.; Kuppermann, A., Chem. Phys. Lett. 1981, 77, 573.
43. Kuppermann, A., Kaye, J. A. J. Phys. Chem., 1981, 85, 1969.
44. Manz, J.; Römelt, J. Chem. Phys. Lett. 1981, 81, 179.
45. Walker, R. B.; Zeiri, Y.; Shapiro, M. J. Chem. Phys. 1981, 74, 1763.
46. Garrett, B. C.; Truhlar, D. G.; Grev, R. S.; Schatz, G. C.; Walker, R. B. J. Chem. Phys. 1981, 85, 3806.
47. Launay, J. M.; Le Dourneuf, M. J. Phys. B 1982, 15, L455.

48. Lee, K. T.; Bowman, J. M.; Wagner, A. F.; Schatz, G. C. J. Chem. Phys. 1982, 76, 3563.
49. Bondi, D. K.; Clary, D. C.; Connor, J. N. L.; Garrett, B. C.; Truhlar, D. G. J Chem. Phys. 1982, 76, 4986.
50. Manz, J.; Pollak, E.; Römelt, J. Chem. Phys. Lett. 1982, 86, 26.
51. Alvariño, J. M.; Gervasi, O.; Laganà, A. Chem. Phys. Lett. 1982, 87, 254.
52. Römelt, J. Chem. Phys. Lett. 1982, 87, 259.
53. Bondi, D. K.; Connor, J. N. L. Chem. Phys. Lett. 1982, 92, 570.
54. Kaye, J. H.; Kuppermann, A. Chem. Phys. Lett. 1982, 92, 574.
55. Shyldkrot, H.; Shapiro, M. J. Chem. Phys. 1983, 79, 5927.
56. Skodje, R. T.; Schwenke, D. W.; Truhlar, D. G.; Garrett, B. C.; J. Phys. Chem. 1984, 88, 628.
57. Skodje, R. T.; Schwenke, D. W.; Truhlar, D. G.; Garrett, B. C. J. Chem. Phys. 1984, 80, 3569.
58. Laganà, A.; Hernandez, M. L.; Alvariño, J. M. Chem. Phys. Lett. 1984, 106, 41.
59. Bondi, D. K.; Connor, J. N. L.; Manz, J.; Römelt, J. Mol. Phys. 1983, 50, 467.
60. Abusalbi, N.; Kouri, D. J.; Lopez, V.; Babamov, V. K.; Marcus, R. A. Chem. Phys. Lett. 1984, 103, 458.
61. Schatz, G. C.; Kuppermann, A. Phys. Rev. Lett. 1975, 35, 1266.
62. Baer, M. J. Chem. Phys. 1976, 65, 493.
63. Redmon, M. J.; Wyatt, R. E. Int. J. Quantum Chem. Symp. 1977, 11, 343.
64. Walker, R. B.; Stechel, E. B.; Light, J. C. J. Chem. Phys. 1978, 69, 2922.
65. Redmon, M. J.; Int. J. Quantum Chem. Symp. 1979, 13, 559.
66. Redmon, M. J.; Wyatt, R. E. Chem. Phys. Lett. 1979, 63, 209.
67. Bowman, J. M.; Lee, K-T. Chem. Phys. Lett. 1979,64,291.
68. Lee, Y. T. Ber. Bunsenges. physik. Chem. 1982, 86, 378.
69. Wyatt, R. E.; McNutt, J. F.; Redmon, M. J. Ber. Bunsenges. physik. Chem. 1982, 86, 437.
70. Gerjuoy, E. In Autoionization (Proc. Symp. At. Interactions and Space Phys.); A. Temkin, Ed.; Mono Book Corp.: Baltimore, 1966, p. 33.
71. Danos, M.; Greiner, W. Phys. Rev. 1966, 145, 708.
72. Nesbet, R. K. Adv. At. Mol. Phys. 1977, 13, 315.
73. Hazi, A. Phys. Rev. A 1979, 19, 920.
74. Truhlar, D. G.; Schwenke, D. W., Chem. Phys. Lett. 1983, 95, 83.
75. Bohm, A. "Quantum Mechanics"; Springer-Verlag: New York, 1979; pp. 478-83.
76. Levine, R. D.; Shapiro, M.; Johnson, B. R. J. Chem. Phys. 1970, 52, 1755.
77. Levine, R. D.; Johnson, B. R. Chem. Phys. Lett. 1969, 4, 365.
78. Levine, R. D.; Bernstein, R. B. J. Chem. Phys. 1970, 53, 686.
79. Smith, F. T. Phys. Rev. 1960, 118, 349.
80. Smith, F. T. In "Kinetic Processes in Gases and Plasmas", Hochstim, A. R., Ed., Academic Press, New York, 1969, p. 321.
81. Celenza, L.; Tobocman, W. Phys. Rev. 1968, 174, 1115.
82. Gien, T. T. Can. J. Phys. 1969, 47, 279.
83. Kinsey, J. L. Chem. Phys. Lett. 1971, 8, 349.
84. LeRoy, R. J.; Bernstein, R. B. J. Chem. Phys. 1971, 54, 5114.
85. Yoshida, S. Annu. Rev. Nucl. Sci. 1974, 24, 1.

86. Korsch, H. J.; Möhlenkamp, R. J. Phys. B 1982, 15, L559.
87. Allison, A. C. Chem. Phys. Lett. 1969, 3, 371.
88. Jackson, J. L.; Wyatt, R. E. Chem. Phys. Lett. 1970, 4, 643.
89. Johnson, B. R.; Balint-Kurti, G. G.; Levine, R. D. Chem. Phys. Lett. 1970, 7, 268.
90. Numrich, R. W.; Kay, K. G.; J. Chem. Phys. 1979, 71, 5352.
91. Reinhardt, W. P. Annu. Rev. Phys. Chem. 1982, 33, 223.
92. Basilevsky, M. V.; Ryaboy, V. M. Int. J. Quan, Chem. 1981, 19, 611.
93. Basilevsky, M. V.; Ryaboy, V. M. Chem. Phys. 1984, 86, 67.
94. Atabek, O.; Lefebvre, R.; Jacom, M. J. Phys. B 1982, 15, 2689.
95. Meyer, H.-D.; Walter, O. J. Phys. B 1982, 15, 3647.
96. Meyer, H.-D. J. Phys. B 1983, 16, 2265.
97. Meyer, H.-D. J. Phys. B 1983, 16, 2785.
98. Brenig, W.; Haag, R. In "Quantum Scattering Theory"; Ross, M., Ed.; Indiana University Press: Bloomington, 1963; p. 13.
99. Newton, R. G. "Scattering Theory"; McGraw-Hill: New York, 1966.
100. Taylor, J. R. "Scattering Theory"; John Wiley & Sons: New York, 1972.
101. Siegert, A. F. Phys. Rev. 1939, 56, 750.
102. Feshbach, H. Ann. Phys. (N.Y.) 1958, 5, 357.
103. Feshbach, H. Ann. Phys. (N.Y.) 1962, 19, 287.
104. Feshbach, H. Ann. Phys. (N.Y.) 1967, 43, 410.
105. Goldberger, M. L.; Watson, K. M. "Collision Theory"; John Wiley & Sons: New York, 1964.
106. Rodberg, L. S.; Thaler, R. M. "Introduction to the Quantum Theory of Scattering"; Academic Press: New York, 1967.
107. McCurdy, C. W.; Rescigno, T. N. Phys. Rev. A 1979, 20, 2346.
108. Stine, J. R.; Marcus, R. A. Chem. Phys. Lett. 1974, 29, 575.
109. Bowman, J. M.; Kuppermann, A.; Adams, J. T.; Truhlar, D. G. Chem. Phys. Lett. 1973, 20, 229.
110. Hirschfelder, J. O.; E. Wigner J. Chem. Phys. 1939, 7, 616.
111. Marcus, R. A. J. Chem. Phys. 1966, 45, 4493.
112. Marcus, R. A. J. Chem. Phys. 1966, 45, 4500.
113. Marcus, R. A. J. Chem. Phys. 1968, 49, 2610.
114. Truhlar, D. G. J. Chem. Phys. 1970, 53, 2041.
115. Miller, W. H.; Handy, N. C.; Adams, J. E. J. Chem. Phys. 1980, 72, 99.
116. Garrett, B. C.; Truhlar, D. G. J. Phys. Chem. 1979, 83, 1079; 1980, 84, 628(E); 1983, 87, 4553(E).
117. Garrett, B. C.; Truhlar, D. G. J. Phys. Chem. 1982, 86, 1136; 1983, 87, 4554(E).
118. Garrett, B. C.; Truhlar, D. G., J. Chem. Phys., 1984, 81, in press.
119. Carrington, T., Jr.; Hubbard, L. M.; Schaefer, H. F., III; Miller, W. H. J. Chem. Phys., 1984, 80, 4347.
120. Skodje, R. T.; Truhlar, D. G.; Garrett, B. C. J. Chem. Phys. 1982, 77, 5955.
121. Dywer, J. P.; A. Kuppermann, unpublished.
122. Dywer, J. P.; Ph.D. Thesis, California Institute of Technology, Pasadena, 1978.
123. Babamov, V. K.; Kuppermann, A., J. Chem. Phys. 1982, 77, 1891.
124. Römelt, J. Chem. Phys. 1983, 79, 197.

125. Aquilanti, V.; Cavalli, S.; Laganà, A. Chem. Phys. Lett. 1982, 93, 179.
126. Pollak, E.; Römelt, J. J. Chem. Phys. 1984, 80, 3613.
127. Holøien, E.; Midtal, J. J. Chem. Phys. 1966, 45, 2209.
128. Taylor, H. S. Adv. Chem. Phys. 1970, 17, 91.
129. Taylor, H. S.; Thomas, L. D. Phys. Rev. Lett. 1972, 17, 1091.
130. Bačić, Z.; Simons, J. J. Phys. Chem. 1982, 86, 1192.
131. Thompson, T. C.; Truhlar, D. G. J. Chem. Phys. 1982, 76, 1790; 1982, 77, 3777(E).
132. Thompson, T. C.; Truhlar, D. G. Chem. Phys. Lett. 1983, 101, 235.
133. Thompson, T. C.; Truhlar, D. G. Chem. Phys. Lett. 1982, 92, 71.
134. Thompson, T. C.; Truhlar, D. G. J. Phys. Chem. 1984, 88, 210.
135. Thirumalai, D.; Thompson, T. C.; Truhlar, D. G. J. Chem. Phys., 1984, 80, in press.
136. Simons, J. J. Chem. Phys. 1981, 75, 2465.
137. McCurdy, C. W.; McNutt, J. F. Chem. Phys. Lett. 1983, 94, 306.
138. Isaacson, A. D.; Truhlar, D. G., to be published.
139. Gerber, R. B.; Ratner, M. A. Chem. Phys. Lett. 1979, 68, 195.
140. Landau, L. D.; Lifshitz, E. M. "Quantum Mechanics", 2nd ed., Pergamon Press, Oxford, 1965, p. 163.
141. Garrett, B. C.; Truhlar, D. G. Chem. Phys. Lett. 1982, 92, 64.
142. Farrelly, D.; Hedges, R. M., Jr.; Reinhardt, W. P. Chem. Phys. Lett. 1983, 96, 599.
143. Smith, A. D.; Liu, W.-K. Chem. Phys. Lett. 1983, 100, 461.
144. Pollak, E.; Child, M. S. Chem. Phys. 1981, 60, 23.
145. Pollak, E.; Wyatt, R. E. J. Chem. Phys. 1982, 77, 2689.
146. Pollak, E. J. Chem. Phys. 1982, 76, 5843.
147. Pollak, E.; Wyatt, R. E. J. Chem. Phys. in press.
148. Taylor, H. S.; Nazaroff, C. V.; A. Golebiewski J. Chem. Phys. 1966, 45, 2822.
149. Clary, D. C.; Connor, J. N. L. Chem. Phys. Lett. 1983, 94, 81.
150. Pollak, E. Chem. Phys. Lett. 1983, 94, 84.
151. Manz, J.; Meyer, R.; Römelt, J. Chem. Phys. Lett. 1983, 96, 607.
152. Yaris, R.; Taylor, H. S. Chem. Phys. Lett. 1979, 66, 505.
153. Atabek, O.; Lefebvre, R. Chem. Phys. 1981, 55, 395.
154. Noro, T.; Taylor, H. S. J. Phys. B 1980, 13, L377.
155. Bačić, Z.; Simons, J. Int. J. Quantum Chem. 1982, 21, 727.
156. Porter, R. N.; Karplus, M. J. Chem. Phys. 1964, 40, 1105.
157. Liu, B. J. Chem. Phys. 1973, 58, 1925.
158. Siegbahn, P.; Liu, B. J. Chem. Phys. 1978, 68, 2457.
159. Truhlar, D. G.; Horowitz, C. J. J. Chem. Phys. 1978, 68, 2466; 1979, 71, 1514E.
160. Gerrity, D. P.; Valentini, J. J. J. Chem. Phys. 1983, 79, 5203.
161. Rettner, C. T.; Marinero, E. E.; Zare, R. N. In "Physics of Electronic and Atomic Collisions: Invited Papers from the XIIIth ICPEAC"; Eichler, J.; Hertel, I. V.; Stolterfoht, N., Eds; North Holland: Amsterdam, in press.
162. Marinero, E. E.; Rettner, C. T.; Zare, R. N. J. Chem. Phys., in press.
163. Gerrity, D. P.; Valentini, J. J. J. Chem. Phys., in press

164. Ashton, C. J.; Child, M. S.; Hutson, J. M. J. Chem. Phys. 1983, 78, 4025.
165. Light, J. C.; Walker, R. B. J. Chem. Phys. 1976, 65, 4272.
166. Macek, J.; Burke, P. G. Proc. Phys. Soc. London 1967, 92, 351.
167. Fels, M. F.; Hazi, A. U. Phys. Rev. A 1972, 5, 1236.
168. Connor, J. N. L. in "Semiclassical Method in Molecular Scattering and Spectroscopy", Child, M. S., Ed.; Reidel, Dordrecht, Holland, 1980.
169. Connor, J. N. L.; Smith, A. D. Mol. Phys. 1981, 43, 347.
170. Marcus, R. A.; Coltrin, M. E. J. Chem. Phys. 1977, 67, 2609.
171. Skodje, R. T.; Truhlar, D. G.; Garrett, B. C. J. Phys. Chem. 1981, 85, 3019.
172. Garrett, B. C.; Truhlar, D. G. J. Phys. Chem. 1979, 83, 1052; 1979, 83, 3058(E).
173. Herzberg, G. "Molecular Spectra and Molecular Structure". Vol. I, "Spectra of Diatomic Molecules", 2nd ed., Van Nostrand, Princeton, NJ, 1950, p. 101.
174. Downing, J. W.; Michl, J. In "Potential Energy Surfaces and Dynamics Calculations"; Truhlar, D. G., Ed.; Plenum: New York, 1981; p. 199.
175. Geiger, L. C.; Schatz, G. C.; Garrett, B. C., contribution included in this volume.
176. Wolf, R. J.; Hase, W. L. J. Chem. Phys. 1980, 73, 3779, 1980, 72, 316.
177. Hedges, R. M.; Reinhardt, W. P. Chem. Phys. Lett. 1982, 91, 241.
178. Kulander, K. C. J. Chem. Phys. 1983, 73, 1279.
179. Shoemaker, C. L.; Wyatt, R. E. J. Chem. Phys. 1982, 77, 4982.
180. Shoemaker, C. L.; Wyatt, R. E. J. Chem. Phys. 1982, 77, 4994.
181. Truhlar, D. G. J. Mol. Spectrosc. 1971, 38, 415.
182. Garrett, B. C.; Truhlar, D. G. J. Phys. Chem. 1979, 83, 1915.

RECEIVED June 11, 1984

21

Atom–Diatom Resonances Within a Many-Body Approach to Reactive Scattering

DAVID A. MICHA[1] and ZEKI C. KURUOGLU[2]

[1]Departments of Chemistry and Physics, University of Florida, Gainesville, FL 32611
[2]Research Institute for Basic Science, Gebze-Kocaeli, Turkey

The resonance scattering of atoms by diatomics can be described within a many-body approach that provides a link between the electronic structure and the quantum dynamics of chemical bonding. Introducing a basis of spin-adapted valence-bond functions, the system hamiltonian matrix is decomposed into sums of atom-atom pairs and intrinsic three-atom contributions. Rearrangement scattering cross sections are obtained from coupled equations for three-atom transition operators, which provide a definition of quasibound states corresponding to resonances. Coupled integral equations in two vector variables are analyzed into angular momentum components and solved to obtain K-matrix elements converged within an s-wave model. This provides positions and widths of resonances for H + H_2, over a wider range of energies than previously investigated.

When the collision energy between an atom A and a diatomic molecule BC is varied, the total energy of the triatomic system may reach values at which a resonance process occurs with formation of a long-lived triatomic state. In this state, the three atoms interact at short distances over times long compared with typical rotation times and eventually break up, for energies below dissociation of BC, into two-body channels (A)(BC), (B)(CA) and (C)(AB) which may be electronically excited or electrically charged as a result of electron transfer. In the language of the chemical dynamics of stationary processes, the above states are neither scattering nor bound states but instead resonances, closely related to the transition states of chemical kinetics. A central task of a quantum theory of reactive scattering is to provide a definition of resonance states that can be used to calculate their properties as observed for example in the energy and angular dependance of cross sections.

We present here a summary of our work on the collision dynamics of three interacting atoms, (1) adding new developments on the theory of resonances in rearrangement collisions. We begin by showing how to go from a description in terms of electrons and nuclei to one

0097-6156/84/0263-0401$06.00/0

involving three interacting atoms by using spin-adapted valence-bond (VB) states (2) in ways related to the VB (3,4) and Diatomics-in-Molecules (5,6) methods. This provides a solid foundation for a model of interacting atomic spins which motivated our earlier study of the reaction dynamics of $H + H_2$ within a theory of three-atom collisions (7). Such a theory was developed to rigorously describe three interacting particles (8) and is being used here in a recent form (9) more convenient for calculations. The theory is readily adapted to discuss permutational symmetry of nuclei and dissociative scattering (2), although we will not touch on the latter. Here we further develop the theory of resonances by first introducing characteristic or Hilbert-Schmidt states by analogy with work on two-particle scattering (10). These states provide the desired equations and normalization conditions to define quasi-bound states physically interpreted as resonances. Within this framework one can introduce integral equations in momentum variables for rearrangement collisions. The procedure is somewhat lengthy and is detailed in two publications which deal with the construction of two-atom transition operators (11) and with the calculation of atom-diatom scattering amplitudes (12). In this work we summarize the procedure followed to obtain scattering amplitudes from the integral equations, with emphasis on the calculation of resonance energies and widths for $H + H_2$ by parametrizing those amplitudes. To avoid overly lengthy equations, we use a compact notation closely related to the one in a review to appear (13), where we also derive many basic results.

The approach we follow here is based on a many-body theory of reactive scattering (1). Other approaches have been reviewed in the literature (14). For example, several of the available results on resonances have been obtained in wave-mechanical approaches that propagate and match wavefunctions (15,16) or expand them in functions of suitable coordinates (17). Another approach which promises to provide accurate resonance cross sections is that of channel-coupling-array equations, also recently reviewed (18). The Feshbach theory of resonances has also been recently applied to molecular reactions (19,20).

From Electrons and Nuclei to Interacting Atoms

The hamiltonian of three interacting atoms A, B and C contains the nuclear and electronic kinetic energies and the sum of all the Coulomb interactions of electrons and nuclei. It would appear that, to describe the formation and breakup of atomic bonds, it would be convenient to distribute the electrons in the hamiltonian among the nuclei, and to consider the interaction of the resulting atomic fragments. This procedure is however defective because it destroys the invariance of the hamiltonian with respect to electronic exchange, as soon as nuclei are removed to large distances. The procedure would also complicate the description of electron transfer among fragments.

The alternative procedure we have followed consists of introducing electronic states $|\alpha_I\rangle$, for atoms I = A, B, C, already symmetrized with respect to electronic exchange. The index α_I gives the electronic configuration and the charge of fragment I. Electronic states of the whole triatomic system may be collected in the

$1 \times N_0$ row matrix

$$|\underset{\sim}{\alpha}_0\rangle = [A_{A,B,C}|\alpha_A\rangle|\alpha_B\rangle|\alpha_C\rangle] \tag{1}$$

where $A_{A,B,C}$ is a supplementary antisymmetrizer. The states collected in $|\underset{\sim}{\alpha}_0\rangle$ constitute a Valence-Bond (VB) set.

For example, a minimal VB set for H_3 includes only a 1s orbital per nucleus, with up or down spins. The VB states are of form $|H_a 1s\rangle|H_b 1s\rangle|H_c 1s\rangle$. with 1s substituted by $\overline{1s}$ for down spins. There is a total of $2^3 = 8$ such states of which only $N_0 = 3$ correspond to projection of the total spin $M_S = 1/2$, the maximum value for the doublet triatomic state.

As a second example consider FH_2, for which one must include also F^-H^+H in the minimal VB set. For FHH we have $|F\ 2p_\xi^2\ 2p_\eta^2\ 3p_\zeta\rangle \times|H_a 1s\rangle|H_b 1s\rangle$ with $(\xi,\eta,\zeta) = (x,y,z)$, or $3 \times 2^3 = 24$ states, while for F^-H^+H we have $|F^-\ 2p_x^2\ 2p_y^2\ 2p_z^2\rangle|\ H_n\ 1s\rangle$, $n = a,b$ or $2^2 = 4$ states. Of this total of 28 states, only half have $M_S > 0$, and among these 3 have $M_S = 3/2$, and 11 have $M_S = 1/2$. In FH_2, spin-orbit coupling plays an important role and all 14 states must be considered in a complete treatment.

The VB set $|\underset{\sim}{\alpha}_0\rangle$ is not orthonormal but instead satisfies

$$\langle\underset{\sim}{\alpha}_0|\underset{\sim}{\alpha}_0\rangle = \underset{\sim}{\Delta} \tag{2}$$

which is a $N_0 \times N_0$ overlap matrix. Furthermore, assuming that the chosen set is sufficient (i.e. complete) to describe the reaction is equivalent to writing the completeness relation

$$|\underset{\sim}{\alpha}_0\rangle\underset{\sim}{\Delta}^{-1}\langle\underset{\sim}{\alpha}_0| = 1. \tag{3}$$

Separating the hamiltonian H as

$$H = K_{nu} + H_{e\ell}, \tag{4}$$

where K_{nu} is the nuclear kinetic energy and $H_{e\ell}$ is the rest, we expand $H_{e\ell}$ in the VB basis so that

$$H_{e\ell} = |\underset{\sim}{\alpha}_0\rangle\underset{\sim}{\Delta}^{-1}\underset{\sim}{H}_{e\ell}\underset{\sim}{\Delta}^{-1}\langle\underset{\sim}{\alpha}_0| \tag{5a}$$

$$\underset{\sim}{H}_{e\ell} = \langle\underset{\sim}{\alpha}_0|H_{e\ell}|\underset{\sim}{\alpha}_0\rangle. \tag{5b}$$

The matrix $\underset{\sim}{H}_{e\ell}$ depends on the position $\vec{r}_n$, of nuclei n = a, b, and c. Next we can define fragment hamiltonians by sequentially removing atoms to infinity. This is conveniently done in terms of Jacobi coordinates which, e.g., for the (a, bc) arrangement will replace

$(\vec{r}_a, \vec{r}_b, \vec{r}_c)$ by $(\vec{R}_{cm}, \vec{R}_1, \vec{r}_1)$, with $\vec{r}_1$ the position of b with respect to c, $\vec{R}_1$ the position of a with respect to the center of mass of bc, and R_{cm} the position of the total center of mass. Working in the center of mass frame (with $\vec{R}_{cm} = 0$) we let $R_1 \to \infty$ and $r_1 \to \infty$ to find

$$\underset{\sim}{H}_{e\ell} \simeq \underset{\sim}{H}'_A + \underset{\sim}{H}_{BC} \quad \quad (6)$$
$$R_1 \to \infty$$

$$\underset{\sim}{H}'_A \simeq \underset{\sim}{H}_A = [E(\alpha'_A,\alpha_A)\ \delta(\alpha'_B,\alpha_B)\ \delta(\alpha'_C,\alpha_C)] \quad \quad (7)$$
$$r_1 \to \infty$$

where $\underset{\sim}{H}_A$ is the hamiltonian of fragment A, and the energy array $E(\alpha'_A,\alpha_A) = E(\alpha_A)\ \delta(\alpha'_A,\alpha_A)$ provided the states $|\alpha_A\rangle$ are constructed to diagonalize it, which we assume in what follows. The diatomic fragment hamiltonian $\underset{\sim}{H}_{BC}$ defines the interaction

$$\underset{\sim}{V}_{BC} = \underset{\sim}{H}_{BC} - \underset{\sim}{H}_B - \underset{\sim}{H}_C \quad \quad (8)$$

and the total $\underset{\sim}{H}_{e\ell}$ can be written as

$$\underset{\sim}{H}_{e\ell} = \sum_I \underset{\sim}{H}_I + \sum_{JK}{}' \underset{\sim}{V}_{JK} + \underset{\sim}{V}_{ABC} \quad \quad (9)$$

where $\underset{\sim}{V}_{ABC}$ is an intrinsic three-atom interaction. Replacing this expression in Equation 5a, one obtains the desired partition of $H_{e\ell}$ into a sum of interacting fragment terms, all of which are invariant under electronic exchange. The nuclear kinetic energy can also be expressed in the $|\underset{\sim}{\alpha}_0\rangle$ basis, keeping in mind that the states depend on the $\vec{r}_n$, so that (21)

$$K_{nu} = |\underset{\sim}{\alpha}_0\rangle \underset{\sim}{\Delta}^{-1} \sum_n (2m_n)^{-1} \left(\frac{\hbar}{i}\frac{\partial}{\partial \vec{r}_n} + \vec{G}_n\right)^2 \underset{\sim}{\Delta}^{-1} \langle \underset{\sim}{\alpha}_0| \quad \quad (10a)$$

$$\vec{G}_n = \langle \underset{\sim}{\alpha}_0 | \frac{\hbar}{i}\frac{\partial}{\partial \vec{r}_n} | \underset{\sim}{\alpha}_0\rangle \quad \quad (10b)$$

where the $\vec{G}_n$ are additions to the kinematic momenta resulting from electron drag by the nuclei. These additional terms can frequently be neglected when one works with VB functions, which change smoothly with nuclear coordinates.

In the absence of spin-orbital coupling in the electronic hamiltonian, one can block-diagonalize the matrix $\underset{\sim}{H}_{e\ell}$ by working with $N^{(i)}$

spin adapted VB functions of total electronic spin quantum numbers S, M_S,

$$|\underset{\sim}{\alpha}^{(i)}> = |\underset{\sim}{\alpha}_0>\underset{\sim}{C}_0^{(i)} \tag{11}$$

where i refers to the spin coupling order I(JK), $\alpha^{(i)}$ stands for $(A^{(i)}s_I(s_Js_K)s_{JK}SM_S)$ with $A^{(i)}$ itself a collection of quantum numbers, and the $N_0 \times N^{(i)}$ matrix $\underset{\sim}{C}_0^{(i)}$ has elements given by angular momentum coupling coefficients. The diatomic interactions $\underset{\sim}{V}_i = \underset{\sim}{V}_{JK}$ in Equation 9 are frequently block diagonal when transformed to the $|\underset{\sim}{\alpha}^{(i)}>$ basis, because they are defined in the limit $\vec{R}_i \rightarrow \infty$ where $\underset{\sim}{\Delta} = \underset{\sim}{I}$. Each block of $\underset{\sim}{V}_i$ may be labelled with an index B_i.

For example, for H_3 and $S = M_S = 1/2$, $\underset{\sim}{V}_1$ is a diagonal 2×2 matrix in the basis $|\underset{\sim}{\alpha}^{(i)}>$, which contains the $(^2S, ^1\Sigma)$ and $(^2S, ^3\Sigma)$ antisymmetrized states in a reference frame attached to H_bH_c. As a result the H_3 electronic hamiltonian is equivalent to the interaction energy of three atomic spins 1/2, (7). For FH_2 and $S = M_S = 1/2$, noting that overlaps between the (F, H_2) and (F^-, H_2^+) sets of states equal zero asymptotically, we have that $V_{(F)(H2)}$ is block diagonal in these two sets. The V_{F,H_2} block is a 6×6 diagonal matrix in the states $(^2P, ^{1,3}\Sigma)$ while $V_{F^-,H2^+}$ is a 2×2 non-diagonal matrix in $(^1S, ^2\Sigma_{g,u})$ states. Further, $V_{(H)(FH)}$ is block diagonal in the states of (H,FH) and (H^+, FH^-). The (H^+, FH^-) block is 1×1 and (H, FH) is 7×7, the latter being made up of a 4×4 block for singlets of FH and a 3×3 block for the triplet.

The described procedure breaks up the full hamiltonian into one-, two- and three-atom contributions. When the three-atom term can be neglected (as in the case of H_3) or considered small, a great simplification arises in the treatment of the interaction dynamics.

Dynamics of Nuclear Rearrangement

The total hamiltonian H, given by Equations 4, 5 and 9, may be partitioned in ways appropriate to each arrangement of the nuclei. The free hamiltonian is of the form

$$H_0 = K_{nu} + \sum_I H_I \tag{12}$$

where H_I refers to the I-th atom, and the total potential is

$$V = \sum_{i=0}^{4} V_i \tag{13}$$

where V_i, i = 1,2,3, refers to pair (JK), V_4 to the intrinsic triatomic potential among A, B and C, and $V_0 = 0$ is introduced to make the following notation consistent. The channel hamiltonian H_i and interaction $V^{(i)}$ are defined by

$$H_i = H_0 + V_i \tag{14}$$

$$V^{(i)} = V - V_i = \sum_{k=0}^{4} \bar{\delta}_{ik} V_k \tag{15}$$

where i = 0 to 4 and $\bar{\delta}_{ik} = 1 - \delta_{ik}$. For i = 1,2,3 it is also convenient to introduce the internal hamiltonian h_i of the (JK) pair and its eigenstates so that

$$[H_I + h_i - E(\beta^{(i)} \nu_i)] \mid \beta^{(i)} \nu_i > = 0 \tag{16}$$

where $|\beta^{(i)}>$ is the asymptotic electronic state of the triatomic (a linear combination of the $|\alpha^{(i)}>$), and ν_i is the collection of nuclear motion quantum numbers of (JK). Indicating with $\vec{p}_i$ and $\vec{P}_i$ the momenta conjugate to $\vec{r}_i$ and $\vec{R}_i$ one finds for the channel hamiltonians that

$$H_i \, |\Phi_i> = E \, |\Phi_i> \qquad , \; i = 1,2,3 \tag{17a}$$

$$|\Phi_i> = |\beta^{(i)} \nu_i \vec{P}_i> \tag{17b}$$

$$H_0 \, |\Phi_0> = E \, |\Phi_0> \tag{18a}$$

$$|\Phi_0> = |\beta^{(0)} \vec{p}_0 \vec{P}_0> \tag{18b}$$

with

$$E = \frac{P_i^2}{2M_i} + E(\beta^{(i)} \nu_i) = \frac{P_0^2}{2M_0} + \frac{p_0^2}{2m_0} + E(\beta^{(0)}) \tag{19}$$

with the 0 index in Equation 18b equal to any of the i = 1,2,3.

The transition amplitudes we want to obtain are

$$A_{i1} = <\Phi_i|V^{(i)}|\Psi_1^{(+)}> \tag{20}$$

where $\Psi_1^{(+)}$ is the full scattering state for incoming waves in channel 1, and outgoing waves in all channels. Differential cross sections follow from

$$d\sigma_{i1}/d\Omega_i = (2\pi)^4 \hbar^2 M_i M_1 (P_i/P_1) |A_{i1}|^2 \qquad (21)$$

when momentum states are normalized to δ functions (7).

Instead of obtaining $\Psi_1^{(+)}$ and then A_{i1} it is more convenient to work with transition operators U_{i1} satisfying $U_{i1}\Phi_1 = V^{(j)}\Psi_1^{(+)}$. Among the several proposed procedures, the most useful is based on the set of coupled equations

$$U_{k1} = U_{k1}^0 + \sum_j U_{ki}^0 \, G_0 \, T_i \, G_0 \, U_{i1} \qquad (22a)$$

$$U_{ki}^0 = \bar{\delta}_{ki} \, G_0^{-1} \qquad (22b)$$

Where G_0 is the resolvent of H_0 and T_i the transition operator for scattering of J and K; in three-body space

$$T_i = V_i + V_i \, G_0 \, T_i . \qquad (23)$$

Equation 22b corresponds physically to a stripping process where, e.g. (A)(BC) rearranges into (AB)(C) on account of favorable momenta relations, without intervention of the potentials V_i (22). These are contained only in the T_i. Hence, when V_4 is negligible $T_4 = 0$ and Equation 22a provides three-atom transition operators in terms of two-atom ones. This is shown graphically in Figure 1 for U_{31}. To include V_4 one can eliminate the i = 4 term in Equation 22a, in which case the equations for U_{k1}, k = 1,2,3, have the same form but with U_{ki}^0 replaced by $U_{ki}^1 = U_{ki}^0 + T_4$. If V_4 is small one can further let $T_4 \approx V_4$.

Permutational Symmetry of Identical Nuclei

When two of the three nuclei are identical it is necessary to work with wavefunctions that are properly symmetrized. This is in fact also computationally advantageous because, as we shall show, the number of coupled equations we have to solve will as a result be decreased.

We introduce the nuclear spins with quantum numbers i_a, i_b and i_c and couple them to a total nuclear spin of quantum numbers (I, M_I). There are three ways of doing this, described by states $|\xi_k\rangle$, k = 1,2,3. Fully symmetrized channel states are given by

$$\Phi_k^{(s)} = (C_k/N!) \sum_P (-1)^{2i_n p} P\,(\Phi_k\ \xi_k) \tag{24}$$

where C_k is a normalization constant, and P is a permutation of parity p in the symmetric group of N identical nuclei with spin i_n. Symmetrized transition amplitudes are now given by (23).

$$\begin{aligned} A_{k1}^{(s)} &= \langle\Phi^{(s)} \mid U_{k1} \mid \Phi_1^{(s)}\rangle \\ &= \langle\Phi_k\ \xi_k \mid U_{k1}^{(s)} \mid \Phi_1\ \xi_1\rangle \end{aligned} \tag{25}$$

which defines the operator $U_{k1}^{(s)}$. This is found to be, for three identical nuclei of spin 1/2, and using Equation 22 for the U_{k1},

$$U_{11}^{(s)} = \frac{1}{2}\,(I - P_{bc})U_s \tag{26a}$$

$$U_s = U_s^1 + U_s^1\,G_0\,T_1\,G_0\,U_s \tag{26b}$$

$$U_s^1 = (P_{abc} + P_{acb})G_0^{-1} + (I + P_{abc} + P_{acb})T_4. \tag{26c}$$

As seen in Equation 26b, only the operator U_s remains to be calculated, even though the leading term U_s^1 is more complicated than before. If the initial and final nuclear spins are not specified by the experimental devices, we must average the cross sections over initial nuclear spins and must sum them over final ones. This procedure however does not eliminate the restrictions over intermediate scattering states imposed by the symmetry which affects, for example, resonance states.

Resonance Scattering

It is convenient in what follows to introduce a compact notation with matrices of operators, so that Equations 22 look like a two-body Lippmann-Schwinger equation. Defining the diagonal matrix $\underset{\sim}{F} = [\delta_{ik}G_0T_iG_0]$, we have

$$\underset{\sim}{U} = \underset{\sim}{U}^0 + \underset{\sim}{U}^0\,\underset{\sim}{F}\,\underset{\sim}{U}. \tag{27}$$

Each element in $\underset{\sim}{F}$ is complex, since we can write $G_0T_iG_0 = G_i - G_0$, where G_i is the resolvent of the hamiltonian H_i. Considering in what follows only energies E below the threshold for break-up into A + B + C, we have that G_0 is real and that

$$G_i = G_i^P - i\pi\Delta_i \tag{30a}$$

$$\Delta_i = \delta(E - H_i) = \sum_{\text{open}} |\Phi_i\rangle\langle\Phi_i|, \tag{30b}$$

where G_i^P is the real part, or principal value, of G_i. The operator Δ_i, expanded in a basis of eigenstates of H_i, is a sum of projection operators over the energetically accessible, or open, states of H_i. This is shown graphically in Figure 2. Consequently we can write

$$\underset{\sim}{F} = \underset{\sim}{F}^P - i\pi\underset{\sim}{\Delta} \tag{31}$$

and derive a Heitler-like equation convenient for the study of resonances (24). Letting

$$\underset{\sim}{U}^P = \underset{\sim}{U}^0 + \underset{\sim}{U}^0 \underset{\sim}{F}^P \underset{\sim}{U}^P, \tag{32}$$

we immediately obtain that equation as

$$\underset{\sim}{U} = \underset{\sim}{U}^P + \underset{\sim}{U}^P (-i\pi\underset{\sim}{\Delta})\underset{\sim}{U}. \tag{33}$$

Next we expand the real operator $\underset{\sim}{U}^P$ in a basis of characteristic, or Hilbert-Schmidt, states as follows (10). Define

$$\overline{\underset{\sim}{U}}^P = (\underset{\sim}{U}^0)^{-1/2}\underset{\sim}{U}^P(\underset{\sim}{U}^0)^{-1/2} = \underset{\sim}{I} + \underset{\sim}{J}\overline{\underset{\sim}{U}}^P \tag{34a}$$

$$\underset{\sim}{J} = (\underset{\sim}{U}^0)^{1/2}\underset{\sim}{F}(\underset{\sim}{U}^0)^{1/2} \tag{34b}$$

where $\underset{\sim}{I}$ is the unit matrix and the transformation has led to a symmetric kernel J, a function of E. Its eigenstates $|\overline{\Theta}_s\rangle$ (each a column of four states) and eigenvalues $\lambda_s(E)$ satisfy

$$\underset{\sim}{J}\ |\underset{\sim}{\overline{\Theta}}_s\rangle = \lambda_s\ |\underset{\sim}{\overline{\Theta}}_s\rangle \tag{35}$$

with $\langle\underset{\sim}{\overline{\Theta}}_s \mid \underset{\sim}{\overline{\Theta}}_{s'}\rangle = \delta_{ss'}$. This gives

$$\overline{\underset{\sim}{U}}^P = \sum_s |\underset{\sim}{\overline{\Theta}}_s\rangle\langle\underset{\sim}{\overline{\Theta}}_s|\ /\ (1 - \lambda_s). \tag{36}$$

More familiar equations result by defining $|\underset{\sim}{\Theta}_s\rangle = (\underset{\sim}{U}^0)^{1/2}|\underset{\sim}{\overline{\Theta}}_s\rangle$, in which case

$$\underset{\sim}{F}^P\underset{\sim}{U}^0\ |\underset{\sim}{\Theta}_s\rangle = \lambda_s\ |\underset{\sim}{\Theta}_s\rangle \tag{37a}$$

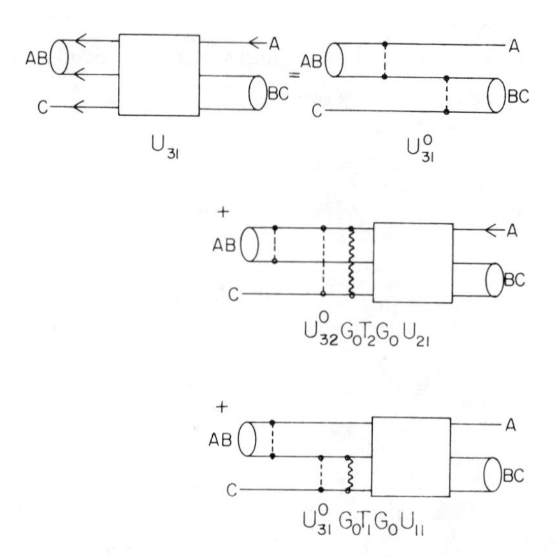

Figure 1. Graphical display of Equation 22 for the amplitude U_{31} of A + BC → AB + C.

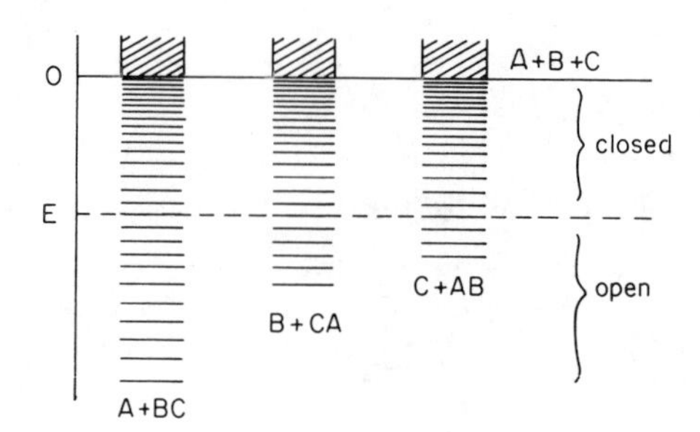

Figure 2. Energetically accessible (open) and inaccessible (closed) states of (A, B, C) in their three arrangements for given total energy E.

$$<\underset{\sim}{\Theta}_s|\underset{\sim}{U}^0|\underset{\sim}{\Theta}_{s'}> = \delta_{ss'} \tag{37b}$$

and

$$U^P = \sum_s (\underset{\sim}{U}^0|\underset{\sim}{\Theta}_s><\underset{\sim}{\Theta}_s|\underset{\sim}{U}^0)/(1 - \lambda_s) \tag{38}$$

Equation 37a may be transformed by using

$$\underset{\sim}{F}^P\underset{\sim}{U}^0 = [\overline{\delta}_{jk}\ G_j^P\ V_j], \tag{39}$$

which shows that the four wavefunctions making up $|\underset{\sim}{\Theta}_s>$ in the coordinate representation satisfy standing-wave boundary conditions. Further, their normalization, from Equation 37b, is neither that of bound states nor of scattering states, but of quasi-bound states, appropriate to the description of resonances. These may be found by letting $\lambda_s = 1$ in Equation 37a, and by finding the energy E_r and state $|\underset{\sim}{\Theta}_r(E_r)>$ which satisfy that equation with the normalization in Equation 37b. Several such resonances may exist, but we assume for simplicity that there is only one. Expanding around $E = E_r$.

$$1 - \lambda_r(E) = (E - E_r)\ N_r + \mathcal{O}(E - E_r)^2 \tag{40}$$

we can rewrite Equation 38 as

$$\underset{\sim}{U}^P = \underset{\sim}{U}^1 + N_r^{-1}(\underset{\sim}{U}^0|\underset{\sim}{\Theta}_r><\underset{\sim}{\Theta}_r|\underset{\sim}{U}^0)_{E_r}\ /\ (E - E_r) \tag{41}$$

where $\underset{\sim}{U}'$ varies smoothly with E around E_r. Then $\underset{\sim}{U}$ may be obtained from direct and resonance terms as follows

$$U = U^{(d)} + U^{(r)} \tag{42a}$$

$$\underset{\sim}{U}^{(d)} = \underset{\sim}{U}'\ (1 + i\pi\underset{\sim}{\Delta}\underset{\sim}{U}')^{-1} \tag{42b}$$

$$U^{(r)} = |\underset{\sim}{\Lambda}_r><\underset{\sim}{\Lambda}_r^*|\ /\ (E - E_r - \Delta_r + i\Gamma_r/2) \tag{42c}$$

where

$$|\underset{\sim}{\Lambda}_r> = (\underset{\sim}{1} - i\pi\underset{\sim}{U}^{(d)}\underset{\sim}{\Delta})N_r^{-1/2}\underset{\sim}{U}^0\ |\underset{\sim}{\Theta}_r> \tag{43}$$

and $|\underset{\sim}{\Lambda}_r^*>$ is its complex conjugate, while the resonance energy shift Δ_r and width Λ_r are given by

$$\Delta_r = \pi\ \mathcal{I}m\ <\underset{\sim}{\Lambda}_r^*|(\underset{\sim}{1} + i\pi\underset{\sim}{\Delta}\underset{\sim}{U}')\underset{\sim}{\Delta}|\underset{\sim}{\Lambda}_r> \tag{44a}$$

$$\Gamma_r = 2\pi\ Re\ \langle \underset{\sim}{\Lambda}^*_r | (\underset{\sim}{1} + i\pi\underset{\sim}{\Delta}\underset{\sim}{U}')\underset{\sim}{\Delta} | \underset{\sim}{\Lambda}_r \rangle \tag{44b}$$

Taking matrix elements of $\underset{\sim}{U}^{(r)}$ between states Φ_1 and Φ_i, and using the sum in Equation 30b for $\underset{\sim}{\Delta}$, one finds for the resonance scattering amplitudes

$$A_{i1}^{(r)} = \langle \Phi_i | U_{i1}^{(r)} | \Phi_1 \rangle = \gamma_{ir}\ \gamma^*_{ir}\ /\ (E - E_r - \Delta_r + i\Gamma_r/2) \tag{45}$$

where, in terms of the components of $|\underset{\sim}{\Lambda}_r\rangle$,

$$\gamma_{ir} = \sum_k \langle \Phi_i | \Lambda_r \rangle_k \tag{46}$$

is the decay amplitude from the resonance r into channel i. A reaction resonance may be defined as a resonance r for which the amplitudes γ_{ir} are substantially different from zero for decay into a single channel, then one is simply dealing with a channel resonance. At energies E very close to E_r, the background term $\underset{\sim}{U}^1$ can be neglected, and the amplitudes and width are simply expressed in terms of the components of the quasibound states $|\underset{\sim}{\Theta}_r\rangle$, since then

$$\gamma_{ir} = \sum_k \langle \Phi_i | \underset{\sim}{U}^0 | \Theta_r \rangle_k\ N_r^{-1/2}, \tag{47}$$

$\Delta_r = 0$, and

$$\begin{aligned} \Gamma_r &= 2\pi \langle \underset{\sim}{\Theta}_r | \underset{\sim}{U}^0 \underset{\sim}{\Delta} \underset{\sim}{U}^0 | \underset{\sim}{\Theta}_r \rangle\ /\ |N_r| \\ &= 2\pi \sum_i \sum_{\text{open } \Phi_i} |\gamma_{ir}|^2 \end{aligned} \tag{48}$$

Hence the resonance width is a sum of partial widths into the several arrangement channels.

Integral Equations for Rearrangement

To calculate cross sections with the previous operator equations, one must further specify a basis set in the space of nuclear degrees of freedom. It is convenient in our approach to use eigenstates of the nuclear momenta because resolvents are then diagonal in the basis. In this basis the operator equations become a set of coupled integral equations in the momentum variables $\vec{P}_i$ and $\vec{p}_i$ for relative and internal degrees of freedom respectively.

To obtain the integral equations we begin by expanding the operators T_i of $\underset{\sim}{F}$ in a basis $|\underset{\sim}{\chi}_i\rangle$ which satisfies

$$\langle \underset{\sim}{\chi}_i | H_I + h_i | \underset{\sim}{\chi}_i \rangle = \underset{\sim}{\varepsilon}_i \tag{49}$$

where $|\underset{\sim}{\chi}_i\rangle$ is a row of states $|\chi_i(n_i)\rangle$ and $\underset{\sim}{\varepsilon}_i$ is a diagonal matrix of elements $\varepsilon_i(n_i)$. Since $\underset{\sim}{V}_i$ and hence $\underset{\sim}{h}_i$ is block diagonal in the basis of electronic states, we can separately diagonalize it for each block B_i so that $n_i = (B_i v_i j_i m_{ji})$. The operator T_i may be expanded in the $|\underset{\sim}{\chi}_i\rangle$ basis using the Lippmann-Schwinger variational principle, which gives (11)

$$T_i(z) = \int d^3\vec{P}_i |P_i\rangle t_i(z_i) \langle \vec{P}_i| \tag{50a}$$

$$t_i = V_i |\underset{\sim}{\chi}_i\rangle \underset{\sim}{\tau}_i \langle \underset{\sim}{\chi}_i | V_i \tag{50b}$$

$$\underset{\sim}{\tau}_i = [\langle \underset{\sim}{\chi}_i | (V_i - V_i g_{0i} V_i) | \underset{\sim}{\chi}_i \rangle]^{-1} \tag{50c}$$

where $\underset{\sim}{\tau}_i$ is a block-diagonal square matrix of elements $\tau_i(n_i, n_j'; z_i)$ and $z_i = z - P_i^2/(2M_i)$. This procedure leads to the correct analytical behavior of τ_i with z_i, since

$$t_i(z_i) \simeq V_i |\chi_i(n_i)\rangle [z_i - \varepsilon_i(n_i)]^{-1} \langle \chi_i(n_i) | V_i \tag{51}$$

for $z_i \to \varepsilon_i(n_i)$. A further important advantage is that the expansion includes contributions of the continuous spectrum of h_i and consequently of virtual transitions to A + B + C during the collision.

The form of Equation 50b makes it advantageous to work with "vertex states"

$$|n_i \vec{P}_i\rangle = V_i |\chi_i(n_i)\rangle \otimes |\vec{P}_i\rangle \tag{52}$$

and to define the amplitudes

$$X_{kl}(\vec{P}_k n_k, \vec{P}_l n_l) = \langle \vec{P}_k n_k | G_0 \, U_{kl} \, G_0 \, | \vec{P}_l n_l \rangle . \tag{53}$$

The reason is that for $\vec{P}_k$ "on-the-energy-shell," i.e. satisfying

$$E = P_k(n_k)^2/(2M_k) + \varepsilon_k(n_k) \tag{54}$$

one finds

$$G_0(z) V_k |\chi_k(n_k), \vec{P}_k(n_k)\rangle = |\chi_k(n_k), \vec{P}_k(n_k)\rangle$$

which is identical to $|\Phi_k\rangle$, so that on-energy-shell

$$X[\vec{P}_k(n_k),n_k;\ \vec{P}_1(n_1),n_1] = A_{k1}, \tag{55}$$

the required transition amplitude. The full set of integral equations is, from Equation 22,

$$X_{k1}(\vec{P}_k n_k,\vec{P}_1 n_1) = X^1_{k1}(\vec{P}_k n_k,\vec{P}_1 n_1) + \sum_i \sum_{n_i n_i'} \int d^3P_i X^1_{ki}(\vec{P}_k n_k,\vec{P}_i n_i) \tag{56a}$$

$$\times\ \tau_i(n_i,n_i';z_i)\ X_{i1}(\vec{P}_i n_i',\vec{P}_1 n_1)>$$

$$X^1_{ki}(\vec{P}_k n_k,\vec{P}_i n_i) = <\vec{P}_k n_k|G_0 U^1_{ki} G_0|\vec{P}_i n_i> \tag{56b}$$

In this expression we have already eliminated the i = 4 channel, including T_4 in U^1_{ki}. The amplitude X^1_{ki} simplifies to

$$Z_{ki} = \bar{\delta}_{ki}\ <\vec{P}_k n_k|G_0|\vec{P}_i n_i> \tag{57}$$

when T_4 can be neglected. The elements Z_{ki} are the basic transfer amplitudes from channel i to k. They describe a process where decoupling occurs in channel i followed by free motion and recoupling in channel k. It is significant that Equations 56 do not contain electronic quantum numbers. All the information on electronic matrix elements is contained in the τ_i factors. If $N_i(B_i)$ is the number of χ_i functions used for block B_i, then the total number of coupled amplitudes is

$$N = \sum_i \sum_{B_i} N_i(B_i).$$

To proceed with a calculation one must expand the $|\vec{P}>$ states in partial waves, and couple angular momenta. This is a lengthy procedure in the general case, because one must e.g. couple to total electronic spin and orbital angular momenta, and these two to the total nuclear angular momenta. Things are somewhat simpler when the electronic angular momenta are null, as is the case in $H + H_2$ (25).

Calculation of $H + H_2$ Resonances

In this section, we summarize the details presented in Reference 12 for $H + H_2$ collision resonances. Since the three atoms are identical we need to refer only to channel 1. The diatomic potential matrix V_1 is diagonal, with elements $B_1 = {}^1\Sigma$ and ${}^3\Sigma$. The triatomic states $\beta^{(1)} = ({}^2S,\ B_1)$ are constructed for $S = M_S = 1/2$, i.e. they are the doublet electronic states with the lowest asymptotic energy. The singlet and triplet potentials $^1V_1$ and $^3V_1$ were taken from the parametrization of an accurate H_3 potential energy surface (26), the so called LSTH potential. Since the molecular states must be antisymmetric with respect to exchange of the protons, one finds the selection rule

$$j_1 + s_1 + i_1 = \text{even}$$

where j, s and i are respectively the nuclear rotational, electronic spin and nuclear spin quantum numbers of H_bH_c. Our treatment shows that this rule applies not only asymptotically but also for all intermediate steps of the collision.

To eliminate the angle variables in Equations 56 we expand the $|\vec{P}>$ state into its partial-wave components $|P\ell m_\ell>$ and then coupled this to the states $|\beta vjm_j>$, where the electronic state $\beta = (s\ i)$ for the specified $S = M_S = 1/2$ and for total nuclear spin quantum numbers $I = M_I = 1/2$. The result of angular momentum coupling is to provide states $|b\ JM_J>$, where (J, M_J), are the total nuclear angular momentum quantum numbers, and

$$b = (s\ i\ v\ j\ \ell).$$

Expanding the amplitudes X_{ki} and X^1_{ki} in the new basis one obtains functions of the momentum length P as coefficients, which are diagonal in J and I and are independent of M_J and M_I due to rotational invariance. Omitting these conserved quantum numbers, Equations 56 simplify to

$$X_{b'b}(P',P_b) = Z_{b'b}(P',P_b) + \sum_{cc'} \int_0^\infty dQ\ Q^2 \times Z_{b'c}(P',Q)\tau_{c'c}(E_Q)X_{c'b}(Q,P_b) \qquad (58)$$

after reducing the number of equations by means of permutational symmetry (see Equations 26 b,c). The symbols c have the same meaning as b; $\tau_{cc'}$ is diagonal in all indices except (v, v'), and depends on $E_Q = E - Q^2/(2M)$. The momentum P_b is an on-energy-shell value.

To obtain cross sections we have followed the steps described before for the operators U_{kl}. First we introduce the solutions $X^P_{b'b}$ of Equation 58 where only the principal value of τ is included. A Heitler equation is then obtained for the on-shell amplitudes $A_{b'b} = X_{b'b}(P_{b'},P_b)$ and $K_{b'b} = X^P_{b'b}(P_{v'},P_b)$ of the form

$$A_{b'b} = K_{b'b} - i\pi M \sum_{b''} P_{b''} K_{b'b''} A_{b''b}. \tag{59}$$

Changing from momentum to energy normalization we recover the usual expressions for the $\underset{\sim}{T}$ and $\underset{\sim}{S}$ matrices,

$$\underset{\sim}{T} = (\underset{\sim}{1} + i\,\underset{\sim}{R})^{-1}\underset{\sim}{R} \tag{60a}$$

$$\underset{\sim}{S} = \underset{\sim}{1} - 2\pi i\underset{\sim}{T} \tag{60b}$$

where $T_{b'b} = M(P_{b'}P_b)^{1/2} A_{b'b}$ and $R_{b'b} = M(P_{b'}P_b)^{1/2} K_{b'b}$. Transition probabilities are given by $P_{b'b} = |S_{b'b}|^2$.

Calculations were done for J = 0 and j = 0 in initial and final channels. Hence initially and finally one has $\ell = 0$ from the triangle rule. Our χ-basis was restricted to j = 0 which implies that $\ell = 0$ also in intermediate channels. As a result ours is an s-wave model calculation, which is however converged in other expansions. The selection rule for this case is

$$s_1 + i_1 = \text{even}$$

and since the diatomics are in singlet states, with $s_1 = 0$, this restricts i_1 to 0 and we are describing <u>para</u> to <u>para</u> transitions. However, triplet states, with $s_1 = 1$, appear at intermediate stages of the collision and require that the diatomic nuclear spins couple to $i_1 = 1$, the ortho state. As results of the s-wave model and of the selection rule, we only need to specify the quantum numbers s and v in each channel, and only v asymptotically. The final results are simply given by $R_{v'v}$ and $P_{v'v}$.

Various aspects of the computational procedure are as follows. (i) $N_T = 15$ basis functions were used for the triplet, and 30 functions for the singlet were contracted to $N_S = 15$ functions by diagonalization of h. (ii) The singular kernel in Equation 58, with the principal value integral, was regularized by adding and subtracting the integrals at the known poles ε_{sv} of $\tau_{sv,s'v'}$. (iii) The matrix $Z(P',P)$ was obtained by numerical integration from the χ_{sv} and was in fact one of the most time-consuming steps in the calculation. To obtain it, the factors $\langle p|V|\chi_{sv}\rangle$ were stored as a Table in the momentum p and were interpolated as needed. (iv) The integral equation with a regular kernel was transformed into coupled linear algebraic equations by discretization of the integral over Q, using Gauss-Legendre quadrature points and weights situated in intervals bounded by the poles of τ. A total of $N_Q = 50$ quadrature points was used. (v) The number of algebraic equations was as a result quite high, in fact equal to $(N_S + N_T)N_Q = 1500$ and required using a modified form of an elimination algorithm suitable to large, nonsymmetric and nonsparse matrices.

The calculated matrix elements $R_{v'v}$ converged to an estimated 5% accuracy, and the range of energies reliably covered went up to the v = 3 threshold, higher than in other works. The R-matrix elements were found to change sign and grow as functions of the total energy E at singularities typical of resonances. Indicating with E_r these singular energies, the elements were parametrized near E_r by

$$R_{v'v} = (2\pi)^{-1} (\Gamma_{v'r} \Gamma_{vr})^{1/2} / (E - E_r). \quad (61)$$

Two such resonances were found in the range of investigated energies. The first one, called here resonance r = a, has energy (in atomic units) $E_a = -0.13808\ E_h = \varepsilon_1 + 0.203$ eV and width $\Gamma_a = 0.167$ eV. The second resonance r = b was found at $E_b = -0.1201\ E_h = \varepsilon_2 + 0.202$ eV with width $\Gamma_b = 0.274$ eV. Hence both resonances are at about the same distance from the vibrational energy thresholds ε_1 and ε_2 for v = 1 and 2, with the higher resonance appreciably wider. Partial widths are also shown in Table I.

Table I. Resonance Energies E_r, Partial Widths Γ_{vr} and Total Widths Γ_r for Resonances r = a and b (All in eV)

r	E_r	Γ_r	Γ_{0r}	Γ_{1r}	Γ_{2r}
a	$\varepsilon_1 + 0.203$	0.167	0.132	0.0351	
	$\varepsilon_1 + 0.191$ [1]	0.035 [1]			
	$\varepsilon_1 + 0.200$ [2]				
b	$\varepsilon_2 + 0.202$	0.275	0.220	0.0502	0.0046

[1] From Reference 15, with the PK potential; [2] From Reference 27, with the full LSTH potential.

Calculating decay times by $\tau_{vr} = \hbar/\Gamma_{vr}$ we see that for r = a decay is about four times shorter into v = 0 than into v = 1. This is also true for resonance b, where decay also occurs into v = 2 but taking ten times longer than into v = 1. Table I shows results of two other calculations of resonance a based on propagation and matching of wavefunctions, with post-symmetrization on the Porter-Karplus (PK) potential (15) and on the full LSTH potential (27). The energies are in excellent agreement, but our width is about five times larger than the one in (15). This discrepancy is not surprising because we used a different potential, only s-waves, and prior symmetrization. Widths are more sensitive to these differences than resonance energies. Resonances covering a wider range of energies have also been recently described by treating H_3 as a bound linear system undergoing normal mode vibrations (28,29). Our resonance a corresponds to the lowest vibrational state with symmetric stretch quantum number $n_{str} = 1$. For $n_{str} = 2$ it is found that several resonances exist, which are classified by asymmetric stretch and bending quantum numbers. Our model does not differentiate among these, because it contains only s waves. Consequently our b resonance must be considered a combination of all those with $n_{str} = 2$. Our energy E_b is the average of the $n_{str} = 2$ resonance energies, and our width Γ_b should be a sum of the corresponding widths.

Although we extracted resonance parameters from the transition amplitudes, our previous discussion of resonances makes clear that they are properties of quasibound states. These states are superpositions of channel contributions, each of which is given in our model by an isotropic distribution of diatomic positions and an isotropic distribution of the third atom with respect to diatomic center of mass. The description underlines the essential difference between our treatment, which allows for all the arrangements present in three-dimensional motion, and treatments of collinear models.

Conclusions

We have shown at the beginning of this contribution how, under very general conditions, one can describe a triatomic system in terms of atom-pair interactions plus an intrinsic three-atom term that is asymptotically null in all arrangements. This leads to a treatment of the dynamics based on the collision theory of three interacting particles with internal structure. Such a theory provides a rigorous mathematical framework for discussing resonances. Thus, a particularly interesting result is that one can associate resonances to quasibound states which satisfy a standard Schrödinger-like equation and are normalized as shown by Equations 37, and for which one must let $\lambda = 1$. This indicates that properties of resonance states could be calculated with methods developed for bound states.

Our calculations of cross sections were based on sets of coupled integral equations in momentum variables. Although this is quite different from wave-mechanical approaches, it leads to results by simply solving large sets of coupled linear equations, in our case for as many as 1,500 variables. How this was done is explained in detail in Reference 12. It can be seen that one of the biggest challenges is to computationally process the large matrices involved. This, however, is very efficiently done by array processors which are becoming increasingly accessible. Hence, there are good prospects for applying the present approach to a variety of collision processes.

Acknowledgments

The present work has been supported by NSF grants CHE 80-19510 and CHE 83-15696.

Literature Cited

1. Micha, D.A. Nucl. Phys. 1981, A353, 309c.
2. Kuruoglu, Z.C.; Micha, D.A. J. Chem. Phys. 1983, 79, 6115.

3. Van Vleck, J.H.; Sherman, A. Rev. Mod. Phys. 1935, 7, 165.
4. Balint-Kurti, G.G. Adv. Chem. Phys. 1975, 30, 137.
5. Tully, J. Adv. Chem. Phys. 1980, 42, 63.
6. Kuntz, P.J. In "Atom-Molecule Collision Theory"; Bernstein, R.B., Ed.; Plenum Press: New York, 1979; Chap. 3.
7. Micha, D.A. J. Chem. Phys. 1972, 57, 2184.
8. Faddeev, L.D. Sov. Phys. JETP 1961, 12, 1014.
9. Alt, E.O.; Grassberger, P.; Sandhas, W. Nucl. Phys. 1967, B2 167.
10. Weinberg, S. Phys. Rev. 1963, 131, 440.
11. Kuruoglu, Z.C.; Micha, D.A. J. Chem. Phys. 1980, 72, 3327.
12. Kuruoglu, Z.C.; Micha, D.A. J. Chem. Phys. 1984, to appear May 1st.
13. Micha, D.A. In "The Theory of Chemical Reaction Dynamics"; Baer, M., Ed.; CRC Press: Boca Raton, FL, to appear; Chap. II.3.
14. Walker, R.B.; Light, J.C. Annu. Rev. Phys. Chem. 1980, 31, 401.
15. Schatz, G.C.; Kuppermann, A. Phys. Rev. Lett.1975, 35, 1266.
16. Kuppermann, A. Theor. Chem. Adv. Perspect. 1981, 6A, 80.
17. Wyatt, R.E.; McNutt, J.F.; Redmon, M.J. Ber. Bunsenges. Phys. Chem. 1982, 86, 437.
18. Baer, M. Adv. Chem. Phys. 1982, 48, 191.
19. Shoemaker, C.L.; Wyatt, R.E. J. Chem. Phys. 1982, 77, 4982.
20. Shyldkrot, H.; Shapiro, M. J. Chem. Phys. 1983, 79, 5927.
21. Smith, F.T. Phys. Rev. 1969, 179, 111.
22. Micha, D.A.; Yuan, J.-M. J. Chem. Phys. 1975, 63, 5462.
23. Micha, D.A. J. Chem. Phys. 1974, 60, 2480.
24. Newton, R.J. "Scattering Theory of Waves and Particles"; McGraw-Hill: New York, 1966; Chap. 7.
25. Kuruoglu, Z.C. Ph.D. Thesis, Chemistry Department, University of Florida, 1978.
26. Truhlar, D.G.; Horowitz, C.J. J. Chem. Phys. 1978, 68, 2466.
27. Walker, R.B.; Stechel, E.B.; Light, J.C. J. Chem. Phys. 1978, 69, 2922.
28. Pollak, E.; Wyatt, R. J. Chem. Phys. in press.
29. Garrett, B.C.; Schwenke, D.W.; Skodje, R.T.; Thirumalai, D.; Thompson, T.C.; Truhlar, D.G., this volume.

RECEIVED June 11, 1984

22

Resonances in the Collisional Excitation of Carbon Monoxide by Fast Hydrogen Atoms

LYNN C. GEIGER[1], GEORGE C. SCHATZ[1], and BRUCE C. GARRETT[2]

[1]Department of Chemistry, Northwestern University, Evanston, IL 60201
[2]Chemical Dynamics Corporation, Columbus, OH 43220

This paper presents a detailed study of resonances in the collisional excitation of CO by H atoms at 1-3 eV translational energy. These resonances are all associated with formation of the metastable species COH during collision. The dynamics is studied using both classical and quantum versions of the infinite order sudden (IOS) approximation, and the resonance energies are characterized using stabilization calculations. We find that the vibrationally inelastic transition probabilities show complex resonance structure at energies above 1.2 eV, with the lowest six resonance states of COH narrow enough to be characterized by stabilization methods. Each resonance energy varies with both the orbital angular momentum ℓ and the atom-diatom orientation angle γ. A complete mapping of the (ℓ,γ) dependence of these six lowest resonance states is determined, and the resonances seen in the scattering calculations are all assigned in terms of their CO and OH stretch quantum numbers. These assignments are confirmed by examining plots of the scattering wavefunctions. The resonance widths are studied and are found to vary with vibrational state, with ℓ, and with γ. The smallest calculated width is about 4×10^{-4} eV corresponding to a lifetime of 2 ps. Integral cross sections for vibrational excitation (rotationally summed) are found to be not very sensitive to the presence of resonances, with good agreement between quantum and classical IOS results and between IOS and quasiclassical trajectory results except close to threshold for each vibrational state. A qualitative study of the γ dependent probabilities suggests that the resonances should measureably influence rotationally inelastic cross sections for high rotational states. They should be most important in studies of the rotationally resolved differential cross sections.

Recent advances in experimental technology have opened up a new area of dynamical processes involving the collision of molecules with

0097-6156/84/0263-0421$06.00/0

fast hydrogen atoms to study by state to state methods (1-8). The hydrogen atoms are produced by excimer laser photolysis of H_2S, HCl, HBr and other species and typically have center of mass energies in the range 1-4 eV. Collisions of H with molecules like CO (3-4), NO (3,5), DCl (1), CO_2 (2), D_2 (6,7), O_2 (8), and H_2O (8) lead to vibrational and rotational excitation of the colliding molecule and in some cases chemical reaction to produce vibrationally and/or rotationally excited products. The nascent distribution of vibration/rotation states in the resulting molecules is then monitored by techniques such as infrared fluorescence, laser induced fluorescence, multiphoton ionization, and CARS, thereby providing information which should be sensitive to the intricate details of the collision dynamics. Because of the high energies involved in these experiments, many features of potential surfaces which are not accessible in thermal collision processes can play an important role in determining the vibration/rotation distributions. For collisions of H with CO, for example, the metastable species COH whose potential minimum is at least 1 eV above H + CO can be formed in these collisions, leading to the possibility of resonances in the scattering process. H + CO is not unusual in this regard. In fact most molecules have either stable or metastable minima associated with addition of a hydrogen atom or they have rearrangement pathways which can support reactive scattering resonances of the type which have been studied in H + H_2 and F + H_2. Thus two important questions in developing a theoretical interpretation of the vibration/rotation distributions seen in fast hydrogen atom collisions are: (1) are resonant states being formed, and (2) how do the resonances influence the vibration/rotation distributions and other dynamical variables?

In this paper we address these questions for the H + CO system, with a detailed study of the quantum scattering dynamics using the infinite order sudden approximation, and with comparisons with analogous classical trajectory calculations. H + CO is a good system for this study for many reasons. First, there now exist two experimental studies of the CO vibration/rotation distributions using the fast hydrogen atom technique (3,4). Second, only a single potential energy surface appears to be important in the dynamics of H + CO, and this potential surface has been fairly well characterized (9), with a global potential surface recently developed (10). Third, two of us have recently studied H + CO using quasiclassical trajectories (10) where we found that formation of the COH species does play an observable role in the measured vibrational distributions. At the same time, the role of the other minimum on the potential surface, the stable HCO (formyl radical) species, can be fairly cleanly separated from that of COH. In this context, the paper by Bowman et al (11) in this volume describes a different energy regime for H + CO where resonances associated with HCO are important. Fourth, the COH potential minimum falls at a convenient energy for studies of resonances in that the number of resonances associated with that minimum varies from zero to a few dozen over the experimentally accessible range of energies. Fifth, the H + CO mass combination is close to optimum for using the infinite order sudden (IOS) approximation (12-15). Without this, the use of quantum scattering methods in this problem would not be feasible. Of course there is an interesting paradox here: the IOS treatment of the scattering dynamics is only accurate for impulsive (nonresonant) collisions, but the corresponding fixed rotor orientation description of the vibrational motions is quite accurate for determining bound state energies provided only that bending is slow compared to the other degrees of freedom. The connection between these two pictures is provided by the process of quantizing the bend in the IOS calculation; we will show how to do this to determine resonance energies.

Perhaps the most severe problem associated with the present calculation is the large number of partial waves which contribute to the vibrational excitation cross sections at most of the energies of interest. This is a consequence of the high velocity of the hydrogen atoms and it makes the characterization of resonances rather complicated. At any given energy, the resonance energies depend on both the orbital angular momentum ℓ and the diatomic orientation angle γ. A complete characterization of each resonance requires mapping out the "trajectory" of the resonance in the ℓ, γ plane. Since many of the resonances are quite narrow, the determination of the resonance values of γ for each ℓ requires searching a very fine grid of γ values, necessitating a very large number of scattering calculations.

To supplement these scattering calculations in determining the ℓ versus γ "trajectories" of each resonance, we also present the results of calculations of the resonance energies by the stabilization method (16-18). Very good correspondence between the stabilization and scattering calculations is found, and this enables us to characterize each resonance in terms of the associated CO and OH stretch quantum numbers. We also present plots of the scattering wavefunctions to further confirm the quantum number assignments, and we present an analysis of adiabatic potential curves which provides much insight concerning resonance decay patterns and lifetimes.

One of the most important questions to be addressed in this study is the importance of quantum effects in determining vibrational excitation probabilities and cross sections. We will study this here by comparing the calculated quantum IOS transition probabilities with the corresponding classical IOS probabilities. Integral cross sections for vibrational excitation will also be compared. The extreme complexity of the γ and ℓ dependence of transition probabilities has prevented us from obtaining rotational and angular distributions, but qualitative information about these distributions will be inferred from the results that we do have.

To summarize the rest of this paper, in the next section we describe the methods used for the quantum and classical IOS calculations and for the stabilization calculations. The results of our calculations are presented and discussed in the following section and a summary of our findings is presented in the Conclusion.

Theory

Potential Surface. We consider collisions of H with CO using the potential energy surface of Ref. 10. This surface was developed using the Sorbie-Murrell method (19) to fit *ab initio* results (9) associated with the HCO and COH minima and with the barriers for addition and isomerization. The HCO portion of the potential was also adjusted to reproduce experimentally derived values of the formyl radical abstraction barrier and energy defect. No such adjustment was possible for the COH part of the potential, and in the trajectory studies of Ref. 10, it was found that better agreement with vibrational excitation data was obtained by lowering the 1.72 eV COH addition barrier by 0.2 eV. We will ignore this problem in the present calculations. Recently, Bowman et al (11) have extended the H + CO ab initio calculations to accurately map out all of the surface which can be accessed in collisions up to a few eV. We did not use this surface in our calculations because the functional form of the resulting surface is so complicated as to make extensive scattering calculations much less practical. Comparison of the Bowman et al surface with the much simpler surface that we did use indicates that our surface is approximately correct except near the isomerization barrier where our surface is much too repulsive. This will not be of major consequence in the present study where the region near the COH minimum is of most interest. Both the Bowman et al surface and the one from Ref. 10 are subject to errors on the order of 0.2 eV in the value of the COH minimum energy, which means that the calculated resonance energies are uncertain by that amount.

IOS Calculations. Both quantum and quasiclassical scattering calculations have been performed using the infinite order sudden approximation. This method has been rather extensively used in quantum mechanical applications to both reactive and nonreactive scattering (12-14) but it has not often been used in studies of resonances (20). Its classical mechanical counterpart has received less attention (15) although it is a straightforward extension of the corresponding quantum calculation.

For H + CO, the IOS Hamiltonian is given by

$$H = \frac{P_R^2}{2\mu_t} + \frac{P_r^2}{2\mu_v} + \frac{\ell^2}{2\mu_t R^2} + \frac{j^2}{2\mu_v r^2} + V(R,r,\gamma) \qquad (1)$$

Here the coordinate R is the H to center of mass of CO distance; P_R and P_r are the radial momenta conjugate to R and r, respectively; μ_t and μ_v are the translational and vibrational reduced masses; ℓ and j are the orbital and rotational angular momenta; and V is the full potential energy function. V depends on the three variables R, r and γ where γ is the angle between R and r such that $\gamma = 0$ for linear COH geometries. In the IOS treatment, the quantities ℓ,j and γ are taken to be parameters, with ℓ chosen in the usual way (from 0 to ∞ in the quantum calculations and 0 to ℓ_{max} in the classical calculations with ℓ_{max} large enough so that all vibrationally inelastic probabilities are converged), and γ is varied from 0 to π. The rotational angular momentum quantum number j has been fixed to the most probable j (i.e., 7) for a 300K Boltzmann distribution. Our earlier trajectory study (10) demonstrated complete insensitivity of the vibrational excitation cross sections to the choice of j for j = 0 - 20. This is the expected result for the collision of a rapidly moving light atom like H with a heavier molecule like CO and is one of the reasons why IOS is expected to be a good approximation here.

The quantum IOS calculations were done using two different independently developed computer programs, both of which gave identical results. The first used a vibrational basis of contracted harmonic oscillator functions and the R matrix propagation method (21). In this approach the vibrational wavefunctions are expanded using a primitive basis of 28 harmonic oscillator functions, and then the 12 lowest eigenfunctions of the asymptotic potential are used to determine interaction potential matrix elements for the R matrix propagation.

The second quantum scattering code uses a basis of 12 vibrational wavefunctions obtained from a 200-point finite difference calculation using the asymptotic potential. This basis is then used to determine matrix elements for use with a modified Gordon integrator (22). This second method is somewhat less efficient than the first, but can be used to determine one piece of information that the first does not: the scattering wavefunction. In this case, a stabilization procedure developed by Schatz et al (23) was used to avoid numerical instabilities in determining the complete wavefunction at all distances. This stabilization procedure, not to be confused with the stabilization method discussed in the next section, insures that the numerical wavefunctions found in the first step of the quantum scattering calculation are sufficiently linearly independent to be combined stably into scattering solutions.

The classical IOS calculations were done using standard integration and sampling methods (15). Consistent with the quasiclassical calculations in (10), a substantial fraction of the trajectories at high translational energies and for γ values where the COH well was sampled involved long-lived "chattering" collisions. This made it difficult to determine vibrational excitation probabilities, since the outcome of each collision varied rapidly with initial vibrational phase and initial translational energy. In the results where this was important we have "smoothed" the probability curves by least squares fitting.

Stabilization Calculations. Stabilization is now a widely used method for determining resonance energies and widths (16-18). In the present context we applied the stabilization method to determine resonances for the IOS Hamiltonian in Equation 1 expanding the wavefunction in a set of two-mode harmonic oscillator functions associated with the normal coordinates of the COH well. Roughly speaking, these two normal coordinates correspond to the CO and OH stretch distances, and we will label the resulting resonance wavefunctions according to the quantum numbers υ_{CO} and υ_{OH} associated with the dominant harmonic configuration in the wavefunction. Typical wavefunction expansions used 34 basic functions, including configurations with all quantum numbers up to $\upsilon_{CO} + \upsilon_{OH} = 7$.

Following a standard procedure for determining stable eigenvalues (see, e.g., 24), we solved the vibrational secular equation at several values of the factor η which scales the frequencies in the harmonic oscillator basis set. Plots of eigenvalue versus η were then constructed and the stable eigenvalues were identified by visual examination of these plots. The identification was straightforward for the resonances corresponding to $\upsilon_{CO} \leq 2$ and $\upsilon_{OH} \leq 1$, and because of this, we will not present the stabilization plots. Most of the resonances with quantum numbers higher than these were quite broad (with widths comparable to the vibrational spacing) so we will not attempt to characterize them. In general, the resonance energies and widths depend on both γ and ℓ, so a complete stabilization analysis requires the construction of stabilization plots at each γ, ℓ of interest. In the present application this full analysis would require construction of hundreds of such plots. To avoid this, the stabilization plots were constructed only at the γ value corresponding to the COH potential energy minimum for each ℓ. Resonance energies at other γ's were then obtained from a single eigenvalue calculation at that γ, by assuming that the configuration with the same dominant harmonic component that gave a stable eigenvalue at the minimum-energy γ (for the same ℓ) would be stable at the γ of interest. A few test calculations at γ's other than the minimum energy γ indicated that this procedure was reliable except at γ's near $\pi/2$ where the potential rapidly becomes repulsive.

Results

Quantum IOS Transition Probabilities and Wavefunctions. For the purpose of characterizing resonance energies and widths, the most easily analyzed result of the IOS calculations is the vibrationally inelastic transition probability $P_{\upsilon\upsilon'}$ between initial CO vibrational state υ and final state υ'. In the IOS treatment, this probability is a function of three variables: the initial translational energy E_0, the orbital angular ℓ and the angle γ. In this paper we will analyze the dependence of $P_{\upsilon\upsilon'}$ on all three of these momentum variables, choosing υ and υ' to be (0,1), (0,2) and (0,3). These vibrational states are the most important in the laser photolysis results of Refs. 3 and 4.

Figure 1 presents the probabilities P_{01}, P_{02} and P_{03} as a function of E_0 for ℓ = 20 and γ = 0.5856 radians. The choice of ℓ = 20 represents a partial wave which is among the largest contributors to the 0 → 2 cross section in the region of energies (1.84 - 2.51 eV) of most relevance to the experiments. The choice γ = 0.5856 is roughly where the effective potential (the sum of the potential energy and the orbital and rotational terms in Equation 1) for ℓ = 20, j = 7 shows a minimum corresponding to the COH well.

Figure 1 shows that the probabilities have a very complex dependence on E_0, with very distinct resonance dips and/or peaks at E_0 = 1.72, 1.88, 2.02, 2.06, 2.18, and 2.28 eV having widths of about 0.0004, 0.0017, 0.031, 0.0015, 0.045, and 0.050 eV respectively. There is also a good deal of additional structure in the 2.30 - 3.00 eV range but it is difficult to identify specific resonances in the results. The direct contributions to the results in Figure 1 are typical of vibrational excitation transition probabilities, with P_{01} already quite large at 1.5 eV then dropping off above 1.8 eV. The other transition probabilities peak at higher energies, roughly in accord with the spacing of about 0.27 eV between the lowest vibrational states of isolated CO.

We will use several methods in this paper to analyze the resonance structure seen in Figure 1 including a stabilization analysis in the next section and an analysis of vibrationally adiabatic potential curves in the subsequent section. In this section we concentrate on what can be learned directly from the IOS calculations.

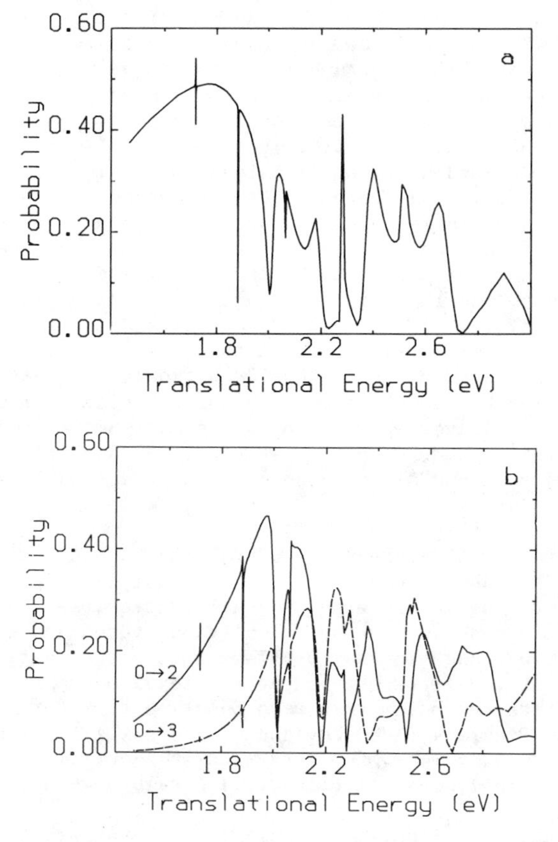

Figure 1. Transition probability versus E_0 in eV for ℓ = 20, γ = 0.5856. Figure 1a depicts P_{01} while Figure 1b shows P_{02} and P_{03}.

In particular let us now examine the scattering wavefunction ψ. Figure 2 plots contours of the real part of ψ versus R and r for translational energies of 1.85 eV, 2.0175 eV and 2.285 eV. Superimposed on these plots are contours of the effective potential. The potential contours clearly show the COH well, and we see that at 1.85 eV where there is no resonant structure in Figure 1, there is no significant penetration of the wavefunction into this well. At 2.0175 and 2.285 eV, on the other hand, we see significant penetration of the scattering wavefunction into the well.

Figures 2b and 2c reveal that the resonance wavefunction inside the well has a very transparent nodal structure. In Figure 2b, there is one node in the R direction inside the well and none in the r direction. According to our previous definitions of vibrational quantum numbers, this resonance thus corresponds to $\upsilon_{CO} = 0$, $\upsilon_{OH} = 1$ or (0,1) in an abbreviated notation which we use from here on. The resonance in Figure 2b is clearly (1,1) in the same notation. Applying the same analysis to all the clearly resolved resonances in Fig. 1 leads to the following assignments: 1.72 eV is (0,0), 1.88 eV is (1,0), 2.02 eV is (0,1), 2.06 eV is (2,0), 2.18 eV is (3,0) and 2.28 eV is (1,1). All of these resonances lie about where they might be expected to based on the COH harmonic frequencies of 0.16 eV for CO stretch and 0.40 eV for OH stretch except (2,0) and (1,1) which seem to be reversed. Although the assignment in Figure 2b seems straightforward, we note that the (2,0) and (1,1) states are nearly degenerate and mix enough to cause a reasonable shift in their energy levels. This near degeneracy will show up in several other results in this paper.

We now study the dependence of P_{02} and P_{03} on γ and ℓ for a given E_0. The energy chosen for this purpose is 1.84 eV. This corresponds to the lowest experimentally accessible energy where resonances are observed in the calculations for many partial waves. It is also low enough so that problems with broad overlapping resonances are not too important at the IOS level of approximation. Fig. 3 presents these probabilities as a function of γ for $\ell = 0$, 10, 20 and 30 while Figure 4 plots P_{02} and P_{03} versus ℓ for $\gamma = 0.5856$. Approximately 100 values of γ were used in generating the results in Figure 3 at each ℓ, and all (integral) ℓ values were considered in Figure 4.

Examination of Figure 3 reveals that the probabilities are all smooth in the range $1.2<\gamma<\pi$. This range corresponds mostly to collisions of H with the carbon end of CO, and even though there is a deep well associated with a slightly bent HCO geometry, no resonance structure is apparent. Presumably this is because the translational energy being considered is much higher than the 0.11 eV barrier to dissociation from this well. Figure 3 does show that the transition probabilities peak near $\gamma = 1.9$ (where penetration into the HCO well is important) and are small near $\gamma = \pi/2$ (corresponding to repulsive perpendicular collisions). For $\gamma < 1.2$, complicated resonance structure is observed in Figures 3a-3c but not in Figure 3d. Figure 4 indicates that Figure 3d is on the tail of the opacity function where centrifugal effects have eliminated all resonances.

The assignment of the resonance structure in Figures 3-4 can be achieved by examining wavefunctions as in Figure 2, but in the present case it is easier to examine the stabilization results to which we now turn.

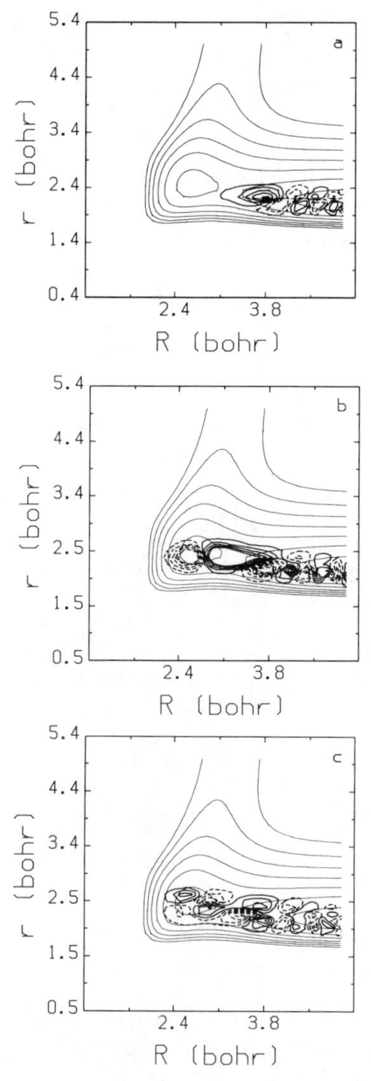

Figure 2. Contours of the real part of IOS scattering wavefunction versus R and r for ℓ = 20, γ = 0.5856 and for E_0 = 1.85 eV (Figure 2a), 2.0175 eV (Figure 2b) and 2.285 eV (Figure 2c). The contours are plotted for the flux normalized wavefunction from -9 a.u. to +9 a.u. with a spacing of 2 a.u. Also plotted are contours of the effective potential (including orbital and rotational terms). The lowest contour plotted is at -11.43 eV (relative to complete dissociation) and the interval between contours is 1.09 eV.

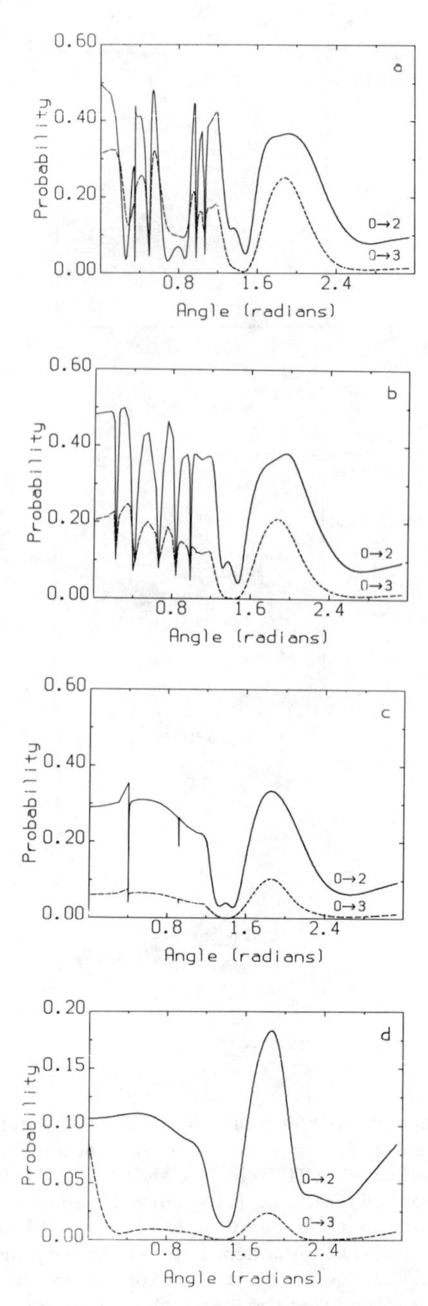

Figure 3. P_{02} and P_{03} versus γ for E_o = 3. P_{02} and P_{03} versus γ for E_o = 1.84 eV and ℓ = 0 (Figure 3a), ℓ = 10 (Figure 3b), ℓ = 20 (Figure 3c) and ℓ = 30 (Figure 3d).

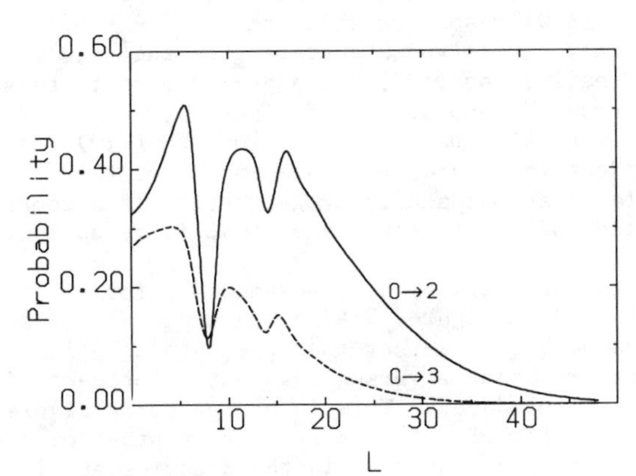

Figure 4. P_{02} and P_{03} versus ℓ for E_o = 1.84 eV and γ = 0.5856.

<u>Stabilization Calculations</u>. Figure 5 presents the lowest six stable eigenvalues obtained from calculations at $\ell = 10$ as a function of γ. The assignments (υCO, υOH) for each of these eigenvalues is indicated, based on the dominant configuration associated with each eigenvalue. As discussed previously, the (2,0) and (0,1) eigenvalues are often very close, and whenever this happens the corresponding eigenfunctions are strongly mixed. Note that the (2,0) and (0,1) eigenvalue curves should not cross as γ is varied. We have ignored this in plotting Figure 5. Comparison of Figure 5 and Figure 3b immediately enables us to assign the resonances in Figure 3b. Specifically, we find that the dips at $\gamma = 0.62$ and 0.82 are both due to the (3,0) resonance while those at $\gamma = 0.40$ and 1.0 involve apparently unresolved structure from the (2,0) and (0,1) resonances. Finally, the (1,0) resonance appears to cause the dip at $\gamma = 0.22$. From the appearance of Figure 5, one might expect to see additional structure near $\gamma = 1.2$ from the (1,0) resonance, but nothing is evident in Figure 3b. This indicates that the rapid rise in the potential near perpendicular geometries is accompanied by enough broadening of this resonance so that it is no longer observable.

Construction of plots similar to Figure 5 for all the ℓ and γ values considered in Figures 3-4 should now enable assignment of all of these resonances. Rather than present such plots, let us instead plot the resonant ℓ versus resonant γ for each of the resonances seen at $E_0 = 1.84$ eV. This is presented in Figure 6, and is one of the key results of this paper since it provides a complete picture of how resonances show up in the dynamics at 1.84 eV. Going back to Figure 3a, we can now read off the assignments of each resonance using Figure 6. In particular, the dip at $\gamma = 0.28$ is due to (0,1), that at 0.35 is (2,0) and that at 0.49 is an incompletely resolved (1,1) plus (3,0). The latter combination appears again at 0.97 while (2,0) plus (0,1) shows up at 1.08. Considering $\ell = 20$ in Figure 3c we find that only the (0,0) resonance is present, appearing at both $\gamma = 0.4$ and 0.92. When the ℓ dependence of P_{02} and P_{03} is considered in Figure 4 we find that the dip at $\ell = 8$ is due to the (1,1) + (3,0) resonance while that at $\ell = 14$ arises from (2,0) + (0,1). The (0,0) and (1,0) resonances do not show up in Figure 4 because the resonant ℓ values for $\gamma = .5856$ are not integers.

To completely specify all the resonance energies as a function of ℓ and γ, we have fit our stabilization eigenvalues to the following multinomial

$$E_{\upsilon CO,\ \upsilon OH}(\gamma,\ell) = \begin{aligned} &C_1 + C_2x + C_3x^2 + C_4y + C_5y^2 + C_6xy + C_7y^3 \\ &+ C_8xy^2 + C_9y^4 + C_{10}y^5 + C_{11}y^6 \end{aligned} \qquad (2)$$

where $x = \ell(\ell + 1)$ and $y = \gamma - 0.75398$. The angle $\gamma = 0.75398$ corresponds to the COH minimum for $\ell = 0$. All of the fits using just the terms indicated in Equation 2 are quite accurate, with an average rms deviation of 0.02 eV.

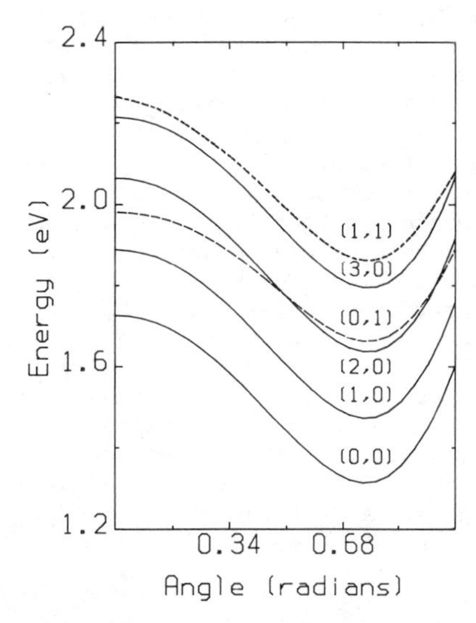

Figure 5. Lowest six stable eigenvalues from a 34-configuration stabilization calculation for ℓ = 10 as a function of orientation angle γ. Stability of each eigenvalue was determined at the orientation γ where the sum of the potential plus centrifugal energy minimizes. The eigenvalues were then determined as a function of γ using single scale factor calculations.

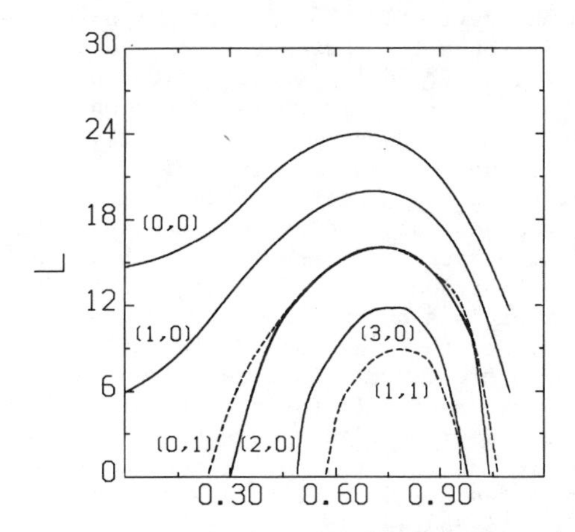

Figure 6. Resonant ℓ value versus resonant γ value for E = 1.84 eV and the lowest six stable eigenvalues.

The coefficients C_1 - C_{11} for each of the 6 lowest resonances are presented in Table I. Two interesting results may be inferred from this Table. First, the dependence of the resonant energies on ℓ is very accurately described by using just the rigid rotor C_2 coefficient, and this is largely independent of υ_{OH} and υ_{CO}. The value of this coefficient is consistent with a moment of inertia $\mu_t R^2$ = 10807a.u. which implies R = 2.47a_o. This is almost identical to the 2.46a_o value of R at the COH minimum (10). Second, the γ dependence of the resonance energies defines an effective "Born Oppenheimer" bending potential which can be used to generate resonance energies for which bending is also quantized. If we assume that the effective moment of inertia for bending motions is $\mu_t R^2$, then the lowest three bending states for the (0,0) resonance for ℓ = 0 are at 1.244, 1.360 and 1.469 eV. Analogous energies may be derived for the other resonances, and these may be used to estimate that the total number of stretch/ bend resonance states which are energetically accessible at 1.84 eV for ℓ = 0 is 19. To estimate the widths of these bend/stretch states would require averaging the IOS width as a function of γ over the probability distribution associated with that state. From limited results available from our scattering calculations, it appears that the width varies slowly with γ near the COH equilibrium, suggesting that the widths seen in Figure 1 are representative of values which would be obtained for the stretch/ bend states.

Analysis of Adiabatic Potential Curves. Another way to characterize the resonances in H + CO is to analyze the potential curves constructed by assuming that the CO motions are vibrationally adiabatic for a given ℓ and γ. We found that this can be done quite accurately for H + CO by fitting a Morse function to the effective potential at each R, using three r values located near the Morse function minimum to determine the Morse parameters. (This is similar to the Morse II model used previously (25) for reactive resonances.) The resulting adiabatic potential curves for υ_{CO} = 0-4 are presented in Figure 7 for ℓ = 20, γ = 0.5856. The positions of the lowest 6 resonance energies from Figure 1 are indicated on the curves and we can see from this plot that υ_{OH} = 1 is apparently the highest bound OH stretch state for υ_{CO} = 0 and 1. In addition, we can infer some information about the decay patterns of these resonances. For example, if each resonant state decays primarily by tunnelling through the adiabatic barrier for that state then the (2,0) resonance would be much narrower than (0,1) since the barrier is larger. This agrees with what is seen in Figure 1. However, many of the resonance states appear to decay more efficiently by first being deactivated to the υ = 0 curve. This is evidenced by the dips rather than peaks that appear in the curves in Figure 1 on resonance. Only the 0 → 0 probability gets enhanced on resonance which presumably is due to the fact that all the resonances except the lowest can decay on the υ = 0 curve in Figure 7 with a smaller barrier than occurs via adiabatic decay.

Classical IOS Results. Figure 8 presents the probabilities P_{02} and P_{03} from the classical IOS calculation for ℓ = 20, γ = 0.5856. These curves are compared with the analogous quantum probabilities from Figure 1, and we find that the nonresonant portions agree quite

Table I. Coefficients in Fit of Equation 2 to Stabilization Calculation Eigenvalues[a]

Coefficient $(\upsilon_{CO},\upsilon_{OH})$ =	(0,0)	(1,0)	(2,0)	(0,1)	(3,0)	(1,1)
C_1	1.184	1.346	1.519	1.543	1.654	1.731
C_2	1.259(-3)	1.227(-3)	1.082(-3)	1.116(-3)	1.228(-3)	1.311(-3)
C_3	-2.237(-7)	-7.905(-8)	3.431(-7)	-6.152(-8)	-2.284(-7)	-5.867(-7)
C_4	-4.457(-2)	-8.198(-2)	-9.951(-2)	-5.509(-2)	-9.917(-2)	-2.743(-1)
C_5	2.872	2.810	2.842	2.554	3.819	2.345
C_6	8.117(-4)	8.279(-4)	8.224(-4)	6.151(-4)	4.147(-4)	1.212(-3)
C_7	3.192	3.434	3.399	2.706	6.026	3.728
C_8	3.318(-4)	4.676(-4)	5.962(-4)	1.038(-4)	-1.175(-4)	1.051(-3)
C_9	4.839(-1)	7.899(-1)	9.013(-2)	-1.688	-3.813	2.261
C_{10}	b	b	-7.588(-1)	-2.354	-1.762(1)	2.129
C_{11}	b	b	b	b	-1.229(1)	1.525
rms deviation	0.024	0.0046	0.015	0.0027	0.035	0.035

(a) All energy levels are in eV's, angles in radians.

(b) Term not used in fit.

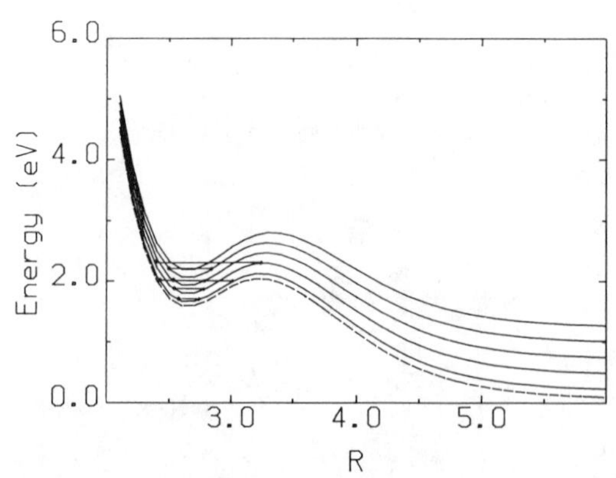

Figure 7. Vibrationally adiabatic potential curves as a function of R for ℓ = 20, γ = 0.5856 for CO vibrational states υ = 0-4. Also plotted is the minimum CO potential energy at each R (the dashed curve) and the lowest six resonant eigenvalues (as obtained from stabilization calculations).

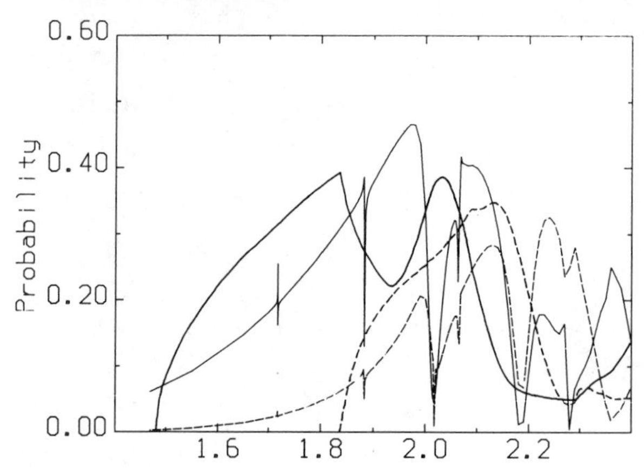

Figure 8. Classical and quantum probabilities P_{02} and P_{03} versus translational energy for ℓ and γ values that are the same as Figure 1b.

well in general. The main differences are found near the threshold for each vibrational state where the quantum probabilities are found to be well above the corresponding classical probabilities. The classical results also do not describe resonances correctly, but because most of the resonances at energies less than 2.4 eV are quite narrow, the overall energy region where resonance effects are important is small.

Integral Cross Sections. Because of the complex resonant structure seen in Figures 1-4, the complete determination of rotationally resolved integral and differential cross sections for H + CO will be a truly formidable task. Even at the relatively low energy of 1.84 eV, we estimate that 50 partial waves would be needed, many of which would require 50-100 values of γ to accurately map out the resonant and nonresonant regions. Such calculations are beyond the scope of our present study, but it is possible to do a more coarse-grained evaluation to determine vibrational excitation cross sections. In this evaluation we have evaluated the quantum IOS probabilities on a grid of 25 equally spaced values in γ at ℓ = 0, 10, 20, 30, ...110. The calculation of rotationally summed integral cross sections involves only the integral of $P_{\upsilon\upsilon'}(\gamma,\ell)$ over γ, and it is apparent from Figure 3 that the resonant contributions to this integral will be minor. As a result, the resulting integrated probabilities are smooth functions of ℓ, which allows us to interpolate between the ℓ's not actually computed.

The resulting cross sections $Q_{0\upsilon}$ from the quantum and classical IOS calculations (QIOS and CIOS) and from the quasiclassical QC calculations of Ref. 10 are summarized in Table II. Comparison of the three sets of cross sections indicates excellent agreement for the lower vibrational states (υ = 1,2). The agreement for the highest vibrational states is not as good, with some of the cross sections differing by factors of 2 or more. Note that the QIOS cross sections for the highest vibrational states are consistently higher than QC or CIOS. Presumably this is due to the threshold error in the classical calculations that was noticed in Figure 8. This result has some bearing on the comparison with experiment given in Ref. 10, since one conclusion of that study was that the QC cross sections for high vibrational states were too small to explain the experimental data without modifying part of the HCO potential surface. The present results suggest that this problem with QC is in fact a quantum threshold effect which can be corrected using QIOS.

The fact that the QCT and CIOS cross sections are in generally excellent agreement attests to the validity of the IOS approximation. The fact that the CIOS and QIOS cross sections are close for all but the highest υ states indicates that the nonresonant part of the scattering is well described by classical mechanics. The one anomaly in this comparison is for υ = 3 at 2.51 eV where the QIOS cross section is significantly higher than either CIOS or QC even though υ = 3 is not close to the highest populated vibrational state.

Conclusion

In this paper we have discovered that the H + CO collision dynamics has a complex resonant structure with at least six COH stretch

Table II. QIOS, CIOS and QC Cross Sections $Q_{0\upsilon}$ (in a_0^2).

(A) E_0 = 1.84 eV	QIOS	CIOS[a]	QC (Ref. 10)
υ = 1	11.9	11.9	14.3 ±0.7
2	3.9	3.0	2.3 ±0.3
3	0.94	0.7	0.4 ±0.1
4	0.13	---	---
(B) E_0 = 2.30 eV			
υ = 1	13.2	12.8	14.6 ±0.8
2	5.4	4.8	5.1 ±0.4
3	2.5	1.7	0.75 ±0.16
4	0.97	0.6	0.52 ±0.12
5	0.25	0.06	0.11 ±0.04
6	0.03	---	---
(C) E_0 = 2.51 eV			
υ = 1	13.6	12.8 ±0.2	15.2 ±0.8
2	5.7	5.1 ±0.4	5.7 ±0.5
3	3.0	1.8 ±0.3	1.7 ±0.2
4	1.5	1.2 ±0.3	0.56 ±0.13
5	0.56	0.3 ±0.1	0.25 ±0.08
6	0.15	0.05	0.04 ±0.02
7	0.02	---	---

a. Statistical uncertainty in the CIOS results is ±1 in the last figure except as otherwise indicated.

states playing an important role in the dynamics at 1.84 eV and above. These resonances have lifetimes that vary from many picoseconds (i.e. hundreds of CO vibrational periods) to 0.01 ps (a few periods), and whose energies vary with ℓ like a rigid rotor and with γ like an anharmonic oscillator. Although the IOS description does not treat bending contributions to the resonance energies correctly, it is easy to correct this by quantizing the bend degree of freedom, and since the bend is still the lowest frequency mode even in the COH well, one expects (26) that the resulting "Born Oppenheimer"like resonance energies will be quite accurate.

While the COH resonances play an important role in the dynamics of H + CO in the 1-3 eV range, so also do direct scattering processes. In fact, the rotationally summed vibrationally inelastic cross sections appear to be completely dominated by direct scattering. This indicates that the conclusions of a previous quasiclassical trajectory study of H + CO concerning measured vibrationally inelastic cross sections will not be significantly altered by the resonances. In fact, the influence of quantum threshold effects was found to be the most important difference between the classical and quantum results. The question then arises as to what experimental measureables will show significant resonance effects. Examination of Figure 3 indicates one possibility, namely, the high j part of the rotational distributions. In the IOS treatment, the rotational distributions are obtained from Legendre moments of the γ dependent scattering matrices. The many resonances seen in Figure 3 should all cause high frequency structure in these scattering matrices, leading to high j contributions to the cross sections. The rotationally resolved differential cross sections should reveal this structure more clearly than the integral cross sections. Unfortunately, the task of evaluating rotationally resolved integral and differential cross sections will be very formidable. The possibility that such information may eventually be measured experimentally suggests that such calculations are worth doing.

Acknowledgments

We thank Dr. Michael Redmon for serving as the catalyst in getting this project started. LCG and GCS were supported by a grant from the National Science Foundation (CHE-8115109).

Literature Cited

1. Magnotta, F.; Nesbitt, D.J.; Leone, S.R., Chem. Phys. Lett. 1981, 83, 21-25.
2. Quick, Jr., C.R.; Weston, Jr., R.E.; Flynn, G.W., Chem. Phys. Lett. 1981, 83, 15-20.
3. Wight, C.A.; Leone, S.R., J. Chem. Phys. 1983, 78, 4875-86.
4. Wood, C.F.; Flynn, G.W.; Weston, Jr., R.E., J. Chem. Phys. 1982, 77, 4776-7.
5. Wight, C.A.; Leone, S.R., to be published.
6. Gerrity, D.P.; Valentini, J.J., J. Chem. Phys. 1983, 79, 5202-3.
7. Marinero, E.E.; Rettner, C.T.; Zare, R.N., J. Chem. Phys., to be published.

8. Kleinermanns, K.; Wolfrum, J., Appl. Phys. B, in press.
9. Dunning, T.H., J. Chem. Phys. 1980, 73, 2304-9.
10. Geiger, L.C.; Schatz, G.C., J. Phys. Chem. 1984, 88, 214-221.
11. Bowman, J.M.; Lee, K.T.; Romanowski, H.; Harding, L.B., in this volume.
12. Kouri, D.J., in "Atom Molecule Collision Theory", ed. by Bernstein, R.B. (Plenum, New York, 1979), p.301.
13. Schinke, R.; Bowman, J.M., in "Molecular Collision Dynamics", (Topics in Current Physics, Vol. 33) ed. by Bowman, J.M. (Springer, Berlin, 1983) p. 61-115.
14. Secrest, D.; Lin, C.S., J. Chem. Phys. 1979, 70, 3420-3 and references therein.
15. Mulloney, T.; Schatz, G.C., Chem. Phys. 1980, 45, 213-223.
16. Holoien, E.; Midtal, J., J. Chem. Phys. 1966, 45, 2209-16.
17. Taylor, H.S.; Williams, J.K.; Eliezer, I., J. Chem. Phys., 1967, 47, 2165-77.
18. Taylor, H.S., Adv. Chem. Phys. 1970, 18, 91-147.
19. Sorbie, K.S., Murrell, J.N., Mol. Phys. 1975, 29, 1387-1407; 1976, 31, 905-920.
20. Zhang, Z.H., AbuSalbi, N., Baer, M., Kouri, D.J., in this volume.
21. Harvey, N.M., Ph.D. Thesis, University of Minnesota, Minneapolis, 1979.
22. Schatz, G.C., Chem. Phys. Lett. 1983, 94, 183-187.
23. Schatz, G.C., Hubbard, L.M., Dardi, P.S., Miller, W.H. J. Chem. Phys., accepted.
24. Thompson, T.C., Truhlar, D.G., Chem. Phys. Lett. 1982, 92, 71-5.
25. Garrett, B.C., Truhlar, D.G., J. Phys. Chem. 1982, 86, 1136-1141; 1983, 87, 4554 (E).
26. Bowman, J.M., Romanowski, H., to be published.

RECEIVED June 11, 1984

Resonant Quasi-periodic and Periodic Orbits

For the Three-Dimensional Reaction of Fluorine Atoms with Hydrogen Molecules

C. C. MARSTON[1] and ROBERT E. WYATT[2]

[1]Departments of Physics and Chemistry and Institute for Theoretical Chemistry, University of Texas, Austin, TX 78712

[2]Department of Chemistry and Institute for Theoretical Chemistry, University of Texas, Austin, TX 78712

Numerical methods are described for locating resonant quasiperiodic and periodic orbits in the 3D $F+H_2$ reaction with J=0. A number of plots of both types of resonant orbit are presented. This is the first time that resonant orbits have been found for a non-collinear reaction. These orbits are then used in the arbitrary trajectory semiclassical quantization scheme of DeLeon and Heller. The lowest resonance energy predicted using this procedure is in good agreement with all available quantal and adiabatic semiclassical results.

Over the past few years, resonances in chemical reactions have been the focus of numerous theoretical (1) and experimental studies (2). On the theoretical side, both quantal and semiclassical methods have been used to calculate resonance energies and widths, principally for collinear reactions, although there are a few studies of 3D reactions. In quantal studies of 3D reactions, some close-coupling calculations on $H+H_2$ have been reported (1f), (1i), but the large number of channels has necessitated approaches based upon the J_z-conserving (3), IOS (4), BCRLM (5), or DWBA (6) approximations. In addition to these quantal studies, several semiclassical approaches have been applied to collinear (7) and 3D reactions (7b), (8). In the Pollak-Child theory, the energies of periodic orbits in the collinear collision complex are adjusted to satisfy integer action quantization conditions (7c). These resonant periodic orbits (RPO's) were first mentioned in an earlier study of trajectories trapped in entrance or exit regions (periodic orbit dividing surfaces - PODS) or in the collision complex (9) of collinear reactions. In an extension to predict resonance energies in the 3D $H+H_2$ and $F+H_2$ reactions, Pollak and Wyatt (8) developed an adiabatic reduction scheme in which semiclassically quantized RPOs at fixed values of the bending angle were com-

0097-6156/84/0263-0441$06.00/0

puted in the first step. In the second step of the reduction scheme, these energies then served as an effective potential for the slower bending motion. Semiclassical quantization of the periodic bending orbits led to time-averaged effective moments of inertia for the slow overall tumbling motion. These and other studies based upon PODS (10) or RPOs have been extensively reviewed by Pollak (11). In related studies, Duchovic, Swamy, and Hase found quasiperiodic orbits above the dissociation threshold for the H-C-C → H+C≡C fragmentation (12). They used an iterative method to semiclassically quantize the energies of some of these orbits.

The present study is concerned with the application of numerical methods to locate orbits for chemical reactions that are not restricted to collinear geometry. In contrast to the Pollak-Wyatt adiabatic reduction scheme (8), the present treatment does not require an adiabatic separation of motions. At a given energy E, the trajectory must be started at a point such that, after a time interval, it nearly (quasiperiodic case) or exactly (periodic case) returns to its starting point. The problem then is to systematically locate these starting points. In practice, it is found that trajectories initiated close to the resonant orbit quickly evolve into the asymptotic reactant or product regions; the initial conditions must be adjusted to minimize the escaping tendency of the trajectory from the collision complex into both the reactant and product channels. Here, the escaping tendency is measured by the atom-molecule relative momentum at a turning point in the motion; this momentum is denoted $\vec{P}_z$ or $\vec{P}_{z'}$, in reactants or products, respectively. To locate RPOs for collinear reactions, Pollak and Child (7c) discussed the turning point (TP) and reactant-product (RP) boundary methods. In the TP method, the trajectory is followed to the first turning point, where the sign of the momentum component perpendicular to $\vec{\nabla}V$ is examined. The starting point of the trajectory in the reactant channel is then adjusted in an attempt to force this momentum component to be zero. On the other hand, the RP method was recently used by Pollak to locate a quasiperiodic orbit near the entrance channel v=1 vibrational adiabatic barrier in the 3D $H+H_2$ reaction (13). Using this method, the trajectory is integrated from the region of the collision complex long enough to see whether it moves toward reactants or products. The initial condition is then adjusted to locate the boundary between orbits decaying into reactants or into products. At the RP boundary, the orbit could be either periodic or quasiperiodic. The method that we describe in the next section is a generalization to noncollinear geometries of the TP method. In the second part of the same section, we will describe a method for locating RPOs for 3D reactions. The method is based upon minimization of an "aperiodicity index," A, by again adjusting the initial conditions.

The Hamiltonian is that of an atom-diatomic molecule collision at total angular momentum J=0,

$$H = \frac{1}{2\mu}[P_R^2 + P_r^2 + \frac{1}{r^2}P_\gamma^2] + V(R,r,\gamma),$$

where R and r are the scaled (14) reactant atom-molecule and molecular vibrational coordinates, respectively, and where γ is the bending angle between R and r (In the next section, we will also use the notation z and ρ for R and r, respectively.) In addition, μ is the effective reduced mass of the three atom system. The Muckerman V potential (15) was used for $V(R,r,\gamma)$.

Having computed quasiperiodic or periodic resonant orbits, we then use them in a semiclassical quantization scheme in order to predict resonance energies. Since these trapped trajectories are bound states embedded in the continuum of the collision complex, one of the semiclassical quantization schemes devised for (truly) bound systems may be used (16). Although the semiclassical prediction of resonance energies has been considered previously for the collinear $H+H_2$ reaction (7a) and for a model atom-diatom inelastic collision (17), where (using quasiperiodic trajectories) the iterative surface-of-section method (18) was successfully employed, we found the noniterative arbitrary-trajectory method of DeLeon and Heller (19) to admirably suit our needs. An important advantage of this method is that "arbitrary" trajectories (relatively close to the quantum energy being sought) may be used, thus eliminating the need for difficult root searches to find the "right" trajectories which satisfy quantum conditions on the action integrals. Later, several $F+H_2$ resonant quasiperiodic orbits will be illustrated. Then, the semiclassically quantized resonance energy, computed from two RPOs, is compared to results from all available quantal and semiclassical studies.

Numerical Methods for Locating Quasiperiodic and Periodic Orbits

Before describing the numerical methods used to locate quasiperiodic or periodic resonant orbits, we will define several sets of coordinates that are useful in specifying the size and shape of the three atom triangle. Let $\vec{R}$ be the (scaled) vector from the center-of-mass of H_2 to the F atom, and let $\vec{r}$ be the (scaled) H_2 separation vector (14). In addition, let γ be the angle between $\vec{R}$ and $\vec{r}$, such that $\gamma=0$ or π denote collinear configurations, while $\gamma=\pi/2$ or $3\pi/2$ denote perpendicular configurations. In order to orient the molecular vector $\vec{r}$ relative to $\vec{R}$, we may also use Cartesian coordinates, $x=r\cos\gamma$ and $y=r\sin\gamma$, so that y measures the deviation from collinearity (y=0 thus defines collinear geometries, $\gamma=0$ or π). Thus, the three Cartesian coordinates (R,x,y), where R is the length of vector $\vec{R}$, or the cylindrical coordinates (R,ρ,γ), where $\rho(=r)$ is the length of vector $\vec{r}$, may be used to specify the size and shape of the FH_2 three-atom triangle.

Continuity of classical dynamics with respect to initial conditions suggests that the search for quasiperiodic trajectories in 3D should begin by selecting initial conditions close to the known collinear RPOs, but rotated slightly out of the collinear plane. To a first approximation, the cylindrical coordinates R_o and ρ_o are set equal to R_o and x_o, respectively, of the known collinear RPO ($y_o=0$) and the orientation of the $\vec{r}$ vector is determined by the value of γ_o (initially 1.0°); that is to say, $y_o=\rho_o\cos\gamma_o$. This approximation to the initial position vector must then be adjusted within the $\gamma=\gamma_o$ plane along the component of $\vec{\nabla}V$ lying in that plane to within 10^{-12}eV of the desired energy. Numerical integration of the equations of motion is allowed to proceed through the first two extremes in the ρ motions, or turning points (the trajectory is again near its starting point) and the value of the dissociative momentum $|\vec{P}_z|$ is compared with that of the previous trajectory so that subsequent displacements (ΔR_o, with γ_o fixed) will be in the direction of decreasing final dissociative momentum. This procedure of advancing the initial position vector in the direction of decreasing final dissociative momentum is continued until the value obtained is no longer less than that of the previous displacement, at which point the displacement direction (ΔR_o) is reversed and the search is continued in a convergent sequence. The integrator is then allowed to proceed to the next turning point in the ρ motion and the minimization procedure (the adjustment of ΔR_o) is repeated as before, but with an appropriate definition of the dissociative momentum direction, depending upon whether the trajectory is temporarily terminated in the entrance or exit channel of the potential energy surface. The procedure of minimizing the dissociative momentum after an increasing number of turning points in the ρ motion and with ever-increasing precision is continued until the accuracy of the numerical integrator is exhausted (typically after 14 turning points). In this way, the escaping tendency of the trajectory (toward $F+H_2$ reactants or FH+H products) is minimized, at a gradually increasing number of turning points.

Having thus located the starting condition R_o for quasiperiodic dynamics initiated at a particular value of γ_o, the initial position vector may be rotated to a higher initial angle and the successive minimization procedure repeated at this new value of γ_o. Proceeding through 2° rotational increments, it was possible to find quasiperiodic orbits at angles below γ_{max}, without encountering the barrier to high angle bending expected on the FH_2 potential energy surface.

In order to locate 3D periodic orbits, a slightly different procedure was used. For quasiperiodic orbits at E=0.4 eV, a comparison of γ vs. ρ plots (as in Figure 3B) for different starting conditions (R_o,γ_o) revealed that the turning points of the trajectory initiated at 24.8° appeared to separate into four distinct sets in the trajectory initiated at γ_o=24.0°. The quaisperiodic trajectory initiated at 21.1° was located and its γ vs. ρ plot was found to be consistent with this trend in that the grouping effect was even more distinct than that observed at higher angles. These orbits are thus becoming increasingly periodic. This progression toward an exact superposition of turning points into four clusters was quantified by dividing the difference of the γ values at turning points 8 and 10 (see Figure 3B) by the initial value of γ to obtain an "aperiodicity index," A, for which a value of zero would imply exact periodicity. By plotting A as a function of γ_o, it was possible to extrapolate to A=0 to obtain successively better estimates of the value of γ_o leading to exact periodicity. At γ_o=17.7°, the value was considered acceptably small to assume essentially exact periodicity of the dynamics. The same method was used to find resonant periodic orbits at other starting angles and energies. An example will be provided later.

Semiclassical Quantization Using Arbitrary Trajectories

The semiclassical quantization procedure of DeLeon and Heller (19) was used to obtain quantized resonance energies because of its capability of yielding accurate results from "arbitrary" trajectories (i.e., root searches for the "right" quantizing trajectories are not required at all). The method recognizes that integrability of the dynamics permits the energy to be expressed as a function of only the N action variables obtainable from a system of N degrees of freedom. A first order expansion of the energy from one set of action variables to another is then possible using the expression:

$$E=E^{o}+\frac{\partial E}{\partial \vec{J}}\cdot\delta\vec{J} \tag{1}$$

In order to obtain an approximate energy eigenvalue, $\delta\vec{J}$ must be selected to be appropriate for an expansion to a set of action variables consistent with the quantization conditions. If we now let $\delta\vec{J}=\vec{J}_v-\vec{J}^{(o)}$, where $\vec{J}_v=\vec{V}\hbar$ denotes the set of quantized action integrals ($\vec{V}$ is a set of quantum numbers), and where $\vec{J}^{(o)}$ denotes the actions of the starting trajectory, and using $\partial E/\partial\vec{J}=\vec{\omega}$ from Hamilton-Jacobi theory, the energy quantization expression becomes

$$E(\vec{V})=E(\vec{J}^{o})+\vec{\omega}\cdot(\vec{V}-\vec{J}^{(o)})\hbar \tag{2}$$

Thus, if $\vec{J}^{(o)}$ and $\vec{\omega}$ can be obtained from trajectories at energy $E^{o}=E(\vec{J}^{(o)})$, then the approximate quantized resonance energy, $E(\vec{V})$, labeled by the set of quantum numbers $\vec{V}$, can be predicted. The actions may be obtained from a consideration of the average phase of the trajectories. For one trajectory,

$$\phi_T=\frac{1}{T}\oint\vec{p}\cdot d\vec{q}=\frac{1}{T}\sum_i n_i\oint_{c_i}\vec{p}\cdot d\vec{q}=\frac{1}{T}\sum\frac{T\omega_i}{2\pi}\oint_{c_i}\vec{p}\cdot d\vec{q}$$

$$=\sum_i\omega_i\left[\oint_{c_i}\frac{\vec{p}\cdot d\vec{q}}{2\pi}\right]=\sum_i\omega_i J_i=\vec{\omega}\cdot\vec{J} \tag{3}$$

where i indexes the topologically distinct paths on the torus manifold, and n_i is the number of windings of the i-th topologically distinct path before (exact or nearly exact)closure. The actual trajectory is assumed to wind back on itself (exactly or approximately) in a time T. Also, in conforming to the notation of DeLeon and Heller, we introduced a denominator of 2π into the definition of the action integral. Introducing the index (k) to specify a particular trajectory within a set, all at energy $E(\vec{J}^{(o)})$, this becomes,

$$\phi_T^{(k)}=\vec{\omega}^{(k)}\cdot\vec{J}^{(k)} \tag{4}$$

Assuming that the values of $\vec{J}^{(k)}$ for the different trajectories are essentially equal (for a fixed value of energy), the set of vectors $\vec{J}^{(k)}$ may be replaced by the single vector $\vec{J}^{(o)}$ to obtain the approximation

$$\phi_T^{(k)}=\vec{\omega}^{(k)}\cdot\vec{J}^{(o)} \tag{5}$$

The set of equations implied by the above notation may be expressed in the single matrix equation

$$\underline{\underline{\Phi}}=\underline{\underline{\Omega}}\,\underline{J}^{(o)} \tag{6}$$

and the action variables are then obtained by inverting the frequency matrix: $\underline{J}^{(o)}=\underline{\underline{\Omega}}^{-1}\underline{\Phi}$. In Equation 6, $\Omega_{ij}=\omega_j^{(i)}$ is the j-th frequency for trajectory i, so that the frequencies for a given trajectory run across a row.

For this reaction, the two actions will be denoted J_a for the high-frequency translation-vibration (asymmetric) motion and J_b for the low-frequency bending motion. Integer quantization conditions are used for both actions,

$$\vec{J}_v=(v_a,v_b)=(n_a+1,n_b+1) \tag{7}$$

For J_a, an integer quantization condition is used by analogy to the collinear RPO studies of Pollak and Child (7c); they found that v_a = 4, 6, 8, ... led to semiclassical resonance energies which were close to the exact quantal values. For J_b, integer quantization is used because we are trying to obtain the ground state energy of a doubly degenerate bending degree of freedom. To predict the lowest resonance energy of 3D $F+H_2$, we will thus use $v_a=4$ and $v_b=1$. Higher energy resonances could be predicted with $v_a=4$, $v_b=2$, $v_a=6$, $v_b=1$, etc.

For the current problem, Equation 2 becomes

$$E(v_a,v_b)=E(J_a^{(o)},J_b^{(o)})+\omega_a(v_a-J_a^{(o)})\hbar+\omega_b(v_b-J_b^{(o)})\hbar. \tag{8}$$

Two trajectories, at the energy $E(J_a^{(o)},J_b^{(o)})$, are required to compute the frequencies and actions: ω_a, $J_a^{(o)}$, and ω_b, $J_b^{(o)}$. In order to simplify the Fourier analysis, trajectory number one will be the collinear RPO, and trajectory number two will be the noncollinear RPO. In Equation 8, the two frequencies and the two actions will both refer to trajectory number two. Recently, Miller has shown how just one trajectory may be used to predict an eigenvalue in the arbitrary trajectory method (20).

Quasiperiodic Resonant Orbits

In this Section, we will illustrate several quasiperiodic resonant orbits for $F+H_2$. Using the numerical methods discussed earlier, the quasiperiodic orbit at E=0.9 eV and γ_o=17° was computed. The total energy, E, is measured from the floor of the entrance valley on the FH_2 potential surface. In Figure 1, this orbit is illustrated in (R,x,y) Cartesian internal coordinate space. In addition, projections are shown of the orbit upon the three coordinate planes (R,x), (R,y), and (x,y). Recall that (R,x) is the collinear plane. In part A of the figure, the orbit has been integrated through 9 turning points, while in part B the integration time was extended to 18 turning points. The projection of the orbital motion in the collinear (R,x) plane is a "blurred" or thickened version of the collinear resonant periodic orbits illustrated by Pollak and Child (6c). This thickening arises because the orbit evolves on a two-dimensional curved surface embedded in the three-dimensional space. Figure 2 shows another quasiperiodic resonant orbit, this time for E=0.4 eV and γ_o=20°. Part A again shows the orbit (integrated through 14 turning points in (R,x,y) space, while part B shows the orbit within the FH_2 potential space. The potential surface in Figure 2B is the locus of points with V=0.4 eV; the reactant region on the left is connected to the product region on the right by the FHH interaction region in the middle of the figure. In reactants or products, the classically allowed region ($V \leq 0.4$ eV) lies between the two tubes; for example, in reactants, H_2 vibration, H_2 rotation about the $F—H_2$ vector, and $F—H_2$ relative translation correspond to radial motion between the cylinders, circular motion between the cylinders about the axis labeled "$F+H_2$," and motion parallel to the "$F+H_2$" axis, respectively. The quasiperiodic orbit in part B of the figure is shown in the interaction region of the potential surface. This gives a clear indication of the amount of configuration space swept out during the orbital motion.

For the quasiperiodic orbit shown in Figure 2, three projections of the orbit in the cylindrical coordinate space (R,ρ,γ) are illustrated in Figure 3. In parts A and B, the direction of motion of the orbit between t=0 and t=end (14 turning points) is also shown. Some of the turning points are numbered. Symmetry of the orbit about the γ=0 (collinear) plane is evident in part B. Part C of this figure again shows that the (ρ,R) motion is similar to a "blurred" collinear resonant periodic orbit.

Quasiperiodic resonant (and periodic ones, too) orbits for $F+H_2$ are unstable in the sense that a small displacement $\delta\vec{r}$ from the starting position $\vec{r}_o$ eventually causes the trajectory to leave the interaction region and move into reactants or products. This behavior is illustrated in Figures 4-6. First, in Figure 4, we see a quasiperiodic orbit at 0.9 eV, with γ_o=17°. Small displacements from the starting vector in Figure 4 lead either to decay into products, Figure 5, or decay into reactants, Figure 6. In Figure 5, rotational excitation of FH is evident from the manner in which the trajectory winds around the product tube. This contrasts with the asymptotic motion shown in Figure 6; H_2 is vibrationally excited, but there is

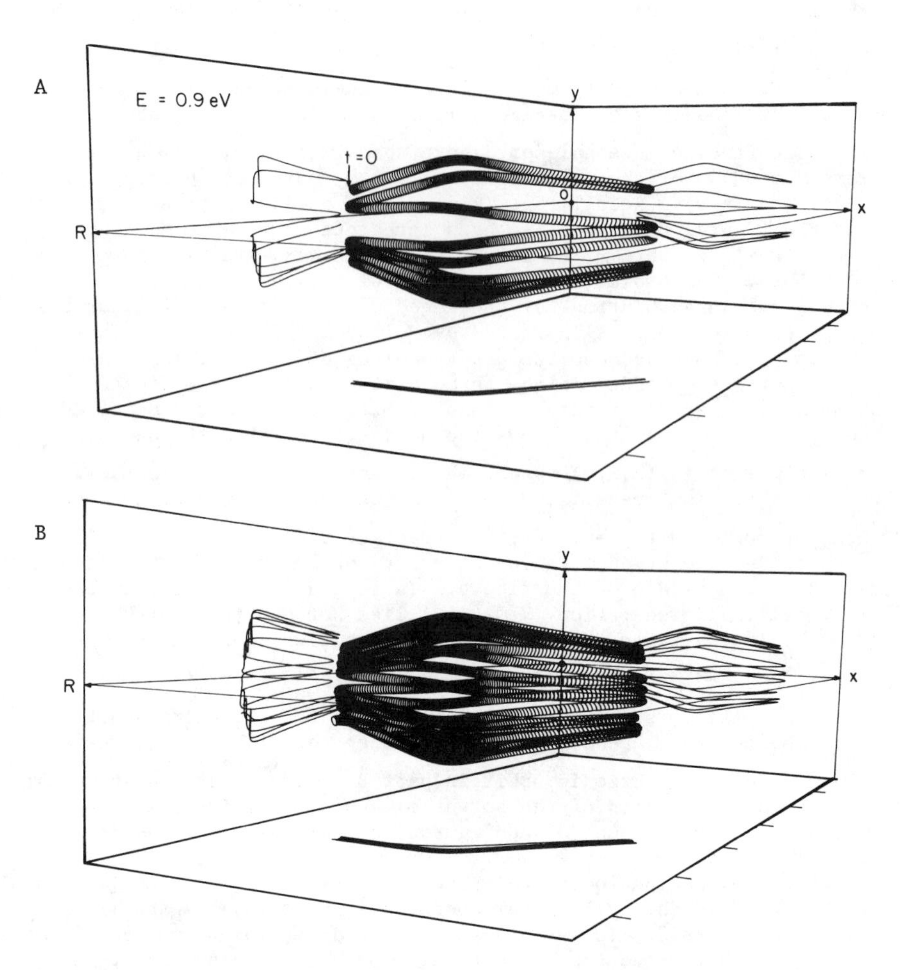

Figure 1. A resonant quasiperiodic orbit displayed in Cartesian coordinate space (R,x,y) at E=0.9 eV with γ_o=17°. The total energy measured from the bottom of the F+H_2 reactant valley is denoted as E. In A, the orbit (thickened, for better visibility) has been integrated to 9 turning points, while in B, the orbit has been extended to 18 turning points. Projections of the orbit onto the Rx, Ry, and xy coordinate planes are also shown; however, for better viewing, the collinear (Rx) projection has been displaced down into the lower horizontal plane.

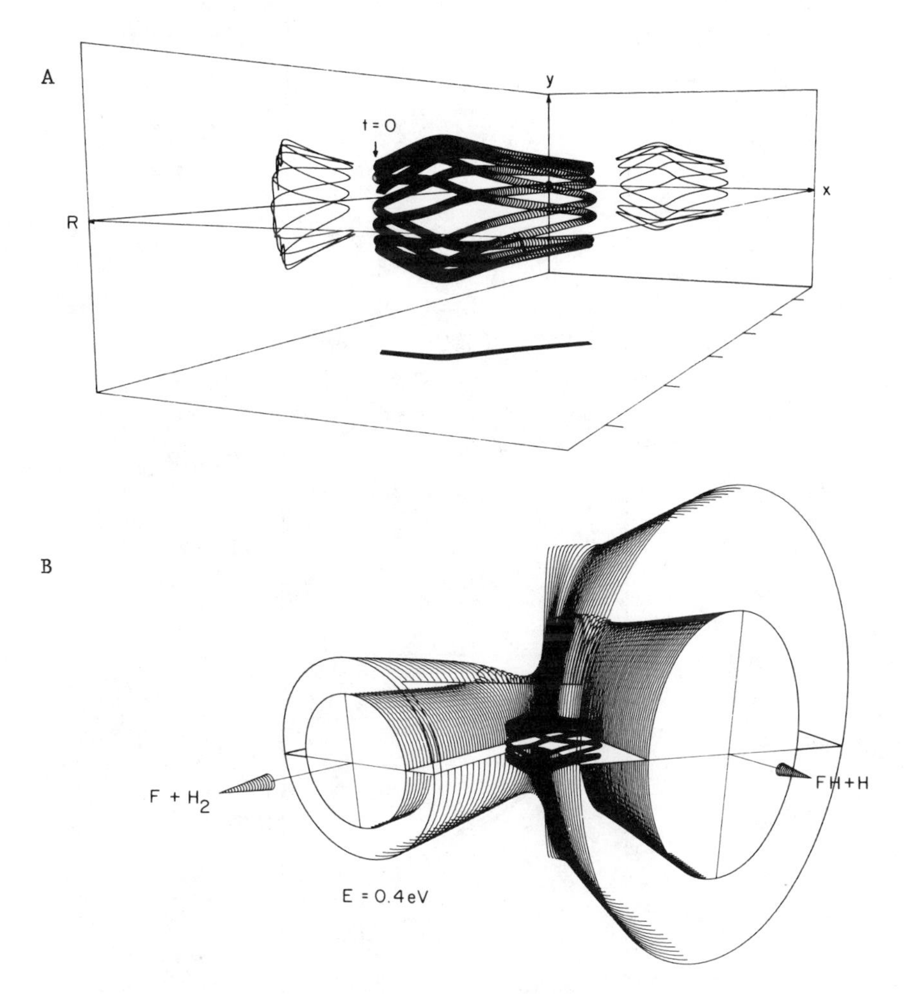

Figure 2. A resonant quasiperiodic orbit at E=0.4 eV with γ_o=20°. In A, the orbit is shown in (R,x,y) internal coordinate space. In B, the orbit is displayed within the V=0.4 eV FH_2 potential energy surface, with the reactant tube extending to the left and the product tube extending to the right.

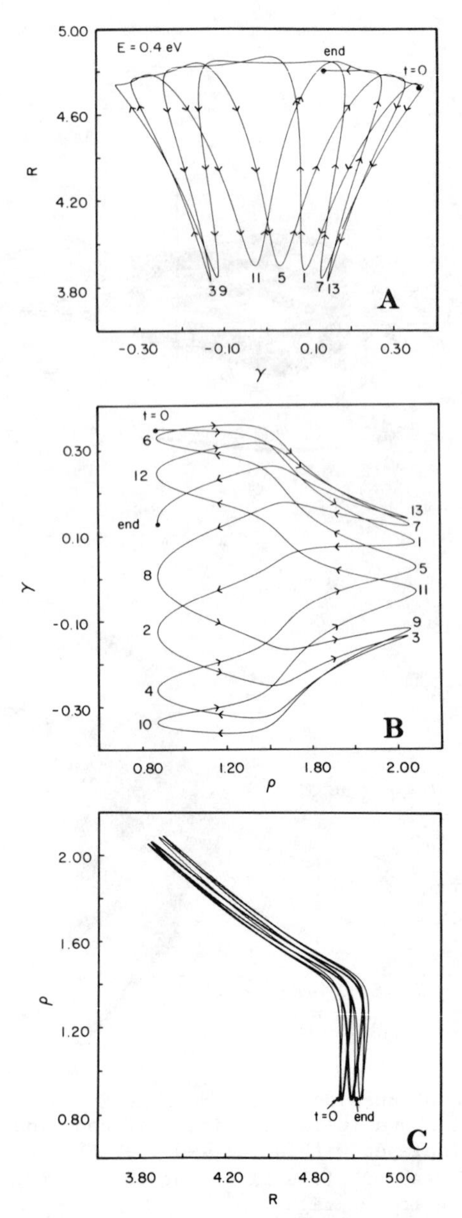

Figure 3. Plots of pairs of polar coordinates for the quasiperiodic trajectory at E=0.4 eV with γ_o=20° (see Figure 2). Recall that R and ρ are (scaled) F—H_2 and H—H distances, while γ measures deviations from collinearity (γ=0 denotes the collinear plane). In A and B, the initial (t=0) and final (end) values are indicated. Some turning points are numbered in A and B.

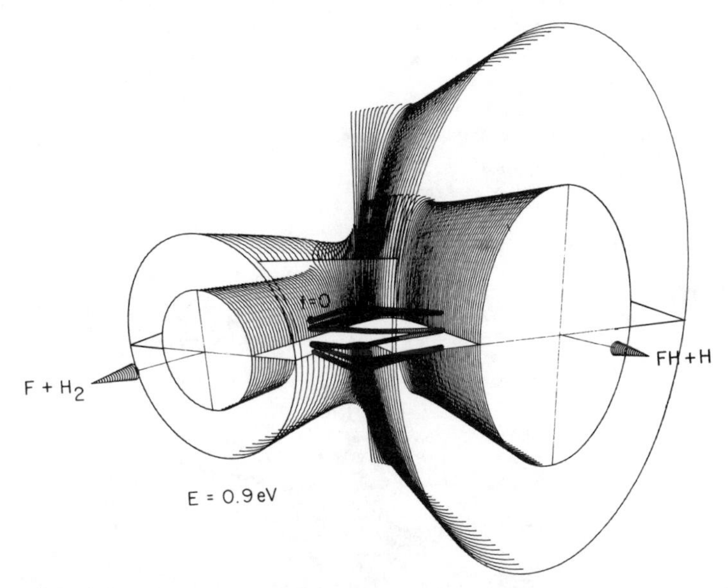

Figure 4. A resonant quasiperiodic orbit at E=0.9 eV, with γ_o=17°. The orbit has been integrated through 6 turning points.

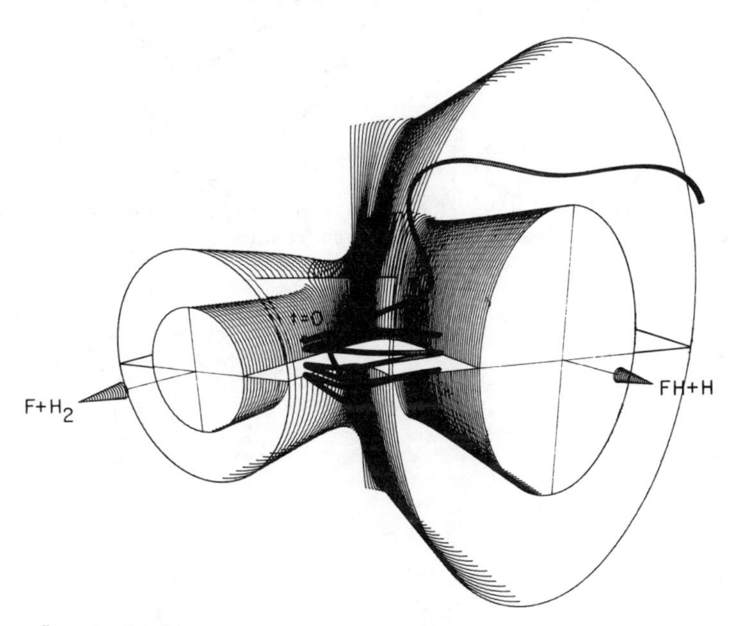

Figure 5. A slight change in the initial condition on the orbit in Figure 4 leads to decay into products.

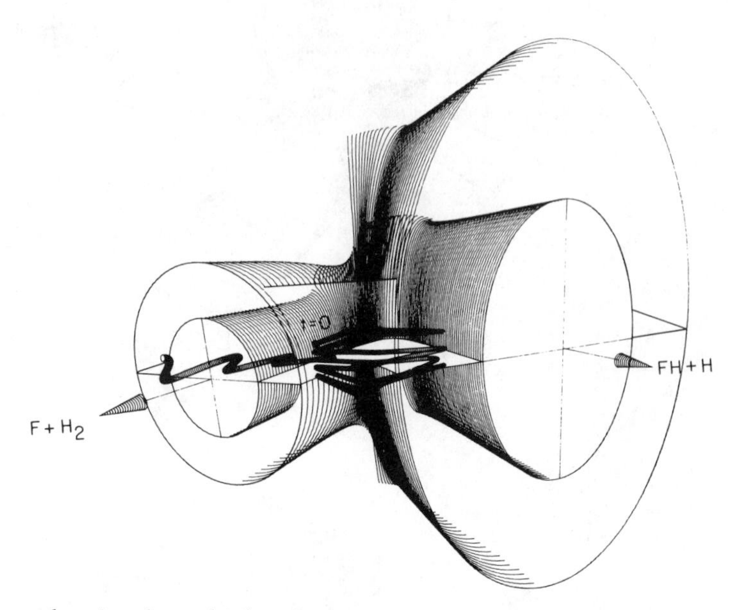

Figure 6. Another slight change in the initial condition on the orbit in Figure 4 leads to decay into reactants.

a relatively small amount of energy in the rotational (γ) degree of freedom.

Resonant Periodic Orbits

Using the numerical method described earlier, a number of periodic resonant orbits have been computed for 3D $F+H_2$. One of them, at E=0.4 eV and γ_o=29.0°, is illustrated in Figure 7. During one complete period, there are three vibrational-translation (approximately FHH asymmetric stretch) cycles within each bending cycle, so this can be referred to as a 3:1 frequency resonance. Clearly, the two frequencies associated with this motion are commensurate, $\omega_a/\omega_b=3$, where "a" and "b" refer to asymmetric and bending motions, respectively. For this orbit, $\omega_a=281.1\ (ps)^{-1}$ and $\omega_b=93.7\ (ps)^{-1}$. This periodic orbit, along with the collinear periodic orbit at the same energy, was then used to compute the $\underline{\phi}$ and $\underline{J}$ vectors and the $\underline{\underline{\Omega}}$ matrix (defined earlier in connection with the arbitrary-trajectory semiclassical quantization method). In this case, we obtain 0.3435 eV for the lowest reso-energy. Repeating the calculation with an analogous pair of trajectories at E=0.34 eV gave essentially the same value (0.3439 eV). These values again compare very well with all available quantal results for this reaction (3a), (5b), (8). An advantage of using periodic orbits, rather than quasiperiodic orbits, in the semiclassical quantization scheme is that the Fourier analysis needed to compute the two frequencies is essentially trivial.

Summary

Numerical methods were described for locating both quasiperiodic and periodic resonant orbits in the noncollinear $F+H_2$ reaction with J=0. Using these methods, several resonant quasiperiodic and periodic orbits were computed and plotted in the internal coordinate space. These orbits exhibit resonant energy transfer between local (dressed) vibration-bend oscillations in the entrance and exit regions of the collision complex. Frequencies and actions from the periodic orbits were then used in the arbitrary-trajectory semiclassical quantization scheme (19). The lowest resonance energy predicted for the J=0 reaction was in good agreement with all available quantal and adiabatic results. Further properties of both types of orbit, including those obtained from a stability analysis, will be presented elsewhere (21).

Acknowledgments

We thank Eli Pollak for many helpful comments. In addition, Don Truhlar contributed many valuable suggestions on an earlier version of this manuscript.

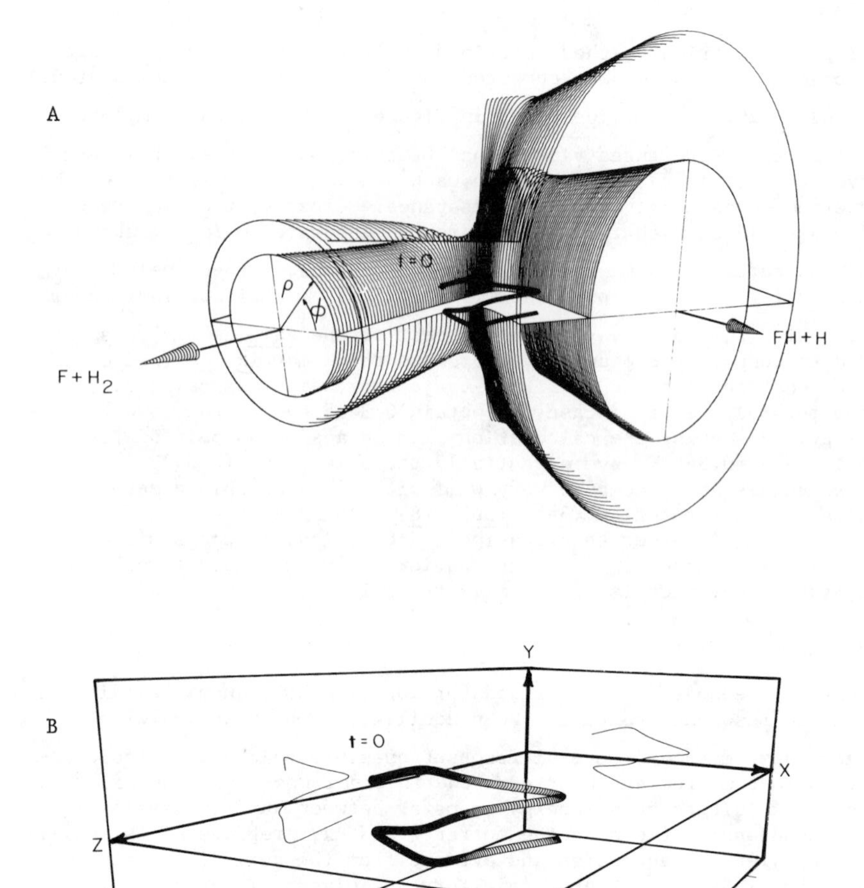

Figure 7. A resonant periodic orbit at E=0.4 eV (γ_o=29°). Part A shows the orbit within the V=0.4 eV FH_2 potential surface. Part B shows the orbit in the internal coordinate space (R,x,y). Once again, projections of the orbit onto the three coordinate planes are also shown.

Literature Cited

1. a. Truhlar, D. G.; Kuppermann. J. Chem. Phys. 1970, 52, 3841; 1972, 56, 2232.
   b. Levine, R. D.; Wu, S. F. Chem. Phys. Lett. 1971, 11, 557.
   c. Wu, S. F.; Johnson, B. R.; Levine, R. D. Mol. Phys. 1973, 25, 609.
   d. Schatz, G. C.; Bowman, J. M.; Kuppermann, A. J. Chem. Phys. 1975, 63, 674.
   e. Chapman, F. M., Jr.; Hayes, E. F. J. Chem. Phys. 1975, 62, 4400; 1976, 65, 1032.
   f. Schatz, G. C.; Kuppermann, A. Phys. Rev. Lett. 1975, 35, 1267.
   g. Chapman, F. M., Jr.; Hayes, E. F. J. Chem. Phys. 1977, 66, 2554.
   h. Latham, S. L.; McNutt, J. F.; Wyatt, R. E.; Redmon, M. J. J. Chem. Phys. 1978, 69, 3746.
   i. Walker, R. B.; Stechel, E. B; Light, J. C. J. Chem. Phys. 1978, 69, 2922.
   j. Connor, J. N. L. Comp. Phys. Comm. 1979, 17, 117.
   k. Kupperman, A. In "Potential Energy Surfaces and Dynamics Calculations"; Truhlar, D. G., Ed.; Plenum: New York, 1981; Ch. 16.
   l. Launay, J. M.; Le Dorneuf, M. J. Phys. 1982, B15, L455.
   m. Shoemaker, C. L.; Wyatt, R. E. Adv. Quantum Chem. 1982, 14, 169.
   n. Shoemaker, C. L.; Wyatt, R. E. J. Chem. Phys. 1982, 17, 4982 and 4994.
   o. Skodje, R. T.; Truhlar, D. G.; Garrett, B. C. J. Chem. Phys. 1982, 77, 5955.
   p. Skodje, R. T.; Schwenke, D. W.; Truhlar, D. G.; Garrett, B. C. J. Phys. Chem. 1984, 88, 628.
2. Neumark, D. M.; Wodtke, A. M.; Robinson, G. N.; Hayden, C. C.; Lee, Y. T., molecular beam studies of the $F+H_2$ reaction; to be published.
3. a. Redmon, M. J.; Wyatt, R. E. Chem. Phys. Lett. 1979, 63, 209.
   b. Wyatt, R. E.; Redmon, M. J. Chem. Phys. Lett. 1983, 96, 284.
   c. Schatz, G. C. Chem. Phys. Lett. 1983, 94, 183.
4. Baer, M.; Jellinek, J.; Kouri, D. J. J. Chem. Phys. 1983, 78, 2962.
5. a. Walker, R. B.; Hayes, E. F. J. Phys. Chem. 1983, 87, 1255.
   b. Hayes, E. F., Jr.; Walker, R. B. J. Phys. Chem. To be published.
6. Emmons, R. W.; Suck, S. H. Phys. Rev. A. 1983, 27, 1803.
7. a. Stine, J. R.; Marcus, R. A. Chem. Phys. Lett. 1974, 29, 575.
   b. Garrett, B. C.; Truhlar, D. G. J. Phys. Chem. 1982, 86, 1136.
   c. Pollak, E.; Child, M. S. Chem. Phys. 1981, 60, 23.
8. Pollak, E.; Wyatt, R. E. J. Chem. Phys. 1982, 77, 2689; Pollak, E.; Wyatt, R. E. To be published.
9. Pollak, E.; Pechukas, P. J. Chem. Phys. 1978, 69, 1218.

10. a. Pollak, E.; Child, M. S.; Pechukas, P. J. Chem. Phys. 1980, 72, 1669.
b. Pollak, E. J. Chem. Phys. 1981, 74, 5586; 1981, 75, 4435.
c. Pollak, E. Chem. Phys. 1981, 61, 305.
d. Pollak, E.; Wyatt, R. E. J. Chem. Phys. 1983, 78, 4464.
e. Costley, J.; Pechukas, P. J. Chem. Phys. 1982, 77, 4957.
11. Pollak, E. In "Theory of Chemical Reaction Dynamics"; CRC Press: New York,1984; to be published.
12. Duchovic, R. J.; Swamy, K. N.; Hase, W. L. J. Chem. Phys. 1984, 80, 1462.
13. Pollak, E. Chem. Phys. Lett. 1982, 91, 27.
14. Hutchinson, J.; Wyatt, R. E. J. Chem. Phys. 1979, 70, 3509.
15. Muckerman, J. T. In "Theoretical Chemistry: Advances and Perpectives"; Eyring, H.; Henderson, D. H., Eds.; Academic: New York, 1981; Vol. 6A, p. 1.
16. Noid, D. W.; Koszykowski, M. L.; Marcus, R. A. Ann. Rev. Phys. Chem. 1981, 32, 267.
17. Noid, D. W.; Koszykowski, M. L. Chem. Phys. Lett. 1980, 73, 114.
18. a. Percival, I. C. Adv. Chem. Phys. 1977, 36, 1.
b. Noid, D. W.; Marcus, R. A. J. Chem. Phys. 1975, 62, 2119; 1977, 67, 559.
19. DeLeon, N.; Heller, E. J. J. Chem. Phys. 1983, 78, 4005.
20. Miller, W. H., to be published.
21. Marston, C. C.; Wyatt, R. E., to be published.

RECEIVED June 11, 1984

24

Resonance Phenomena in Quantal Reactive Infinite-Order Sudden Calculations

Z. H. ZHANG[1], N. ABUSALBI[1], M. BAER[2], D. J. KOURI[1], and J. JELLINEK[3]

[1]Department of Chemistry and Department of Physics, University of Houston–University Park, Houston, TX 77004
[2]Applied Mathematics, Soreq Nuclear Research Center, Yavne, Israel 70600
[3]Department of Chemical Physics, Weizmann Institute of Science, Rehovot, Israel 76100

In this paper we discuss the resonance tuning hypothesis as an important mechanism whereby resonances are spread in the $F+H_2$ and $F+D_2$ reaction systems and examine whether the shift from backward to sideways scattering of the HF and DF products is a resonance signature. All results are obtained using the Muckerman 5 potential surface.

The development of nonperturbative quantal approximations for 3-dimensional reactive scattering is very important and is currently being pursued by several groups.(1-12) Of the available methods, one of the most widely applied has been the reactive infinite order sudden (RIOS) approximation.(4-7,10,12) This method has now been applied to the $H+H_2$, $F+H_2$, $F+D_2$, $D+H_2$ and D+HCl reactions. In the case of $H+H_2$, qualitatively correct results were obtained for angular distributions and good quantitative agreement was obtained for total reactive cross sections compared to exact close coupling (CC) results.(4-5,13) In the case of the $F+H_2$ and $F+D_2$ systems,(6-7) the ℓ-av RIOS results for the total integral reactive cross sections agreed well with Muckerman's classical trajectory(CT) results,(14) as well as with CT results of Ron, Pollak and Baer.(15) Evidence suggested that the total integral reactive cross section is not subject to large quantum effects(6-7) and thus the RIOS appeared to give acceptable accuracy for such state summed quantities. The branching ratios for different product vibrational states appeared however to show large quantum effects,(6-7) in qualitative agreement with the j_z-conserving results of Redmon and Wyatt(3) (obtained with a somewhat different potential surface). Of particular interest were the angular distributions of the HF and DF product molecules as a function of final vibrational state v_f. This is due to the fact

0097–6156/84/0263–0457$06.50/0

that only for two systems ($F+H_2$ and $F+D_2$) is such state-resolved data currently available from experiment.(16) The experiments show a fascinating behavior in which at the lowest energy, all HF and DF products were backward scattered while at a slightly higher energy, all HF and DF product molecules continued to be backward scattered except the HF $v_f=2$ and DF $v_f=3$ products, which were sideways scattered. More recent studies suggest, however, that there is a significant forward peak in the $v_f=3$ HF and $v_f=4$ DF products.(17) The behavior for $F+H_2$ was tentatively interpreted in terms of a mechanism in which the well known $v_f=2$ collinear resonance(18) for the $F+H_2$ system was shifted to higher orbital angular momentum as the energy was increased until ultimately, a shift to sideways scattering occurred.(16,3) In the case of $F+H_2$ the RIOS calculations using the Muckerman 5 potential(14) were in qualitative agreement with this shift from backwards to sideways scattering of the HF($v_f=2$) product but did not unambiguously indicate that a resonance mechanism was responsible.(6) Thus, while there was evidence that resonance effects could be present, it was not necessarily clear that they were solely responsible for the shift in the angular pattern for the HF($v_f=2$) products. Other model calculations have recently been reported that also show sideways peaking of the $v_f=2$ HF product.(11b,19) In addition, they show backward scattering for the $v_f=3$ HF product at both the low and higher energy. Of these, only the Bowman, et.al.(17) results included a sum over all final HF rotational states. Classical trajectory (CT) calculations have been performed by Ron, Baer and Pollak (RBP)(20a) for a single energy $E_{tot}=0.5$ eV (which is somewhat higher than the experimental energies) where E_{tot} is the total energy of the system relative to the $H_2(D_2)$ diatomic well. The RBP study consisted of both forward and reverse CT calculations. The forward study showed only backwards scattering for all four final vibrational states, and in this sense confirmed the earlier forward CT results obtained by Blais and Truhlar.(20b) In the case of the reverse CT calculations, a different picture was obtained. There are many accessible rotational states of the HF molecule which can be populated at the nominal experimental energies. Therefore, the reverse CT studies had to be done for a variety of rotational states. It was found that all rotational states for HF in the v=3 state yielded backwards scattering, but several of the most probable rotational states for HF with v=2 yielded sideways scattering. It was felt by RPB that these results point up a technical problem with the forward CT calculations having to do with boxing of the HF rotational-vibrational states. In the reverse CT calculations, one boxes on the final H_2 states and begins the calculation in a well defined HF rotational-vibrational state. Thus, RPB contend that the CT results taken as a whole cannot be considered to support the existence of quantum effects in the vibrational selected angular distributions.

In addition to the $F+H_2$ system, Sparks, et.al. (16,17) also performed measurements on the $F+D_2$ system. Collinear calculations for this system also show interesting resonance behavior, although the resonance appears to be significantly broader(21) than in the case of collinear $F+H_2$. The experimental results for this system showed pure backward scattering for DF(v_f) for all v_f at a relative kinetic energy of 2.34 kcal/mole while at 4.51 kcal/mole, the products are scattered progressively more forward. The results show the v_f=4 DF product is more strongly forward peaked than backward at 4.51 kcal/mole, with the sequence v_f=3,2,1 as one goes to larger scattering angle. Thus, the DF experimental results agree qualitatively with the newer HF experimental results which appear to show the v_f=3 HF product with significant forward scattering, followed by the v_f=2 and then the v_f=1 HF products. Recently, Ron, Pollak, and Baer(15) have performed a forward CT calculation also for the $F+D_2$ system at the two nominal experimental energies. Again, in contrast to the experimental results and the quantum RIOS results,(7) all the state-to-state differential cross sections were found to be scattered backward. AbuSalbi et al.(7b) have carried out quantum RIOS calculations for $F+D_2$ and found that this system (using the Muckerman 5 potential) behaves in a manner completely analogous to $F+H_2$ so far as the fixed-γ reaction probabilities and state-resolved angular distributions are concerned. Thus, the v_f=3 state of DF plays the role of the v_f=2 state of HF, while v_f=3 HF corresponds to v_f=4 DF. These highest states behave different from experiment in both FH_2 and FD_2 since they are backward scattered at both low and higher energies. The question of resonances in the FD_2 system was answered unequivocably using the lifetime matrix method of Smith,(22) and it was shown that the energy of the resonance did tune as γ or ℓ was changed where γ is the internal angle between the vector from the diatom center of mass to the atom (F) and the diatom (H_2 or D_2) axis and ℓ is the CS orbital parameter (more details about ℓ will be given below). However, the grid in γ and ℓ was not fine enough to thoroughly study how the resonance shifted and changed in intensity as γ and ℓ were varied. In addition, resonances in FH_2 are expected to be stronger than in FD_2 and it is of interest to examine in more detail how sensitive the resonance is to tuning in γ and ℓ. In particular, it is important to examine just how rapidly the lifetime of the collinear resonance decreases as γ and/or ℓ are increased.

In this paper, we wish to examine the results of earlier RIOS calculations for the FH_2 and FD_2 systems. Our purpose in this is to show how the two systems behave in a parallel fashion. We also summarize the results of the lifetime studies of the FD_2 system. Next we report new, more detailed lifetime calculations carried out for the FH_2 system. These results show in much more detail how the energy of the resonance shifts with changes in γ and ℓ. Also, we show how the magnitude of the delay time changes with γ and ℓ. Finally, we compare the FH_2 delay times for various γ, ℓ with the characteristic times of molecular rotation and vibration in order to get a better feeling for how strongly delayed the system is. All these results show in detail the tuning mechanism. Finally, the question of whether the resonance phenomenon is responsible for the sideways shift of the $FH_2(v_f=2)$ angular distribution is addressed.

Summary of Theory

The theory of the quantual RIOS was introduced in an earlier series of papers.(4,6) Here we simply summarize some of the salient points of the formalism.

Differential and integral cross sections. The degeneracy averaged differential cross section from a given initial state $(v_\lambda j_\lambda m_\lambda)$ to a final vibrational state v_ν and all rotational states is given by

$$\frac{d\sigma(v_\nu|v_\lambda j_\lambda m_\lambda|\theta,\psi)}{d\omega} = \frac{1}{(2j_\lambda + 1)} \sum_{j_\lambda m_\lambda} |f(v_\nu j_\nu m_\nu|v_\lambda j_\lambda m_\lambda|\theta,\psi)|^2 \tag{1}$$

where λ and ν denote the initial and final arrangements, ω is a solid angle comprised of the scattering angel θ, and the azimuthal scattering angle ψ; v_α, j_α and m_α, $\alpha=\lambda,\nu$, are the vibrational, rotational and p-helicity quantum numbers, respectively, and $f(v_\nu, j_\nu m_\nu|v_\lambda, j_\lambda m_\lambda|\theta, \psi)$ is the scattering amplitude. In what follows, we drop the v_λ and v_ν indices to simplify the notation.

The scattering amplitude is related to the body-fixed scattering matrix element according to

$$f(j_\nu m_\nu|j_\lambda m_\lambda|\theta,\psi) = \frac{i^{j_\lambda - j_\nu + 1}}{2k_{j_\lambda}} \sum_J [J] d^J_{m_\lambda m_\nu}(\theta) S^{j_\nu m_\nu}_{Jj_\lambda m_\lambda} , \tag{2}$$

where [X] stands for (2X + 1) and $d^J_{m_\lambda m_\nu}(\theta)$ is the Wigner rotation matrix (the notion of Rose is employed (23)). The body-fixed S matrix elements are related to the Arthurs-Dalgarno (AD)(24) matrix elements by

$$S^{j_\nu m_\nu}_{Jj_\lambda m_\lambda} = \sum_{\ell_\nu \ell_\lambda} i^{\ell_\lambda - \ell_\nu} \frac{\sqrt{[\ell_\lambda][\ell_\nu]}}{[J]} \langle \ell_\lambda 0 j_\lambda m_\lambda | J m_\lambda \rangle \langle \ell_\nu 0 j_\nu m_\nu | J m_\nu \rangle S^{j_\nu \ell_\nu}_{Jj_\lambda \ell_\lambda} \tag{3}$$

and the AD S matrix elements are approximated within the RIOS as(4,6)

$$S^{j_\nu \ell_\nu}_{Jj_\lambda \ell_\lambda} = i^{\ell_\lambda + \ell_\nu - 2\ell} \sum_{\Omega_\lambda \Omega_\nu} \frac{\sqrt{[\ell_\nu][\ell_\lambda]}}{[J]} \langle \ell_\lambda 0 j_\lambda \Omega_\lambda | J\Omega_\lambda \rangle \langle \ell_\nu 0 j_\nu \Omega_\nu | J\Omega_\nu \rangle \, 2\pi \int_0^\pi d\gamma_\lambda \sin\gamma_\lambda Y^*_{j_\nu \Omega_\nu}[\gamma_\nu(B,\gamma_\lambda),0] \, d^J_{\Omega_\lambda \Omega_\nu}[\Delta(\gamma_\lambda,B)] S_\ell(\gamma_\lambda) Y_{j_\lambda \Omega_\lambda}(\gamma_\lambda,0) , \tag{4}$$

where γ_α, $\alpha=\lambda,\nu$ is defined as

$$\gamma_\alpha = \cos^{-1}(\hat{r}_\alpha, \hat{R}_\alpha); \; \alpha=\lambda,\nu \tag{5}$$

and Δ is an angle defined as

$$\Delta = \cos^{-1}(\hat{R}_\lambda, \hat{R}_\nu) . \tag{6}$$

B is a parameter which originates from the matching of the λ

channel wave function with the ν channel wave function (more details about B will be given in the next section) and ℓ is the CS orbital parameter(25) which is identified either as the initial value of the angular momentum quantum number in the λ arrangement channel, namely ℓ_λ, or as the average value; i.e.,

$$\ell = (\ell_\lambda + \ell_\nu)/2 \ . \tag{7}$$

In the first case the ℓ labeling is known as the ℓ-initial labeling and in the second as the ℓ-average labeling. In the integral we have also a collinear-type γ-dependent (and also B-and ℓ-dependent) S matrix element.

Lifetime Matrix. In the analysis of our RIOS results for both FH_2 and FD_2, we shall employ the lifetime matrix analysis due to Smith.(22) In this approach, the lifetimes are given by the diagonal elements of the time delay matrix

$$Q_{ij}\ (E) = i\hbar \sum_k S_{ik} dS^*_{jk}/dE \ . \tag{8}$$

where E is either the total or relative translational energy. By going to a diagonal representation of the $\underline{\underline{Q}}(E)$ matrix, one obtains the most compact description of the system. The eigendelays can be negative (corresponding to a process in which the collision is accellerated by the potential, relative to no potential at all) or positive (in which the potential causes a delay relative to the time required to traverse the collision region in the absence of a potential). When one or more eigendelay time is positive and substantial, one may speak of a resonant process occurring. In our calculations reported herein, a 5-point numerical differentation formula has been used. The $\underline{\underline{Q}}(E)$ matrix was found to be diagonally dominant so that following the eigenvalues did not prove at all difficult for FH_2 and FD_2.

Comparison of FH_2 and FD_2

We shall begin by first comparing a variety of RIOS results for the FH_2 and FD_2 systems. One of the advantages of the RIOS method is the fact that it permits us to examine how the reaction responds to changes in the relative configuration of the system, as well as in the relative orbital angular momentum. Although not of direct physical significance, the so called "primitive reaction probabilities" obtained from the γ-dependent S-matrix elements

$$S^{\lambda v_\lambda}_{\ell \nu v_\nu}(\gamma)$$

afford insight into the "tuning" of the reaction as a function of ℓ and γ.

In Figures 1-2 we give the reactive state-to-state probabilities

$$P^{\lambda v_\lambda}_{\ell \nu v_\nu}(\gamma)$$

as a function of γ for different ℓ values and E_{tot}, for the system FH_2(6b). In Figures 1(a) and 2(a) are shown the results for v_f=1, in Figures 1(b) and 2(b) for v_f=2 and in Figures 1(c) and 2(c) for v_f=3. In Figure 1 are shown the results for E_{tot}=0.34 eV and in Figure 2 for E_{tot}=0.423 eV.

The main features to be noticed are: (a) the probabilities for v_f=1,2 and for v_f=3 exhibit a different γ dependence for most of the ℓ values. Probabilities for v_f=1 and v_f=2 tend to peak at angles which are far removed from the collinear arrangement while the v_f=3 probabilities all have a maximum at γ=0, and then slowly decay to zero as γ increases. (b) The differences between the results in Figures 1 and 2 are minor and both figures exhibit very similar patterns. The only noticeable changes are that all the probabilities, as a function of γ, extend to larger γ values when the energy is increased from E_{tot}=0.34 eV to E_{tot}=0.423 eV. Also the v_f=1,2 results tend to shift their peak region to higher angles. As for v_f=3, most of the highest probabilities are still concentrated around the collinear arrangement.

The FD_2 reactive state-to-state probabilities as functions of γ for different ℓ values are analogous. As in the FH_2 system, the main features are: (a) the probabilities for v_f=1,2,3 are qualitatively different as functions of γ and ℓ from those for v_f=4. The probabilities for v_f=1,2,3 tend to peak at angles away from the collinear while those for v_f=4 tend to have their maximum at =0°. Further, the qualitative features are essentially the same at both the low and higher energy. (b) Any differences between the results at E_{tot}=0.291 eV and E_{tot}=0.385 eV do not appear to be of an essential nature. Once again, the main changes are that all probabilities extend to larger γ values at the higher energy. (The v_f=4 results with ℓ=0 and ℓ=5 do show a slight maximum away from the ℓ=0 angle but higher ℓ's remain peaked at γ=0.) Thus, comparing the $F+D_2$ results with the corresponding $F+H_2$ results(6b), it is noted that qualitatively, the two sets of results are the same, with the v_f=3 HF and v_f=4 DF in correspondence with each other and the v_f=2 HF and v_f=3 DF in correspondence with each other.

Very similar results are obtained for the reactive probability for fixed γ as a function of ℓ,

$$P_{\gamma\nu v_\nu}^{\lambda v_\lambda}(\ell),$$

for both FH_2 and FD_2. In general, the HF (v_f=2) and DF(v_f=3) results for

$$P_{\gamma\nu v_\nu}^{\lambda v_\lambda}(\ell)$$

peak at ℓ values significantly larger than ℓ=0, while those for HF(v_f=3) and DF(v_f=4) peak at ℓ=0. These qualitative features again are the same at energies both below and above the collision energy at which the DF(v_f=3) and HF(v_f=2) products shift from backward to sideways peaking. However, it is found that the values of ℓ for which

$$P_{\gamma\nu v_\nu}^{\lambda v_\lambda}(\ell)$$

has its maximum (for HF(v_f=2) and DF(v_f=3)) are larger for the higher energy than for the low energy. These results suggest that the

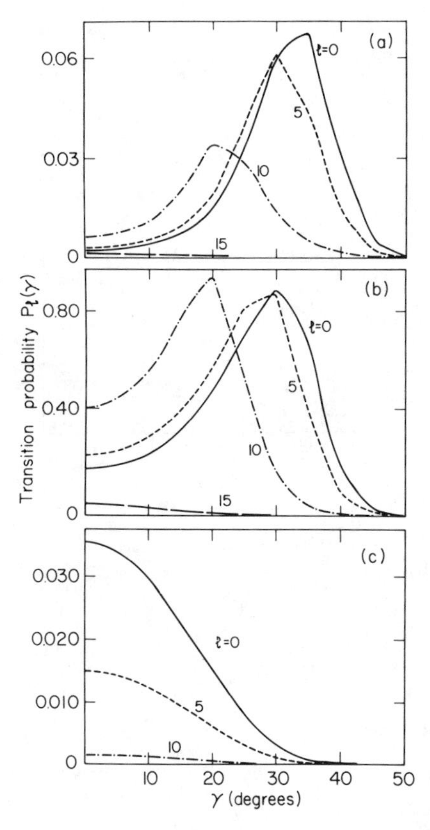

Figure 1. Primitive $F+H_2$ reactive state-to-state probabilities $P(v=o\rightarrow v')$ as a function of γ_λ for different ℓ-values. The energy is 0.34 eV. (a) $v_i=0\rightarrow v_f=1$ (b) $v_i=0\rightarrow v_f=2$ (c) $v_i=0\rightarrow v_f=3$. Reproduced by permission from Ref. 6b, Copyright 1983, American Institute of Physics.

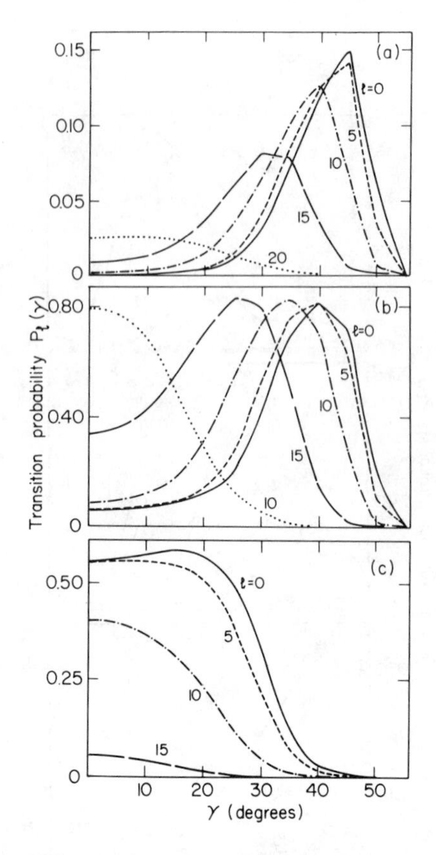

Figure 2. Same as Figure 1 except E_{tot}=0.423 eV. Reproduced by permission from Ref. 6b, Copyright 1983, American Institute of Physics.

resonance tuning spreads any resonance effects over a range of energies, but there is still a critical energy region where the scattering will shift from backwards to sideways.

In Figure 3 are shown FH_2 reactive transition probabilities as a function of energy (E_{tot}=0.28-0.7 eV) for a few values of γ and ℓ. The state-to-state $P(v_i=0 \rightarrow v_f=2)$ probabilities are shown in Figures 3(a), 3(c) and 3(e), and the state-to-state $P(v_i=0 \rightarrow v_f=3)$ probabilities are shown in Figures 3(b), 3(d), and 3(f). In each figure, three curves which correspond to three different ℓ values are given. We first call attention to the clear presence of the (γ=0, ℓ=0) resonance at E_{tot} 0.28 eV that was detected long ago.(26) This (γ=0, ℓ=0) resonance is characterized by a steep rise in probability in the threshold region. For γ=0, v_f=2 [Figure 3(a)], increasing causes a significant broadening of the resonance peak.

By contrast, the γ=0, v_f=3 [Figure 3(b)] results show the normal rise at threshold followed by a steady decline in the probability. In Figure 3(c) we have the results for v_f=2, γ=30° and we observe that the ℓ=0 case is qualitatively like the γ=0, v_f=2, ℓ=10 or 20 cases in Figure 3(a). One has the resonance peak which is again broadened and shifted to higher energy compared to the γ=0, ℓ=0, v_f=2 case. In fact, for ℓ=10 and 20, γ=30°, and v_f=2, one sees the broadened resonance peak followed by the shoulder produced by the background nonresonant reaction probability. The v_f=3, γ=30° results again show no resonant peak. Finally, when γ=50°, the v_f=2 results do not seem to show a resonance feature. Regarding the (v_i=0 $\rightarrow$ v_f=3) results, an important thing to note is the appearance of a resonance for the collinear arrangement and ℓ=0 at the vicinty of E_{tot}=0.7 eV (see also Reference (26)). Another feature is the relatively high threshold energies as compared with those of (v_i=0 $\rightarrow$ v_f=2) transitions once the angle γ becomes large.

In Figure 3 the primitive probability functions clearly show indications of the resonance tuning effect alluded to above. It is also seen that the resonant peak broadens and shifts to higher energy as the value is increased (for fixed γ). However, for γ=50°, there does not appear to be a resonance. Finally, we note that the height of the γ=0, ℓ=10 result suggests an even longer lived resonance than in ℓ=0,γ=0 (however, we shall see that in fact, the time delay is less for γ=0, ℓ=10).

For $F+D_2$, approximately identical results are obtained for the energy dependence of the fixed γ FD_2 probabilities for the v_i=0 $\rightarrow$ v_f=3 and v_i=0 $\rightarrow$ v_f=4 reactive transitions for the orbital angular momentum quantum numbers ℓ=0, 10 and 20. Just as for FH_2, the FD_2 results show the presence of a resonance which has previously been found in γ=0, ℓ=0 calculations.(21) Furthermore as γ increases, the threshold for reaction goes up in energy and also as ℓ increases, the threshold energy increases. Again in analogy to FH_2, we find that initially as γ increases, the maximum value of the $P^{\ell}_{3,0}$ reaction probability increases. The same effect is seen when ℓ is increased (at least at lower γangles). In the case of the v_i=0 $\rightarrow$ v_f=4 reaction, however, increasing ℓ causes a decrease in the maximum reaction probability, as does also increasing γ for fixed ℓ.

In Figure 4, we give Argand plots of the fixed-γ FD_2 reactive S matrix elements for the v_i=0 $\rightarrow$ v_f=3 reaction process (we include results for ℓ=0, 10, and 20). Results for γ=0° and γ=15° are shown.

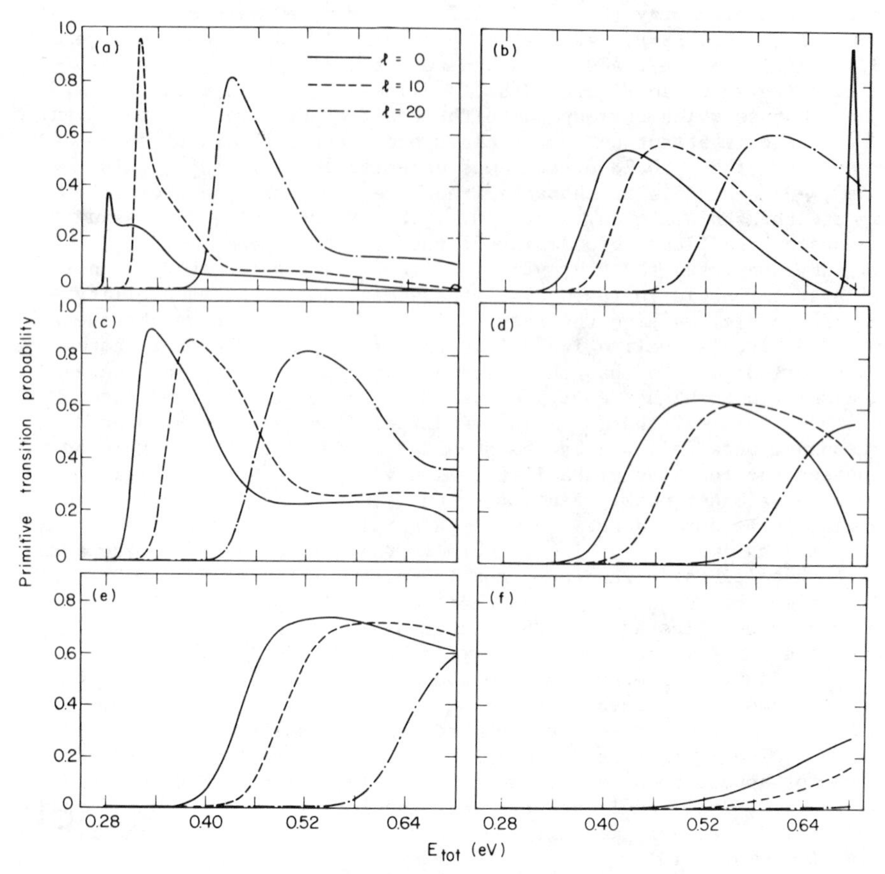

Figure 3. Primitive reactive state-to-state probabilities $P(v_i=0 \rightarrow v_f)$ as a function of energy for different γ_λ and ℓ values. (a) $\gamma_\lambda=0°$, $v_f=2$ (b) $\gamma_\lambda=0°$, $v_f=3$ (c) $\gamma_\lambda=30°$, $v_f=2$ (d) $\gamma_\lambda=30°$, $v_f=3$ (e) $\gamma_\lambda=50°$, $v_f=2$ (f) $_\lambda=50°$, $v_f=3$. Reproduced by permission from Ref. 6b, Copyright 1983, American Institute of Physics.

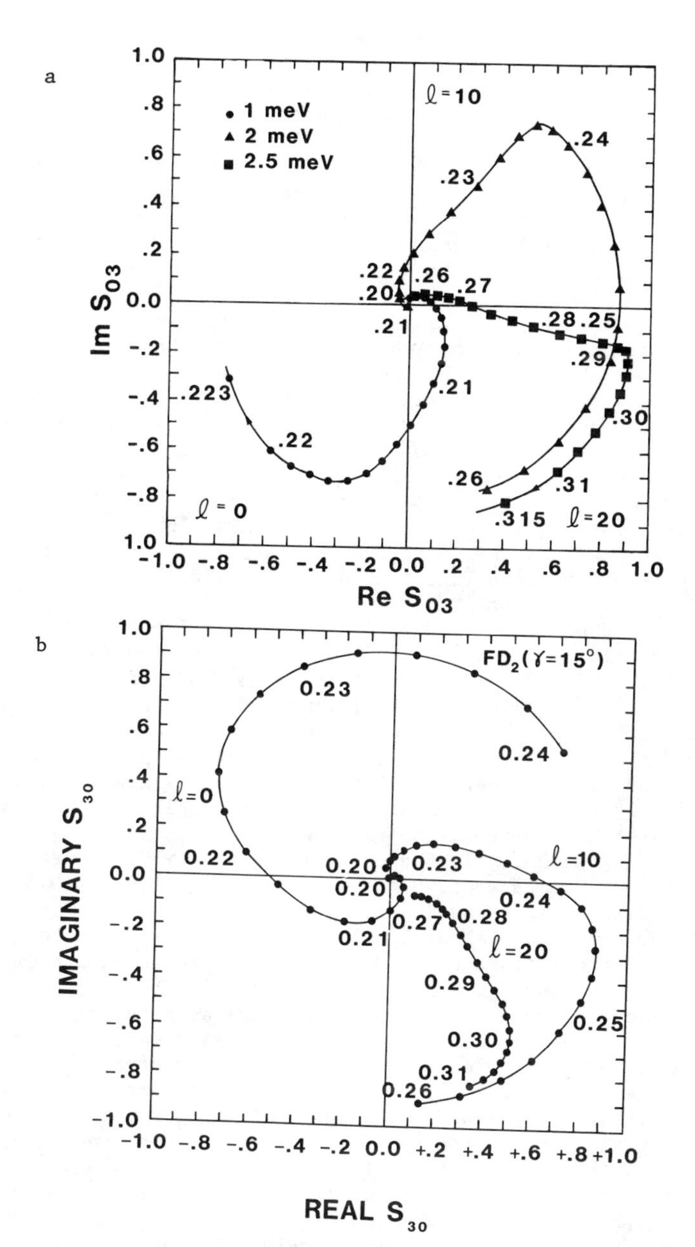

Figure 4. Argand plot of the fixed-γ S-matrix versus total energy for the process $F+D_2$ $(v_i=0)\rightarrow DF(v_f=3)+D$ for various ℓ-values. (a) γ=0° (b) γ=15°. Circles represent points spaced by 1 meV; triangles represent points spaced by 2 meV; and squares represent points spaced by 2.5 meV. Reproduced by permission from Ref. 7b, Copyright 1984, American Institute of Physics.

The characteristic feature of these results is the occurrence of a flattened portion of the Argand plot. Earlier collinear (γ=0°) studies of $F+D_2$ with ℓ=0 showed this same behavior(21) and it was found in that case to signal a resonance in the presence of strong background scattering.

In Figure 5, we give Argand plots of the fixed-γ FH_2 reactive S matrix elements for the $v_i=0 \rightarrow v_f=2$ reaction process. Here, the resonance for ℓ=0 causes a greater flattening of the Argand plot while for ℓ=10, the plot actually shows a complete loop (counterclockwise). One might be tempted to conclude from this, plus the fact that the maximum in the γ=0, ℓ=10 reactive probability curve in Figure 3a is much higher than for γ=0, ℓ=0, that the γ=0, ℓ=10 resonance is stronger than for γ=0, ℓ=0. (Similar behavior is seen for FD_2.) As we shall see, this is not true and the resonances for both the FH_2 and FD_2 systems are the longest lived for γ=0, ℓ=0. The reason the resonance in γ=0, ℓ=0 shows up only as a flattening of the Argand plot is due to the stronger background for this case compared to that for γ=0, ℓ=10.

The most precise characterization of the resonances is obtained using the lifetime matrix analysis of Smith.(22) In Figure 6 we give the eigendelay times for FH_2 with γ=0, ℓ=0 and γ=0, ℓ=10. It is immediately recognized that the time delay for γ=0, ℓ=0 is over twice as large as that for γ=0, ℓ=10. This is so in spite of the fact that the Argand plots in Figure 5 show a complete counterclockwise loop for γ=10, ℓ=0 and not for γ=0, ℓ=0. The time delay for FH_2 γ=0, ℓ=0 is about 3.4×10^{-13} s. This corresponds to about 45 vibrations of the (free) H_2 molecule. The γ=0, ℓ=10 resonance lasts about 20 vibrations of the (free) H_2 molecule. These are very substantial lifetimes. Similarly, FD_2 calculations for given γ and ℓ values as a function of energy also show that there is a significant delay indicating the occurrence of a resonance complex being formed in the $F+D_2$ system. However, the resonance delay is not as great as that occurring in FH_2. In the case of FD_2, the γ=0, ℓ=0 resonance lasts 1.4×10^{-13} s (compared to 3.4×10^{-13} s for FH_2 with γ=ℓ=0). This corresponds to about 13 vibrations of a free D_2 molecule. The γ=0, ℓ =10 FD_2 resonance lasts about 11 vibrations so it appears that the lifetime of the resonance in FD_2 does not decrease as rapidly with ℓ as does that in FH_2. When γ=0 , it is found that the time delay experienced decreases (relatively slowly) as a function of orbital angular momentum. Similar behavior is also shown for $\gamma \neq 0$. Furthermore, the magnitude of the time delay also decreases as increases. These results for both FH_2 and FD_2 show unequivocably that resonances are occurring over a range of γ and ℓ values. We also point out that the nonresonant time delays are all negative, which indicates that the duration of the reactive collision is shorter than the transit time that would occur if no interaction were present. This is likely a reflection of the large exothermicity of the HF exit channel which strongly accelerates the product molecules.

Differential and Integral Cross Sections. In Figure 7 are shown the state-to-state ℓ-average reactive differential cross sections for $F+H_2$ (degeneracy-averaged over m_λ, summed over m_ν and j_ν but resolved with respect to the final vibrational state v_ν). In each figure,

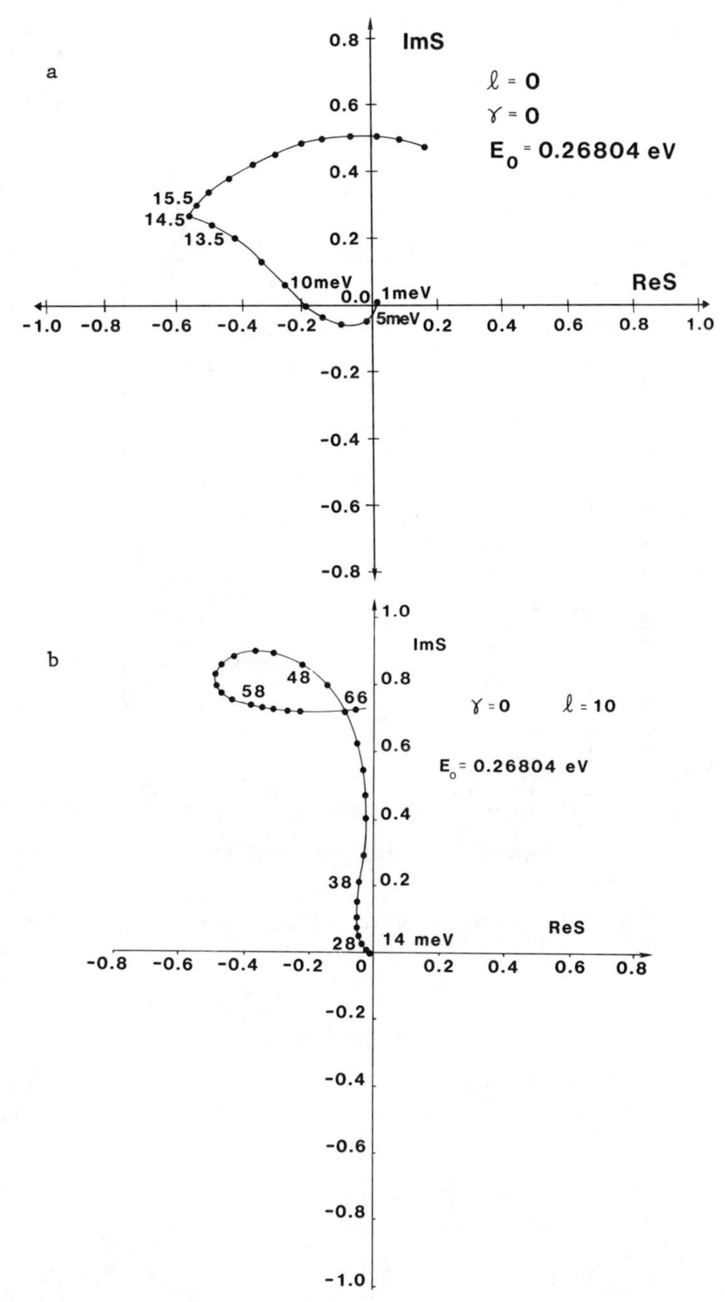

Figure 5. Argand plot of the fixed γ S-matrix versus total energy for the process $F+H_2(v_i=0)\rightarrow HF(v_f=2)+H$ for $\gamma=0°$, (a) $\ell=0$ (b) $\ell=10$.

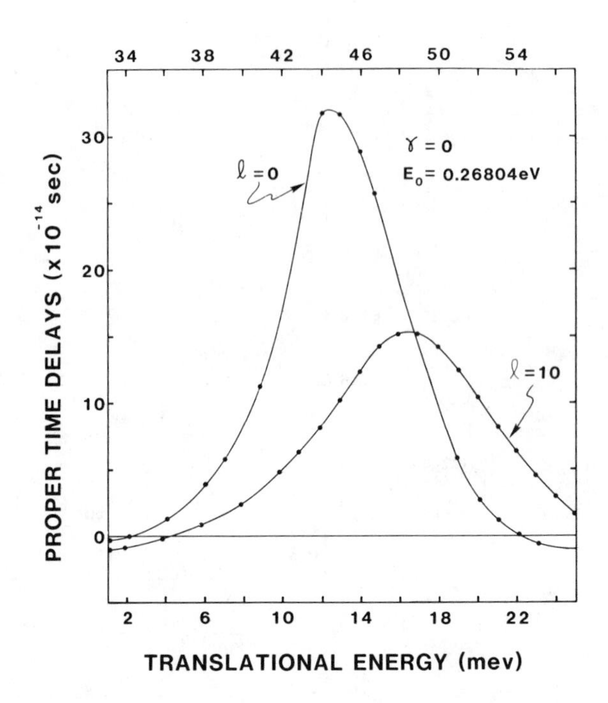

Figure 6. Eigendelay times in units of 10^{-14}sec for γ=0°, ℓ=0 and ℓ=10 for $F+H_2(v_i=0)\rightarrow HF(v_f=2)+H$.

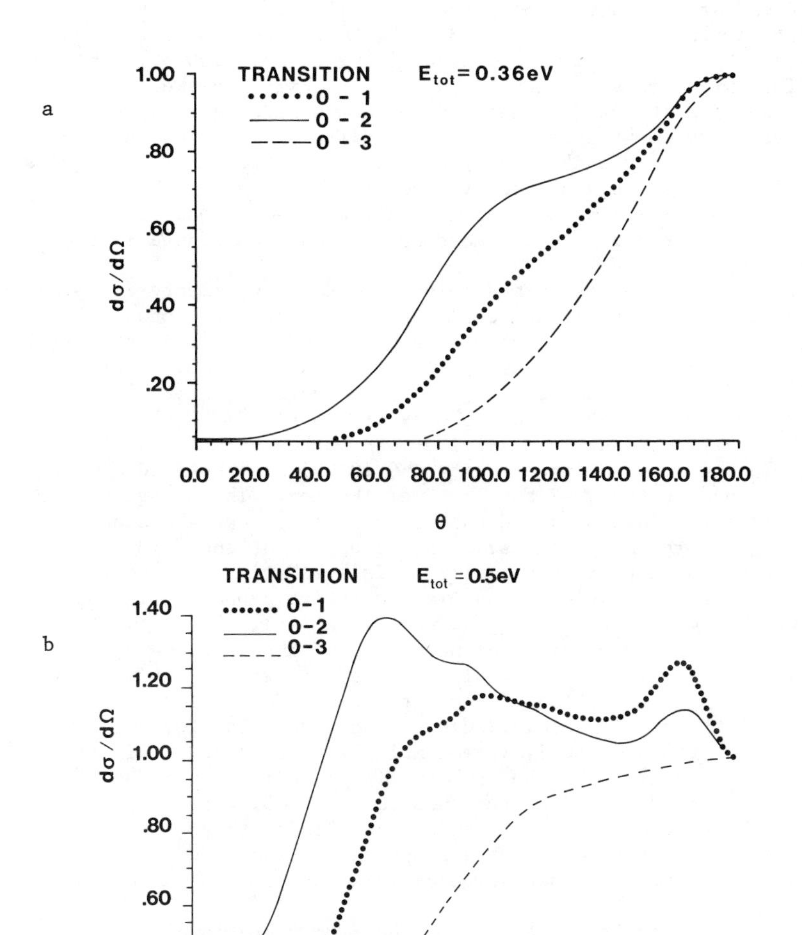

Figure 7. Vibrational State resolved $F+H_2 \rightarrow HF+H$ differential cross section. All cross sections are normalized to one at $\theta=180°$. (a) $E_{tot}=0.36$ eV (b) $E_{tot}=0.5$ eV. Reproduced with permission from Ref. 6b, Copyright 1983, American Institute of Physics.

three curves are given for the reactive cross sections into v_f=1, 2, and 3 at a given total energy. All the results are scaled to be equal to one at $\theta=\pi$, and display the following features: (a) The differential cross section for v_f=3 is backwards peaked at $\theta=\pi$ for all energies studied (E_{tot}=0.31, 0.34, 0.36, 0.423, 0.50 eV) (b) The differential cross section for v_f=1 is backwards peaked ($\theta\cong\pi$) for all energies studied but has a significant sideways component ($\theta\sim\pi/2$) at the highest energy. (c) The differential cross section for v_f=2 is backwards peaked ($\theta=\pi$) only for the three lowest energies; at the fourth it shows a secondary peak at $\theta\sim100°$ and at the fifth it peaks at $\theta\sim\pi/3$. (d) In addition, the differential cross section for v_f=2 has oscillations, with the most pronounced occurring a little forward of π.

Sparks et.al.(16) performed molecular beam experiments in which they measured the state-to-state differential cross sections and found the v_f=2 differential cross sections to behave differently from the other two (i.e., v_f=1,3) differential cross sections. Whereas for a low energy measurement (E_{trans}=0.1 eV where $E_{trans}=E_{tot}-E_0$. E_0 is the zero-point vibrational energy of the diatom (H_2 or D_2)) they found that all three angular distributions peaked backwards, at a higher relative translational energy (∿0.15 eV) they noticed a dramatic change in the v_f=2 angular distribution. The other two distributions remained backward peaked but the v_f=2 peaked sideways (∿120°). (More recently Sparks, et.al. report that the v_f=3 HF product is found to scatter forward.(17)) This behavior was consistent with the difference in the partial wave reactive probabilities of RW(3) for the two energies E_{trans}∿2 and 3 kcal/mol. That is, at low energy, RW found a monotonic decline with ℓ of the reactive v_f=2 probability while at E_{trans}∿3 kcal/mol, they observed a maximum for v_f=2 at ℓ=10. Our QM-IOS angular distributions for v_f=1,2,3 qualitatively fit the experimental findings. In particular for the three energies below or equal to 2.14 kcal/mol all three angular distributions peak backwards as can be seen in Figure 7a for E_{tot}=0.36 eV but for E_{trans}=0.156 eV, which corresponds to E_{trans}=3.54 kcal/mol a change in the v_f=2 distribution is noticed. In contrast to the two others, it shows a strong sideways component at about 80°. At the energy E_{tot}=0.5 eV, the sideways peaking is very evident as can be seen in Figure 7b.

Similar results are obtained for the RIOS vibrational-state-resolved F+D_2 angular distributions. We may summarize the results at 2.91 kcal/mol as follows: (a) the DF molecules are back scattered for v_f=1,2,3,4; (b) the least backward peaked is v_f=3 and the most is v_f=4; and (c) interesting undulations are observed. The general features of the 4.51 kcal/mol results are: (a) the v_f=1,2,3 DF molecules have all developed pronounced sideways peaks; (b) the order of the peaks (as the scattering angle is increased from zero toward π) is first v_f=3, then v_f=2 and finally v_f=1; and (c) the v_f=4 DF product continues to show very strong backward peaking and shows a pronounced minimum in the forward direction. Thus, the v_f=3 HF results are analogous to the v_f=4 DF results and similarly for the v_f=2,1 HF and v_f=3,2,1 DF results.

Thus, we again find strong resemblances between the angular distributions for the DF and HF products, with the v_f=4 DF and v_f=3 HF behaving similarly and the v_f=3,2,1 DF and v_f=2,1 HF products behaving similarly. In the case of the experimental results, it also would appear that the $F+D_2$ and $F+H_2$ systems behave qualitatively the same. Thus in $F+H_2$, the most recent results indicate that the $v_f=3$ is forward peaked while in $F+D_2$, it is the $v_f=4$ which is forward peaked.(17) Then as one moves away from the forward direction, the $v_f=2$ HF and $v_f=3$ DF are next in order of increasing scattering angle, followed by the $v_f=1$ HF and $v_f=2$ DF.(17) Thus, the latest experimental data shows a similar correspondence between the v_f state of HF with the v_f+1 state of DF. The present theoretical results agree qualitatively with the experimental results so far as the $v_f=3,2,1$ DF product and $v_f=2,1$ HF product angular distributions but disagree with experiment for the $v_f=4$ DF and $v_f=3$ HF products. However, even in this instance, the theoretical and experimental results agree that these products behave similarly. Regarding the backward angular distribution of the $v_f=4$ DF product (and analogously the $v_f=3$ HF product), there seems to be little chance, in our opinion, that the Muckerman 5 surface can result in behavior other than this. If the experimental findings are indeed correct, then we believe this is clear evidence that the Muckerman 5 surface is not capable of fully describing the vibrational-state-resolved angular distributions.

Resonance Tuning in the $F+H_2$ System

It would appear on the basis of the results discussed above that there are definitely resonances predicted within the RIOS method for both the FD_2 and FH_2 systems. Here, we wish to examine these resonances for $F+H_2$ in more detail with the particular aim of exploring the resonance tuning concept. We have chosen the $F+H_2$ system rather than $F+D_2$ because the former appears to show longer lived resonances and the tuning effect seems to be accentuated. In Figure 8 we give a two-dimensional plot of the maximum time delay as a function of γ and ℓ. It is very clearly seen that increasing either γ or ℓ causes a decrease in the time delay experienced in the $F+H_2$ collision. However, the rate of decrease of the maximum delay time is not such as to completely eliminate the resonance before fairly substantial values of γ and/or ℓ are attained. Of course, in an actual collision, the angle γ is not in fact constant (as is assumed in the RIOS). In Figure 9 we give a two dimensional plot showing how the total energy E_{res} of the maximum time delay varies with γ and ℓ. This shows very clearly the tuning mechanism referred to earlier.(6-7) It is easily seen that increasing either γ or ℓ causes a shift to higher energy of the maximum delay time, thereby implying that the resonance occurs at higher collision energy. This is the essence of the resonance tuning hypothesis and it implies that as the collision energy is raised, the reaction will be influenced by the resonance occurring in higher orbital angular momenta and for larger γ values. Very similar behavior is found for the $F+D_2$ system.

Role of the Resonance in the $F+H_2$ and $F+D_2$ Angular Distributions

The central remaining question to be addressed is what is the role of resonance tuning in the angular distribution of the $F+H_2$ reaction. In part, this question also involves the question of the relevance of the Muckerman 5 potential surface for the $F+H_2$ reaction.

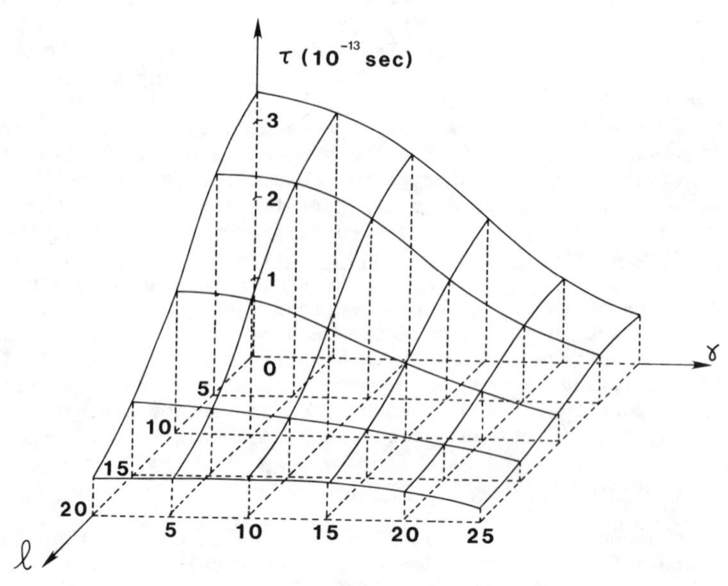

Figure 8. The maximum delay time for $F+H_2$ is shown versus γ and ℓ.

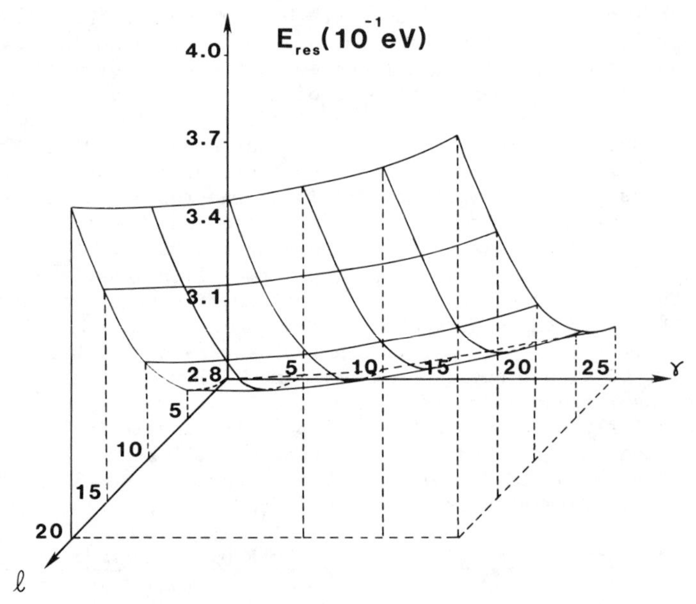

Figure 9. The resonance energy E_{res} (defined as the total energy at which the time delay is maximal) as a function of γ and ℓ.

To an extent, the Muckerman 5 surface does give results that are in substantial agreement with experiment. As was discussed earlier in this and previous papers,(6,7) both theory and experiment indicate that the FH_2 and FD_2 systems are completely analogous, with the state v_f in HF corresponding in behavior to state $v_f + 1$ of DF. However, we strongly believe that the fact that experiment shows the v_f=3 HF and v_f=4 DF products strongly forward scattered(17) must be taken as pointing up a central defect in the Muckerman 5 surface. We believe the RIOS results for the state-resolved angular distributions do represent at least qualitatively (and perhaps even quantitatively) what results from the Muckerman 5 surface. In the case of the RIOS state-resolved angular distributions, we also believe that the resonance tuning does have an important influence. However, its influence is primarily to accelerate (as a function of energy) the shift from backward to forward scattering which normally occurs whether or not there is any resonance. That is, any reaction will show a shift from backward to forward scattering as the collision energy is increased from low to higher values. However, this shift normally occurs over a broader energy range than is experienced in FH_2 and FD_2. We believe the resonance tuning is responsible for sharpening the energy range over which this shift occurs. We also believe that the most significant aspect of the comparison between experiment and the RIOS results is the fact that experiment shows the v_f=3 HF and v_f=4 DF products more forward scattered than any other product. This suggests that it is these product states which are most strongly influenced by resonant phenomena and this should prove extremely useful in obtaining further refinements of the FH_2 and FD_2 potentials(28) (beyond the Muckerman 5 surface).

Acknowledgments

Partial support of this research under National Science Foundation Grant CHE82-151317 is gratefully acknowledged. Acknowledgment is made to the donors of the Petroleum Research Fund, administered by the American Chemical Society, for partial support of this research.

Literature Cited

1. Baer, M. Adv. Chem. Phys. 1982, 49, p. 191; Wyatt, R.E. In Atom-Molecule Collision Theory: A Guide for the Experimentalist, Bernstein, R.B., Ed. Plenum: N.Y., 1979 p. 447.
2. Elkowitz, A.B., Wyatt, R.E. Mol. Phys., 1976, 31, 189.
3. Redmon, M.J.; Wyatt, R.E. Int. J. Quant. Chem. 1977, 11, 343; Chem. Phys. Lett 1979, 63, 209.
4. Khare, V.; Kouri, D.J.; Baer, M. J. Chem. Phys. 1979, 71, 1188; Baer, M.; Khare, V.; Kouri, D.J. Chem. Phys. Lett. 1979, 68, 378; Baer, M.; Mayne, H.; Khare, V.; Kouri, D.J.; ibid. 1980, 72, 269; Jellinek, J.; Baer, M.; Khare, V.; Kouri, D.J. ibid. 1980, 75, 460; Khare, V.; Kouri, D.J.; Jellinek, J.; Baer, M. In Potential Energy Surfaces and Dynamics Calculations, Truhlar, D.G. Ed. Plenum: N.Y., 1981 p. 475; Baer, M.; Kouri, D.J.; Jellinek, J. J. Chem. Phys. (in press).
5. Bowman, J.M.; Lee, K.T. J. Chem. Phys. 1978, 68, 3940; Chem. Phys. Lett. 1979, 64, 29; J. Chem. Phys. 1980, 72, 5071.

6. (a) Jellinek, J.; Baer, M.; Kouri, D.J. Phys. Rev. Lett. 1981, 47, 1588; (b) Baer, M. Jellinek, J.; Kouri, D.J. J. Chem. Phys. 1983, 78, 2962; Jellinek, J.; Baer, M.; Kouri, D.J. J. Phys. Chem. 1983, 87, 3370.
7. (a) Shoemaker, C.L.; Kouri, D.J., Jellinek, J; Baer, M. Chem. Phys. Lett. 1983, 94, 359; (b) AbuSalbi, N.; Shoemaker, C.L.; Kouri, D.J.; Jellinek, J.; Baer, M. J. Chem. Phys. 1984, 80, 3210.
8. Barg, G.D.; Drolshagen, G. Chem. Phys. 1980, 47, 209.
9. (a) Clary, D.C.; Mol. Phys. 1981, 44, 1067 and 1083; (b) Clary, D.C.; Garrett, B.C.; Truhlar, D.G. J. Chem. Phys., 1983, 78, 777.
10. Clary, D.C.; Drolshagen, G. J. Chem. Phys. 1982, 76, 5027.
11. (a) Walker, R.B.; Hayes, E.F. J. Phys. Chem. 1982, 86, 85; (b) Walker, R.B.; Hayes, E.F.; J. Phys. Chem. 1983, 87, 1255.
12. Jellinek, J.; Kouri, D.J. In Theory of Chemical Reaction Dynamics, Baer, M., Ed. CRC Press; Boca Raton, Fla., 1983; Jellinek, J., Ph.D. Thesis, Weizmann Institute of Science, Rehovot, Israel, 1983,
13. Schatz, G.C.; Kuppermann, A. J. Chem. Phys. 1976, 65, 4624, 4642, 4668.
14. Muckerman, J.T. In Theoretical Chemistry: Advances and Perspectives, Vol. 6A, Henderson, D.H., Eds.; Academic, N.Y., 1981, p.1.
15. Ron, S.; Pollak, E.; Baer, M. J. Chem. Phys. (in press).
16. Sparks, R.K.; Hayden, C.C.; Shobatake, D.; Neumark, D.M.; Lee, Y.T. In Horizons in Chemistry, Fukui, K.; Pullman, B.; Eds.; D. Reidel, N.Y., 1980, p. 91.
17. Sparks, R.F.; Shobatake, K.; Lee, T.Y.; personal communication.
18. Schatz, G.C.; Bowman, J.M.; Kuppermann, A.; J. Chem. Phys. 1973, 56, 1024; ibid. 1975, 63, 674.
19. Bowman, J.M.; Lee, K.-T.; Ju, G.-Z. Chem. Phys. Lett. 1982, 86, 384.
20. (a) Ron, S.; Baer, M.; Pollak, E. J. Chem. Phys. 1983, 78, 4414; (b) Blais, N.C., Truhlar, D.G. J. Chem. Phys. 1982, 76, 4490.
21. Kuppermann, A. In Potential Energy Surfaces and Dynamics Calculations, Truhlar, D.G. Ed. Plenum, N.Y., 1981, p. 375.
22. Smith, F.T. Phys. Rev. 1960, 118, 349.
23. Rose, M.E. Elementary Theory of Angular Momentum, Wiley, J. and Sons, N.Y., 1963, p.52.
24. Arthurs, A.M.; Dalgarno, A. Proc. Roy. Soc. 1960, A256, 540.
25. McGuire, P.; Kouri, D.J. J. Chem. Phys. 1974, 60, 2488; Pack, R.T. ibid. 1974, 60, 633; Secrest, D. ibid. 1975, 62, 710; Shimoni, Y.; Kouri, D.J. ibid. 1976, 65, 3372 and 3958 and 1977, 66, 675 and 2841; Truhlar, D.G.; Poling, R.E.; Brandt, M.A. ibid. 1976, 64, 26; Parker, G.A.; Pack, R.T. ibid. 66, 2850; Kouri, D.J.; Shimoni, Y. ibid. 1977, 67, 86; Goldflam, R.; Kouri, D.J. ibid. 1977, 66, 542; Khare, V. ibid. 1977, 67, 3897; Monchick, L. ibid. 1977, 67, 4534; Fitz, D.E. Chem. Phys. 1977, 24,133; Khare, V.; Kouri, D.J.; Pack, R.T. J. Chem. Phys. 1978, 69, 449; Fitz, D.E. Chem. Phys. Lett. 1978, 55, 202; Schinke, R.; McGuire, P. Chem. Phys. 1978, 28, 129; Monchick, L.; Kouri, D.J. J. Chem. Phys. 1978, 69, 3262; Coombe, D.A.; Snider, R.F. ibid. 1979, 72, 4284; Stolte, S.; Reuss, J. In Atom-Molecule Collision Theory: A Guide for the Experimentalist, Bernstein, R.B. Plenum, N.Y., 1979; Kouri, D.J. ibid.; Khare, V.; Kouri,

D.J.; Hoffman, D.K. J. Chem. Phys. 1981, 74, 2275; Khare, V.; Fitz, D.E.; Kouri, D.J. ibid. 1980, 73, 2802and 4148; Khare, V.; Kouri, D.J. Chem. Phys. Lett. 1981, 80, 262; Khare, V.; Kouri, D.J.; Hoffman, D.K. J. Chem. Phys. 1981, 74, 2656.
26. Wu, S.F.; Johnson, B.R.; Levine, R.D. Mol. Phys. 1973, 25, 839; Schatz, G.C.; Bowman, J.M.; Kuppemann, A. J. Chem. Phys. 1973, 56, 1024 and 1975, 63, 674.
27. Siegbahn, P.; Liu, B. J. Chem. Phys. 1978, 68, 2457; Truhlar, D.G.; Horowitz, C.J. ibid. 1978, 68, 2466 and 1979, 71, 1514.
28. Truhlar, D.G.; Garrett, B.C.; Blais, N.C. J. Chem. Phys. 1984, 80, 232.

RECEIVED June 11, 1984

25

Dynamic Resonances in the Reaction of Fluorine Atoms with Hydrogen Molecules

D. M. NEUMARK, A. M. WODTKE, G. N. ROBINSON, C. C. HAYDEN[1], and Y. T. LEE

Materials and Molecular Research Division, Lawrence Berkeley Laboratory and Department of Chemistry, University of California, Berkeley, CA 94720

The reactions of F + H_2, HD and D_2 were studied in high resolution crossed molecular beams experiments. Center-of-mass translational energy and angular distributions were determined for each product vibrational state. In the F + H_2 reaction, the v=3 product showed intense forward scattering while the v=2 product was backward-peaked. These results, in contrast to the backward scattering of all DF product vibrational states from F + D_2 at the same collision energy, suggest that dynamical resonances play an important role in the reaction dynamics of this system. In the F + HD reaction, the strong forward scattering of HF products and backward scattering of DF products is in agreement with the prediction of a stronger resonance effect for HF formation. The effect of the H_2 rotational excitation and the reactivity of $F(^2P_{1/2})$ are also discussed.

Understanding the potential energy surface governing the formation and destruction of chemical bonds during a chemical reaction continues to be an important goal of experimental reaction dynamics. The region of the potential energy surface near the transition state, or strong coupling region, critically influences many of the observable features of a chemical reaction but has, in general, proved inaccessible to direct study. One approach to solving this problem has been to develop increasingly state-selective techniques with the ultimate goal of obtaining state-to-state rate constants for chemical reactions. Although the results of these experiments certainly reflect the properties of the strong coupling region, it is difficult, working backwards, to derive quantitative information about the structure of the potential energy surface near the transition state. Another method which is, in principle, more direct involves probing the transition state by

[1]Current address: Department of Chemistry, University of Wisconsin, Madison, WI 53706.

0097-6156/84/0263-0479$06.00/0

photon absorption and emission (1,2). However, the continuous spectra obtained from this technique represent electronic transitions between two unknown surfaces and are therefore difficult to interpret.

A more promising method is suggested by quantum mechanical scattering calculations on several elementary chemical reactions (3-5). Calculations of the energy dependence of reaction cross sections predict the existence of resonances arising from quasi-bound states in the strong coupling region (6-8). The nature of these resonances is very sensitive to the detailed structure of the potential energy surface in this region (9-10). The experimental observation of reactive resonances should therefore provide a far more sensitive probe of the strong coupling region than has previously been available. Resonances appear to be most prominent in hydrogen transfer reactions, and the two most thoroughly studied examples are the $H + H_2$ (3) and $F + H_2$ (4-5) reactions. Of these two, the $F + H_2$ reaction is considerably more amenable to experimental investigation. We have performed a high resolution crossed molecular beam study of this reaction in an attempt to observe resonance effects in reactive scattering.

In light of previous experimental and theoretical work on the $F + H_2$ reaction, it can be seen why an experiment of this complexity is necessary in order to observe dynamic resonances in this reaction. The energetics for this reaction and its isotopic variants are displayed in Figure 1. Chemical laser (11) and infrared chemiluminescence (12) studies have shown that the HF product vibrational distribution is highly inverted, with most of the population in v=2 and v=3. A previous crossed molecular beam study of the $F + D_2$ reaction showed predominantly back-scattered DF product (13). These observations were combined with the temperature dependence of the rate constants from an early kinetics experiment (14) in the derivation of the semiempirical Muckerman 5 (M5) potential energy surface (15) using classical trajectory methods. Although an ab initio surface has been calculated (16), M5 has been the most widely used surface for the $F + H_2$ reaction over the last several years.

Collinear quantal calculations on M5 show sharp resonances in the reaction probability as a function of collision energy (5-6). The resonance energies have been shown to correspond to the energies of quasi-bound FH_2 states (6,8). The resonance widths are typically 0.01 eV, indicating that the lifetime of the quasi-bound states is on the order of several vibrational periods. Only the lowest energy resonance is readily accessible in our experiment; on the M5 potential energy surface, this decays exclusively to HF(v=2) product.

The collinear results imply that resonances might be observed in an experiment which measures the energy dependence of the total reaction cross section, but three-dimensional quantal calculations (17,18) on $F + H_2$ show that this approach will not succeed. These calculations reveal no sharp structure in the reaction cross section as a function of collision energy, a result which can be understood by considering the effects of angular momentum. The orbital angular momentum, L, of the reactants becomes rotational angular momentum of the FH_2 quasi-bound state. If E_o is the energy of a non-rotating quasi-bound state, then at approximately $E_o + BL(L+1)$, a rotating quasi-bound state can be formed by a collision of orbital angular momentum $L\hbar$, where B is an appropriate rotational constant for the quasi-bound state. Consequently, as the collision energy is increased beyond E_0, collisions with progressively larger values of orbital angular momentum will be brought into resonance. The large number of partial waves contributing to reactive scattering allows the resonance to be accessed over a wide energy range, and resonances appear as broad, smooth features in the collision energy dependence of the reaction cross section.

On the other hand, a comparison of classical (15) and three-dimensional quantal (17) calculations of the reaction probability as a function of reactant orbital angular momentum, that is, the opacity function, shows that a resonance enhances the contribution of high L collisions to the production of HF(v=2) over a range of collision energies above an L=0 resonance, while for the production of HF(v=3), classical and quantal calculations give similar results. If the M5 surface were correct then at collision energies somewhat above the lowest energy resonance the HF(v=2) product angular distribution should exhibit more sideways and forward scattering than the other HF vibrational states which are formed primarily by low orbital angular momentum collisions. The only experiment that appears likely to reveal dynamic resonances in this reaction is the measurement of vibrationally state resolved differential cross sections of the HF product at appropriate translational energies.

Our previous experimental studies on this reaction (19-20) have shown that vibrationally state resolved differential cross-section of HF products can indeed be obtained in a crossed molecular beams experiment by measuring both the laboratory angular and velocity distributions. The high HF background near the F beam and the lack of sufficient velocity resolution due to the 20% spread in the F beam velocity prevented us from obtaining the complete picture, although hints of quantum effects were observed. The experimental results reported here were obtained with an entirely new experimental arrangement designed to overcome all of the difficulties encountered in our previous experiment.

Experimental

The major features of the crossed molecular beams apparatus used in these studies have been described elsewhere (21-22). However, several important modifications were made specifically for these studies. The major objectives were to reduce the velocity spread of the reactant beams in order to resolve the product vibrational states as distinct peaks in time-of-flight measurements and to reduce the background of mass 20 in the detector, especially near the F atom beam. A schematic top cross sectional view of the experimental arrangement is shown in Figure 2.

An effusive beam of F atoms was produced by thermally dissociating F_2 at 2.0 torr and 920° K in a resistively heated nickel oven. The F beam was velocity selected with a FWHM velocity spread of 11%. The H_2 beam was produced by a supersonic expansion of 80 psig through a 70 micron orifice at variable temperatures with a FWHM spread of 3%. Rotational state distributions of H_2 in the beam were studied previously using molecular beam photoelectron

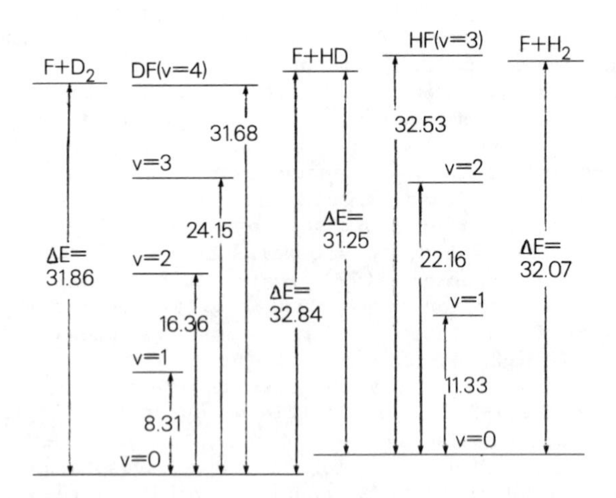

Figure 1. Energetics of the F + H_2, F + D_2, and F + HD reactions. All values are in kcal/mol. H_2, D_2, and HD are in their lowest internal states (v=0, J=0).

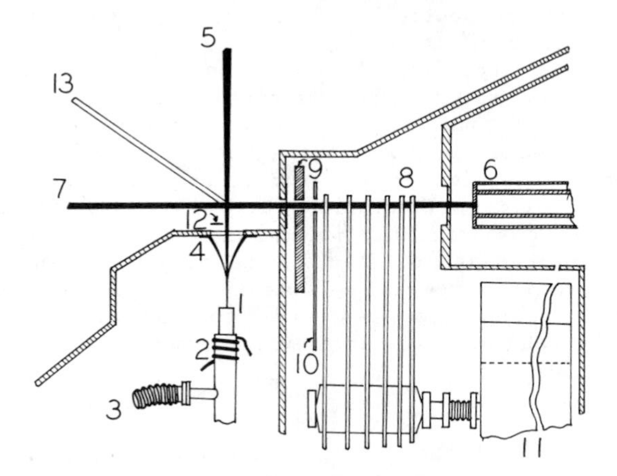

Figure 2. Top cross-section view of the beam sources and collision region. 1) H_2 source (70 micron orifice) 2) coaxial heater cable 3) liquid nitrogen contact 4) 18 mil skimmer 5) supersonic H_2 beam 6) F oven 7) F beam 8) velocity selector 9) differential chamber cold shield 10) radiation shield 11) mounting block for velocity selector 12) 150 Hz tuning fork chopper 13) UHV rotatable mass spectrometric detector.

spectroscopy which showed that about 80% of the p-H_2 will be in J=0 under these conditions (23). The FWHM reactant kinetic energy spread in the center-of-mass was only 0.1 kcal/mol. The HF product was detected with a rotatatable, ultrahigh-vacuum mass spectrometer. Angular scans were taken by modulating the H_2 beam at 150 Hz with a tuning fork chopper and recording the modulated HF signal as a function of angle. Product velocity distributions were obtained at 19 angles between 8° and 54° from the F beam by the cross-correlation time-of-flight technique (24). The flight length was nominally 30 cm.

Results and Analysis

The angular distribution for the HF product from F + p-H_2 at 1.84 kcal/mol is shown in Figure 3. The LAB angle Θ is measured from the F beam. The Newton diagram below the figure aids in relating features in the laboratory angular distribution to the center-of-mass(CM) distributions. ($\underset{\sim}{v}_F$, $\underset{\sim}{v}_{H2}$) and ($\underset{\sim}{u}_F$, $\underset{\sim}{u}_{H2}$) are the LAB and CM velocities, respectively, of the reactants. Θ_{CM} is the angle of the velocity vector for the center-of-mass of the colliding pair in the LAB frame. The tip of this vector defines the origin of the center-of-mass coordinate system. In the CM coordinate system, θ=0° is defined as the direction of the incident F beam, $\underset{\sim}{u}_F$. The radii of the "Newton circles" represent the maximum center-of-mass speeds for HF product formed in the indicated vibrational state. The broad peaks in the angular distribution around 28° and 45° are from back-scattered v=3 and v=2 product, respectively. The sharp peak at 8° is from forward-scattered v=3. This prominent feature was obscured by a high m/e=20 background near the F beam in the earlier studies. No product signal was detected on the other side of the F beam.

Whereas LAB product angular distributions alone yield only a qualitative picture of the reaction, time-of-flight measurements of the product velocity distributions allow one to quantitatively determine the contribution from each vibrational state to the total signal at a LAB angle. High angular and velocity resolution combined with the small amount of product rotational excitation relative to the HF vibrational spacing results in discrete peaks in the TOF spectra from the various product vibrational states. Sample TOF spectra with their vibrational state assignments are shown in Figure 4. The spectrum at Θ=18° shows three distinct peaks. The fastest peak is from v=2 product, and the two slower peaks are from v=3. The two v=3 peaks merge at Θ=30° and at other LAB angles which are nearly tangent to the v=3 Newton circle. The spectrum at Θ=8° confirms that the forward peak in the angular scan is from v=3 product.

The translational energy and angular distributions in the CM coordinate system were determined for each product vibrational state by forward convolution. A trial CM distribution for each state was input to a computer program which averaged over beam velocity spreads and detector resolution. The program generated a LAB angular distribution and TOF spectra, and the trial distributions were adjusted until the computer-generated results matched the data. The lines in Figures 3 and 4 are the LAB distributions generated by the best-fit CM parameters.

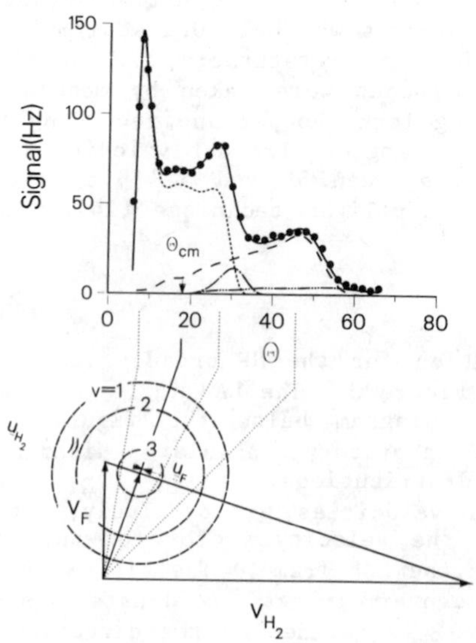

Figure 3. LAB angular distribution for F + p-H_2, 1.84 kcal/mol, and Newton diagram. Both the data and calculated LAB distributions are shown (• data, ———— total calculated, —— --- —— v = 1, —— —— —— v=2, - - - - - v=3, —— — —— v=3').

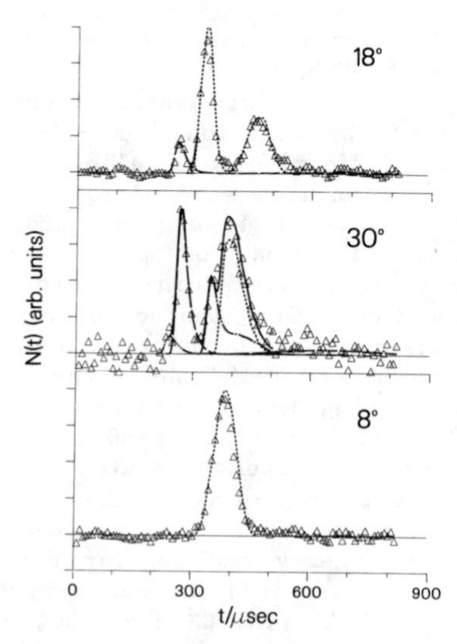

Figure 4. Time-of-flight spectra at LAB angles 18°, 30°, and 8° with vibrational state assignments (data Δ, total calculated dist. ————, vibrational state same as Figure 3, solid line not shown when it obscures a vibrational state).

The slow peak at Θ=30° has a fast shoulder. A similar feature appears in other TOF spectra that sample v=3 product near Θ=180°, and this could be fit only by assuming it was due to HF(v=3) from reactants with approximately 1 kcal/mol internal excitation. This product, designated as v=3', could originate from spin-orbit excited $F(^2P_{1/2})$ which lies 1.16 kcal/mol above the $^2P_{3/2}$ ground state and constitutes 21% of the F beam at 920°K, but the reaction between $F(^2P_{1/2})$ and H_2 can only occur by an electronically non-adiabatic process and is expected to be inefficient (25-26). It is more likely that the v=3' product is from the reaction of $F(^2P_{3/2})$ with H_2(J=2) which is 1.03 kcal/mol above H_2(J=0) and makes up about 20% of the para-H_2 beam (23).

The CM distributions for the HF products are summarized graphically in Figure 5, a contour map of the velocity flux distribution as a function of the CM scattering angle θ. The v=1 contours are not reliable due to the low intensity of that state. The v=2 state is backward-peaked and drops off slowly for θ<180°. The v=3 state has a broad maximum around θ=80° and, in contrast to the v=2 state, has a sharp, intense peak at θ=0°. The v=3' state scatters entirely into the backward hemisphere.

Similar results obtained for F + D_2 at 1.82 kcal/mol, as well as F + HD product angular distributions and the contour map for DF products at 1.98 kcal/mol are shown in Figures 6, 7, and 8. The effect of rotational excitation was also examined for F + H_2 by comparing the angular distributions from para and normal hydrogen at 1.84 kcal/mol. Angular distributions for F + H_2(J=0) and F + H_2(J=1) are shown in Figure 9.

Discussion

When contrasted with the strong back-scattering of HF(v=2) in the reaction of F + p-H_2 at 1.84 kcal/mol (Figure 5), the sharp forward peak and pronounced sideways scattering of the v=3 product is the most compelling evidence to date for quantum mechanical dynamic resonance effects in reactive scattering. The shape of the distribution is consistent with what one expects when collisions at relatively high impact parameters contribute to the formation of a quasi-bound state followed by selective decay to v=3 products. The intense forward peak results from the strong correlation between the direction of $\underline{L}$ and $\underline{L}'$, the final orbital angular momentum vector, for the collisions that form the quasi-bound state. In truly long-lived complexes that survive for several rotational periods, this correlation results in a symmetric angular distribution peaking at 0° and 180° in the center-of-mass frame (27). The much weaker intensity at 180° in this experiment shows that the quasi-bound state only lives a fraction of a rotational period which is about 5×10^{-13} s for bound F-H-H rotating with 10 $\hbar$ of angular momentum.

Our studies on the effect of isotopic substitution on this reaction further support our identification of the v=3 forward peak as a resonance effect. Dynamical resonances are predicted to be highly isotope-dependent; they should be much weaker in F + D_2 and in F + DH forming DF than in F + H_2 (28) and much stronger in F + HD forming HF. The contour map for F + D_2 at 1.82 kcal/mol displayed in Figure 6 shows no forward peak, although the v=4 product is slightly sideways-peaked. The predominant forward

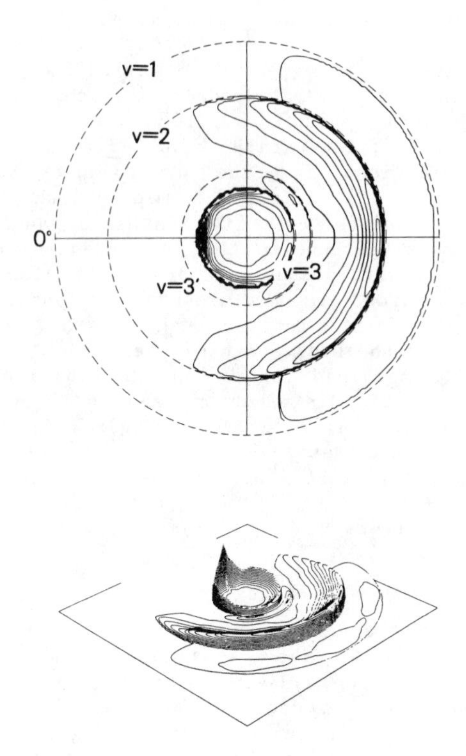

Figure 5. Center-of-mass velocity flux contour map for F + n-H_2, 1.84 kcal/mol, with three-dimensional perspective.

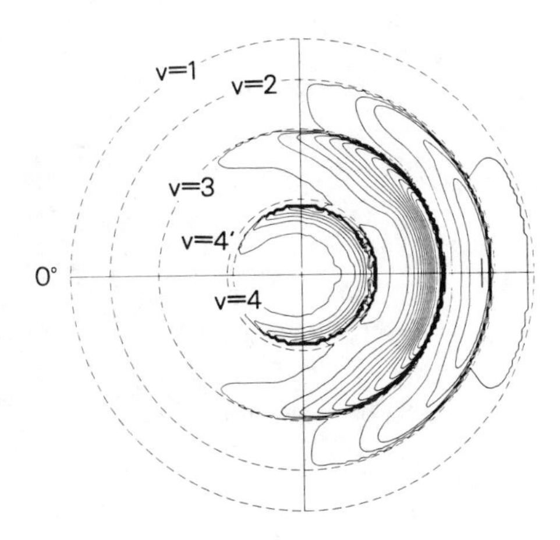

Figure 6. Center-of-mass velocity flux contour map for F + D_2, 1.82 kcal/mol.

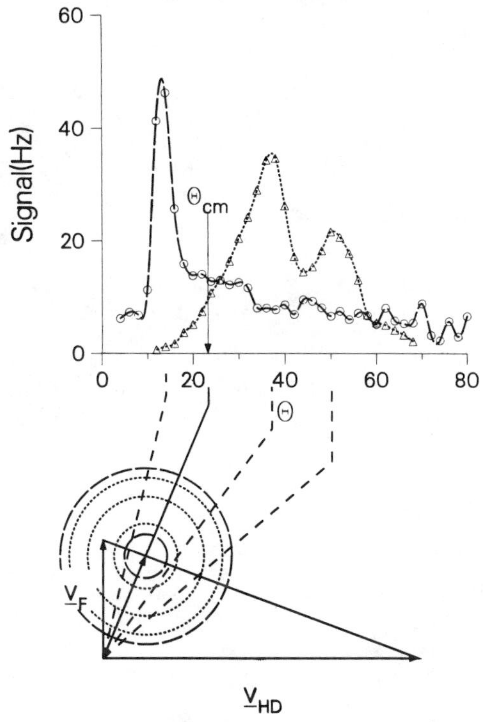

Figure 7. LAB angular distribution for F + HD, 1.98 kcal/mol (o ——— ——— HF product, ----- DF product). The Newton circles corresponding to HF and DF product are drawn with the same texture as the lines in the angular distributions. The HF(v=3) and v=2 circles are shown, as are the v=4, 3, and 2 circles for DF.

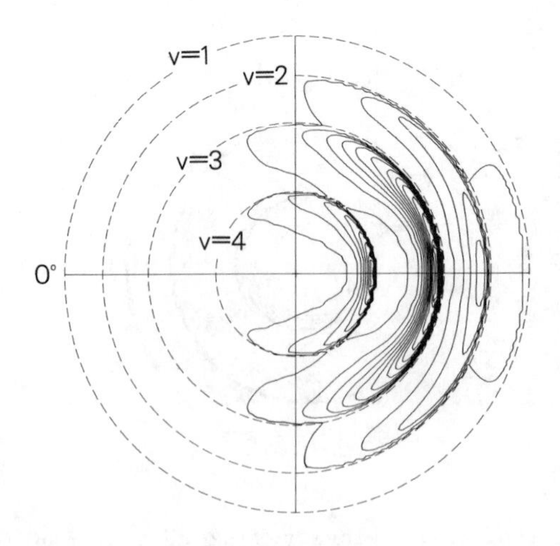

Figure 8. Center-of-mass velocity flux contour map for F + HD, DF + H, 1.98 kcal/mol.

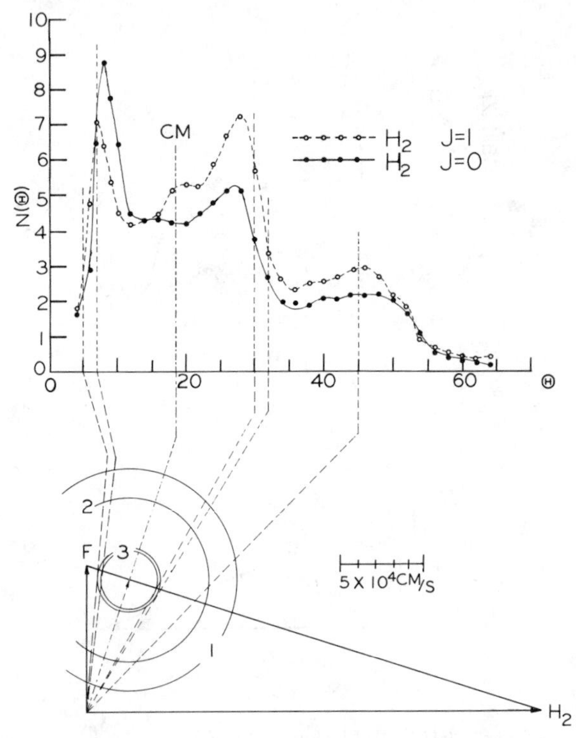

Figure 9. LAB angular distributions for F + H_2(J=0) and J=1. The innermost and next smallest Newton circles are for HF(v=3) product from H_2(J=0) and J=1, respectively.

scattering of HF products and back-scattering of DF products in the reaction of F + HD at 1.98 kcal/mol shown in Figure 7 also agrees with the resonance hierarchy predicted in quantal reactive scattering calculations. Classical trajectory calculations do not show such isotope-specific behavior in the angular distributions for these reactions.

Our results indicate that the quasi-bound state of FH_2 decays to v=3 product, in disagreement with calculations on M5 which predict decay exclusively to v=2 at energies near threshold. This discrepancy, as well as other recent developments (15,29-30), suggest that the M5 surface is an inadequate representation of the F + H_2 potential energy surface. The exact nature of the modifications to be made are beyond the scope of this article and will be discussed in a future publication. However, it should be noted that quantal collinear calculations on other model surfaces do show the corresponding resonance decaying to v=3 (10). Although these surfaces are inferior to M5 in terms of exothermicity and entrance channel barrier height, the calculations illustrate that small changes in the interaction region markedly affect the role of resonances in this reaction.

LAB angular distributions for F + H_2(J=0) and F + H_2(J=1) at 1.84 kcal/mol in Figure 9 show that there is considerably less forward peaking of the HF(v=3) product from H_2(J=1) than from H_2(J=0). It appears that resonance effects are less pronounced when starting from H_2(J=1). This may suggest how reactant rotation is coupled to the degrees of freedom of the reaction intermediate. If the H_2 rotational energy goes into overall rotation of the quasi-bound state, a large change in the strength of resonance effects would not be expected. However, if H_2 rotation were coupled to bending motion of the reaction intermediate, one might expect resonance effects to become less pronounced if only because the additional internal energy would shorten the lifetime of the resulting quasi-bound state. In addition Ron et al (31) have shown that on the M5 surface, the lowest energy quasi-bound state does not exist for F-H-H angles greater than 20°. These two considerations suggest that there is at least some coupling between reactant rotation and bending motion in the strong coupling region.

Since 20% of F atoms in the beam are in the excited $^2P_{1/2}$ state which is about 1 kcal/mol higher in energy than the ground $^2P_{3/2}$ state, questions naturally arise about the reactivity of $F(^2P_{1/2})$. The reaction of $F(^2P_{1/2})$ with H_2 does not correlate to ground state products (25-26), but semi-classical calculations have predicted a small contribution to reaction via a non-adiabatic transition in the entrance channel (32-33). Close-coupling calculations of inelastic $F(^2P_{1/2})$ + H_2 scattering have shown that the near resonant process

$$F(^2P_{1/2}) + H_2(J=0) \rightarrow F(^2P_{3/2}) + H_2(J=2)$$

is about an order of magnitude more efficient than any other electronic quenching process (34,35). This might lead one to propose that the v=3' state results from quenching followed by reaction on the ground state surface with the newly formed H_2(J=2). The reactivity of $F(^2P_{1/2})$ should then be higher with p-H_2 than with n-H_2 because of the larger J=0 population in

$p-H_2$. However, it is unlikely that this process can lead to more product than the reaction between $F(^2P_{3/2})$ and $H_2(J=2)$ already in the beam. Thus the previous assignment of the v=3' state to rotationally excited H_2 is more convincing, though slightly less exotic. As a result our data can be fit without assuming any contribution to the reaction from $F(^2P_{1/2})$. If the reactivity of $H_2(J=2)$ with $F(^2P_{3/2})$ is similar to that of $H_2(J=0)$, the contribution from $F(^2P_{1/2})$ must be very small.

Conclusion

The vibrationally state-resolved differential cross sections obtained in this experiment show a dramatic effect from dynamical resonance phenomena and represent an important step in our ability to characterize a chemical reaction completely. The comparison of these results with state-of-the-art reactive scattering calculations, nearly all of which have been performed on the Muckerman 5 potential energy surface, indicates that an improved potential energy surface is needed in order to achieve better agreement with the experimental results. It now appears that the critical region of the potential energy surface can be more accurately characterized by varying the surface until the results of scattering calculations agree with the experimentally observed reactive resonance phenomena.

Acknowledgment

This work was supported by the Director, Office of Energy Research, Office of Basic Energy Sciences, Chemical Sciences Division of the U.S. Department of Energy under Contract No. DE-AC03-76SF00098. The authors acknowledge the contributions of R.K. Sparks and K. Shobatake in the early phases of this work. We would especially like to thank R. Wyatt, A. Kuppermann, R. Walker, and E. Pollak for several enlightening discussions on dynamical resonances during the course of these studies.

Literature Cited

1. Brooks, P. R.; Curl, R. F.; Maguire, T. C. Ber. Bunsenges. Phys. Chem. 1982, 86, 401.
2. Foth, H.-J.; Polanyi, J. C.; Telle, H. H. J. Phys. Chem. 1982, 86, 5027.
3. Truhlar, D. G.; Kuppermann, A. J. Chem. Phys. 1970, 52, 3841.
4. Wu, S.-F.; Levine, R. D. Mol. Phys. 1971, 22, 881.
5. Schatz, G. C.; Bowman, J. M.; Kupperman, A. J. Chem. Phys. 1975, 63, 674.
6. Launay, J. M.; LeDourneuf, M. J. Phys. B 1982, 15, 1455.
7. Pollak, E.; Child, M. S.; Pechukas, P. J. Chem. Phys. 1980, 72, 1669.
8. Pollak, E. Chem. Phys. 1981, 60, 23.
9. Kuppermann, A. In "Potential Energy Surfaces and Dynamics Calculation"; Truhlar, D. G., Ed.; Plenum: New York, 1981, p.
10. Connor, J. N. L.; Jakubetz, W.; Manz, J. Mol. Phys. 1978, 35, 1301.
11. Coombe, R. D.; Pimentel, G. C. J. Chem. Phys. 1973, 59, 251.

12. Polanyi, J. C.; Woodall, K. B. J. Chem. Phys. 1972, 57, 1574.
13. Schafer, T. P.; Siska, P. E.; Parson, J. M.; Wong, Y. C.; Lee, Y. T. J. Chem. Phys. 1970, 53, 3385.
14. Mercer, P.D.; Pritchard, H.O. J. Phys. Chem. 1959, 63, 1468.
15. Muckerman, J. T. In "Theoretical Chemistry:Advances and Perspectives", Eyring, H.; Henderson, D., Eds.; Academic: New York, 1981, Vol. 6A, p.1-77.
16. Bender, C. F.; Pearson, P. K.; O'Neal, S. V.; Schaefer III, H. F. Science, 1972, 176, 1412.
17. Redmon, M. J.; Wyatt, R. E. Chem. Phys. Lett. 1979, 63, 209.
18. Baer, J.; Jellinek, J.; Kouri, D. J. J. Chem. Phys. 1983, 78, 2962.
19. Sparks, R. K.; Hayden, C. C.; Shobatake, K.; Neumark, D. M.; Lee, Y. T. In "Horizons in Quantums Chemistry", Fukui, K; Pullman, B., Eds. Reidell, Boston, 1980, p.91-105.
20. Hayden, C. C. Ph.D. Thesis, University of California, Berkeley, California, 1982.
21. Lee, Y. T.; McDonald, J. D.; LeBreton, P. R.; Herschbach, D. R. Rev. Sci. Instrum. 1969, 40, 1402.
22. Sparks, R. K. Ph.D. Thesis, University of California, Berkeley, California, 1980.
23. Pollard, J. E.; Trevor, D. J.; Lee, Y. T.: Shirley, D. A. J. Chem. Phys. 1982, 77, 4818.
24. Skold, K. Nucl. Instrum. Methods 1968, 63, 114.
25. Truhlar, D. G. J. Chem. Phys. 1972, 56, 3189, 1974, 61 440(E).
26. Muckerman, J. T.; Newton, M. D. J. Chem. Phys. 1972, 56, 3191.
27. Miller, W. B.; Safron, S. A.; Herschbach, D. R. Disc. Faraday Soc. 1967, 44, 108.
28. Schatz, G. C.; Bowman, J. M.; Kuppermann, A. J. Chem. Phys. 1975, 63, 685.
29. Wurzberg, E.; Houston, P. L. J. Chem. Phys. 1980, 72, 4811.
30. Heidner III, R. F.; Bott, J. F.; Gardner, C. E.; Melzer, J. E. J. Chem. Phys. 1980, 72, 4815.
31. Ron, S.; Baer, M.; Pollak, E. J. Chem. Phys. 1983, 78, 4414.
32. Tully, J. C. J. Chem. Phys. 1974, 60, 3042.
33. Komornicki, A.; Morokuma, K.; George, T. F. J. Chem. Phys. 1977, 67, 5012.
34. Rebentrost, F.; Lester, Jr., W. A. J. Chem. Phys. 1971, 67, 3367.
35. Wyatt, R. E.; Walker, R. B. J. Chem. Phys. 1979, 70, 1501.

RECEIVED June 11, 1984

26

Reactive Resonances and Angular Distributions in the Rotating Linear Model

EDWARD F. HAYES[1] and ROBERT B. WALKER[2]

[1]National Science Foundation, Washington, DC 20550
[2]Los Alamos National Laboratory, Los Alamos, NM 87545

We use the Bending-Corrected Rotating Linear Model (BCRLM) to investigate in detail the way in which resonances may affect the angular distribution of reaction products. Using a lifetime matrix method, we separate the resonant and direct parts of the S matrix, and from the direct part we obtain angular distributions in the absence of the resonance. When applied to the $F+H_2(v=0) \rightarrow HF(v'=2)+H$ reaction on the M5 surface, at an energy for which the angular distribution is strongly sideways peaked, we find the resonance enhances the intensity of the sideways scattering, and shifts the sideways peak toward the forward direction. We also describe conditions for which angular distributions have a significant forward scattering component at the reaction threshold. For the $F+H_2(v=0) \rightarrow HF(v'=2)+H$ reaction on an improved potential surface, there is forward scattering at threshold because a resonance builds in at large partial waves. For the $He+H_2^+(v=3) \rightarrow HeH^+(v'=0)+H$ reaction, there are many overlapping resonances in all partial waves, and the reaction proceeds without a barrier. Reactivity from many partial waves at the reaction threshold results in significant forward scattering.

In this paper, we will present a detailed analysis of the way in which resonances may affect the angular distribution of the products of reactive collisions. To do this, we have used an approximate three-dimensional (3D) quantum theory of reactive scattering (the Bending-Corrected Rotating Linear Model, or BCRLM) to generate the detailed scattering information (S matrices) needed to compute the angular distribution of reaction products. We also employ a variety of tools, notably lifetime matrix analysis, to characterize the importance of a resonance mechanism to the dynamics of reactions. As a result, we hope to gain insight into how the resonant component of the scattering mechanism is manifested in the angular distribution of reaction products. Applications of these techniques to two reactive systems, $F+H_2$ and $He+H_2^+$, will be reviewed.

0097-6156/84/0263-0493$06.50/0

The Bending-Corrected Rotating Linear Model

The study of quantum effects such as resonances in atom-molecule reactions has been largely confined to coupled-channel calculations for collisions constrained to collinear geometries. Progress in quantum reactive scattering techniques is reviewed periodically (1-4). A few 3D quantum calculations of simple reactions, some more approximate (5-17) than others (18-19), have been concerned with resonance features in the reaction dynamics, and with the increasing sophistication and sensitivity of molecular beam experiments (20-23), it has become evident that the angular distribution of reaction products is likely to be the most sensitive observable manifestation of resonant contributions to reaction mechanisms.

Coupled channel methods for collinear quantum reactive calculations are sufficiently well developed that calculations can be performed routinely. Unfortunately, collinear calculations cannot provide any insight into the angular distribution of reaction products, because the impact parameter dependence of reaction probabilities is undefined. On the other hand, the best approximate 3D methods for atom-molecule reactions are computationally very intensive, and for this reason, it is impractical to use most 3D approximate methods to make a systematic study of the effects of potential surfaces on resonances, and therefore the effects of surfaces on reactive angular distributions. For this reason, we have become interested in an approximate model of reaction dynamics which was proposed many years ago by Child (24), Connor and Child (25), and Wyatt (26). They proposed the Rotating Linear Model (RLM), which is in some sense a 3D theory of reactions, because the line upon which reaction occurs is allowed to tumble freely in space. A full three-dimensional theory would treat motion of the six coordinates (in the center of mass) associated with the two vectors $\vec{r}$ and $\vec{R}$ which specify the internuclear diatomic axis and the atom-molecule separation respectively. In a space-fixed frame, these vectors have polar coordinates (r,θ',ϕ') and (R,θ,ϕ). In the RLM, two degrees of freedom are eliminated by requiring the polar coordinates of the $\vec{r}$ and $\vec{R}$ vectors to coincide. The Hamiltonian for this system treats the four degrees of freedom associated with the coordinates (R,r,θ,ϕ). As a model for angular distributions for reactive collision dynamics, the RLM possesses two attractive features. First, the angular degrees of freedom are handled in a partial wave expansion in the total angular momentum index ℓ, which introduces an impact parameter into the theory. Second, the RLM coupled-channel equations for each partial wave, when expressed in either Cartesian, polar, or hyperspherical coordinates, are identical to those of a purely collinear formulation, with the addition of an effective centrifugal potential

$$V^{\ell}_{RLM}(r,R) = \hbar^2 \frac{[\ell(\ell+1)+1]}{2\mu(R^2+r^2)} \tag{1}$$

to the collinear potential energy surface for reaction. Consequently, the solution of the reaction dynamics in the RLM is only as difficult in practice as solving a family of collinear problems, one for each partial wave. In enforcing asymptotic boundary conditions, we use spherical Bessel functions and ignore the centrifugal zero-point motion $1/(R^2+r^2)$ in Equation 1.

An obvious defect of the RLM is that it neglects motion of the $\vec{r}$ vector relative to the $\vec{R}$ vector. Asymptotically, the neglected degrees of freedom describe the rotational motion of the diatomic, and in the interaction region, they describe the internal bending of the collision complex. Since the model only samples the collinear projection of the entire potential hypersurface, one would expect the RLM to be most reliable for those reactions which proceed through linear collision intermediates with strongly hindered bending motion. In order to improve the quantitative aspects of the dynamics, especially in the reaction threshold region, we have added a bending correction to the RLM, in the same spirit as described earlier (27-30) for other models. The model thus generated is called the Bending-Corrected Rotating Linear Model (BCRLM), and the results we describe here will all be derived from this model. The details of the bending correction are described in an earlier publication (14).

In a series of recent papers (14-17), we have used the BCRLM to examine the manifestation of resonances for several reactions, including

$$F+H_2(v) \rightarrow HF(v') + H$$

$$H+H_2(v) \rightarrow H_2(v') + H$$

and several of their isotopic variants. In these papers, we have considered the importance of quantum effects, especially resonances, upon integral and differential cross sections and upon rate constants for these reactions. The hope is that these approximate results contain many of the important features of more accurate theoretical methods.

At any total scattering energy E, elements of the multichannel S matrix in the RLM are labelled by the total angular momentum index ℓ, and by the initial and final vibrational quantum numbers v and v'. Equations for physical observables in the BCRLM have been given previously (24-26), and we only summarize the final results here, in order to establish a common notation. The opacity function gives the impact parameter dependence of the reaction probabilities, $P^{\ell}_{v'v}(E)$, at fixed E, so that

$$P^{\ell}_{v'v}(E) = \left|S^{\ell}_{v'v}(E)\right|^2 \quad , \tag{2}$$

where the impact parameter is related to the angular momentum index by the semiclassical expression $bk \simeq (\ell+1/2)$, and k is the translational wavenumber. The angular distribution $I_{v\rightarrow v'}(\theta)$ of products is given by the differential cross section

$$d\sigma_{v\rightarrow v'}(\theta;E)/d\Omega = (k_{v'}/k_v)\left|f_{v'v}(\theta;E)\right|^2 = I_{v\rightarrow v'}(\theta) \tag{3}$$

where the scattering amplitude is

$$f_{v'v}(\theta;E) = \frac{1}{2i(k_v k_{v'})^{1/2}} \sum_{\ell} (2\ell+1)(S^{\ell}_{v'v}(E)-\delta_{v'v})P_{\ell}(\cos\theta) \quad (4)$$

The integrated cross section is

$$\sigma_{v\to v'}(E) = 2\pi \int I_{v\to v'}(\theta) \sin\theta d\theta = \pi k_v^{-2} \sum_{\ell} (2\ell+1)P^{\ell}_{v'v}(E) \quad (5)$$

All angular distributions reported in this paper are calculated using Equations 3 and 4 above. However, in interpreting features of the angular distribution, it is often instructive to appeal to the elements of the classical formula for angular distributions (35)

$$I_{v\to v'}(\theta) = \frac{b\, P_{v'v}(b)}{\sin\theta \left|\frac{d\theta(b)}{db}\right|} \quad (6)$$

In Equation 6, $\theta(b)$ is the classical deflection function, and specifies the angle θ at which the particles separate after a collision at an initial impact parameter b. The deflection function has the quantum analog (35)

$$\theta_{v'v}(\ell;E) \equiv \frac{d\, \mathrm{Arg}(S^{\ell}_{v'v}(E))}{d\ell} = \mathrm{Re}\left(-i[S^{\ell}_{v'v}(E)]^{-1} \frac{dS^{\ell}_{v'v}(E)}{d\ell}\right) \quad (7)$$

where Arg(z) is the phase angle of the complex number z. Equations 3, 4, 6, and 7 make it clear that the shape of the angular distribution of reaction products depends on both the magnitudes and phases of the elements of the S matrix.

Analysis of Resonances

Resonances modify the "direct" or slowly varying energy and impact parameter dependence of the phases and magnitudes of the S-matrix elements. Resonances therefore alter the shape of the angular distribution function. Kuppermann (31) has recently reviewed many of the tools which can be used to characterize resonances in single and multichannel problems. The analysis of resonances in a multichannel problem is simplest if one assumes (33-38) the isolated narrow resonance conditions (INR), in which a resonance is due to a single simple pole of the S matrix located at energy $E = E_R - i\Gamma/2$ in the complex plane. It is assumed that the background or direct component of the scattering has a much slower energy dependence than the resonant component, which is equivalent to requiring that the resonance width Γ be small, or equivalently that the pole lie close to the real axis. The slowly varying background contribution condition will be compromised if there is either a new channel threshold near the resonance energy E_R, or if there are other nearby (overlapping) resonances.

The effect of resonances on angular distributions in reactive scattering can be effectively modeled if one can successfully separate the multichannel S matrix into its background and resonant components (33,36), so that (in matrix form),

$$\underline{\underline{S}} = \underline{\underline{S}}_0 + \underline{\underline{S}}_R \quad , \tag{8}$$

where it will be understood in what follows that this analysis is done at each partial wave index ℓ. Invoking the INR conditions, Equation 8 can be written in a generalization of the single-channel Breit-Wigner form (36)

$$\underline{\underline{S}} = \underline{\underline{S}}_0^{1/2}(\underline{\underline{1}} - \frac{i\underline{\gamma}\underline{\gamma}^T}{E - z})\, \underline{\underline{S}}_0^{1/2} \tag{9}$$

with $z=E_R-i\Gamma/2$. For N open channels, the column vector $\underline{\gamma}$ is composed of N real numbers, such that $\underline{\gamma}^T\underline{\gamma}=\Gamma$. Comparing Equations 8 and 9 makes it clear that the complete characterization of the resonant and background parts of the S matrix requires knowledge of the resonant partial widths $(\underline{\underline{S}}_0^{1/2}\underline{\gamma})$, where

$$\underline{\underline{S}}_R = \frac{-i(\underline{\underline{S}}_0^{1/2}\underline{\gamma})(\underline{\underline{S}}_0^{1/2}\underline{\gamma})^T}{E - z} \tag{10}$$

All resonances are necessarily characterized by the lifetime of the compound state, and Smith's (39) definition of the lifetime matrix is

$$\underline{\underline{Q}} = i\hbar\underline{\underline{S}}\,\frac{d\underline{\underline{S}}^\dagger}{dE} \tag{11}$$

If we substitute Equation 9 into Equation 11 and make the simplifying assumption that $\underline{\underline{S}}_0$ is independent of energy, we obtain a direct expression for $\underline{\underline{Q}}$ in the vicinity of the resonance

$$\underline{\underline{Q}} = \frac{\hbar(\underline{\underline{S}}_0^{1/2}\underline{\gamma})(\underline{\underline{S}}_0^{1/2}\underline{\gamma})^\dagger}{(E-E_R)^2 + \Gamma^2/4} = \frac{\hbar\Gamma}{(E-E_R)^2+\Gamma^2/4}\,\underline{\omega}\underline{\omega}^\dagger \quad , \tag{12}$$

where the unit vector $\underline{\omega}$ is

$$\underline{\omega} = \Gamma^{-1/2}\underline{\underline{S}}_0^{1/2}\underline{\gamma} \quad . \tag{13}$$

Equation 12 tells us three things about the lifetime matrix near an INR when the energy dependence of $\underline{\underline{S}}_0$ is negligible: (i) the trace of $\underline{\underline{Q}}$ has a Lorentzian form

$$\text{Trace}\ (\underline{\underline{Q}}) = \frac{\hbar\Gamma}{(E-E_R)^2 + \Gamma^2/4} \quad , \tag{14}$$

and (ii) all the eigenvalues $[q_i]$ of $\underline{\underline{Q}}$ are zero except for one ($\underline{\underline{Q}}$ is a matrix of rank 1), and (iii), the eigenvector of $\underline{\underline{Q}}$ corresponding to the nonzero eigenvalue is the unit vector $\underline{\omega}$ defined in Equation 13. Combining Equations 13 and 10 gives a direct expression for $\underline{\underline{S}}_R$ in terms of the appropriate eigenvector of $\underline{\underline{Q}}$, namely

$$\underline{\underline{S}}_R = \frac{-i\Gamma}{E-E_R + i\Gamma/2}\ \underline{\omega}\underline{\omega}^T \tag{15}$$

Combining Equation 15 with Equation 8 allows a determination of the background S matrix $\underline{\underline{S}}_0$, and from $\underline{\underline{S}}_0$ at each partial wave one can calculate the angular distribution in the absence of a resonance, Equation 3. In practice, the energy dependence of $\underline{\underline{S}}_0$ compromises the practical application of the preceding discussion, but as we hope to show, the scheme outlined nevertheless seems quite useful in analyzing resonances.

Resonance Effects On Quantum Deflection Functions

In a recent paper (16) we explored the effects of resonances on the phase behavior of individual elements of the S matrix, with particular attention paid to the quantum deflection function, defined in Equation 7. Following in the spirit of Child's (35) analysis, we assume the partial wave dependence of the resonance energy to be

$$E_R(\ell) = E_R(0) + B\ell(\ell+1) \tag{16}$$

and further assume the resonance width depends quadratically on ℓ

$$\Gamma_\ell = \Gamma_0(1+a\ell+b\ell^2) \quad . \tag{17}$$

Consider now the phase behavior predicted by Equation 9 for the elements of the S matrix. For simplicity, we assume the background contribution provided by $\underline{\underline{S}}_0$ is weak compared to $\underline{\underline{S}}$, and we concentrate on off-diagonal transitions. In this case, the phase behavior of individual S-matrix elements is

$$S^\ell_{v'v} = \left|S^\ell_{v'v}\right| e^{i\eta^\ell_{v'v}} \quad , \tag{18}$$

$$\eta^{\ell}_{v'v} = \eta^{\ell}_{v'v}(\text{background}) + \eta^{\ell}(\text{resonant}) \quad , \tag{19}$$

$$\tan \eta^{\ell}(\text{resonant}) = \frac{\Gamma_{\ell}/2}{E_R(\ell)-E} \quad . \tag{20}$$

Equations 18-20 resemble the single-channel case, but differ by a factor of two, since we are concerned here only with off-diagonal elements. Note that it also follows from Equation 9 that the resonant part of the phase shift is independent of the channel labels v and v'. Equations 16-20 then lead to the following expression for the behavior of the quantum deflection function near a resonance,

$$\Theta_{v'v}(\ell) = \Theta^{0}_{v'v}(\ell) + \frac{1}{2} \; \frac{\Gamma_0(a+2b\ell)(E_R-E) - B\Gamma_{\ell}(2\ell+1)}{(E_R(\ell)-E)^2 + \Gamma^2_{\ell}/4} \quad . \tag{21}$$

In an earlier paper (16), we showed that Equation 21 (with a=b=0) was remarkably good in modeling the resonant behavior of the deflection function.

Application to Angular Distributions

In this section, we present an application of the techniques described in the previous sections to the angular distribution for the reaction

$$F+H_2(v=0) \rightarrow HF(v'=2)+H$$

A series of calculations, using the BCRLM, was performed on the M5 surface (40), as reported earlier (16). Here we will concentrate on the angular distribution at a fixed total energy E=1.807 eV [measured from the minimum of the HF vibrational potential for asymptotic H+HF geometries]. The initial relative kinetic energy of ground state reactants at this total energy is 0.162 eV [or 3.73 kcal/mol]. We show the impact parameter dependence of the magnitudes and phases of the appropriate elements of the S matrix in Figures 1 and 2. Figure 1 shows the opacity function, P^{ℓ}_{20} vs. ℓ [see Equation 1], and Figure 2 shows the quantum deflection function Θ^{ℓ}_{20} vs. ℓ [see Equation 7]. The presence of a resonant contribution to the reaction mechanism is evident as a peak in the opacity function near ℓ=14 and as a dip in the deflection function near ℓ = 16. The dip in the deflection function is a consequence of the enhanced lifetime of the resonant complex, which rotates toward more forward angles before it decays. The angular distribution which results from these calculations is shown in Figure 3.

Let us now analyze the resonance contribution to this angular distribution more closely. Since Figure 2 shows that at E=1.807 eV the resonant contribution to the deflection of products is greatest near ℓ=16, we show in Figure 4 a plot of the eigenvalues of the ℓ=16 lifetime matrix [Equation 11] as a function of total energy E. Over

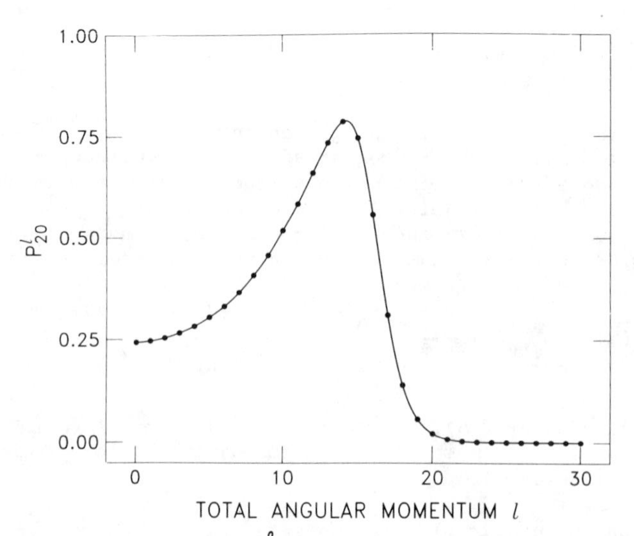

Figure 1. Opacity function, P^{ℓ}_{20} vs. ℓ, for the reaction $F+H_2(v=0) \rightarrow HF(v'=2)+H$ at total energy E = 1.807 eV.

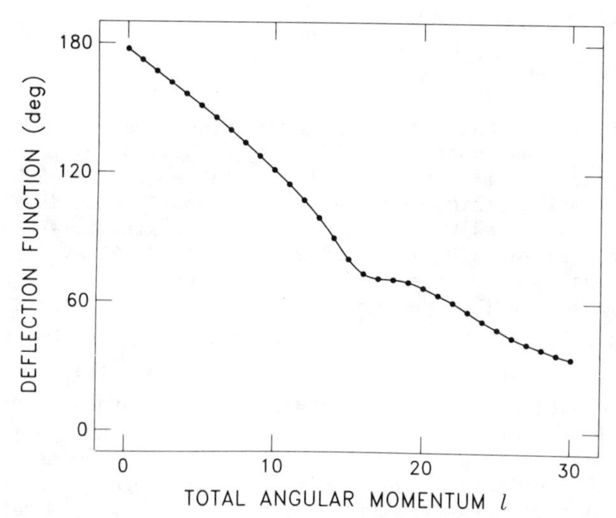

Figure 2. Quantum deflection function, $\Theta(\ell)$, as defined by Equation 7, for the reaction $F+H_2(v=0) \rightarrow HF(v'=2)+H$ at total E = 1.807 eV.

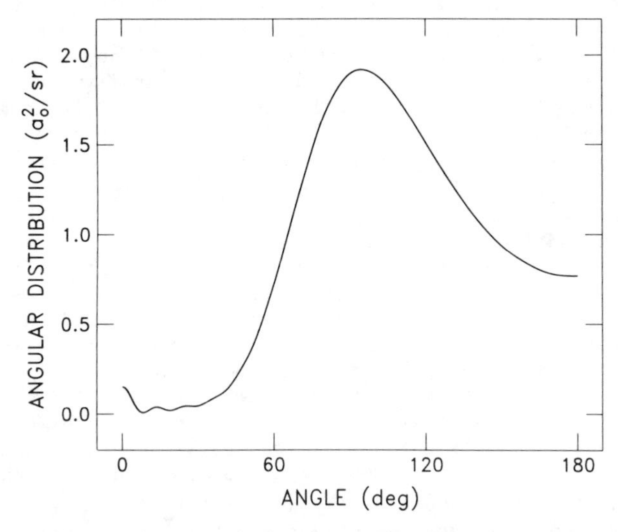

Figure 3. Differential cross section per unit solid angle for the reaction $F+H_2(v=0) \rightarrow HF(v'=2)+H$ at total energy E = 1.807 eV.

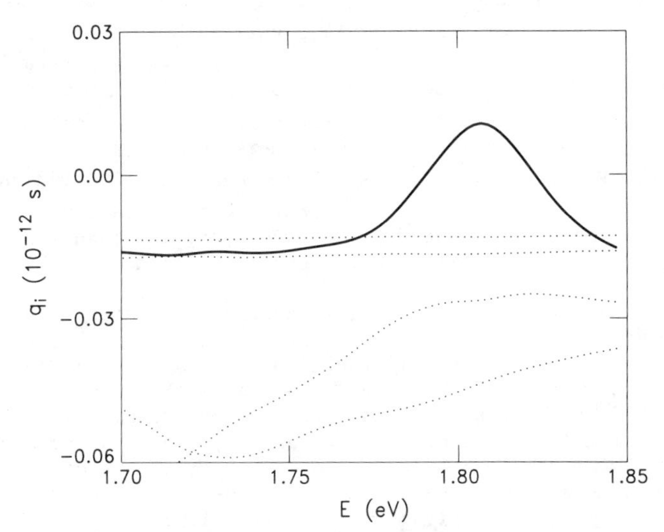

Figure 4. Eigenvalues of the collision lifetime matrix $\underline{\underline{Q}}$ vs. total energy for the ℓ=16 partial wave of the $F+H_2$ reaction. The eigenvalue for the resonant eigenlifetime channel is drawn as a solid curve, and all the other eigenvalue curves are dotted.

this range of total energies, there are are five open channels in the S matrix [v=0 of reactants and v'=0,1,2,3 of products]. It is clear in Figure 4 that a Lorentzian shaped feature is indeed centered near E=1.807 eV in one of the eigenlifetimes. If this feature is fitted to the form of Equation 16, with a quadratic background behavior, the resonance energy $E_R(16)$ and width Γ_{16} can be determined. Values of $E_R(\ell)$ and Γ_ℓ for other ℓ values are determined similarly. In Figure 5, we show a plot of the energy dependence of the magnitudes of the elements of the eigenvector of $\underline{\underline{Q}}$ corresponding to the eigenvalue with the Lorentzian profile. These are related to the partial widths by Equation 13. Figures 4 and 5 give an indication of the degree of validity of the assumptions discussed from Equations 8 to 15. Although one of the eigenlifetimes shows the resonance feature much more strongly than the others, the other eigenlifetimes are neither zero over the width of the resonance, nor are they even constant. Similarly, in Figure 5, we see that the resonant partial widths are not necessarily slowly varying with energy over the width of the resonance.

Using the data of Figure 5 to generate the partial widths, we calculate the resonant part of the S matrix using Equation 16, and the background S matrix using Equation 8. Because the elements of the $\underline{\omega}$ vector in Equation 16 come from the numerical diagonalization of $\underline{\underline{Q}}$ calculated using Equation 11, we obtain $\underline{\omega}$ with an undetermined phase $e^{-i\phi}$. Possible choices of ϕ are determined iteratively by requiring the background S matrix $\underline{\underline{S}}_0$ be unitary [in addition to being symmetric]. At energies $E=E_R(\ell)$, only one value of ϕ makes $\underline{\underline{S}}_0$ unitary, and at other values of E, a unitary $\underline{\underline{S}}_0$ is obtained from either of two choices of ϕ. The physically satisfactory choice is the one for which the resonant behavior of $\underline{\underline{S}}_0$ is eliminated (the other choice tends to accentuate the resonance). An example of this procedure is demonstrated in Figure 6, where we have plotted $|S_{20}^{16}(E)|$ vs. E. The solid line shows the magnitudes of the full S-matrix elements, the solid circles show the background S-matrix elements corresponding to the physical root, and the open circles show background S-matrix elements corresponding to the nonphysical root.

Assembling the magnitudes of the background S-matrix elements for the physical roots at all partial waves from ℓ=0 to 18, we construct the background opacity function shown as the solid curve in Figure 7, at E = 1.807 eV. The large dip in this opacity function near ℓ=12 is the result of the crossing of two eigenvalues of the lifetime matrix in this range of ℓ and E. The crossing eigenvalues induce interactions between the eigenvectors and perturb the calculation of the background S matrix. Consequently, we have also used the opacity function given by the dashed line in Figure 7 to produce an estimate of the background angular distribution. Furthermore, because of these practical difficulties with the determination of the magnitudes of the background S-matrix elements, we elected not to use the phases determined in this manner, but to use phases from the deflection function analysis described earlier. The set of resonance energies is reasonably described by the form of

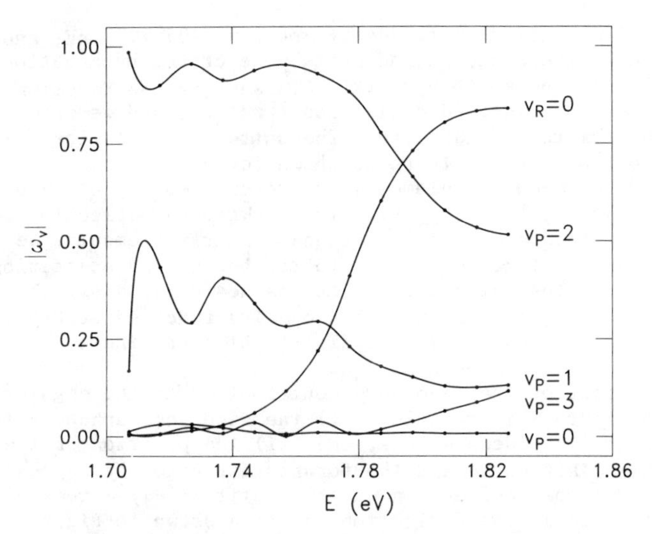

Figure 5. Energy dependence of the magnitudes of the elements of the eigenvector of $\underline{\underline{Q}}$ corresponding to the resonant eigenvalue. Near the resonant energy E = 1.806 eV, the channels which participate most in the resonance dynamics are clearly $v_R=0$ of the H_2 reactants and $v_P=2$ of the HF products.

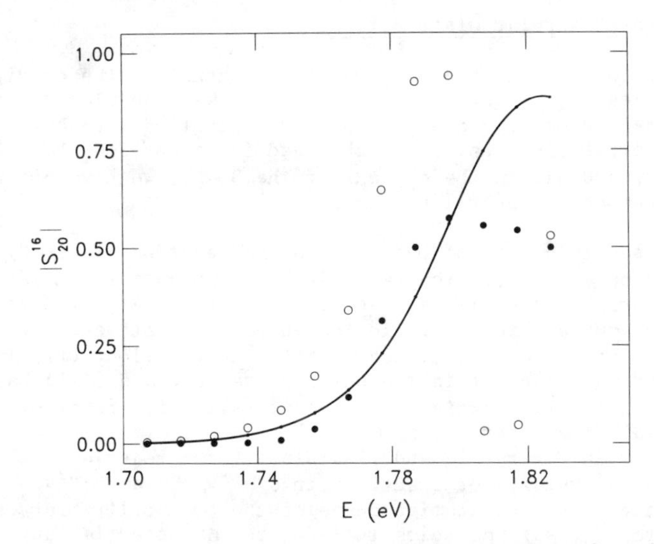

Figure 6. Magnitudes of background S-matrix elements for the reaction $F+H_2(v=0) \rightarrow HF(v'=2)+H$, partial wave $\ell=16$. The solid curve shows the full S-matrix elements, and the open and filled circles show two choices of background elements corresponding to the two phase choices for the partial widths discussed in the text. The open circles are elements corresponding to a nonphysical root, the closed circles correspond to the physical root.

Equation 18 with $E_R(0) = 1.7234$ eV and $B = 3.05\times10^{-4}$ eV, and the ℓ dependence of the widths is adequately described by Equation 17 with $\Gamma_0 = 0.031$ eV, and $a = b = 3.11\times10^{-3}$. We then use Equation 21 to produce the background deflection function, and we show results in Figure 8 as the solid curve. The dotted curve is the full deflection function of Figure 2, shown for comparison. Phases of the background S-matrix elements are determined, up to an unimportant constant, by integrating the background deflection function. Having determined a set of background S-matrix elements, we calculate an angular distribution due to the background scattering, shown in Figure 9 as the solid curve. The dashed curve shows the angular distribution of using the smoothed opacity function of Figure 7, and the dotted curve reproduces the distribution of Figure 3 for comparison.

The effect of the resonance contribution to the angular distribution in Figure 9 is twofold: (i) the resonance enhances the intensity of the sideways peak, and (ii) the position of the sideway speak is shifted toward the forward direction by approximately 20°, an additional deflection which quantitatively agrees with the size of the dip in the deflection function shown in Figure 8. There is also a small reduction in the interference oscillations in the forward direction. It is significant to note, however, that the background angular distribution is sideways peaked, as a consequence of the generally increasing behavior of the opacity functions of Figure 7 at low ℓ.

Forward Peaked Angular Distributions

In the remainder of this paper, we wish to consider the conditions which may lead to broad, or even forward peaked, angular distributions of products for energies near the reaction threshold. Such angular distributions have been observed in recent $F+H_2$ beam experiments (23), and within the context of the BCRLM, we have observed this phenomenon in several systems.

$F+H_2$. We see this sort of behavior in the reaction $F+H_2(v=0) \rightarrow HF(v'=2)+H$ on a modification (41) of the M5 potential surface. This modified surface is called surface 3 in reference 41; it differs from the M5 surface in the following three qualitative ways: (i) the new surface has a lower collinear barrier to reaction, (ii) it has a softer bending potential in the entrance valley, and (iii) has a much lower adiabatic barrier in the exit valley for reaction into the HF(v'=3) final state.

We show in Figures 10 and 11 a plot of the reaction probabilities for this system, for reactions into HF(v'=2) and HF(v'=3) respectively. The qualitative shape of the probability curves for reaction to HF(v'=3) resembles that for the M5 potential surface, except for a reduced threshold delay at low partial waves [a consequence of the lower adiabatic barrier]. There is a much larger difference in the dynamics leading to HF(v'=2) products. At low partial waves, the reaction into HF(v'=2) is suppressed relative to the M5 case, and at larger partial waves, there is an obvious onset of a resonance feature. Presumably, the exit-valley adiabatic

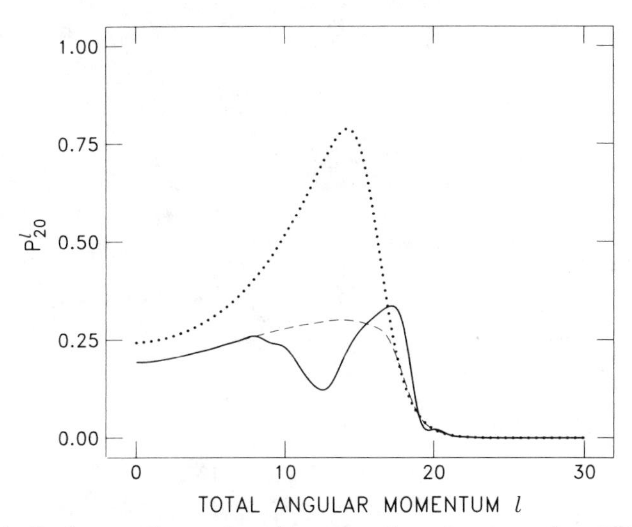

Figure 7. Background opacity function for the reaction $F+H_2(v=0) \rightarrow HF(v'=2)+H$ at E = 1.807 eV. The solid curve is the direct result of the extraction procedure discussed in the text, and the dashed curve is a smoothed version. The dotted curve reproduces the opacity function of Figure 1 for comparison.

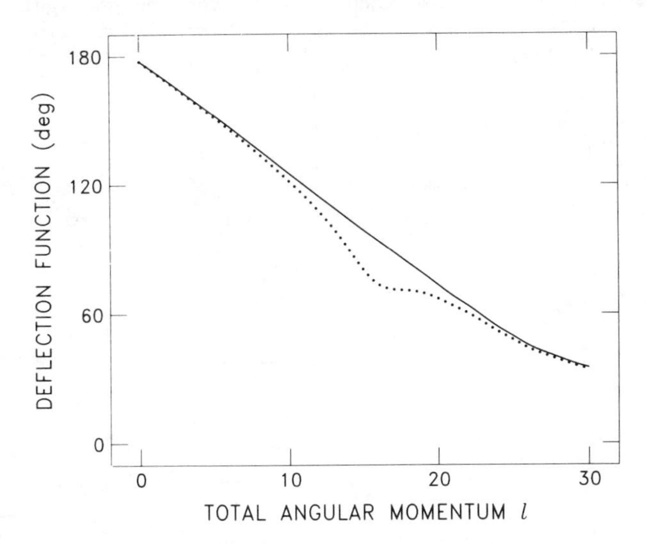

Figure 8. Background deflection function for the reaction $F+H_2 \rightarrow HF(v'=2)+H$ at E = 1.807 eV. The solid curve removes the resonance contribution using Equation 21. The dotted curve reproduces the deflection function of Figure 2 for comparison.

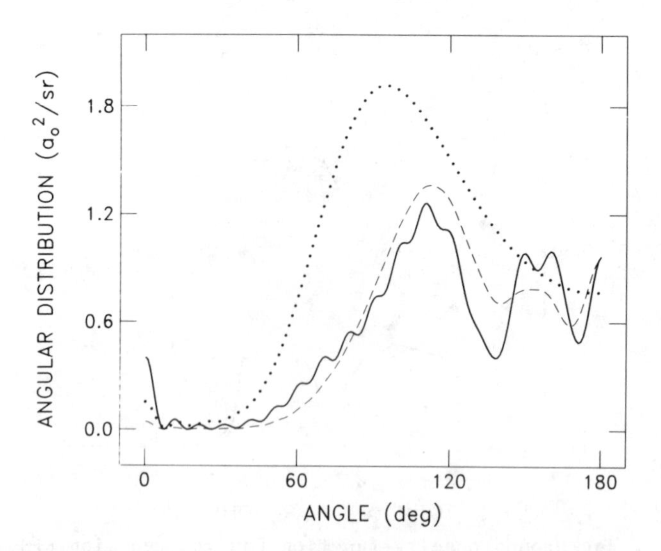

Figure 9. Background angular distributions for the reaction $F+H_2(v=0) \rightarrow HF(v'=2)+H$ at E = 1.807 eV. The solid curve is based on the opacity function shown in Figure 7 as a solid curve, and the smoother, dashed curve shows the angular distribution which results from the smoothed opacity function of Figure 7. The dotted curve reproduces the full angular distribution of Figure 3 for comparison.

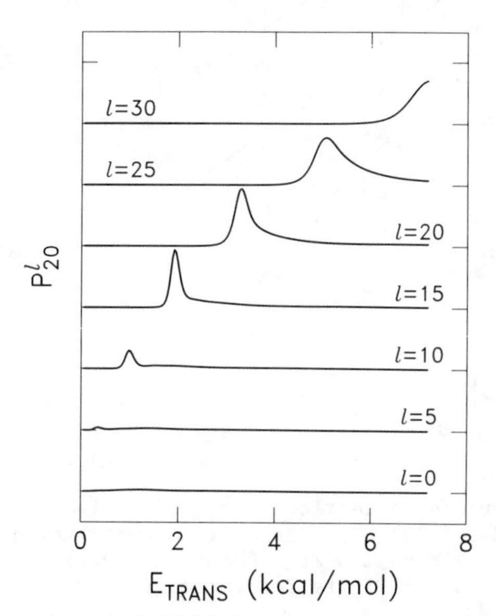

Figure 10. Reaction probabilities vs. relative translational energy for the reaction $F+H_2(v=0) \rightarrow HF(v'=2)+H$ on surface 3 of reference 41.

barrier which is present in the M5 surface, but has been reduced in this newer surface, is replaced by a similar [but centrifugal] barrier at intermediate partial waves, thereby restoring the resonance mechanism. The effect on angular distributions is obvious. Figure 12 shows opacity function plots at several energies, and Figure 13 shows the corresponding angular distributions. The resonant mechanism in this case is considerably stronger than for the corresponding M5 calculation, and the effect on the angular distributions is to induce much stronger interference oscillations, since there are many fewer partial waves contributing significantly to the reaction. In the case of the reaction to produce HF(v'=3) on this surface, it is evident that the v'=3 state of products participates more strongly in the resonance mechanism, as can be seen in the angular distribution plots of Figure 14. The angular distribution shifts from backward to sideways peaked, and has noticeable interference oscillations, indicative of resonance dynamics, at all energies.

The $F+H_2$ reaction on this newer potential surface demonstrates one way in which the BCRLM will produce an angular distribution at the reaction threshold which is not smooth and backward-peaked. In this case, the absence of a significant reaction probability for low partial waves, and the appearance of a resonance feature at larger partial waves, combine to produce an opacity function which peaks at large partial waves, and hence an angular distribution which has a predominant forward distribution of reaction products. We have seen results similar to these (41) in the angular distribution for the reactions $F+D_2(v=0) \rightarrow DF(v'=3,4)+D$ on this same surface.

$He+H_2^+$. We performed a set of BCRLM calculations for the reaction $He+H_2^+ \rightarrow HeH^+ + H$ using the DIM representation (42) of the potential surface. A great many resonance features have been seen (43) in the collinear reactive dynamics of this system, and many have been identified (44) with the attractive well which is present on this surface in the entrance valley. Because the bending potential predicted by the DIM surface is very weak, the BCRLM dynamics (for the $\ell=0$ partial wave) resembles that of the collinear calculation. Reaction probabilities for the reaction $He+H_2^+(v=3) \rightarrow HeH^+(v'=0)+H$ are shown in Figure 15 for several partial waves. The predominance of many resonances in each partial wave is evident in the figure. As written, this reaction is endothermic by 0.81 eV, but there is no barrier over and above the endothermicity of reaction. Consequently, reaction occurs from many partial waves at the reaction threshold. There is little or no evidence of a centrifugal barrier to reaction appearing in the dynamics until the $\ell=15$ partial wave. For this reason the opacity function, at the reaction threshold, contains contributions from a large number of partial waves. The resulting angular distributions of products are therefore quite broad, and show the effects of the multitude of resonances through the very oscillatory nature of the distributions, as is shown in Figure 16. Results similar to these, except for reaction from other initial vibrational states of the H_2^+ reactant molecule, are qualitatively the same -- reaction from many partial waves at threshold makes possible a very broad angular distribution of products.

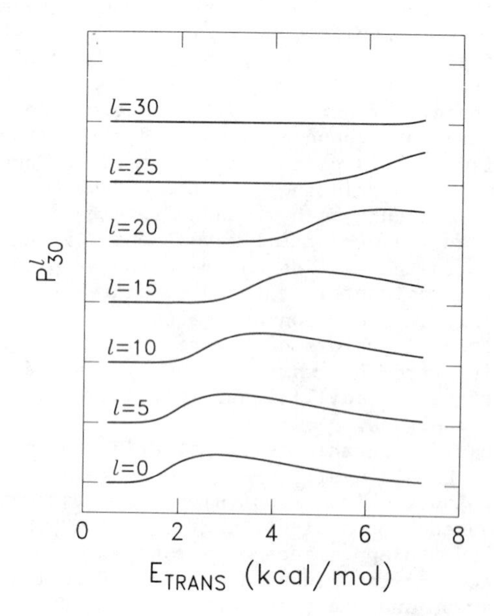

Figure 11. As in Figure 10, except v'=3.

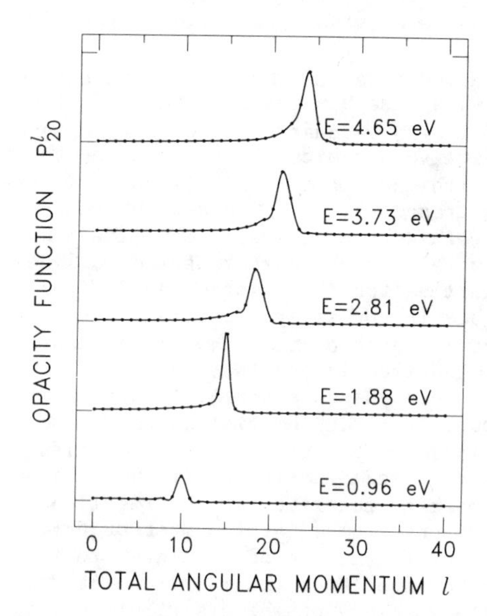

Figure 12. Opacity functions at several energies for the reaction $F+H_2(v=0) \rightarrow HF(v'=2)+H$ on surface 3 of reference 41.

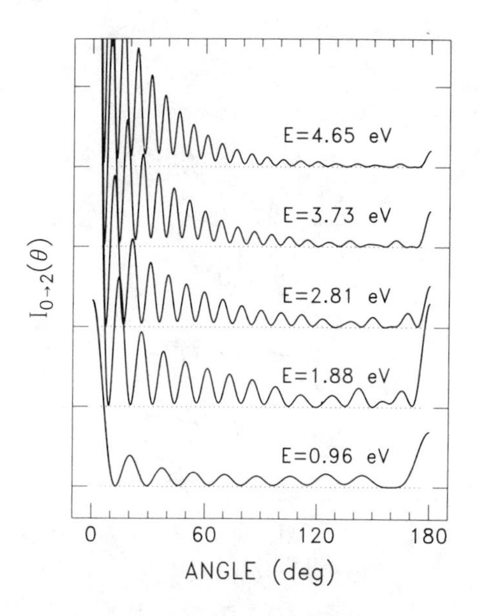

Figure 13. Angular distributions at the same energies as Figure 12 for the reaction $F+H_2(v=0) \rightarrow HF(v'=2)+H$ on surface 3 of reference 41.

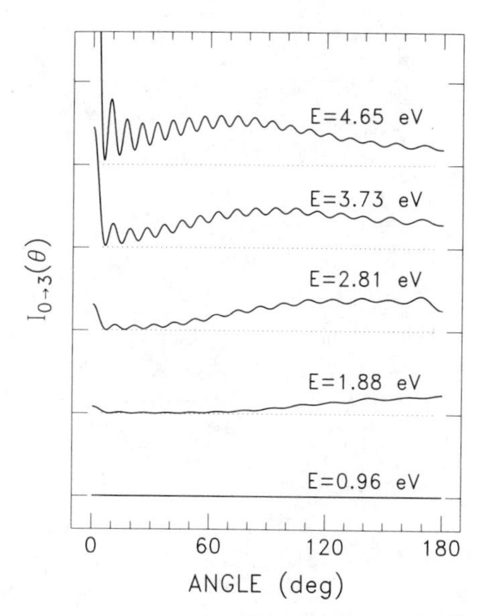

Figure 14. As in Figure 13, except v'=3.

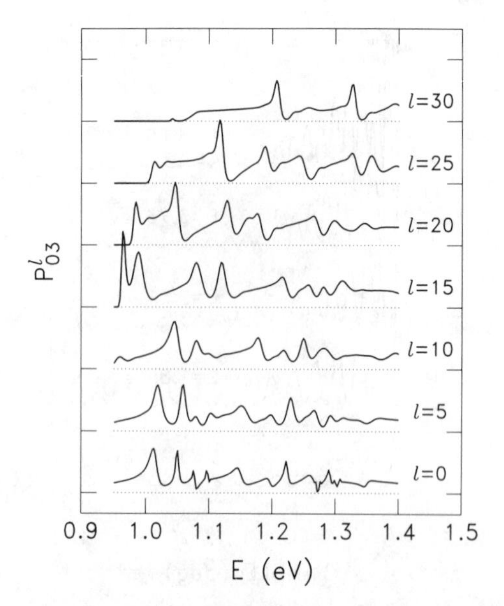

Figure 15. Reaction probabilities vs. energy for the reaction $He+H_2^+(v=3) \rightarrow HeH^+(v'=0)+H$ on the DIM surface of reference 42. Total energies are measured from the minimum of the H_2^+ entrance valley well.

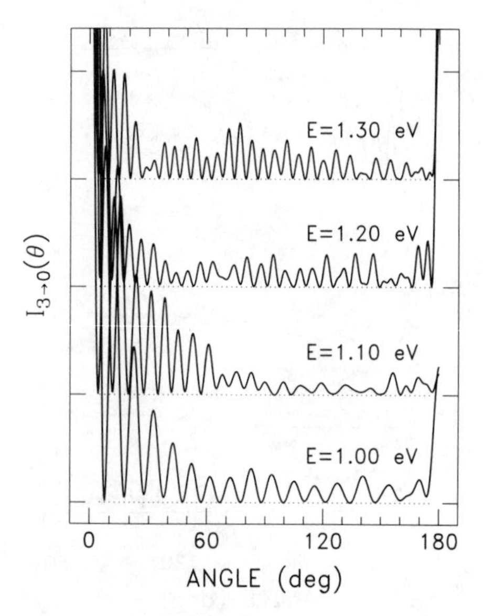

Figure 16. Angular distributions for the reaction $He+H_2^+(v=3) \rightarrow HeH^+(v'=0)+H$. Total energies are measured from the minimum of the H_2^+ entrance valley well.

Summary

We have attempted to describe here the way in which resonances in reactive scattering may affect the angular distribution of the products of reaction. To do this, we have employed a simple 3D model of reactions, the Bending-Corrected Rotating Linear Model, and have computed angular distributions for several reactive systems in which resonances contribute to the scattering dynamics.

The most straightforward way to determine the effect of resonances on angular distributions is to define a partition of the angular distribution into its direct and resonant components. We described an attempt to do this for the reaction $F+H_2(v=0) \rightarrow HF(v'=2)+H$ at an energy where the calculated angular distribution is decidedly sideways peaked. In this analysis, we found the effect of the resonant part of the scattering was to enhance the intensity of the sideways peak and to shift it to a more forward angle. However, even without the resonant contribution, we find the background angular distribution to be sideways peaked. This observation need not be surprising, however, for it has been observed for many years that a direct scattering mechanism can lead to sideways peaked angular distributions as the collision energy is increased (45-52, 15,16).

We have also sought to describe the conditions under which the angular distribution of reaction products could be forward-peaked at the reaction threshold, within the context of the BCRLM. Two sources of this effect have been seen in our calculations, and the essential feature which they have in common is that the reaction must, at threshold, contain contributions from scattering at large partial waves. In the example shown for the $F+H_2$ reaction on a modification (41) of the M5 surface, a forward peaked distribution at threshold was seen to result from a resonant mechanism which is absent in low partial waves, and becomes significant for larger partial waves. In the second example, for the $He+H_2^+$ reaction, forward peaked angular distributions at threshold result from the fact that the reaction proceeds without a barrier (other than that due to the overall endothermicity of reaction). In the $He+H_2^+$ reaction, there are indeed many resonances which contribute to the dynamics, but it is possible that the angular distribution would have been peaked in the forward direction even without the resonances.

Literature Cited

1. Walker, R. B.; Light, J. C. Ann. Rev. Phys. Chem. 1980, 31, 401.
2. Schatz, G. C. In "Potential Energy Surfaces and Dynamics Calculation"; Truhlar, D. G., Ed.; Plenum: New York, 1981; Chap. 12.
3. Kuppermann, A. In "Theoretical Chemistry: Advances and Perspectives, Vol. 6A"; Academic: New York, 1981; p. 79-164.
4. Baer, M. Adv. Chem. Phys. 1982, 49, 191.
5. Wyatt, R. E. in "Horizons in Quantum Chemistry"; Fukui, K.; Pullman, B., eds.; Reidel: Dordrecht, 1980; p. 63.
6. Redmon, M. J.; Wyatt, R. E. Chem. Phys. Letters 1979, 63, 209.
7. Wyatt, R. E.; McNutt, J. F; Redmon, M. J. Ber. Bunsenges. Phys. Chem. 1982, 86, 437.

8. Wyatt, R. E.; Redmon, M. J. Chem. Phys. Letters 1983, 96, 284.
9. Bowman, J. M.; Lee, K. T. Chem. Phys. Letters 1979, 64, 291.
10. Bowman, J. M.; Lee, K. T. J. Chem. Phys. 1980, 72, 5071.
11. Baer, M.; Jellinek, J.; Kouri, D. J. J. Chem. Phys. 1983, 78, 2962.
12. Shoemaker, C. L.; Kouri, D. J.; Jellinek, J.; Baer, M. Chem. Phys. Letters 1983, 94, 359.
13. Schatz, G. C. Chem. Phys. Letters 1983, 94, 183.
14. Walker, R. B.; Hayes, E. F. J. Phys. Chem. 1983, 87, 1255.
15. Walker, R. B.; Hayes, E. F. J. Phys. Chem. 1984, 88, 1194.
16. Hayes, E. F.; Walker, R. B. J. Phys. Chem., accepted.
17. Walker, R. B.; Blais, N. C.; Truhlar, D. G. J. Chem. Phys. 1983, 80, 246.
18. Schatz, G. C.; Kuppermann, A. Phys. Rev. Letters 1975, 35, 1266.
19. Walker, R. B.; Stechel, E. B.; Light, J. C. J. Chem. Phys. 1978, 69, 2922.
20. Sparks, R. K.; Hayden, C. C.; Shobatake, K.; Neumark, D. M.; Lee, Y. T. in "Horizons in Quantum Chemistry"; Fukui, K.; Pullman, B., Eds.; Reidel: Dordrecht, 1980; p. 91.
21. Lee, Y. T. Ber. Bunsenges. Phys. Chem. 1982, 86, 378.
22. Hayden, C. C. PhD. Thesis, Lawrence Berkeley Laboratory, LBL-13660, Berkeley, California, 1982.
23. Neumark, D. M.; Wodtke, A. M.; Robinson, G. N.; Lee, Y. T. See contribution in this volume.
24. Child, M. S. Mol. Phys. 1967, 12, 401.
25. Connor, J.N.L.; Child, M. S. Mol. Phys. 1970, 18, 653.
26. Wyatt, R. E. J. Chem. Phys. 1969, 51, 3489.
27. Mortensen, E. M. J. Chem. Phys. 1968, 48, 4029.
28. Mortensen, E. M.; Pitzer, K. S. Chem. Soc. Spec. Publ. 1962, 16, 57.
29. Garrett, B. C.; Truhlar, D. G. J. Am. Chem. Soc. 1979, 101, 4534.
30. Bowman, J. M.; Ju, G.-Z.; Lee, K. T. J. Phys. Chem. 1982, 86, 2232.
31. Kuppermann, A. In "Potential Energy Surfaces and Dynamics Calculations"; Truhlar, D. G., Ed.; Plenum: New York, 1981; Chap. 16.
32. Kaye, J.; Kuppermann, A. J. Phys. Chem. 1981, 85, 1969.
33. Taylor, J. R. "Scattering Theory"; Wiley: New York, 1972; p. 407-417.
34. Wu, T.-Y.; Ohmura, T. "Quantum Theory of Scattering"; Prentice-Hall: Englewood Cliffs, N.J., 1962; pp. 20-22, 427-429.
35. Child, M. S. "Molecular Collision Theory"; Academic: New York, 1974; pp. 53-56, 73-75.
36. Ashton, C. J.; Child, M. S.; Hutson, J. M. J. Chem. Phys. 1983, 78, 4025.
37. Skodje, R. T.; Schwenke, D. W.; Truhlar, D. G.; Garrett, B. C. J. Phys. Chem. 1984, 88, 628.
38. Skodje, R. T.; Schwenke, D. W.; Truhlar, D. G.; Garrett, B. C. J. Chem. Phys. 1984, to be published.
39. Smith, F. T. Phys. Rev. 1960, 118, 349.
40. Muckerman, J. T. In "Theoretical Chemistry: Advances and Perspectives, Vol. 6A"; Academic: New York, 1981; p. 1.

41. Steckler, R.; Truhlar, D. G.; Garrett, B. C.; Walker, R. B.; Blais, N. C., to be published. This modified surface is called new surface number 3; it is obtained by a slight modification of new surface number 2 of Truhlar, D. G.; Garrett, B. C.; Blais, N. C. J. Chem. Phys. 1984, 80, 232; the value for the Sato parameter is increased from 0.106 to 0.168. The new value removes the exit-channel adiabatic barrier for FD(v'=4)+D.
42. Kuntz, P. J. Chem. Phys. Letters 1972, 16, 581.
43. Adams, J. T. Chem. Phys. Letters 1975, 33, 275.
44. Chapman, F. M., Jr.; Hayes, E. F. J. Chem. Phys. 1975, 62, 4400.
45. Tang, K. T. Ph.D. Thesis, Columbia University, New York, 1965.
46. Karplus, M.; Tang, K. T. Disc. Faraday Soc. 1967, 44, 56.
47. Tang, K. T.; Karplus, M. Phys. Rev. A 1971, 4, 1844.
48. Truhlar, D. G.; Dixon, D. A. in "Atom-Molecule Collision Theory"; Bernstein, R. B., Ed.; Plenum: New York, 1979; Chap. 18.
49. Agmon, N. Chem. Phys. 1981, 61, 189.
50. Suck, S. H.; Emmons, R. W. Phys. Rev. A 1981, 24, 129.
51. Emmons, R. W.; Suck, S. H. Phys. Rev. A 1982, 25, 178.
52. Salk, S. H. Suck; Emmons, R. W.; Klein, C. R. Phys. Rev. A 1984, 29, 1135.

RECEIVED June 19, 1984

Author Index

Abusalbi, N., 457
Baer, M., 457
Borondo, F., 323
Bowman, Joel M., 43
Burrow, P. D., 165
Casassa, M. P., 305
Chu, Shih-I, 263
Collins, L. A., 65
Davis, Michael J., 337
Dehmer, J. L., 139
Garrett, Bruce C., 375,421
Geiger, Lynn C., 421
Gentry, W. Ronald, 289
Giordan, Judith C., 193
Harding, Lawrence B., 43
Hayden, C. C., 479
Hayes, Edward F., 493
Hedges, R. M., Jr., 323
Howard, Allison E., 183
Janda, Kenneth C., 305
Jellinek, J., 457
Jordan, K. D., 165
Kouri, D. J., 457
Kuruoglu, Zeki C., 401
Langhoff, P. W., 113
Le Roy, Robert J., 231
Lee, Ki Tung, 43
Lee, Y. T., 479
Lefebvre, R., 35
Lucchese, R. R., 89
Lynch, D. L., 89
Marston, C. C., 441
McCurdy, C. William, 17
McKoy, V., 89
Micha, David A., 401
Michaud, M., 211
Moore, John H., 193
Neumark, D. M., 479
Pollak, Eli, 353
Reinhardt, W. P., 323
Robinson, G. N., 479
Romanowski, Hubert, 43
Romelt, Joachim, 353
Sanche, L., 211
Schatz, George C., 421
Schneider, B. I., 65
Schwenke, David W., 375
Simons, Jack, 3
Skodje, R. T., 323,375
Staley, Stuart W., 183
Thirumalai, Devarajan, 375
Thompson, Todd C., 375
Tossell, John A., 193
Truhlar, Donald G., 375
Wagner, Albert F., 337
Walker, Robert B., 493
Western, Colin M., 305
Wodtke, A. M., 479
Wyatt, Robert E., 441
Zhang, Z. H., 457

Subject Index

A

Adiabatic eigenenergies
  collinear $F+H_2$ reaction, 358,359f
  collinear $H+H_2$ reaction, 356,357f
Adiabatic generation of semiclassical resonance eigenvalues, 331-35
Adiabatic potential curve
  atom-diatom reactions with collinear minimum energy paths, 391
  collisional excitation of CO by fast H atoms, 434,435f
AE--See Attachment energy
Angular distributions
  $F+H_2$ reaction, 504,506f,507,509f
  forward peaked, 504,506-10
  $He+H_2^+$ reaction, 507,510f
  photoelectrons, 75
  and reactive resonances in the rotating linear model, 493-511
Angular momentum eigenfunction, 268
Anion resonance states of organometallic molecules, 193-208
Anisotropy strength functions
  $Ar-H_2$ potential surface, 244,246f
  Ar-HD, 284f

Anisotropy-induced splitting of metastable levels of H_2-Ar, 247,248f
Antibonding orbitals for cyclopropane, 188,191f
Atom-diatom complexes, Hamiltonian, 235
Atom-diatom potential energy surfaces, 239-47
Atom-diatom predissociation, schematic representation, 233f
Atom-diatom reactions
  adiabatic potential curve, 391
  resonances, 390-94
Atom-diatom resonances within a many-body approach to reactive scattering, 401-18
Atom-electron interaction, 168
Atom-rigid rotor anisotropic potentials, 273
Attachment energy
  definition, 195
  metallocenes, 198f
  orbitals of Fe, Co, and Ni compounds, 197
  $Si(H,Me)_4$-(H,Me)Cl, 205,207f
  substituted benzenes and substituent parent compounds, 206,208f
Average adiabatic surfaces for the IHI system, 365,366f

B

Background deflection function for the $F+H_2$ reaction, 504,505f
Background opacity function for the $F+H_2$ reaction, 502,505f
Background S-matrix elements for the $F+H_2$ reaction, 502,503f
Backscattered current distribution, 214-15
Band profiles, vdW molecules, 309,311
Basis functions, 270-71
Basis set expansion of electron exchange, 72-73
Bending-corrected rotating linear model (BCRLM), 494-96
Benzene, shape and core-excited resonances, 174-76
Bimolecular reactive collisions, 375-95
Bond energy, polyatomic vdW molecules, 295-96
Bound-state energy levels of collinear IHI, 362
Branching probabilities, definition, 378

C

Carbonyl sulfide, 337-48
  dissociation, 346-48
  energy relaxation, 339,341-42,343f
  exponential separation, 338-39
  individual trajectory dynamics, 340f,342-44,345f
CCCC methods--See Complex-coordinate coupled-channel methods
Center-of-mass coordinate systems for atom-diatom vdW molecules, 266,267f
Center-of-mass velocity flux contour map
  $F+D_2$ reaction, 485,487f
  $F+H_2$ reaction, 485,486f
  F+HD reaction, 485,488f
Centrifugal barrier of photoelectron, 142-44
Chloro-substituted methanes, ETS, 176-78
Classical dynamics above dissociation, local-mode systems with two degrees of freedom, 325
Collinear exact quantum (CEQ) scattering matrix, 46
Collinear reactive scattering resonances, 355-64
Collisional resonances
  excitation of CO by fast H atoms, 421-39
  relevance of polyatomic vdW molecules, 301-2
Collisions and half collisions, metastable states, 4-5
Complex adiabatic energies, comparison with resonance energies, 41
Complex adiabatic paths, 40-41
Complex configuration interaction (CCI) calculations, 24
Complex coordinate resonances for a model vdW system, 51-52
Complex coordinate studies, quantum dynamics, 325,327,329f
Complex energy quantization for molecular systems, 35-41
Complex scaling method for numerical and piecewise potentials, 265-66
Complex self-consistent field (CSCF) calculations, 20-22
Complex stabilization calculations, 24-25
Complex variational principles for resonances, 18-19
Complex-coordinate coupled-channel methods
  advantages, 283,287
  predissociating resonances in vdW molecules, 263-87

Constants of motion and exponential separation, 344,346,347f
Continuum MS-Xα cross section
decomposition, 200,204f
elastic electron scattering, 206,207f
Continuum orbital expansion, 92
Contour diagram for a BC_3 potential for H_2-Ar, 244,246f
Core-excited resonances
benzene, 174-76
widths, 379
Coupled-channel approach to determination of resonances, 36-38
Coupling operator approximation, 385
Cross sections, differential and integral, 460-61,468,471-73
Cyclopropane, antibonding orbitals, 188,191f
Cyclopropene and cyclobutene, temporary negative ion states, 184-88

D

Diagonal-corrected vibrational adiabatic hyperspherical (DIVAH) model, 355-64
Diatom stretching dependence, 250-51
Differential cross section
$F+H_2$ reaction, 499,501f
$H+H_2$ reaction, 44-51
Direct variational methods
complex resonance energies, 17-32
resonance calculations, 20-30
Dissociation of planar OCS, 346-48
Distorted wave Born approximation
calculation of resonances, 58-60
resonance energies and widths for two-mode HCO, 58,60
Distorted wave Born resonances for a model vdW system, 51-52
Dynamic resonances in $F+H_2$ reaction, 479-90
Dynamics
molecular photoionization processes, 89-110
nuclear rearrangement, 405-7,410f
vibrational states of triatoms, 323-35

E

Effective potentials
collinear $F+H_2$ reaction, 358,359f
collinear $H+H_2$ reaction, 356,357f
predissociation of H_2-Ar, 258-60
Eigenchannel contour maps of unbound electrons, 149-51
Eigenphase sums, 80f,381
Electron + H_2 scattering, 78-79,80f
Electron + H_2^+ scattering, 79,80f
Electron + N_2 scattering, 76-78
Electron energy loss spectra, 216-22
Electron exchange interactions, 71-73
Electron excitation function of quasi-elastic and vibrational energy losses, 222,224f
Electron impact energy vs. ion current on $Cr(CO)_6$, 200,201f
Electron scattering theory, 213-16
Electron transmission spectroscopy
chloro-substituted methanes, 176-78
$Cr(CO)_6$, 200,201f
ethylene, butadiene, and hexatriene, 169-73
general discussion, 166-67
metallocenes, 196f
negative ion states of three- and four-membered ring hydrocarbons, 183-91
recent developments, 178-79
Electron-atom interaction, 168
Electron-molecule collisions, polarization and correlation, 73-74
Electron-molecule scattering, resonances, 65-86,155-57
Electronic resonances, 6,8-10
Energetics of the $F+H_2$, $F+D_2$, and F+HD reactions, 480,482f
Energies and symmetries of gas- and solid-phase transient anions, 225
Energy dependence
reflectivity and excitation processes, 222,223f
vibrational excitation intensity of N_2, 222,223f
Energy levels for states involved in the $Q_1(0)$ transitions of H_2-Ar, 240,241f
Energy relaxation in planar OCS, 339,341-42,343f
Energy variation of resonance energies for $F+H_2$ reaction, 368,369f
Equipotential energy contours, HCO system, 56
Ethylene complex photodissociation
dynamics, 314-16
parameters, 309
spectra, 308f,310f
Ethylene dimer, vdW interactions, 312-14
ETS--See Electron transmission spectroscopy
Excitation functions, 222-26
Exponential separation in collinear OCS, 338-39
Extended X-ray absorption fine structure (EXAFS), 147

F

Feshbach golden rule, 388-89
Film thickness vs. intensity of electron energy losses, 220,221f
Force constants for resonances in $F+H_2$ reaction, 368,369f
Forward peaked angular distributions, 504,506-10
Frontier orbitals of $M(Cp)_2$, qualitative diagram, 198f

G

Green's function, 69-70,91
Ground electronic states of metallocenes, 197

H

H_2 + electron scattering, 78-79,80f
H_2^+ + electron scattering, 79,80f
Half collisions and collisions, metastable states, 4-5
Hamiltonian
- atom-diatom complexes, 235
- atom-diatom molecular collision, 355,442-43
- body-fixed, 270-71
- and CCCC formulation in body-fixed coordinates, 270-72
- and CCCC formulation in space-fixed coordinates, 266,268-70
- channel, 406
- partial wave, 47
- reduced-dimensionality, 58
- scattering path, 55,57
- target electronic, 67
- two local-mode oscillators, 325
- zero-order, 58

Hartree-Fock and Stieltjes orbitals
- ionization of CO, 129,131f
- ionization of H_2CO, 129,133f
- ionization of NO, 126,127f

Hartree-Fock valence orbitals of NO, 123,124f
Hartree-Fock wavefunction, 74,75
Heavy-particle resonances, 10-12,13t
Hilbert-Schmidt states, 409
Hydrocarbons and their derivatives, negative ion states, 165-80

I

Improved virtual orbital (IVO) equation, 76
Integral cross sections, collisional excitation of CO by fast H atoms, 437
Integral equations for rearrangement, 412-14
Intensity of electron energy losses vs. film thickness, 220,221f
Interacting atoms, 402-5
Intramolecular dynamics of highly excited OCS, 337-48
Ion current vs. electron impact energy on $Cr(CO)_6$, 200,201f
IOS calculations, resonances in collisional excitation of CO by fast H atoms, 424-25,434-37
Ionization potential (IP) of substituted benzenes and substituent parent compounds, 206,208f
IR photodissociation of polyatomic vdW molecules, 290-95
Isolated narrow resonance (INR), 376-77
Isotopic substitution, effect on $F+H_2$ reaction, 485,489
Isotropic-channel potential curves, $Ar-N_2$ system, 268,269f

K

K-shell photoelectron asymmetry parameters, N_2, 103,104f
K-shell photoionization of N_2, 101-3,145-47,148f
Kinetic energy and product momentum, polyatomic vdW molecules, 296

L

LAB angular distribution
- $F+H_2$ reaction, 483,484f,485,488f
- F+HD reaction, 485,487f

LaGrange multipliers, 74-75
Lambert's law, 214
LAMOPT formalism, 75
Lanczos functions, 117,119-20,121f
Lifetime matrix, 461
Lineshape functions for Stieltjes orbitals, 122f
Linewidth-lifetime correlations with structure and dynamics of vdW molecules, 294
Linewidths and unimolecular decay rates, polyatomic vdW molecules, 296-301
Lippman-Schwinger equation, 91
Local-mode systems with two degrees of freedom, classical dynamics above dissociation, 325
Lorentzian transition probability, 377

M

Main group compounds, 203,205-208
Many-body approach to reactive scattering, 401-18
Metallocenes
anion resonance states, 195-99
correlation diagram of AE, 198f
ETS, 196f
Metastable states, 3-15
Molecular absorption and ionization, 115
Molecular continuum, theoretical methods, 66-71
Molecular fields, shape resonances, 139-60
Molecular hydrogen-inert gas potential energy surfaces, 242-47
Molecular photoionization resonances, 113-36,155-57
Molecular systems
complex energy quantization, 35-41
solution of equations, 93-95
Mulliken, 114

N

N_2 + electron scattering, 76-78
Negative ion states
hydrocarbons and their derivatives, 165-80
three- and four-membered ring hydrocarbons, ETS, 183-91
Non-Franck-Condon effects, shape resonance induced, 151-55
Nonadiabatic Feshbach calculations, 384-85
Nonconjugated dienes, interactions, 173-74
Nuclear rearrangement dynamics, 405-7,410f
Numerical comparison of methods for collinear reactions, 386-90
Numerical methods for locating quasi-periodic and periodic orbits, 443-44

O

OCS--*See* Carbonyl sulfide
Opacity function for F+H_2 reaction, 499,500f,507,508f
Optical potential formalism, 73-74
Orbital energies in the ground states of $Mo(CO)_6$ and CO, 202
Orbital occupations of metallocenes, 197
Orbital symmetry and temporary anions, 168
Orbiting resonances, 5,7f
Organometallic molecules, anion resonance states, 193-208
Orthonormalized square-integrable basis functions, 270-71

P

Partial channel photoionization cross section in NO, 123,125f
Partial wave rotationally cumulative reaction probability, 45
Partial wave scattering functions, 92
Partial widths, collinear atom-diatom reaction, 388-89
Partitioning degrees of freedom, two nonlinear polyatomics, 299
Periodic orbits
F+H_2 reaction, 441-54
numerical methods, 443-44
Permutational symmetry of identical nuclei, 407-8
Photoabsorption spectra, H_2S and SF_6, 143f
Photodissociation
ethylene complexes, 305-16
molecules on excited electronic surfaces, 316-17, 318f
multilevel system, 311
vdW molecules, 290-95,305-18
Photoelectron
angular distribution above the K-shell IP of N_2, 147-49
asymmetry parameters, 97,98f,107,109f
centrifugal barrier, 142-44
Photoexcitation and ionization, N_2, NO, and N_2O molecules, 123-29
Photoionization
CO, H_2CO, and CO_2 molecules, 129-34
CO_2, 81,84
molecular processes, 89-110
N_2, 79,81,152-55
NO, 81
resonances, 65-86,145-51
Photoionization cross sections
C_2H_2, 107,110f
$C_2N_2^+$, 107,108f
CO^+, 103,105,106f
CO_2, 85f,86f
definition, 90
H_2S, 134,135f
N_2, 82f,83f,97,98f,100f,101
N_2^+, 97,98f,99f
NO, 82f,83f

Photon energy and vibrational mode, polyatomic vdW molecules, 294
Poincare surfaces of section, 325,326f
Polarization and correlation, electron-molecule collisions, 73-74
Polarization components of photoionization cross sections of CO and H2CO, 129,130f
Pole location in the complex energy plane, 376
Polyatomic vdW molecules
  bond energy, 295-96
  IR photodissociation, 290-95
  linewidths and unimolecular decay rates, 296-301
  photon energy and vibrational mode, 294
  product momentum and kinetic energy, 296
  relevance to collisional resonances, 301-2
  vibrational density of states, 295
Polyatomic vdW molecules--See also VdW molecules
Potential barrier, effect on an unbound wave function, 142,143f,144-45
Potential curves for states involved in the $Q_1(0)$ transitions of H_2-Ar, 240,241f
Potential energy
  HF, 118f
  resonances in $F+H_2$ reaction, 368,369f
  symmetric IHI and IDI stretching mode, 370,371f
Potential matrix elements for HD-Ar, 256,257f
Potential surface, collisional excitation of CO by fast H atoms, 423-24
Potentials, predictions, comparisons, and implications, 251-53
Predissociation
  approximation methods, 237-39
  by internal rotation, 247-60
  vdW molecules, CCCC methods, 263-87
Product momentum and kinetic energy, polyatomic vdW molecules, 296
Propagation along a mixed path, 38-40

Q

Quantal RIOS calculations, theory, 459-61
Quantum deflection functions
  $F+H_2$ reaction, 499,500f
  resonance effects, 498-99
Quantum dynamics, complex coordinate studies, 325,327,329f
Quantum mechanical scattering, 380-82
Quantum mechanical stabilization calculations, 385-86
Quantum RIOS transition probabilities and wavefunctions, 426-31
Quantum studies, zeroth order golden rule, 327-30
Quasi-bound state of $F+H_2$ decays, 489
Quasi-periodic orbits
  $F+H_2$ reaction, 441-54
  IHI system, 368,371f
  numerical methods, 443-44

R

R-matrix formalism, 66-71
Radial strength functions of a BC_3 potential for H_2-Ar, 244,245f
Reaction probabilities
  vs. energy for the $He+H_2^+$ reaction, 507,510f
  vs. relative translational energy for $F+H_2$ reaction, 504,506f,508f
  vs. total energy for the collinear Cl+HCl reaction, 358,360f
Reactive infinite order sudden calculations
  $F+H_2$ and $F+D_2$ systems, 461-68
  resonance phenomena, 457-75
Reactive resonances
  and angular distributions in the rotating linear model, 493-511
  $H+D_2$ reaction, 394
  $H+H_2$ reaction, 390-94
Reactive scattering, atom-diatom resonances within a many-body approach, 401-18
Reactive state-to-state probabilities for the $F+H_2$ and $F+D_2$ reactions, 461-65
Rearrangement, integral equations, 412-14
Reduced mass, 46
Resonance energies
  calculations, 235-37
  collinear atom-diatom reactions, 386-88
  collinear Cl+HCl exchange reaction, 361
  collinear $H+H_2$ reaction, 356
  comparison with complex adiabatic energies, 41
  direct variational methods, 17-32
  threshold, 379
Resonance scattering, 408-12
Resonance states, principal categories, 5-6
Resonance tuning, $F+H_2$ system, 473,474f

Resonance widths
calculations, 235-37
collinear atom-diatom reactions, 390
DWBA, 51-52,53f
Resonances
analysis, 496-98
associations with σ^* orbitals, 176-78
atom-diatom reactions, 390-94
benzene, shape and core-excited, 174-76
calculation for $H+H_2$ reaction, 414-18
characteristics, 167-69
collisional excitation of CO by fast H atoms, 421-39
complex variational principles, 18-19
coupled-channel approach to determination, 36-38
definition, 376
DWBA calculation, 58-60
effects on quantum deflection functions, 498-99
electron-molecule scattering and photoionization, 65-86
electronic, 6,8-10
$H+D_2$ reactions, 394
$H+H_2$ reactions, 390-94
HCO system, 52,54-60
heavy-particle, 10-12,13t
many-body approach to reactive scattering, 401-18
orbiting, 5,7f
physical effects, 144-45
predissociating, vdW molecules, 232-35,263-87
quantal RIOS calculations, 457-75
reactive and nonreactive scattering, 43-61
role in $F+H_2$ and $F+D_2$ angular distributions, 473,475
target-excited, 6,7f,12-15
three- and four-membered rings, 188-91
and threshold energies for the collinear $F+H_2$ reaction, 358
and vibrational bonding, 3D treatment, 367-70,371f
Resonant periodic orbits
collinear IHI system, 364-65,366f
definition, 379
vibrationally bonded molecules, 364-67
Resonant quasi-periodic and periodic orbits for $F+H_2$ reaction, 441-54
RIOS calculations--See Reactive infinite order sudden calculations
Rotational predissociation of $Ar-H_2$ and Ar-HD vdW complexes, 272-83
Rotationally predissociating levels of HD-Ar, calculated widths, 253,257f
Rotationally resolved laser-excited fluorescence spectrum of $NeBr_2$, 318f
Rydberg and valence transitions, 115-17

S

Saddle point coordinate-rotation calculations, 25-26
Scattering matrix, 45-46
Schrodinger equation
body-fixed, 46
one-electron, 91
radial, 268
Semiclassical quantization using arbitrary trajectories, 445-46
Semiclassical resonance eigenvalues, adiabatic generation, 331-35
Semiclassical SCF calculations, 386
Shape resonances
basic properties, 142,51
behavior, 141-42
benzene, 174-76
definition, 90,139
electron-molecule scattering, 155-57
molecular fields, 139-60
molecular photoionization and related connections, 155-57
non-Franck-Condon effects, 151-55
photoionization, 145-51
physical effects, 144-45
Siegert ratio, 39-40
Single-center expansions, maximum values of parameters,94-95
Space-fixed CCCC α-trajectory, 277f,278f,282f,285f
Space-fixed CCCC calculations
$Ar-H_2$ atom-rigid rotor vdW complex, 273-78
$Ar-H_2$ atom-vibrating rotor vdW complex, 280-81,282f
Ar-HD vdW complex, 281,283-86
Spectral lineshapes for Stieltjes orbital representation of cross section of H_2CO, 129,133f
Spectroscopic constants, IHI and IHI^- systems, 372
Spectroscopic properties, $H+H_2$ reactive resonances, 392-93
Spherical Gaussian functions, 94
Stabilization calculations, collisional excitation of CO by fast H atoms, 425-26,432-34,435t
Stabilization graphs, analytic continuation, 26-30,31f

Stieltjes
orbitals, 117-20,121f,122f,126-29
Symmetries and energies of gas- and solid-phase transient anions, 225

T

Target-excited
resonances, 6,7f,12-15
Temporary negative ion states
cyclopropene and cyclobutene, 184-88
hydrocarbons and their derivatives, 165-80
and orbital symmetry, 168
polyatomic molecules, 179-80
previous studies, 165-66
Three- and four-membered rings, resonances, 188-91
Trajectories, semiclassical quantization, 445-46
Trajectory dynamics in planar OCS, 342-44,345f
Transient anions, energies and symmetries, 225
Transition metal
hexacarbonyls, 199-203
Triatoms, dynamics of vibrational states, 323-35

V

Valence and Rydberg
transitions, 115-17
Valence-bond (VB) set, 403
Valence-shell photoabsorption cross section in HF, 118f
Van der Waals (vdW) interactions
Ar atoms, 10,11f
ethylene complexes, 305-16
Van der Waals (vdW) molecules
band profiles, 309,311
CCCC methods for predissociating resonances, 263-87
center-of-mass coordinate systems, 266,267f
definition, 264
distorted wave Born and complex coordinate resonances, 51-52
Van der Waals molecules--Continued
linewidth-lifetime correlation with structure and dynamics, 294
photodissociation, 290-95,305-18
predissociation mechanisms, 264
vibrational predissociation, 231-60
Van der Waals (vdW) molecules--
See also Polyatomic vdW molecules
Vibrational bonding in the collinear IHI and IDI systems, 362-64
Vibrational density of states, polyatomic vdW molecules, 295
Vibrational mode and photon energy, polyatomic vdW molecules, 294
Vibrational predissociation of small vdW molecules, 231-60
with $\Delta v=-1$, 254-60
with $\Delta v=0$, 247-53
Vibrational states of triatoms, dynamics, 323-35
Vibrational-librational excitation and shape resonances in electron scattering, 211-17
Vibrational-rotational state-to-state reaction probability, 45
Vibrationally adiabatic models, bimolecular reactive collisions, 378,382-84
Vibrationally bonded molecules, 353-72
experimental evidence, 370,372
Vibrationally excited states of polyatomic vdW molecules, lifetimes and decay mechanisms, 289-302

W

Wavefunction
continuum state, 385
expansion in body-fixed coordinates, 271
low partial waves, 71
predissociation of H_2-Ar, 258-60
resonance state, 384-85

Z

Zeroth order golden rule, quantum studies, 327-30

Production and indexing by Karen McCeney
Jacket design by Pamela Lewis

Elements typeset by Hot Type Ltd., Washington, D.C.
Printed and bound by Maple Press Co., York, Pa.